Challenges and Future Prospects of Antibacterial Therapy

Challenges and Future Prospects of Antibacterial Therapy

Topic Editors

Kwang-Sun Kim
Zehra Edis

MDPI

Basel • Beijing • Wuhan • Barcelona • Belgrade • Novi Sad • Cluj • Manchester

Topic Editors
Kwang-Sun Kim
Department of Chemistry
Pusan National University
Busan
Republic of Korea

Zehra Edis
Department of
Pharmaceutical Sciences
Ajman University
Ajman
United Arab Emirates

Editorial Office
MDPI AG
Grosspeteranlage 5
4052 Basel, Switzerland

This is a reprint of the Topic, published open access by the journals *Antibiotics* (ISSN 2079-6382), *Journal of Clinical Medicine* (ISSN 2077-0383), *Pharmaceuticals* (ISSN 1424-8247), and *Pharmaceutics* (ISSN 1999-4923), freely accessible at: https://www.mdpi.com/topics/2LY4E47782.

For citation purposes, cite each article independently as indicated on the article page online and as indicated below:

Lastname, A.A.; Lastname, B.B. Article Title. *Journal Name* **Year**, *Volume Number*, Page Range.

ISBN 978-3-7258-7334-0 (Hbk)
ISBN 978-3-7258-7335-7 (PDF)
https://doi.org/10.3390/books978-3-7258-7335-7

Contents

About the Editors . . . vii

Rosesathorn Soontarach, Potjanee Srimanote, Supayang Piyawan Voravuthikunchai and Sarunyou Chusri
Antibacterial and Anti-Biofilm Efficacy of Endolysin LysAB1245 against a Panel of Important Pathogens
Reprinted from: *Pharmaceuticals* **2024**, *17*, 155, https://doi.org/10.3390/ph17020155 . . . 1

Aleen Clare Golla, Jeanne Chaumontet, Rebecca Vande Voorde and Lia Danelishvili
Discovery of *Mycobacterium avium* subsp. *paratuberculosis* Lytic Phages with Extensive Host Range Across Rapid- and Slow-Growing Pathogenic Mycobacterial Species
Reprinted from: *Antibiotics* **2024**, *13*, 1009, https://doi.org/10.3390/antibiotics13111009 . . . 14

Heba Barnawi, Husam Qanash, Abdu Aldarhami, Ghaida Alsaif, Bandar Alharbi, Majed N. Almashjary, et al.
Antimicrobial, Anticancer, and Antioxidant Activities of Maize and Clover Pollen Grains Extracts: A Comparative Study with Phytochemical Characterizations
Reprinted from: *Pharmaceuticals* **2023**, *16*, 1731, https://doi.org/10.3390/ph16121731 . . . 30

Akbar Rijia, Raman Krishnamoorthi, Madhusoodhanan Rasmi, Pambayan Ulagan Mahalingam and Kwang-sun Kim
Comprehensive Analysis of Bioactive Compounds in Wild *Ganoderma applanatum* Mushroom from Kerala, South India: Insights into Dietary Nutritional, Mineral, Antimicrobial, and Antioxidant Activities
Reprinted from: *Pharmaceuticals* **2024**, *17*, 509, https://doi.org/10.3390/ph17040509 . . . 45

Yang Yue, Zhenbo Xu, Thanapop Soteyome, Mahesh Premarathna, Xiaomao Yin and Junyan Liu
Phage Encapsulation and Delivery Technology: A Strategy for Treating Drug-Resistant Pathogenic Microorganisms
Reprinted from: *Pharmaceuticals* **2025**, *18*, 1688, https://doi.org/10.3390/ph18111688 . . . 67

Abrar K. Thabit, Arwa A. Alghamdi, Afnan K. Alsaeed, Nesereen M. Magbool, Yazed S. Alsowaida, Ahmad J. Mahrous, et al.
Linezolid-Associated Thrombocytopenia: Assessment of Risk Factors in Patients without Hemato-Oncologic Diseases
Reprinted from: *J. Clin. Med.* **2024**, *13*, 2380, https://doi.org/10.3390/jcm13082380 . . . 93

Andrea Piccin, Allameddine Allameddine, Gilbert Spizzo, Katrina M. Lappin and Daniele Prati
Platelet Pathogen Reduction Technology—Should We Stay or Should We Go...?
Reprinted from: *J. Clin. Med.* **2024**, *13*, 5359, https://doi.org/10.3390/jcm13185359 . . . 104

Dzhevdet Chakarov, Elena Hadzhieva, Yordan Kalchev and Dimitar Hadzhiev
Aerobic Microbiological Spectrum and Antibiotic Resistance in Children Operated for Anorectal Abscesses
Reprinted from: *J. Clin. Med.* **2024**, *13*, 2414, https://doi.org/10.3390/jcm13082414 . . . 121

Giedrė Valdonė Sakalauskienė, Lina Malcienė, Edgaras Stankevičius and Aurelija Radzevičienė
Unseen Enemy: Mechanisms of Multidrug Antimicrobial Resistance in Gram-Negative ESKAPE Pathogens
Reprinted from: *Antibiotics* **2025**, *14*, 63, https://doi.org/10.3390/antibiotics14010063 . . . 131

Asmaa S. Salman, Shaza N. Alkhatib, Fatimah M. Ahmed and Ragaa A. Hamouda
Chitosan Nanoparticles Loaded with *Capparis cartilaginea* Decne Extract: Insights into Characterization and Antigenotoxicity In Vivo
Reprinted from: *Pharmaceutics* **2023**, *15*, 2551, https://doi.org/10.3390/pharmaceutics15112551 . **174**

Xiancai Li, Yongqing Li, Binghong Xiong and Shengxiang Qiu
Progress of Antimicrobial Mechanisms of Stilbenoids
Reprinted from: *Pharmaceutics* **2024**, *16*, 663, https://doi.org/10.3390/pharmaceutics16050663 . **193**

Lan Zhang, Xiaoyuan Tian, Lei Sun, Kun Mi, Ru Wang, Fengying Gong and Lingli Huang
Bacterial Efflux Pump Inhibitors Reduce Antibiotic Resistance
Reprinted from: *Pharmaceutics* **2024**, *16*, 170, https://doi.org/10.3390/pharmaceutics16020170 . **214**

Ahmad Khalid Madadi and Moon-Jun Sohn
Comprehensive Therapeutic Approaches to Tuberculous Meningitis: Pharmacokinetics, Combined Dosing, and Advanced Intrathecal Therapies
Reprinted from: *Pharmaceutics* **2024**, *16*, 540, https://doi.org/10.3390/pharmaceutics16040540 . **237**

Elizaveta Dzalamidze, Mylene Gorzynski, Rebecca Vande Voorde, Dylan Nelson and Lia Danelishvili
Discovery of Biofilm-Inhibiting Compounds to Enhance Antibiotic Effectiveness Against *M. abscessus* Infections
Reprinted from: *Pharmaceuticals* **2025**, *18*, 225, https://doi.org/10.3390/ph18020225 . . . . **271**

About the Editors

Kwang-Sun Kim

Kwang-sun Kim is a Professor in the Department of Chemistry at Pusan National University, where he leads the RNA Biochemistry Laboratory. His research sits at the intersection of biochemistry, molecular microbiology, genetics, and bioengineering, with a primary focus on the regulatory mechanisms of antibacterial mechanisms and the development of next-generation antimicrobial strategies.

Zehra Edis

Zehra Edis obtained her BSc., MSc. and PhD. from the University of Cologne, Germany. She conducted her PhD. under the Scholarship of the University of Cologne. She worked as a scientific employee at Akzo Nobel Chemicals and a research intern in Bayer AG, Leverkusen, Germany. She joined Ajman University in 2014. Since August 2021, she has been an Associate Professor in Ajman University, UAE. She has published more than 41 papers and received from the Deanship of Graduate Studies and Research (DGSR), Ajman University, six research grants as Principal Investigator. Her research is multidisciplinary, spanning polymer science, iodine and medicinal chemistry for the synthesis of antimicrobial agents.

pharmaceuticals

MDPI

Article

Antibacterial and Anti-Biofilm Efficacy of Endolysin LysAB1245 against a Panel of Important Pathogens

Rosesathorn Soontarach [1,2], Potjanee Srimanote [3], Supayang Piyawan Voravuthikunchai [1] and Sarunyou Chusri [2,*]

[1] Center of Antimicrobial Biomaterial Innovation-Southeast Asia, Faculty of Science, Prince of Songkla University, Songkhla 90110, Thailand; rosesathron_pim@hotmail.com (R.S.); supayang.v@psu.ac.th (S.P.V.)

[2] Division of Infectious Diseases, Department of Internal Medicine, Faculty of Medicine, Prince of Songkla University, Songkhla 90110, Thailand

[3] Graduate in Biomedical Sciences, Faculty of Allied Health Sciences, Thammasat University, Pathum Thani 12121, Thailand; psrimanote01@yahoo.com.au

* Correspondence: sarunyouchusri@hotmail.com

Abstract: Infections caused by antibiotic-resistant bacteria pose a significant global challenge. This study explores the antibacterial effects of a bacteriophage-derived endolysin, LysAB1245, against important pathogens, including *Acinetobacter baumannii, Escherichia coli, Klebsiella pneumoniae, Pseudomonas aeruginosa,* and *Staphylococcus aureus.* We determined the minimal inhibitory concentration (MIC) and minimal bactericidal concentration (MBC) for all tested isolates. A time–kill study was conducted to evaluate the reduction in bacterial survival following treatment with LysAB1245. Additionally, the effects of LysAB1245 on *P. aeruginosa* K1455 and methicillin-resistant *S. aureus* (MRSA) NPRC 001R-formed biofilms were investigated. The MIC and MBC of LysAB1245 against all the tested isolates ranged from 4.68 to 9.36 μg/mL and 4.68 to 18.72 μg/mL, respectively. The time–kill study demonstrated more than a 4 log CFU/mL (99.99%) reduction in bacterial survival within 6 h of LysAB1245 treatment at 2MIC. LysAB1245 (1/8–1/2MIC) treatment significantly reduced biofilms formed by *P. aeruginosa* and MRSA in a concentration-dependent manner. Furthermore, scanning electron and confocal laser scanning microscopy confirmed the potential inhibition effects on 3-day established biofilms formed on abiotic surfaces upon treatment with LysAB1245 at 2MIC. The findings indicate that endolysin LysAB1245 could be employed as a new alternative therapeutic antibacterial and anti-biofilm agent for combating biofilm-related infections.

Keywords: antibiotic-resistant bacteria; endolysin; antibacterial agent; anti-biofilm agent

1. Introduction

The emergence of antimicrobial-resistant pathogens has become one of the most pressing public health threats [1]. Infections caused by these bacteria are often associated with inappropriate and insufficient use of broad-spectrum antibiotics, leading to elevated morbidity and mortality rates, prolonged hospitalization, and increased healthcare costs [2,3]. Gram-positive and Gram-negative bacteria with multiple antibiotic resistances pose a considerable risk for hospital-acquired infections. In 2017, strains of *Staphylococcus aureus, Klebsiella pneumoniae, Acinetobacter baumannii,* and *Pseudomonas aeruginosa* were designated as global priority pathogens, necessitating urgent research and the development of new antibiotics [4]. These pathogens exhibit a high degree of antibiotic resistance through various mechanisms [5].

Biofilms represent a crucial virulence factor in microbial survival under extreme environmental conditions and severe chronic bacterial infections [6,7]. The structural composition of polysaccharides and cellular polymeric substances in biofilms provides an effective diffusion barrier, limiting the penetration of antibacterial agents [8,9]. Numerous studies have demonstrated the heightened antibiotic resistance of bacteria associated with a biofilm-producing phenotype [10–12]. Given the escalating antibiotic resistance among pathogenic bacteria, bacteriophage-encoded endolysins (lysins) are emerging as promising treatment modalities featuring alternative antibacterial agents [13–18].

Endolysins are phage-encoded peptidoglycan-degrading enzymes typically used by most dsDNA bacteriophages to lyse the peptidoglycan layer of their hosts and facilitate progeny release at the end of lytic cycle. Recently, novel lytic phages against multidrug-resistant *A. baumannii* strains were isolated from sewage water [19]. The study suggested the phages encoding important enzymes, including endolysin, could serve as potential alternative antibacterial agents against *A. baumannii* infections [19]. Subsequently, a novel phage endolysin, LysAB1245, was successfully expressed, purified, and characterized [20]. This study demonstrated the lytic activity of LysAB1245 against *A. baumannii* isolates with distinct capsular types; however, its activity in other species has not been explored. We hypothesized that exploring the lytic activity of LysAB1245 in other closely related species could enhance the significance of these findings. Therefore, this study aimed to evaluate the antibacterial and antibiofilm effects of LysAB1245 on a panel of pathogenic Gram-negative and Gram-positive bacteria.

2. Results and Discussion

2.1. Antibacterial Activity Assays

Minimal inhibitory concentration (MIC) and minimal bactericidal concentration (MBC) values of LysAB1245 are presented in Table 1. Endolysin LysAB1245 exhibited broad-spectrum antibacterial effects, with MIC values ranging from 4.68 to 9.36 μg/mL. LysAB1245 exhibited bactericidal activity at MBC range of 4.68 to 18.72 μg/mL. The results revealed that LysAB1245 from a Gram-negative background exhibited potent broad-spectrum antibacterial effects on various Gram-negative and Gram-positive bacteria. In a previous study, endolysin HY-133 displayed MIC50/90 and MBC50/90 values ranging from 0.12 to 0.5 mg/L against methicillin-susceptible *S. aureus* and MRSA isolates [21]. The MIC and MBC values of endolysin LysSS against reference strains of common hospital pathogens, including *A. baumannii, K. pneumoniae, P. aeruginosa, E. coli,* and *S. aureus*, ranged from 0.25 to more than 2 mg/mL [22]. Although our study, consistent with these previous studies, revealed the efficacy of an endolysin against different bacterial strains, the variation in the MIC and MBC values could be attributed to the variations in the secondary and tertiary conformation of the protein and antibacterial efficacy resulting from the unique structural domains of LysAB1245 [20].

Table 1. Minimal inhibitory concentrations (MICs) and minimal bactericidal concentrations (MBCs) of LysAB1245 against pathogenic bacteria.

Pathogenic Organisms	Values of MIC/MBC (μg/mL)
Gram-negative	
Acinetobacter baumannii	
NPRCoE 160516	4.68/4.68
ATCC 19606	4.68/4.68
Escherichia coli	
NPRCoE 161012	9.36/18.72
ATCC 25922	4.68/9.36
Klebsiella pneumoniae	
NPRCoE 160611	9.36/18.72
ATCC 700603	4.68/9.36
Pseudomonas aeruginosa	
K767 (PA01)	4.68/9.36
K1455 (PA01-*nalB*)	4.68/9.36
K1523 (PA01-Δ*mexB*)	4.68/9.36
Gram-positive	
Staphylococcus aureus	
MRSA NPRC 001R	9.36/18.72
ATCC 25923	4.68/9.36

2.2. Time–Kill Assay

The time–kill study was performed to examine the bactericidal kinetics of LysAB1245 on *P. aeruginosa* (PA01 and K1455) and *S. aureus* (MRSA NPRC 001R and ATCC 25923) strains. The results revealed that LysAB1245 showed the concentration-dependent killing of all tested strains (Figure 1). LysAB1245 at 2MIC (9.36 μg/mL) reduced the bacterial viability of both *P. aeruginosa* PA01 and K1455 strains by more than 4 log CFU/mL or 99.99% within 4 h (Figure 1A,B). In addition, LysAB1245 at 2MIC (18.72 μg/mL) exhibited bactericidal effects on MRSA R001, resulting in an approximately 4 log CFU/mL decrease in the number of viable cells within 6 h (Figure 1C). For *S. aureus* ATCC 25923, a reduction of over 5 log CFU/mL in the viable bacterial count was observed after 4 h of incubation with LysAB1245 at 2MIC (9.36 μg/mL) (Figure 1D).

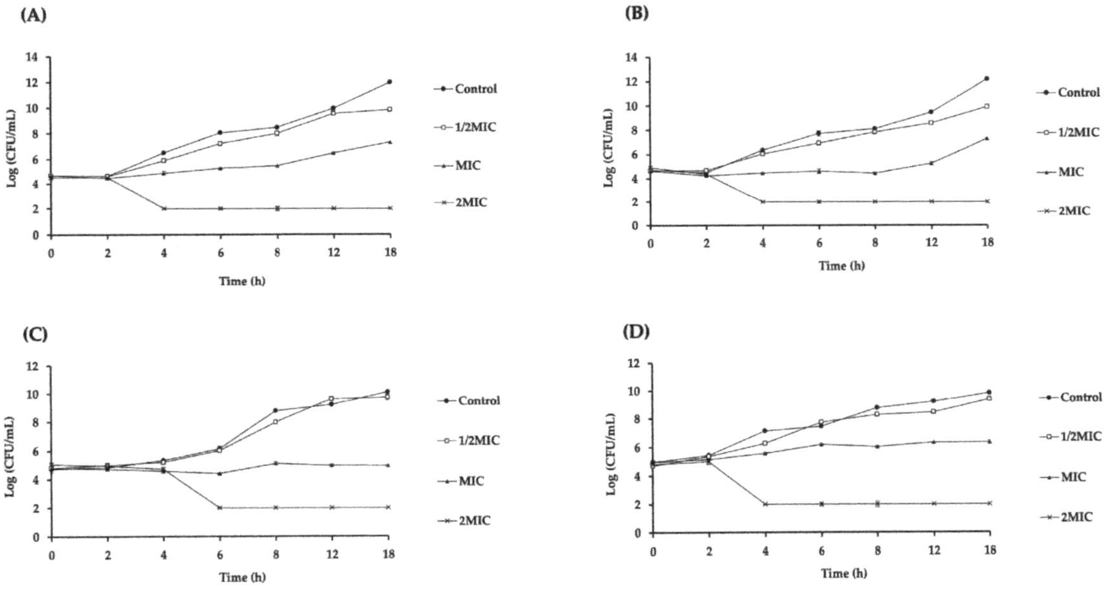

Figure 1. Time–kill curves of Gram-negative and Gram-positive bacteria after treatment with endolysin LysAB1245. Bacterial cultures of *Pseudomonas aeruginosa* PA01 (**A**), *P. aeruginosa* K1455 (**B**), MRSA NPRC 001R (**C**), and *S. aureus* ATCC 25923 (**D**) at 10^6 CFU/mL were incubated with LysAB1245 at 1/2-2MIC for 18 h. Results obtained from two independent experiments performed in triplicate are expressed as mean ± standard deviation (SD).

In this study, two virulent bacterial strains, including *P. aeruginosa* with an overexpressed MexAB-OprM efflux pump (K1455) and MRSA with mecA gene (NPRC 001R), were used to investigate the bactericidal kinetics of LysAB1245. According to a previous study, the N-terminal catalytic domain of phage endolysin LysAB1245 comprises glycosidase domains (N-acetyl-β-D-muramidases) [20]. LysAB1245 disrupts the peptidoglycan cell wall, leading to cell lysis by cleaving the β-1,4 glycosidic bonds between glycan strands [23]. Similar to endolysin SPN9CC, LysAB1245 exhibited exogenous bacteriolytic activity against Gram-negative bacteria without pretreatment with outer membrane permeabilization [24]. These findings suggest that endolysin LysAB1245 could be employed as a potential therapeutic agent effective against both pathogenic Gram-negative and Gram-positive bacteria.

Previous studies have reported that even after several cycles of repeated exposure to phage endolysin at 1/2MIC or MIC, no resistance development was observed in *S. aureus* strains [25,26]. Moreover, endolysin LysAB1245 was derived from a lytic phage T1245. Together, these studies suggest that, within the host strain, endolysin is consistently expressed, and host cells are continually exposed to endolysin without compromising their susceptibility to it. Consequently, the likelihood of bacterial cells developing resistance against endolysin is deemed improbable or occurs at extremely low frequencies.

2.3. Effects of LysAB1245 on A. baumannii Cell Surface by Scanning Electron Microscope

The morphology of *A. baumannii* cells after exposure to LysAB1245 at subinhibitory concentration was visualized using a scanning electron microscope (SEM). The scanning electron micrographs demonstrated shrinking deformation of the LysAB1245-exposed cells, whereas untreated cells displayed normal rod-shaped morphology with smooth and intact surfaces (Figure 2). Together with previous studies, our study suggests that a lysozyme-like catalytic domain of bacteriophage endolysin is pivotal in disrupting the integrity of the bacterial outer membrane, resulting in the leakage of cytoplasmic content [27,28].

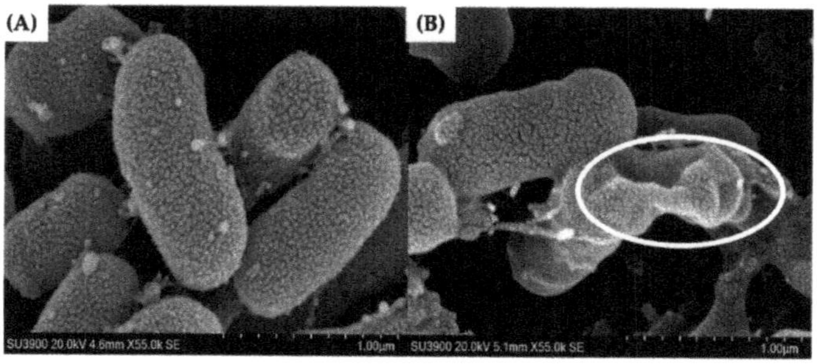

Figure 2. Scanning electron micrographs of untreated (**A**) and LysAB1245-treated *A. baumannii* ABMYH-1245 cells (**B**) at a magnification of 55,000×. The circle indicates the presence of morphological changes.

2.4. Inhibition of Biofilm Formation

Biofilms are a survival strategy of several microorganisms against host defense mechanisms or stressed conditions. The biofilm-forming ability of all the clinically tested isolates was investigated using crystal violet staining. The results revealed that all strains of *P. aeruginosa* evaluated in this study were strong biofilm-producers. The pathogenic isolates of *P. aeruginosa* are commonly used as model organisms to study the formation of biofilm due to their potent biofilm-producing capacity [29]. Therefore, MexAB-OprM overexpressed strain K1455, which is a high biofilm-producing strain of *P. aeruginosa*, was selected as a representative strain to observe the inhibitory effects of LysAB1245 on bacterial biofilm formation. Furthermore, MRSA NPRC 001R, a moderate level of biofilm-forming ability, was also selected as a representative Gram-positive bacterium to assess the anti-biofilm efficacy of LysAB1245.

Endolysin is a bacteriophage-associated peptidoglycan hydrolase with the ability to prevent bacterial biofilm formation. This study revealed that the anti-biofilm effects of LysAB1245 against biofilm formed by *P. aeruginosa* and MRSA strains were concentration-dependent. Furthermore, the crystal violet assay demonstrated a significant reduction in biofilm biomass with a maximum reduction at 1/2MIC of LysAB1245 ($p < 0.0001$; Figure 3). For *P. aeruginosa* strain K1455, treatment with LysAB1245 at concentrations of 1/2MIC, 1/4MIC, 1/8MIC, 1/16MIC, and 1/32MIC led to biofilm reductions of approximately 80.54 ± 6.43%, 67.65 ± 2.0%, 52.91 ± 4.22%, 34.81 ± 10.64%, and 20.05 ± 7.45%, respectively, compared with untreated biofilms (Figure 3A). Similarly, in MRSA NPRC 001R, LysAB1245 treatment (at 1/2MIC, 1/4MIC, 1/8MIC, 1/16MIC, and 1/32MIC) led to biofilm reductions of approximately 76.48 ± 2.89%, 62.23 ± 3.55%, 48.70 ± 5.47%, 33.25 ± 9.02%, and 27.87 ± 4.82%, respectively, compared with untreated biofilms (Figure 3B).

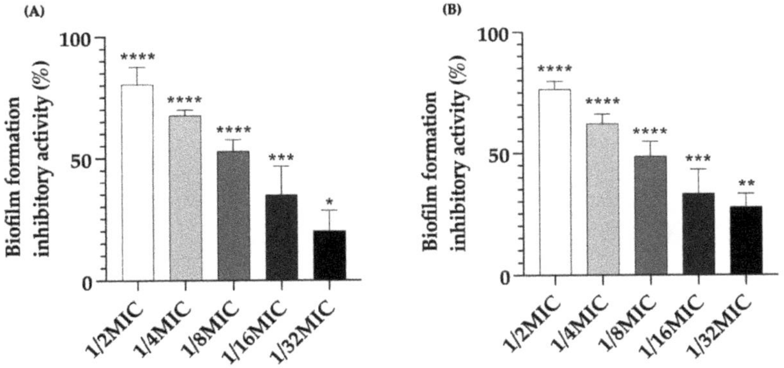

Figure 3. Anti-biofilm effects of LysAB1245 at 1/32-1/2MIC on biofilm formed by *P. aeruginosa* K1455 (**A**) and MRSA NPRC 001R (**B**). Significant reduction in biofilm biomass, compared with control, * $p < 0.05$, ** $p < 0.01$ *** $p < 0.001$, and **** $p < 0.0001$. Results obtained from two independent experiments performed in triplicate are expressed as mean ± standard deviation (SD).

2.5. Effects of LysAB1245 on Established Biofilm Biomass and Viability of Biofilm Cells

After the formation of a biofilm, bacterial cells exhibit remarkable resistance to various antibiotics. Therefore, the effects of LysAB1245 on 3-day established bacterial biofilms were investigated by measuring biofilm biomass (Figure 4). For *P. aeruginosa* K1455 established biofilms, a significant reduction in biofilm formation was observed, with approximately 5.89 ± 1.81%, 21.03 ± 1.62%, 26.32 ± 1.76%, and 75.26 ± 5.33% of biofilm formation after treatment with LysAB1245 at 8MIC, 4MIC, 2MIC, and MIC, respectively (Figure 4A). Additionally, treatments with LysAB1245 at 8MIC, 4MIC, 2MIC, and MIC significantly reduced established biofilms of MRSA NPRC 001R, with biofilm formation percentages of 11.26 ± 2.56%, 26.14 ± 6.08%, 53.31 ± 2.07%, and 61.06 ± 2.45%, respectively (Figure 4B).

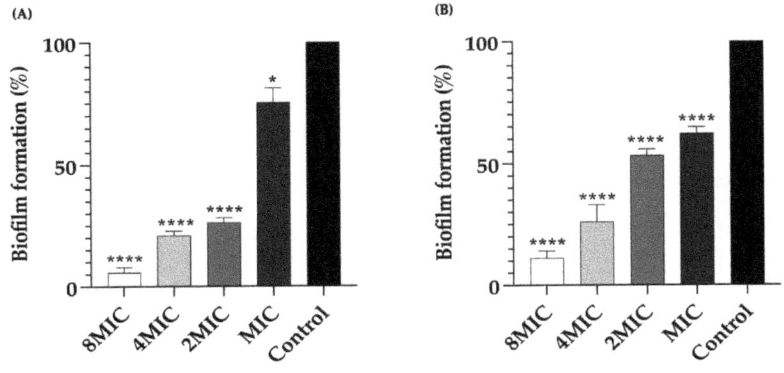

Figure 4. Removal of 3-day established biofilms formed by *P. aeruginosa* K1455 and MRSA NPRC 001R. Biofilms of *P. aeruginosa* K1455 (**A**) and MRSA NPRC 001R (**B**) were treated with LysAB1245 at 1-8MICs for 24 h. Significant reduction in biofilm biomass, compared with control, * $p < 0.05$, and **** $p < 0.0001$. Results obtained from two independent experiments performed in triplicate are expressed as mean ± standard deviation (SD).

Furthermore, the viability of bacterial cells in biofilms was determined using the MTT colorimetric method (Figure 5). As shown in Figure 5A, treatment with LysAB1245 at concentrations ranging from MIC to 8MIC significantly reduced cell viability of *P. aeruginosa* K1455 and MRSA NPRC 001R 3-day established biofilms in a dose-dependent manner. LysAB1245 at 8MIC exhibited a maximum significant reduction in cell viability ($p < 0.0001$).

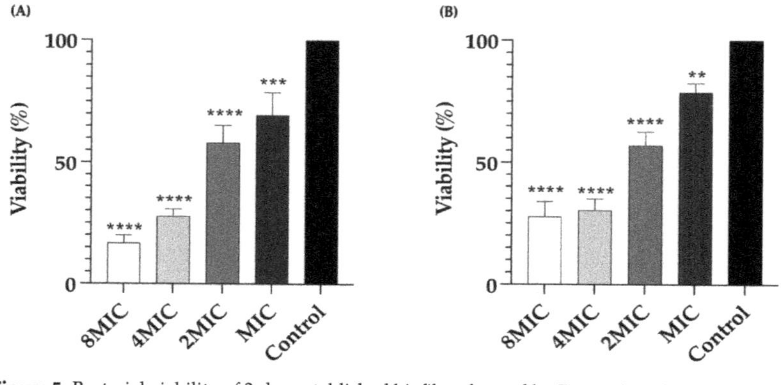

Figure 5. Bacterial viability of 3-day established biofilms formed by *P. aeruginosa* K1455 (**A**) and MRSA NPRC 001R (**B**) after treatment with LysAB1245 at 1–8MICs for 24 h. Significant reduction in biofilm biomass, compared with control, ** $p < 0.01$ *** $p < 0.001$, and **** $p < 0.0001$. Results obtained from two independent experiments performed in triplicate are expressed as mean ± standard deviation (SD).

Biofilms constitute intricate structures of bacterial communities in which cells are embedded within a self-produced matrix. Bacteria living in a biofilm exhibit greater resistance to antibacterial agents compared to their planktonic cells [30–32]. Biofilms formed by *P. aeruginosa* play a significant role in the persistence and endemic spread of nosocomial infections [33,34]. Furthermore, persistent MRSA infections are frequently identified in isolates from patients with cystic fibrosis [35,36]. The formation of biofilms on living tissues or medical devices emerges as a crucial virulence factor in many pathogenic bacteria, leading to chronic infections that significantly complicate the treatment process [37,38]. Due to the association between biofilm formation and antibacterial resistance in many pathogenic bacteria, the prevention and eradication of biofilm formation have become preferred strategies for addressing biofilm-related challenges. This study highlights the importance of the anti-biofilm properties of LysAB1245 in combating bacteria growing in biofilms. The results demonstrated that LysAB1245 induces a significant reduction in both biomass and viability of biofilm cells, established over three days, in both Gram-negative and Gram-positive strains. Therefore, these findings suggest that endolysin LysAB1245 could serve as a promising novel alternative therapeutic approach for tackling biofilm formation and eradication.

2.6. Scanning Electron Micrographs of P. aeruginosa K1455 and MRSA NPRC 001R 3-Day Established Biofilms after Exposure to LysAB1245

Scanning electron microscopy was performed to confirm the effects of LysAB1245 on the morphology of established biofilm cells formed on glass slide surfaces. As shown in Figure 6, treatment with LysAB1245 at 2MIC demonstrated effective removal of 3-days established biofilms for both *P. aeruginosa* K1455 and MRSA NPRC 001R strains, compared with untreated biofilms. In addition, cell membrane deformation and poration were observed after exposure to LysAB1245 at 2MIC for 24 h (Figure 7).

In this study, endolysin LysAB1245 demonstrated promising potential as a novel alternative therapeutic for degrading preformed biofilms produced by *P. aeruginosa* K1455 and MRSA NPRC 001R strains on abiotic surfaces. A previous study has shown that treatment with LysSTG2 at 1 mg/mL for 2 h effectively removed over 99.9% of biofilms formed by *P. aeruginosa* and *P. putida* on a stainless-steel surface [39]. Additionally, the activity of endolysin XZ.700 against MRSA biofilms formed on titanium discs was observed, with no cytotoxic effects on human bone cells after 48 h of exposure at 50 μg/mL [40]. Moreover, the lytic activities of endolysin LysH5 have been reported not only against staphylococcal biofilms but also against rifampicin and ciprofloxacin-induced persister cells obtained by treatment with rifampicin [41].

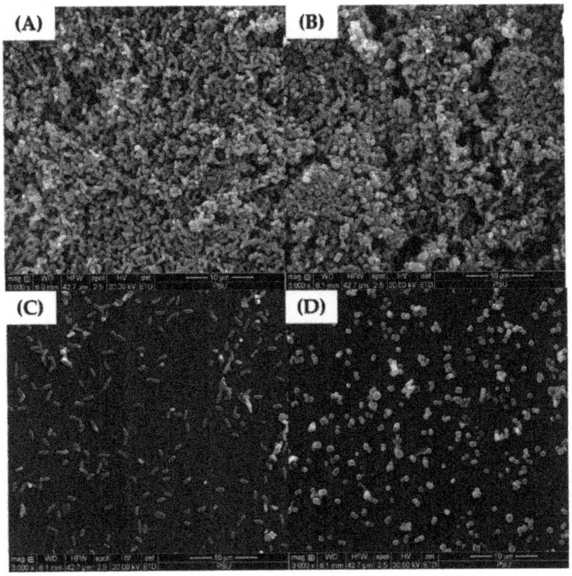

Figure 6. Scanning electron micrographs of 3-day established biofilms on glass slide surfaces at a magnification of 3000×. Scanning electron micrographs of untreated established biofilms formed by *P. aeruginosa* K1455 (**A**) and MRSA NPRC 001R (**B**) present thick bacterial aggregations. Biofilms after treatment with LysAB1245 at 2MIC for 24 h were observed on *P. aeruginosa* K1455 (**C**) and MRSA NPRC 001R (**D**).

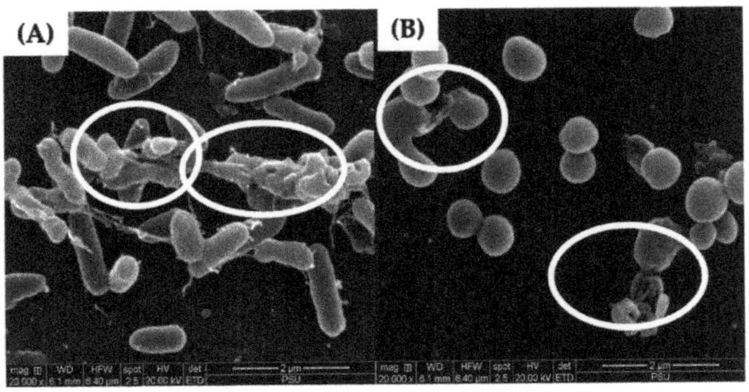

Figure 7. Scanning electron micrographs of 3-day established biofilms formed on glass slide surfaces at a magnification of 20,000×. After treatment with LysAB1245 at 2MIC for 24 h, the destruction of 3-day established biofilms was observed on *P. aeruginosa* K1455 (**A**) and MRSA NPRC 001R (**B**) as indicated by the circle symbol.

2.7. *Confocal of P. aeruginosa K1455 3-Day Established Biofilms after Exposure to LysAB1245*

A three-dimensional (3D) structural image of live and dead cells within the matrix of established biofilms formed by a representative strain of *P. aeruginosa* K1455 on glass surfaces was visualized using confocal laser scanning microscopy (CLSM; Figure 8). The results revealed that LysAB1245 at 4MIC effectively disrupted preformed biofilms (Figure 8A). At 2MIC, a reduction in cell viability was observed, with live cells fluorescing green (Syto 9 dye) and dead cells stained red (propidium iodide) (Figure 8B). In contrast, untreated biofilms displayed dense aggregations of living bacterial cells with green fluorescence (Figure 8C).

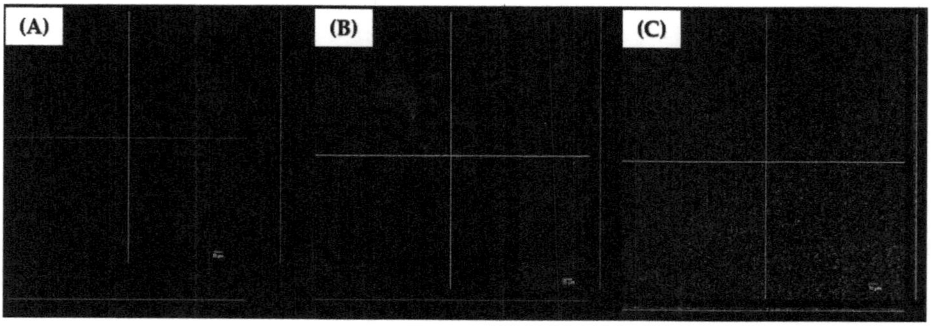

Figure 8. Confocal laser scanning microscopy (CLSM) images of 3-day established biofilms formed by *P. aeruginosa* K1455 on glass slide surfaces. Biofilms were treated with LysAB1245 at 4MIC (**A**), 2MIC (**B**), and control (**C**) for 24 h. All treatments were stained with live/dead staining reagents and subsequent fluorescence microscopy imaging. Living cells are indicated by green Syto 9 staining, and dead bacterial cells are shown in red due to propidium iodide staining.

Based on the findings of this study, the phage-derived endolysins LysAB1245 is considered a potential antibacterial agent with antibiofilm properties for controlling both Gram-negative and Gram-positive bacteria. Furthermore, the topical skin application of the phage endolysin formulated as cream has been proven a safe alternative to antibiotic therapy without causing long-term effects and disrupting the normal flora through preclinical and clinical trials [42]. In 2017, successful treatment was reported for three cases of atopic dermatitis caused by *S. aureus* infections using long-term daily topical Staphefekt SA.100 therapy [43]. Furthermore, the use of endolysins as effective biocontrol agents against various food-borne pathogens holds promise for future applications in the food industry [39,44–46]. However, endolysin LysAB1245 is still in the early stages of development. Considerations should be given to limitations such as pharmacokinetics, route of delivery, and optimal doses.

3. Materials and Methods

3.1. Bacterial Strains and Culture Conditions

Pseudomonas aeruginosa PAO1 strain K767 (wild-type), MexAB-OprM overexpressed strain K1455 (PAO1-nalB), and MexB deletion strain K1523 (PAO1-ΔmexB) were provided by Professor Dr. R. Keith Poole, Queen's University, Kingston, Ontario, Canada. All clinical isolates, including *Acinetobacter baumannii*, *Escherichia coli*, *Klebsiella pneumoniae*, and methicillin-resistant *Staphylococcus aureus* (MRSA) with mecA gene, were obtained from Songklanagarind Hospital (Ethical Approval No. REC 59–241–19–6). Reference strains of *A. baumannii* ATCC 19606, *E. coli* ATCC 25922, *K. pneumoniae* ATCC 700603, and *S. aureus* ATCC 25923 were used in this study. The bacteria were cultured in Luria Bertani (LB) (Difco Laboratories, Detroit, MI, USA) at 37 °C for 18 h and maintained in 20% glycerol at −80 °C.

3.2. Determination of MIC and MBC

The antibacterial efficacy of endolysin LysAB1245 was determined by standard broth microdilution assay according to Clinical and Laboratory Standard Institute (CLSI) M100-ED33 methods [47]. Briefly, LysAB1245 was two-fold serially diluted in concentrations ranging from 1.05 to 33.68 μg/mL in Mueller Hinton broth (MHB, Difco) in 96-well polystyrene microtiter plates. Subsequently, an equal volume of logarithmic phase bacterial cultures at a density of 1×10^6 CFU/mL was added to each well. After incubation at 37 °C for 18 h, the lowest concentration that inhibited bacterial growth was determined as MIC values.

The MBC value is defined as the lowest concentration, which had no bacterial growth on the agar plates. The results were obtained from two independent experiments performed in triplicate.

3.3. Kinetics Analysis

A time–kill kinetic assay was performed on two representative Gram-negative and Gram-positive bacteria (*P. aeruginosa* and *S. aureus*). Bacterial cultures at 1×10^6 CFU/mL were added to MHB containing LysAB1245 at 1/2MIC, MIC, and 2MIC and incubated at 37 °C for 2, 4, 6, 8, 12, and 16 h. Samples from each time interval were collected, serially diluted 10-fold, and plated on Mueller Hinton agar to determine the number of log CFU/mL. The results were obtained from two independent experiments performed in triplicate and were represented as mean ± SD (standard deviation).

3.4. Effects of LysAB1245 on A. baumannii Cell Surfaces Using Scanning Electron Microscope

The potential effects of LysAB1245 on *A. baumannii* ABMYSP-1245, a primary host for phage T1245, were observed. Briefly, bacterial culture at 1×10^6 CFU/mL was treated with LysAB1245 at 1/2MIC (2.34 μg/mL). The mixture of bacterial cells and MHB served as a control. After incubation at 37 °C for 18 h, the supernatant was discarded by centrifugation at 10,000× *g* for 5 min. The pellets were resuspended with 10 mM phosphate-buffered saline (PBS, pH 7.4), fixed with 2.5% (*w*/*v*) glutaraldehyde (Sigma-Aldrich, St Louis, MO, USA), and dehydrated in a graded series of ethanol. After drying with CO_2 and coating with gold, the specimens were examined using a HITACHI SU3900 instrument.

3.5. Screening of Biofilm Formation

All tested clinical isolates were used to evaluate the biofilm-forming ability as described in a previous study [48]. All bacterial isolates were grown overnight at 37 °C in trypticase soy broth (TSB, Difco) with 0.25% glucose. The cultures were then adjusted to 1×10^8 CFU/mL, inoculated into 96-well microtitre plates (200 μL/well), and incubated at 37 °C for 24 h. The contents of the wells were discarded and gently washed twice with 200 μL PBS. The plates were then dried, stained with 0.1% crystal violet for 30 min, and washed twice with distilled water. After drying of the plates, stained biofilm was dissolved in 100 μL dimethyl sulphoxide (DMSO; Merck, Elkton, VA, USA) and the absorbance was measured at 595 nm using a microplate spectrophotometer. All isolates were assigned into four categories according to the following criteria: optical density at (OD595) < 1, non-biofilm forming; 1 < OD595 < 2, weak; 2 < OD595 < 3, medium; and 3 < OD595, strong.

3.6. Effects of LysAB1245 on Biofilm Formation

The anti-biofilm efficacy of LysAB1245 against two representative bacterial strains (*P. aeruginosa* K1455) was investigated [49]. Fifty microliters of bacterial suspensions (at 1×10^6 CFU/mL) were added to the wells containing 50 μL diluted LysAB145 (at 1/32–1/2MIC) or TSB-0.25% glucose (control). After incubation at 37 °C for 18 h, all treatment and control wells were discarded and washed twice with PBS. The amounts of biofilm masses formed on the 96-well plate were measured using the crystal violet staining assay as described in the above method. The results were obtained from two independent experiments performed in triplicate. The following formula was used to calculate the percentage of biofilm formation inhibitory activity:

$$\text{Biofilm formation inhibitory activity } (\%) = \frac{(OD_{control} - OD_{blank}) - (OD_{treated} - OD_{blank})}{(OD_{control} - OD_{blank})} \times 100$$

3.7. Established Biofilms Eradication Ability of LysAB1245 and Viability of Biofilm Cells

Established biofilms of *P. aeruginosa* K1455 and MRSA NPRC 001R were grown as described in a previous study with slight modifications [50]. Briefly, 200 μL bacterial suspension in TSB-0.25% glucose at 1×10^8 CFU/mL was added to 96-well plates and incubated at 37 °C. For 3-day biofilms, planktonic cells were removed, and fresh TSB-0.25%

glucose was added daily. After 3 days of biofilm growth, non-adherent cells were removed, and biofilms were subsequently treated with 100 µL LysAB1245 at different concentrations (1–8MIC) and incubated at 37 °C for 24 h. To evaluate the ability of LysAB1245 to eradicate established bacterial biofilms, crystal violet staining was employed as described in the above method.

To analyze the bacterial cell viability in biofilms, the MTT reduction assay was performed [38]. Briefly, the biofilms were treated with LysAB1245 for 24 h, and the contents of the wells were removed, replaced with 100 µL MTT solution (0.5 mg/mL in PBS; Sigma-Aldrich), and incubated at 37 °C for 2 h. DMSO was used to solubilize the insoluble formazan crystals, and the cell viability was assessed by measuring OD570 using a microplate spectrophotometer. The results were obtained from two independent experiments performed in triplicate. The following formula was used to calculate the percentage of biofilm formation and viability of biofilm cells:

$$\text{Biofilm formation and viability of biofilm cells } (\%) = \frac{(OD_{treated} - OD_{blank})}{(OD_{control} - OD_{blank})} \times 100$$

3.8. Scanning Electron Micrographs of *P. aeruginosa* K1455 3-Day Established Biofilms after Exposure to LysAB1245

The effects of LysAB1245 on established biofilms formed by *P. aeruginosa* K1455 and MRSA NPRC 001R were examined on glass slide surfaces following a previously described method with slight modifications [51]. Briefly, 1 mL of bacterial suspension (at 1×10^8 CFU/mL) was added to 24 wells, with each well containing a glass piece (1 × 1 cm). After incubation for 3 days, the biofilms on glass slides were treated with 500 µL LysAB1245 (at 2MIC) and further incubated at 37 °C for 24 h. The established biofilms on glass slides were incubated with 2.5% glutaraldehyde (Sigma-Aldrich) for 2 h at room temperature. Samples were dehydrated in a graded series of ethanol (20–99.9%), mounted on aluminum stubs, critical point dried, coated with gold, and examined by SEM (Quanta 400 FEG; FEI).

3.9. Confocal Laser Scanning Microscope (CLSM)

The effects of LysAB1245 on 3D visualization of biofilm structure were monitored using CLSM. Established biofilms of *P. aeruginosa* K1455 were developed on glass slide surfaces and treated with LysAB1245 (at 2 and 4MIC). The established biofilms on glass slides were stained with a Live/Dead BacLight bacterial viability kit (Invitrogen, Carlsbad, CA, USA) and incubated for 15 min in the dark. The samples were observed under a CLSM (OLYMPUS FV3000).

3.10. Statistical Analysis

Statistical analysis was performed using GraphPad Prism v8. Dunnett's test was used for comparison of means at a 95% confidential interval.

4. Conclusions

This study highlights the potential of bacteriophage-derived endolysin LysAB1245 as a versatile antibacterial agent with significant applications in controlling both Gram-negative and Gram-positive bacteria. Treatment with LyAB1245 not only significantly reduced biofilm formation but also eradicated preformed biofilms. In addition, results of SEM and CLSM revealed the disruption of biofilm structure and cell morphology of bacterial 3-day established biofilms on glass slide surfaces. Therefore, LysAB1245 is considered a new promising agent to prevent and control clinically relevant biofilms. However, further investigation into the development of LysAB1245-based therapy for suitable applications should be considered.

Author Contributions: Conceptualization, R.S., S.P.V. and S.C.; methodology, R.S.; validation, R.S. and P.S.; formal analysis, R.S.; investigation, S.P.V.; resources, P.S.; data curation, R.S.; writing—original draft preparation, R.S.; writing—review and editing, R.S., S.P.V. and S.C.; visualization, R.S. and P.S.; supervision, P.S. and S.P.V.; project administration, S.P.V. and S.C.; funding acquisition, S.P.V. and S.C. All authors have read and agreed to the published version of the manuscript.

Funding: This research was funded by the National Research Council of Thailand (Grant No. N41A640071) and the Postdoctoral Fellowship from Prince of Songkla University, Thailand.

Institutional Review Board Statement: The study was conducted in accordance with the Declaration of Helsinki and approved by the Human Research Ethics Committee (HREC) of Prince of Songkla University (protocol code: 59-241-19-6, date of approval: 8 August 2016).

Informed Consent Statement: Not applicable.

Data Availability Statement: Data are available in a publicly accessible repository and within the article.

Conflicts of Interest: The authors declare no conflicts of interest.

References

1. Hou, J.; Long, X.; Wang, X.; Li, L.; Mao, D.; Luo, Y.; Ren, H. Global trend of antimicrobial resistance in common bacterial pathogens in response to antibiotic consumption. *J. Hazard. Mater.* **2023**, *442*, 130042. [CrossRef] [PubMed]
2. Peters, L.; Olson, L.; Khu, D.T.; Linnros, S.; Le, N.K.; Hanberger, H.; Hoang, N.T.; Tran, D.M.; Larsson, M. Multiple antibiotic resistance as a risk factor for mortality and prolonged hospital stay: A cohort study among neonatal intensive care patients with hospital-acquired infections caused by gram-negative bacteria in Vietnam. *PLoS ONE* **2019**, *14*, e0215666. [CrossRef] [PubMed]
3. Touat, M.; Opatowski, M.; Brun-Buisson, C.; Cosker, K.; Guillemot, D.; Salomon, J.; Tuppin, P.; De Lagasnerie, G.; Watier, L. A payer perspective of the hospital inpatient additional care costs of antimicrobial resistance in France: A matched case–control study. *Appl. Health Econ. Health Policy* **2019**, *17*, 381–389. [CrossRef] [PubMed]
4. Asokan, G.V.; Ramadhan, T.; Ahmed, E.; Sanad, H. WHO global priority pathogens list: A bibliometric analysis of Medline-PubMed for knowledge mobilization to infection prevention and control practices in Bahrain. *Oman Med. J.* **2019**, *34*, 184. [CrossRef] [PubMed]
5. De Oliveira, D.M.; Forde, B.M.; Kidd, T.J.; Harris, P.N.; Schembri, M.A.; Beatson, S.A.; Paterson, D.L.; Walker, M.J. Antimicrobial resistance in ESKAPE pathogens. *Clin. Microbiol. Rev.* **2020**, *33*, e00181-19. [CrossRef] [PubMed]
6. Akinduti, P.A.; George, O.W.; Ohore, H.U.; Ariyo, O.E.; Popoola, S.T.; Adeleye, A.I.; Akinwande, K.S.; Popoola, J.O.; Rotimi, S.O.; Olufemi, F.O. Evaluation of efflux-mediated resistance and biofilm formation in virulent *Pseudomonas aeruginosa* associated with healthcare infections. *Antibiotics* **2023**, *12*, 626. [CrossRef] [PubMed]
7. Schapira, A.-J.; Drame, M.; Olive, C.; Marion-Sanchez, K. Bacterial viability in dry surface biofilms in healthcare facilities–A systematic review. *J. Hosp. Infect.* **2023**, *144*, 94–110. [CrossRef]
8. Anderl, J.N.; Franklin, M.J.; Stewart, P.S. Role of antibiotic penetration limitation in *Klebsiella pneumoniae* biofilm resistance to ampicillin and ciprofloxacin. *Antimicrob. Agents Chemother.* **2000**, *44*, 1818–1824. [CrossRef]
9. Singh, R.; Ray, P.; Das, A.; Sharma, M. Penetration of antibiotics through *Staphylococcus aureus* and *Staphylococcus epidermidis* biofilms. *J. Antimicrob. Chemother.* **2010**, *65*, 1955–1958. [CrossRef]
10. Parastan, R.; Kargar, M.; Solhjoo, K.; Kafilzadeh, F. *Staphylococcus aureus* biofilms: Structures, antibiotic resistance, inhibition, and vaccines. *Gene Rep.* **2020**, *20*, 100739. [CrossRef]
11. Shenkutie, A.M.; Yao, M.Z.; Siu, G.K.-h.; Wong, B.K.C.; Leung, P.H.-m. Biofilm-induced antibiotic resistance in clinical *Acinetobacter baumannii* isolates. *Antibiotics* **2020**, *9*, 817. [CrossRef]
12. Senobar Tahaei, S.A.; Stájer, A.; Barrak, I.; Ostorházi, E.; Szabó, D.; Gajdács, M. Correlation between biofilm-formation and the antibiotic resistant phenotype in *Staphylococcus aureus* isolates: A laboratory-based study in Hungary and a review of the literature. *Infect. Drug Resist.* **2021**, *14*, 1155–1168. [CrossRef]
13. Nakamura, T.; Kitana, J.; Fujiki, J.; Takase, M.; Iyori, K.; Simoike, K.; Iwano, H. Lytic activity of polyvalent staphylococcal bacteriophage PhiSA012 and its endolysin Lys-PhiSA012 against antibiotic-resistant staphylococcal clinical isolates from canine skin infection sites. *Front. Med.* **2020**, *7*, 234. [CrossRef]
14. Basit, A.; Qadir, S.; Qureshi, S.; Rehman, S.U. Cloning and expression analysis of fused holin-endolysin from RL bacteriophage; Exhibits broad activity against multi drug resistant pathogens. *Enzym. Microb. Technol.* **2021**, *149*, 109846. [CrossRef] [PubMed]
15. Murray, E.; Draper, L.A.; Ross, R.P.; Hill, C. The advantages and challenges of using endolysins in a clinical setting. *Viruses* **2021**, *13*, 680. [CrossRef] [PubMed]
16. Nirwati, H.; Sinanjung, K.; Fahrunissa, F.; Wijaya, F.; Napitupulu, S.; Hati, V.P.; Hakim, M.S.; Meliala, A.; Aman, A.T.; Nuryastuti, T. Biofilm formation and antibiotic resistance of *Klebsiella pneumoniae* isolated from clinical samples in a tertiary care hospital, Klaten, Indonesia. *BMC Proc.* **2019**, *13* (Suppl. 11), 20. [CrossRef] [PubMed]
17. Kim, J.; Kim, J.-C.; Ahn, J. Assessment of bacteriophage-encoded endolysin as a potent antimicrobial agent against antibiotic-resistant Salmonella Typhimurium. *Microb. Pathog.* **2022**, *168*, 105576. [CrossRef] [PubMed]

18. Abdelrahman, F.; Gangakhedkar, R.; Nair, G.; El-Didamony, G.; Askora, A.; Jain, V.; El-Shibiny, A. *Pseudomonas* phage ZCPS1 endolysin as a potential therapeutic agent. *Viruses* **2023**, *15*, 520. [CrossRef] [PubMed]
19. Soontarach, R.; Srimanote, P.; Enright, M.C.; Blundell-Hunter, G.; Dorman, M.J.; Thomson, N.R.; Taylor, P.W.; Voravuthikunchai, S.P. Isolation and characterisation of bacteriophage selective for key *Acinetobacter baumannii* capsule chemotypes. *Pharmaceuticals* **2022**, *15*, 443. [CrossRef] [PubMed]
20. Soontarach, R.; Srimanote, P.; Arechanajan, B.; Nakkaew, A.; Voravuthikunchai, S.P.; Chusri, S. Characterization of a novel bacteriophage endolysin (LysAB1245) with extended lytic activity against distinct capsular types associated with *Acinetobacter baumannii* resistance. *PLoS ONE* **2024**, *19*, e0296453. [CrossRef]
21. Knaack, D.; Idelevich, E.A.; Schleimer, N.; Molinaro, S.; Kriegeskorte, A.; Peters, G.; Becker, K. Bactericidal activity of bacteriophage endolysin HY-133 against *Staphylococcus aureus* in comparison to other antibiotics as determined by minimum bactericidal concentrations and time-kill analysis. *Diagn. Microbiol. Infect. Dis.* **2019**, *93*, 362–368. [CrossRef] [PubMed]
22. Kim, S.; Lee, D.-W.; Jin, J.-S.; Kim, J. Antimicrobial activity of LysSS, a novel phage endolysin, against *Acinetobacter baumannii* and *Pseudomonas aeruginosa*. *J. Glob. Antimicrob. Resist.* **2020**, *22*, 32–39. [CrossRef] [PubMed]
23. Broendum, S.S.; Buckle, A.M.; McGowan, S. Catalytic diversity and cell wall binding repeats in the phage-encoded endolysins. *Mol. Microbiol.* **2018**, *110*, 879–896. [CrossRef] [PubMed]
24. Lim, J.-A.; Shin, H.; Heu, S.; Ryu, S. Exogenous lytic activity of SPN9CC endolysin against gram-negative bacteria. *J. Microbiol. Biotechnol.* **2014**, *24*, 803–811. [CrossRef]
25. Singh, P.K.; Donovan, D.M.; Kumar, A. Intravitreal injection of the chimeric phage endolysin Ply187 protects mice from *Staphylococcus aureus* endophthalmitis. *Antimicrob. Agents. Chemother.* **2014**, *58*, 4621–4629. [CrossRef] [PubMed]
26. Eichenseher, F.; Herpers, B.L.; Badoux, P.; Leyva-Castillo, J.M.; Geha, R.S.; van der Zwart, M.; McKellar, J.; Janssen, F.; de Rooij, B.; Selvakumar, L. Linker-improved chimeric endolysin selectively kills *Staphylococcus aureus* in vitro, on reconstituted human epidermis, and in a murine model of skin infection. *Antimicrob. Agents Chemother.* **2022**, *66*, e02273-21. [CrossRef]
27. Li, X.; Zhang, C.; Wei, F.; Yu, F.; Zhao, Z. Bactericidal activity of a holin-endolysin system derived from *Vibrio alginolyticus* phage HH109. *Microb. Pathog.* **2021**, *159*, 105135. [CrossRef]
28. Zhang, M.; Wang, Y.; Chen, J.; Hong, X.; Xu, X.; Wu, Z.; Ahmed, T.; Loh, B.; Leptihn, S.; Hassan, S. Identification and Characterization of a new type of holin-endolysin lysis cassette in *Acidovorax oryzae* phage AP1. *Viruses* **2022**, *14*, 167. [CrossRef]
29. Sharma, G.; Rao, S.; Bansal, A.; Dang, S.; Gupta, S.; Gabrani, R. *Pseudomonas aeruginosa* biofilm: Potential therapeutic targets. *Biologicals* **2014**, *42*, 1–7. [CrossRef]
30. Ciofu, O.; Tolker-Nielsen, T. Tolerance and resistance of *Pseudomonas aeruginosa* biofilms to antimicrobial agents—How *P. aeruginosa* can escape antibiotics. *Front. Microbiol.* **2019**, *10*, 913. [CrossRef]
31. Craft, K.M.; Nguyen, J.M.; Berg, L.J.; Townsend, S.D. Methicillin-resistant *Staphylococcus aureus* (MRSA): Antibiotic-resistance and the biofilm phenotype. *Medchemcomm.* **2019**, *10*, 1231–1241. [CrossRef]
32. Bowler, P.; Murphy, C.; Wolcott, R. Biofilm exacerbates antibiotic resistance: Is this a current oversight in antimicrobial stewardship? *Antimicrob. Resist. Infect. Control* **2020**, *9*, 162. [CrossRef] [PubMed]
33. Kaiser, S.; Mutters, N.; DeRosa, A.; Ewers, C.; Frank, U.; Günther, F. Determinants for persistence of *Pseudomonas aeruginosa* in hospitals: Interplay between resistance, virulence and biofilm formation. *Eur. J. Clin. Microbiol. Infect. Dis.* **2017**, *36*, 243–253. [CrossRef] [PubMed]
34. Qu, J.; Cai, Z.; Liu, Y.; Duan, X.; Han, S.; Liu, J.; Zhu, Y.; Jiang, Z.; Zhang, Y.; Zhuo, C. Persistent bacterial coinfection of a COVID-19 patient caused by a genetically adapted *Pseudomonas aeruginosa* chronic colonizer. *Front. Cell. Infect. Microbiol.* **2021**, *11*, 641920. [CrossRef] [PubMed]
35. Boudet, A.; Sorlin, P.; Pouget, C.; Chiron, R.; Lavigne, J.-P.; Dunyach-Remy, C.; Marchandin, H. Biofilm formation in methicillin-resistant *Staphylococcus aureus* isolated in cystic fibrosis patients is strain-dependent and differentially influenced by antibiotics. *Front. Microbiol.* **2021**, *12*, 750489. [CrossRef] [PubMed]
36. Gilpin, D.; Hoffman, L.R.; Ceppe, A.; Muhlebach, M.S. Phenotypic characteristics of incident and chronic MRSA isolates in cystic fibrosis. *J. Cyst. Fibros.* **2021**, *20*, 692–698. [CrossRef] [PubMed]
37. Versey, Z.; da Cruz Nizer, W.S.; Russell, E.; Zigic, S.; DeZeeuw, K.G.; Marek, J.E.; Overhage, J.; Cassol, E. Biofilm-innate immune interface: Contribution to chronic wound formation. *Front. Immunol.* **2021**, *12*, 648554. [CrossRef]
38. Kwiatek, M.; Parasion, S.; Rutyna, P.; Mizak, L.; Gryko, R.; Niemcewicz, M.; Olender, A.; Łobocka, M. Isolation of bacteriophages and their application to control *Pseudomonas aeruginosa* in planktonic and biofilm models. *Res. Microbiol.* **2017**, *168*, 194–207. [CrossRef]
39. Zhang, Y.; Huang, H.-H.; Duc, H.M.; Masuda, Y.; Honjoh, K.-i.; Miyamoto, T. Application of endolysin LysSTG2 as a potential biocontrol agent against planktonic and biofilm cells of *Pseudomonas* on various food and food contact surfaces. *Food Control* **2022**, *131*, 108460. [CrossRef]
40. Kuiper, J.W.P.; Hogervorst, J.M.A.; Herpers, B.L.; Bakker, A.D.; Klein-Nulend, J.; Nolte, P.A.; Krom, B.P. The novel endolysin XZ. 700 effectively treats MRSA biofilms in two biofilm models without showing toxicity on human bone cells *in vitro*. *Biofouling* **2021**, *37*, 184–193. [CrossRef]
41. Gutierrez, D.; Ruas-Madiedo, P.; Martinez, B.; Rodríguez, A.; Garcia, P. Effective removal of staphylococcal biofilms by the endolysin LysH5. *PLoS ONE* **2014**, *9*, e107307. [CrossRef]

42. Totté, J.; de Wit, J.; Pardo, L.; Schuren, F.; van Doorn, M.; Pasmans, S. Targeted anti-staphylococcal therapy with endolysins in atopic dermatitis and the effect on steroid use, disease severity and the microbiome: Study protocol for a randomized controlled trial (MAAS trial). *Trials* **2017**, *18*, 404. [CrossRef]
43. Totté, J.E.; van Doorn, M.B.; Pasmans, S.G. Successful treatment of chronic *Staphylococcus aureus*-related dermatoses with the topical endolysin Staphefekt SA. 100: A report of 3 cases. *Case Rep. Dermatol.* **2017**, *9*, 19–25. [CrossRef]
44. Cha, Y.; Son, B.; Ryu, S. Effective removal of staphylococcal biofilms on various food contact surfaces by *Staphylococcus aureus* phage endolysin LysCSA13. *Food Microbiol.* **2019**, *84*, 103245. [CrossRef]
45. Chang, Y. Bacteriophage-derived endolysins applied as potent biocontrol agents to enhance food safety. *Microorganisms* **2020**, *8*, 724. [CrossRef]
46. Lee, C.; Kim, H.; Ryu, S. Bacteriophage and endolysin engineering for biocontrol of food pathogens/pathogens in the food: Recent advances and future trends. *Crit. Rev. Food Sci. Nutr.* **2023**, *63*, 8919–8938. [CrossRef] [PubMed]
47. Clinical and Laboratory Standards Institute (CLSI). *Performance Standards for Antimicrobial Susceptibility Testing*; CLSI: Wayne, PA, USA, 2023.
48. Cevahir, N.; Demir, M.; Kaleli, I.; Gurbuz, M.; Tikvesli, S. Evaluation of biofilm production, gelatinase activity, and mannose-resistant hemagglutination *in Acinetobacter baumannii* strains. *J. Microbiol. Immunol. Infect.* **2008**, *41*, 513–518. [PubMed]
49. Tsukatani, T.; Sakata, F.; Kuroda, R. A rapid and simple measurement method for biofilm formation inhibitory activity using 96-pin microtiter plate lids. *World J. Microbiol. Biotechnol.* **2020**, *36*, 189. [CrossRef] [PubMed]
50. Cendra, M.d.M.; Blanco-Cabra, N.; Pedraz, L.; Torrents, E. Optimal environmental and culture conditions allow the in vitro coexistence of *Pseudomonas aeruginosa* and *Staphylococcus aureus* in stable biofilms. *Sci. Rep.* **2019**, *9*, 16284. [CrossRef] [PubMed]
51. Nithya, C.; Begum, M.F.; Pandian, S.K. Marine bacterial isolates inhibit biofilm formation and disrupt mature biofilms of *Pseudomonas aeruginosa* PAO1. *Appl. Microbiol. Biotechnol.* **2010**, *88*, 341–358. [CrossRef] [PubMed]

antibiotics

MDPI

Article

Discovery of *Mycobacterium avium* subsp. *paratuberculosis* Lytic Phages with Extensive Host Range Across Rapid- and Slow-Growing Pathogenic Mycobacterial Species

Aleen Clare Golla [1], Jeanne Chaumontet [2], Rebecca Vande Voorde [3] and Lia Danelishvili [1,3,*]

[1] Department of Microbiology, College of Science, Oregon State University, Corvallis, OR 97331, USA
[2] National Veterinary School of Toulouse, 31300 Toulouse, France
[3] Department of Biomedical Sciences, Carlson College of Veterinary Medicine, Oregon State University, Corvallis, OR 97331, USA
* Correspondence: lia.danelishvili@oregonstate.edu; Tel.: +1-(541)-737-6544; Fax: +1-(541)-737-2730

Abstract: Background/Objectives: Developing interventions for Johne's disease, which focuses on controlling *Mycobacterium avium* subsp. *paratuberculosis* (MAP) in contaminated environments by treating infected cows and preventing transmission from diseased animals, is a critical priority. Bacteriophage (phage) therapy, an emerging biological intervention, offers a promising alternative for the treatment and management of MAP infections. Methods: In this study, we generated an MAP-specific lytic phage library aimed at characterizing the therapeutic potential of phages under environmental and biological conditions that mimic those encountered in infected cattle such as ruminal fluid, milk, colostrum, and the bovine intestinal epithelium, a key site of MAP colonization and, later, transmission. Results: Our library contains a diverse collection of phages that have demonstrated robust lytic activity against MAP. The host range of these phages was thoroughly assessed, revealing that several isolates produce clear plaques on a range of MAP strains, as well as other pathogenic non-tuberculous mycobacterial (NTM) species and *M. tuberculosis* strains. This broad host range expands the therapeutic potential of the phage collection, positioning it as a potential cross-species antimicrobial tool. In vitro tests under conditions replicating the rumen, milk, and colostrum environments show that selected phages maintain stability and lytic efficacy, even in the presence of complex biological fluids. Furthermore, a subset of these phages was capable of preventing MAP colonization and invasion in cultured bovine epithelial cells, suggesting their potential for direct prophylactic application in cattle. Conclusions. Our collection of MAP phages represents a valuable source that can be developed into probiotic-like preparations, offering a cost-effective solution for prophylaxis and control of Johne's disease.

Keywords: *Mycobacterium avium* subsp. *paratuberculosis*; Johne's disease; lytic phages

1. Introduction

Mycobacterium avium subsp. *paratuberculosis* (MAP) is a highly prevalent and contagious infection in dairy cows, commonly referred to as Johne's disease. Due to MAP broad geographic spread in nearly every country, it poses a global threat to farmers. While the early infection stages of Johne's disease make MAP diagnosis very difficult, the economic trade and distribution of silently infected animals further contribute to MAP dissemination in healthy herds. It is estimated that in the dairy industry, approximately 68% of herds are affected with Johne's disease, and the economic loss due to increased cow replacement costs and reduction in milk production is estimated to be as high as $1.5 billion annually in the United States alone [1,2]. Additionally, while Crohn's disease in humans is traditionally viewed as autoimmune in nature, mounting evidence suggests that it might have an infectious etiology, with MAP being implicated [3–7]. While the association between MAP and Crohn's disease remains highly controversial, the mere possibility raises concerns regarding the potential risks to human health posed by the presence of MAP in milk and beef, as well as its contamination of natural water sources [8].

The current approach to deal Johne's disease is very limited and has unsatisfactory results. Being unpreventable and incurable, there is a need for alternative interventions for the control, treatment, and prevention of Johne's disease. Bacteriophages (phages) present one of the most promising new opportunities to fight infectious bacterial diseases [9,10]. Phages function like traditional antibiotics in a way that they exhibit powerful antimicrobial effect by infecting and lysing host bacteria. However, unlike traditional antibiotics, phages are highly selective and only kill organisms within a narrow taxonomic range. In recent years, phage therapy has gained attention as a promising treatment option for refractory, persistent, and disseminated infections in humans where antibiotics fail. Both in vivo studies and human clinical trials, including those involving pulmonary *P. aeruginosa* [11], and compassionate use in patients against pulmonary non-tuberculous mycobacterial infections [12] indicate that phage therapy is a safe, and often more effective, alternative to conventional antibiotic treatment [13–16].

Phages have been proposed for use in animal health, including veterinary applications [17], animal production environments [18], and more specifically, in the dairy production chain for prophylaxis, therapeutics, environmental decontamination, and dairy product bio-preservation [18]. Phage use in livestock has been mainly directed towards foodborne zoonotic infectious agents such as *Campylobacter*, *Salmonella*, *E. coli*, and *Listeria*. The landmark studies by Dr. Smith and his colleagues at Houghton Institute for Animal Disease Research (UK) that were carried out in young farm animals (calves, piglets, and lambs) led to a successful validation of the therapeutic use of bacteriophages against diarrheal diseases [19–22], demonstrating that severe experimental diarrhea in calves could be cured by a single dose of 10^5 phage particles administered via an oral route, and, most importantly, can be prevented by a dose as low as 100 particles [22]. Multiple studies have explored the use of phages against *Campylobacter* in the poultry industry [23] and against *Salmonella* and pathogenic *E. coli* in the pig industry [24], providing strong evidence for significant reductions in intestinal colonized pathogen counts compared to mock-treated controls without causing any collateral effects on the gut microbiome [25].

Furthermore, phage preparations are used in food safety applications and in the animal production and processing industries to prevent pathogenic bacterial contaminations. Lytic phages effective against *Listeria monocytogenes* are used in ready-to-eat meats, processed fish, cheeses, frozen vegetables, and food contact surfaces [26]. While *Salmonella* phages have been applied in chicken and pork postharvest processing, *E. coli*-specific phages are used in cattle hide wash. Although phage-based products are not yet available for administration to animals, the FDA has granted provisional Generally Recognized as Safe (GRAS) status to food safety treatments against *E. coli* O157:H7, Shiga-toxigenic *E. coli*, Listeria, *Salmonella*, and *Shigella* using a single phage or phage cocktail [27]. Phage-based products such as Agriphage in the US and Biolyse in the UK are currently on the market. Also in the US, a company called Intralytics has launched bacteriophage products for poultry health: PLSV-1™, effective against Salmonella; and INT-401™, effective against *Clostridium perfringens* [28].

The majority of mycobacteriophages have been isolated in *Mycobacterium smegmatis* host [29,30]. However, limited studies report the discovery of phages active against MAP [31], as isolating such phages is extremely challenging due to the complexity of working with slow-growing bacteria and their tendency to clump. Consequently, the potential use of these phages in livestock and veterinary medicine has not yet been explored. In this study, we aimed to uncover MAP phages from diverse geographical regions and to create a lytic phage bank exhibiting activity against a range of animal and human MAP isolates. Our research objectives included characterizing these phages for the host range and virulence, assessing their stability under various environmental stress conditions and in the rumen environment. Additionally, we cross-examined the selected lytic MAP phages against a wide range of fast- and slow-growing mycobacterial strains and collected data on phage receptors. The potent phages of this study can be used toward the development of highly effective phage cocktails or probiotic-like additives for prophylaxis, early therapeutic interventions, and possibly use as bio-disinfectants in dairy farms. The discovery of MAP phages creates an opportunity to develop low-cost tools for Johne's disease control and to support the health and welfare of agricultural animals.

2. Materials and Methods

2.1. Bacterial Strains and Culture Conditions

M. avium subsp. *paratuberculosis* K10 (MAP K10) was obtained from the American Type Culture Collection (ATCC). Additional MAP clinical isolates (Table 1) were sourced from ATCC or Dr. John Bannantine at the National Centers for Animal Health, IA. *M. smegmatis* mc^2 155 and *M. tuberculosis* (Mtb) auxotrophic $H_{37}R_v$ strains mc^2 6206 and mc^2 7901 were provided by Dr. William Jacobs, Albert Einstein College of Medicine, NY. *M. abscessus* clinical strains came from the Cystic Fibrosis Research Program at National Jewish Health, CO. Non-tuberculous mycobacterial (NTM) strains were supplied by Dr. Bermudez at Oregon State University, OR. *M. smegmatis* mc^2 155 was used for phage propagation, and NTM and Mtb strains were used for host range studies. All mycobacterial isolates were grown on Middlebrook 7H10 agar or in 7H9 broth with 5% OADC at 37 °C. For MAP strains, media were supplemented with 2 mg/L ferric mycobactin J.

Table 1. The list of *M. avium* subsp. *paratuberculosis* clinical isolates used in this study.

Strain Designation	Source	Isolation Source
MAP K10	ATCC BAA-968	Animal feces
MAP "Ben" [CIP 103966]	ATCC 43544	Intestinal tissue of Crohn's disease patient
MAP 97R0816	ATCC 700535	Mesenteric lymph node extracted at necropsy from 6-year-old Guernsey cow with a 3-year history of Johne's disease
MAP "Linda" [CIP 103965]	ATCC 43015	Ileum of girl with Crohn's disease
MAP19698	Dr. John Bannantine	Animal feces
MAP "Dominic" [CIP 103967]	ATCC 43545	Intestinal tissue of Crohn's disease patient

2.2. Sample Collection and Preparation

Soil, soil–manure mix, and water samples were collected from North and South America and Asia under USDA permit P526-190513-011. Specimens were placed in 50 mL tubes filled to 25 mL. For solid samples, 20 mL of MP buffer (10 mM Tris/HCl pH 7.6, 100 mM $MgSO_4$, 1 mM $CaCl_2$) was added, followed by 24 h of gentle shaking. Samples were then centrifuged at 2500× *g* for 30 min, and the supernatant was transferred and centrifuged again. Sequential filtration through 1 µm, 0.45 µm, and 0.2 µm syringe filters followed. Water samples were similarly centrifuged and filtered. All cleared samples were spot-plated on 7H10 agar to check for contamination. Filtrates were stored at 4 °C.

2.3. Creation of MAP-Specific Phage Library

MAP K10 (ATCC BAA-968) was cultured at 37 °C on 7H10 Middlebrook agar supplemented with OADC and 2 mg/L ferric mycobactin J for 4–5 weeks. *M. smegmatis* mc^2 155 was grown to the mid-log phase over 3 days. Prior to phage isolation, bacterial suspensions were prepared in MP buffer, sonicated, and allowed to settle for 10 min. The single-cell preparation was then used for phage isolation experiments. Approximately, 100 μL of 3×10^8 CFU (Colony Forming Units)/mL bacteria were incubated with 500 μL of cleared samples at room temperature (24 h for MAP, 30 min for *M. smegmatis*). Afterward, 4 mL of Top Agar was added and overlaid onto Middlebrook 7H10 agar plates (with or without Mycobactin J). Plates were incubated at 37 °C for 4–6 weeks for MAP and 3–5 days for *M. smegmatis* until plaques appeared. Plaques were excised, incubated in MP buffer, and re-propagated in *M. smegmatis* for higher titer recovery. Phages were purified using 100 kDa ultrafiltration membranes. Final phage titers were determined, and lysates ($\geq 10^{10}$ PFU/mL) were stored at 4 °C.

Specimens not producing plaques on MAP K10 but yielding mycobacteriophages in *M. smegmatis* were re-tested on MAP using 10^{10} PFU titrations via spot testing and liquid culturing. In spot testing, 5 μL of phage lysates were applied to MAP lawns, while in liquid culturing, 50 μL of phage preparations were mixed with 10 μL MAP inoculum in 7H9 broth. Cultures were incubated for 7 days, with bacterial growth assessed via OD_{600} and CFU counts. The percentage reduction in bacterial growth was calculated relative to phage-free controls.

2.4. Phage Host Range in MAP Clinical Isolates

A phage library generated in MAP K10 was screened against additional clinical isolates of Johne's and Crohn's disease (Table 1). The host specificity was tested using a phage spot assay with standardized MAP inoculum (3×10^8 CFU/mL). Two hundred microliters of inoculum were mixed with top agar and overlaid onto 7H10 agar plates supplemented with mycobactin J. Phage preparations ($\geq 10^{10}$ PFU) were spotted (5 μL) onto the surface, and plates were incubated for 4–8 weeks or until plaque formation. Phages producing clear plaques were considered lytic against the MAP host. The assay was conducted alongside the MAP K10 reference strain. Additionally, a liquid culture test exposed 50 μL of phage ($\geq 10^{10}$ PFU) to 10 μL of MAP inoculum (3×10^5 CFU) in 200 μL 7H9 broth. After 7 days at 37 °C, bacterial growth was recorded at OD_{600}, and CFU counts were determined. Bacterial reduction was compared to controls without phage, and the percentage of reduction was calculated.

2.5. The Efficiency of Plating (EOP)

The EOP was assessed for phages forming clear zones on MAP K10 across various clinical isolates. Phage preparations were standardized to 10^{10} PFU/mL, with 5 μL of each phage in 10–10^8 titrations spotted onto agar overlays containing 200 μL of MAP at 3×10^8 CFU/mL. Plates were incubated at 37 °C for 4 to 8 weeks until plaque formation was visible. EOP was calculated as the ratio of the phage titer on the clinical isolate to that on MAP K10. Phages were classified based on virulence: highly virulent ($0.1 \leq \text{EOP} \leq 1.00$), moderately virulent ($0.001 < \text{EOP} < 0.1$), and low virulence ($\text{EOP} < 0.001$).

2.6. Lytic Activity of MAP-Specific Phages Across Different Mycobacterial Species

The lytic activity of MAP-specific phages was evaluated against various mycobacterial species, including *M. abscessus*, *M. avium* subsp. *hominissuis*, *M. chelonae*, *M. fortuitum*, *M. kansasii*, *M. xenopi*, *M. gordonae*, *M. mucogenicum*, *M. bovis* BCG, and *M. tuberculosis* $H_{37}R_a$ and $H_{37}R_v$ auxotrophic strains. A standard plaque assay was conducted using phages at a titration of 10^{10} PFU. Two hundred microliters of each bacterial strain at 3×10^8 CFU/mL was overlayed with 4 mL of top agar on 7H10 Middlebrook plates, and 5 μL of phages was stamped on the surface. Plates were incubated until clear plaques formed, taking approximately 3–4 days for fast growers and up to 3 weeks for slow growers.

2.7. Phage Temperature Stability

Active phages (10^{10} PFU) were serially diluted and placed in 96-well PCR plates. These plates were incubated at 43 °C or 55 °C for 5 h, with duplicate plates kept at room temperature as controls. An inoculum of *M. smegmatis* was prepared to a concentration of 3×10^8 CFU/mL using McFarland Standard 1.0. Two hundred microliters of this inoculum was mixed with 4 mL of top agar and overlaid onto Middlebrook 7H10 plates. Five microliters of each phage dilution was stamped onto the agar surface. The plates were incubated until clear zones appeared, typically 3-4 days. Phage activity was evaluated based on the formation of clear zones relative to control plates, with lytic activity assessed by comparing phage reductions to control titration data.

2.8. Phage pH Stability

One hundred microliters of MP buffer at pH 5.0 was added to each well of a 96-well plate, followed by 10 µL of phage preparations (10^4–10^{10}). The phages were incubated in this acidic environment for 24 h. An inoculum of *M. smegmatis* was prepared at 3×10^8 CFU/mL using McFarland Standard 1.0. Four milliliters of top agar was combined with 200 µL of the inoculum and overlaid onto Middlebrook 7H10 plates. Five microliters of the pH 5.0-exposed phages, as well as control phages incubated at pH 7.0, was spotted onto the agar surface. Plates were incubated until plaques formed. Lytic activity was determined by comparing plaque formation between the pH 5.0-exposed phages and the control phages, assessing stability and infectivity under acidic conditions.

2.9. Phage Rumen Stability

Ten microliters of phage preparations (10^4–10^{10}) was placed into wells of a 96-well plate, followed by 100 µL of freshly sterilized rumen fluid. The phages were incubated in rumen fluid for 24 h at 37 °C without agitation. Two hundred microliters of *M. smegmatis* inoculum at 3×10^8 CFU/mL was mixed with 4 mL of top agar and overlaid on Middlebrook 7H10 plates. Five microliters of each phage preparation, including rumen-exposed and control phages, was spot-plated onto the agar. Plates were incubated until plaques appeared. Phage stability and efficacy were evaluated by comparing the plaque formation between rumen-exposed phages and control phages, determining the impact of rumen fluid on phage activity and infectivity.

2.10. Phage Stability in Milk and Colostrum

Freshly collected bovine milk and colostrum samples from Holstein Friesian cows were obtained. Phage preparations with titers ranging from 10^4 to 10^10 PFU were dispensed in 10 µL aliquots into a 96-well plate. Each well received 100 µL of either milk or colostrum, and the plates were incubated for 24 h at 37 °C without agitation. After incubation, 200 µL of *M. smegmatis* inoculum (3×10^8 CFU/mL) was mixed with 4 mL of top agar and overlaid onto Middlebrook 7H10 plates. Five microliters of each phage preparation, along with control phages incubated in MP buffer, was spot-plated. Plates were incubated for 3–4 days, after which phage stability and efficacy were assessed by comparing plaque formation between milk- and colostrum-exposed phages and control phages.

2.11. Mammalian Cell Culture and Colonization/Invasion Studies

Madin–Darby bovine kidney (MDBK) epithelial cells (CCL-22) obtained from ATCC were cultivated in Dulbecco's Modified Eagle Medium (DMEM) supplemented with 10% heat-inactivated fetal bovine serum (FBS; Gemini Bio-Products, Sacramento, CA, USA) at 37 °C in 5% CO_2. The bovine mammary epithelial cell line (MAC-T) was provided by Dr. Lewis Sheffield from the Department of Dairy Science at the University of Wisconsin. MAC-T cells were cultured in DMEM supplemented with 10% heat-inactivated fetal bovine serum, 5 µg/mL insulin, and 1 µg/mL hydrocortisone. MDBK and MAC-T cell monolayers (10^5/well) in 96-well plates were exposed to selected phage at 10^{10} PFU for 30 min before adding MAP K10 inoculum (10^6 CFU/well) at a ratio of 10 bacteria to 1 cell. The wells were lysed at 8 h and 24 h timepoints of MAP post-infection using 0.1% Triton X-100 (Sigma-Aldrich, St. Louis, MO, USA). The lysates were serially diluted in MP buffer and plated on 7H10 agar enriched with mycobactin J to quantify surviving MAP with CFUs in comparison to the MAP growth control of non-phage treatment over time.

2.12. Scanning Electron Microscopy

Approximately 10^8 CFU/mL of MAP K10 strain was incubated with lytic phages (10^{10} PFU/mL) for 30 min or 4 h at room temperature, followed by microcentrifugation at 8000× g to pellet the bacteria. The supernatant was carefully removed, and the samples were fixed with 4% glutaraldehyde and 2% paraformaldehyde (Alfa Aesar, Ward Hill, MA, USA) for 30 min at room temperature. After fixation, the samples were microcentrifuged, the fixative was removed, and the bacterial pellets were resuspended in PBS. A 5 µL drop of each sample was placed onto a freshly glow-discharged, collodion, and carbon-coated grid. After 1 min, the grid was rinsed with 10 mM ammonium acetate and stained with 0.75% uranyl formate for 30 s. The samples were examined and imaged using a Helios 650 Ultra-Resolution Dual Beam FEG scanning transmission electron microscope at the Oregon State University Electron Microscopy Facility.

2.13. Phage DNA Preparation, Sequencing, and Genomic Analysis

Prior to DNA isolation, phage preparations were subjected to bacterial DNA and RNA removal. A 270 µL aliquot of phage samples (10^{10} PFU) was combined with 30 µL of DNase I 10× buffer, 1 µL of DNase I, and 1 µL of RNase. Samples were incubated at 37 °C for 1.5 h. To inactivate DNase I, 10 µL of 0.5 M EDTA was added, and the solution was incubated at 75 °C for 5 min. After eliminating host genetic material, the phage capsids were digested by adding 2.0 µL of Proteinase K (20 mg/mL) and incubating at 65 °C for 1.5 h. Phage DNAs were then purified using the GenElute Bacterial Genomic DNA Kit (Sigma-Aldrich, St. Louis, MO, USA) according to the manufacturer's instructions. Phage DNAs were submitted to the Oregon State University Center for Quantitative Life Sciences Sequencing and Biocomputing facility and sequenced using the Illumina HiSeq 3000 system.

Gene prediction was performed using GeneMarkS and Glimmer, which provided the coordinates and FASTA amino acid sequences of predicted genes. These sequences were then compared against the NCBI BLASTp database to identify homologous genes. Full genome annotation was carried out using the DNA Master software 5.0.2 provided by the SEA-PHAGES program, which automated annotation based on BLASTn results from the NCBI database. A circular map was created using a Circular Genome Viewer (CGView) at http://wishart.biology.ualberta.ca/cgview (accessed on 25 September 2024) [32].

2.14. Phage Susceptibility to M. abscessus (MAB) Gene Knockout Mutants Involved in Surface-Exposed Glycolipid Transport and Synthesis

The MAB_0937c and MAB_0939 knockout mutants were screened for susceptibility via plaque formation assays using MAP phages. Inoculums of the MAB gene knockout mutants, along with the wild-type reference strain MAB 19977, were adjusted to 3 × 10^8 CFU/mL. Then, 200 µL aliquots of the inoculums were mixed with 4 mL of top agar and overlaid onto Middlebrook 7H10 agar plates. Five µL of each phage preparation were spot-plated onto the soft agar surface and incubated for 4–5 days or until clear plaques became visible.

2.15. Statistical Analysis

To ensure reproducibility, most experiments were conducted for three independent trials, unless stated otherwise. Statistical comparisons between experimental and control groups were performed using Student's t-test, analyzed with Prism 10 software (GraphPad). Significance levels were denoted as * $p < 0.05$ and ** $p < 0.01$.

3. Results

3.1. Isolation of MAP Specific Phages

To maximize phage discovery and diversity, specimens were sourced from dairy farms with a known history of MAP infections and from various geographical locations, climates, seasons, and environments, such as water and decaying soils exposed to manure. Through collaborations with veterinarians, veterinary technicians, DVM scientists, and students, we received over 1000 samples from the United States (Colorado, Florida, Hawaii, Minnesota, Nevada, North Carolina, New Mexico, Oregon, Texas, and Virginia), British Columbia, Chile, India, and the Philippines. Cleared samples were processed using the phage lysate spotting method on soft agar overlays of the MAP K10 strain, while the same filtrates were screened against *M. smegmatis* mc^2 155 to efficiently isolate mycobacteriophages. We identified eight lytic phages capable of forming clear plaques on MAP K10 (Figure 1) and 278 phages in *M. smegmatis* [33].

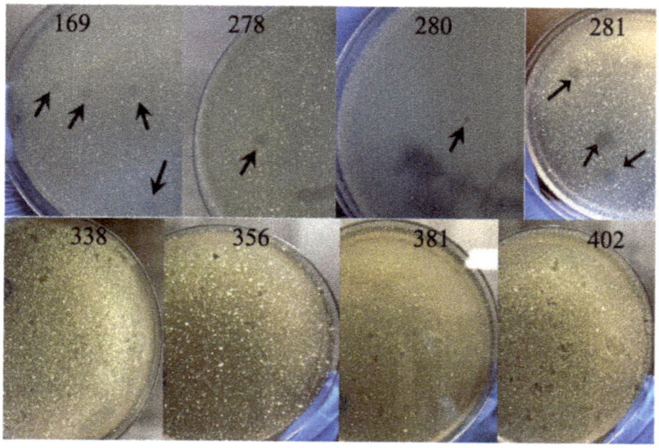

Figure 1. Identification of MAP phages with the double-layer agar plate method directly from environmental sample filtrates. The micrographs, captured at five weeks, display distinct and well-defined single plaques, some highlighted with arrows. These plaques are the result of lytic activity by eight different mycobacteriophages against MAP K10. Individual phage plaques were excised from the gel for subsequent propagation in *M. smegmatis*, achieving high titers of MAP-targeting phages ($\geq 10^{10}$ PFU/mL).

Given the possibility that the original samples contained very low concentrations of phages and that some active MAP phages may have been missed during direct screening, we re-tested the full library of 278 mycobacteriophages against MAP K10 at higher titers. Phage lytic activity was assessed using the spot testing method (Figure 2A) and liquid culturing through the CFU counts of viable MAP after 7 days of phage exposure (Figure 2B). Out of the 278 phages, 23 were found to significantly reduce viable MAP counts ($\geq$50%). However, only 11 phages, including those originally isolated through direct screening on MAP, successfully produced clear plaques on the top agar overlays of the MAP K10 strain and/or fully cleared bacteria in liquid media (Figure 2A,B).

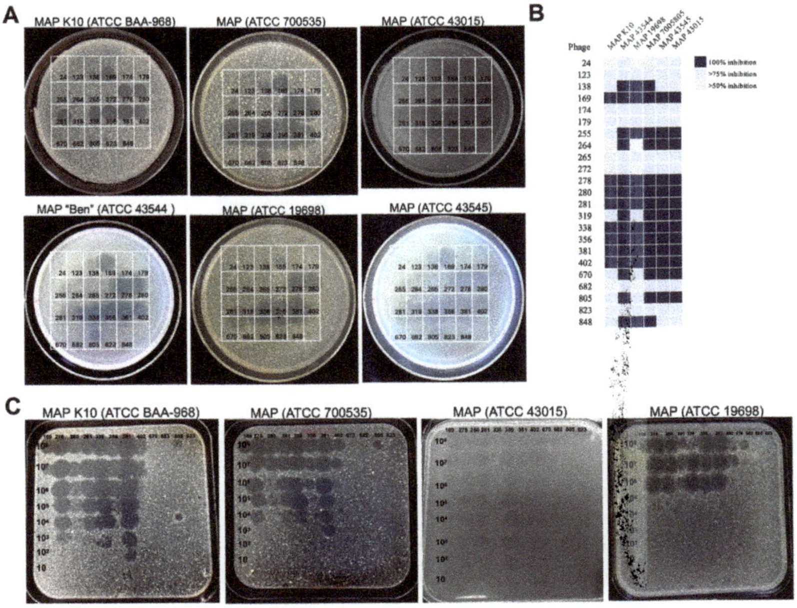

Figure 2. Phage host range and lytic efficiency in MAP strains. (**A**) Twenty-three phages were re-tested against MAP K10 and five additional clinical isolates using the double-layer agar plate method. While the bacteria formed a dense lawn, selected MAP-specific phages produced clear plaques and growth inhibition zones, confirming their lytic activity. Plates were incubated for up to eight weeks or until distinct plaques were observed. (**B**) The heat map summarizes the activity of twenty-three phages against six MAP isolates, assessed through liquid culturing methods. Phage efficacy was quantified by viable bacterial CFU counts, and the percentage of bacterial survival was calculated relative to control cultures without phage exposure. (**C**) Selected phages were further tested through titration to assess their virulence in titrations across different MAP isolates.

3.2. Characterizing the Host Range of MAP-Specific Phages

Twenty-three active MAP-specific phages were tested for their host range using a collection of animal and human MAP isolates, employing both spot testing and liquid culturing methods (Figure 2). Phages were classified as lytic if they completely lysed the host on soft agar (Figure 2A) and showed full clearance of bacteria in liquid culture (Figure 2B). The heat map illustrates the lytic activity and host specificity across six MAP strains. Most tested phages displayed broad activity against MAP isolates, with fifteen identified as lytic. For phage isolates that produced plaques on 7H10 agar overlays of MAP lawns, phage virulence was assessed through titration and compared to the sensitivity of MAP K10 (Figure 2C).

Additionally, selected phages were tested against various mycobacterial species, including *M. abscessus*, *M. avium* subsp. *hominissuis*, *M. intracellulare*, *M. chelonae*, *M. fortuitum*, *M. kansasii*, *M. xenopi*, *M. gordonae*, *M. mucogenicum*, *M. bovis* strain BCG, and *M. tuberculosis* H37Ra and auxotrophic H37Rv strains. Most non-tuberculous mycobacterial species and *M. tuberculosis* strains used in this study displayed susceptibility to a subset of phages, as illustrated in the Figure 3.

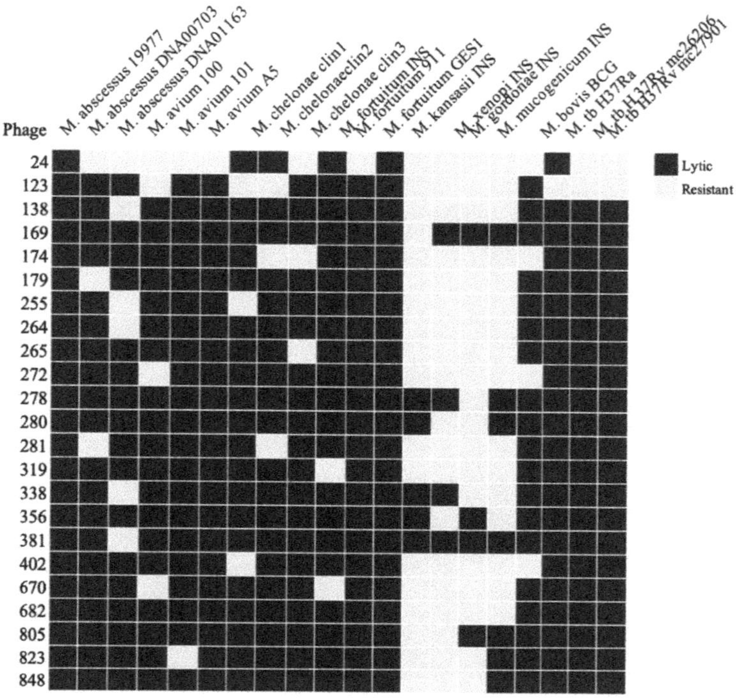

Figure 3. The lytic activity of MAP-specific phages across clinically relevant mycobacterial species. The heat map summarizes the lytic activity of twenty-three phages tested against 17 clinical isolates of various non-tuberculous mycobacterial species, as well as *M. bovis* BCG and *M. tuberculosis* (M. tb) strains. Lytic activity was determined using a plaque assay with the double-layer agar method at a phage concentration of 10^7 PFU.

3.3. Evaluating Phage Stability Across Diverse Environmental Conditions of the Rumen

Evaluating phage stability in the cow rumen is essential due to its unique environment, characterized by temperatures exceeding 37 °C, low pH, and a complex mixture of anaerobic microbiota, ammonia nitrogen, fatty acids, and other metabolites. These factors can significantly impact phage activity, making it crucial to assess phage stability when administered orally through water or feed. We evaluated the stability of 23 phages under these conditions to determine their effectiveness in the rumen. Phage lytic activity was assessed on the fast-growing *M. smegmatis* host after 24 h of exposure to cow rumen fluid. Phage susceptibility was determined via spot testing, alongside a control group of *M. smegmatis* exposed to phages without prior conditioning. The results indicate that while a few phages lost lytic activity under certain conditions, most retained comparable activity to the control group (Figure 4).

3.4. Assessment of Phage Stability in Milk and Colostrum

Given the susceptibility of young animals to MAP infection through colostrum and milk from infected mothers, adding MAP phage cocktails could be a promising preventive strategy to protect calves from infection during feeding. Since milk and colostrum are rich in proteins, fats, carbohydrates, and other macronutrients and minerals, it is important to evaluate how these components might affect phage lytic activity. We incubated 23 selected phages in these environments and studied their activity in a titration range of 10^4–10^{10} PFU in *M. smegmatis*. The results indicate that phage exposure to milk and colostrum did not affect its lytic activity when compared to the control group in MP buffer (Figure 4).

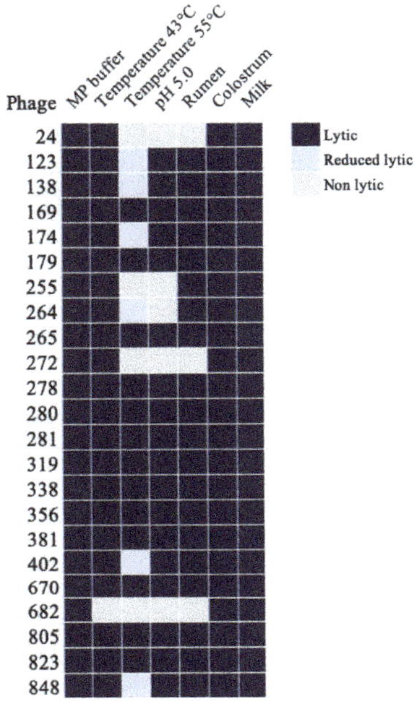

Figure 4. The impact of various environmental conditions on phage stability. The heat maps illustrate the lytic activity of phages under different conditions measured within a titration range of 10^4 to 10^9 PFU on *M. smegmatis* host. Phages were exposed to high temperatures or a low pH for 5 h or incubated in rumen fluid, milk, or colostrum for up to 24 h at 37 °C without agitation. After incubation, 5 μL of each phage preparation was spot-plated onto *M. smegmatis* overlays on Middlebrook 7H10 agar plates. The lytic activity of phages was evaluated by comparing plaque formation between phages exposed to various environmental conditions and control phages incubated in MP buffer. Lytic activity was defined as no significant difference in titers between the treated and control conditions. A reduction in lytic activity was defined as at least a 1 log decrease in titer, while non-lytic activity was characterized by a complete loss of plaque formation.

3.5. MAP-Specific Phages Effectively Prevent MAP Colonization and Invasion in Tissue Culture

This assay examined the ability of phages to inhibit MAP colonization and entry into bovine epithelial cells. MDBK cell monolayers were pre-exposed to selected phages at 10^{10} PFU for 30 min, followed by the addition of MAP K10 at an MOI of 10. Control wells without phage served as MAP growth controls. To account for the potential internalization of MAP by epithelial cells, MDBK cells were lysed at 8 h and 24 h, and bacterial CFUs were assessed. The results indicated varying degrees of inhibition, with three phages demonstrating significant early inhibition and complete clearance of MAP by 24 h when compared with control groups at corresponding times (Figure 5A). Five additional phages, while not completely eliminating MAP, still achieved a significant reduction in bacterial numbers at 8 h and 24 h timepoints. Additionally, the assay was performed on MAC-T bovine mammary epithelial cells (Figure 5B), yielding results consistent with those observed in MDBK cells.

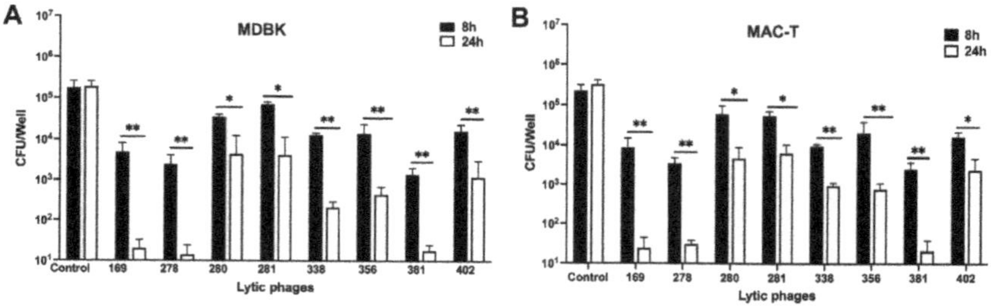

Figure 5. Phages effectively inhibit MAP colonization and invasion in epithelial cells. (**A**) MDBK or (**B**) MAC-T cell monolayers were pre-treated with phages at a concentration of 10^9 PFU for 30 min before infection with MAP at an MOI of 10. Control wells, without phage pre-exposure, were included for comparison. At 8 h and 24 h post-infection, cells were lysed with 0.1% Triton X-100, serially diluted, and MAP counts were assessed by CFU enumeration following 4–5 weeks of incubation on Middlebrook 7H10 agar plates. The significance was determined using Student's *t*-test (* $p < 0.05$, ** $p < 0.01$) within the treatment groups over time and between control and experimental groups. Data represent the mean ± SD of three independent biological replicates, each performed in quadruplicate.

3.6. Assessing Phage Attachment on MAP with Scanning Electron Microscopy

MAP K10 was incubated with four different phages, three of which demonstrated the complete clearance of MAP in both in vitro assays and a bovine tissue culture model. SEM was used to visualize phage adsorption on the cell wall of MAP. The micrographs shown in Figure 6 reveal that selected phages began adsorbing to the bacterial surfaces as early as 30 min. In addition, SEM revealed that all four phages depicted in micrographs A through D belong to the Siphoviridae family, characterized by isometric heads and long, flexible, non-contractile tails, and are known to carry a double-stranded DNA genome.

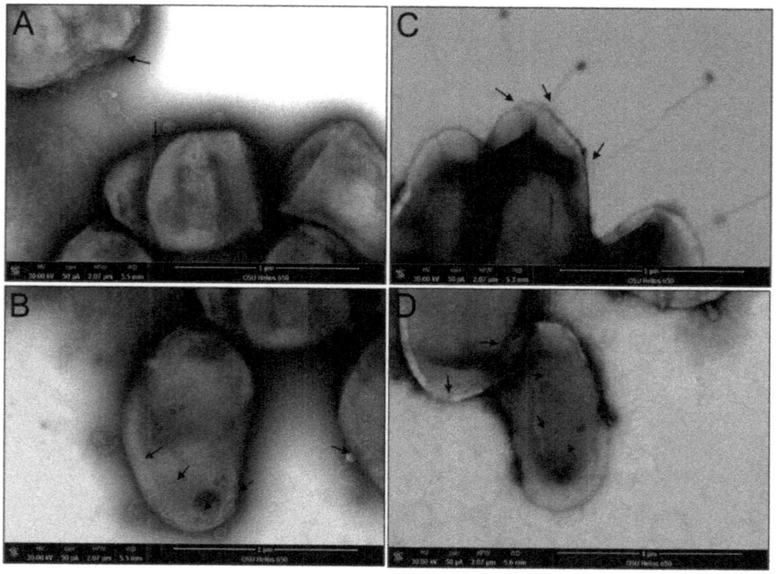

Figure 6. Scanning electron microscopy micrographs visualizing phage absorption on the cell wall of MAP. Panels (**A–D**) display four distinct phages (169, 278, 381, and 402, respectively) binding to the bacterial surface, as indicated by arrows. These phages can be identified by plaque formation on MAP, feature long, flexible tails, and are classified within the Siphoviridae family. They are characterized by their isometric heads and carry double-stranded DNA genomes.

3.7. Phage Genomic Analysis

A comparative analysis of MAP-specific phages 169, 278, and 381 was performed using the complete nucleotide sequences through BLASTn in GenBank. The genetic and physical maps of sequenced phages are presented in Figure 7. Phage L169 has a genome size of 69.7 Kb, with 131 identified ORFs, 102 of which are annotated as hypothetical proteins. It exhibits a GC content of 48.6% and shares 98% sequence identity with phage MrMiyagi (GenBank Accession MT776806). Phage 278 has a genome size of 71.6Kb with a GC content of 56%, containing 118 ORFs, 84 of which are hypothetical proteins. It shares 99% sequence identity with phage Send513 (GenBank Accession JF704112). Phage 381 has a genome size of 56.6 Kb with a GC content of 61% and contains 115 ORFs, with 75 encoding hypothetical proteins. It shares 95% sequence identity with mycobacteriophage JalFarm20 (GenBank Accession MZ958744).

Figure 7. The genetic and physical maps of sequenced phages.

3.8. Establishing the Subset of Phages Utilizing Trehalose Polyphleates as Surface Receptors

Our previous research identified the surface-exposed glycolipids 2,3-diacyl trehalose (DAT) and polyacyltrehalose (PAT) as key targets for phage adsorption [33]. The MAB_0937c gene encodes the MmpL10 efflux pump, which exports these glycolipids to the bacterial surface, while the MAB_0939 gene encodes the polyketide synthase (Pks3/4) involved in DAT synthesis. Given that DAT and PAT are crucial components of the mycobacterial outer cell wall and essential in phage adsorption, this study utilized MAB_0937c and MAB_0939 gene knockout mutants to assess their susceptibility to MAP phages. Of the 23 phages tested, the absence of DAT/PAT resulted in partial resistance to 13 phages and full resistance to 10 phages (Figure 8), underscoring the importance of these glycolipids in phage adsorption mechanisms.

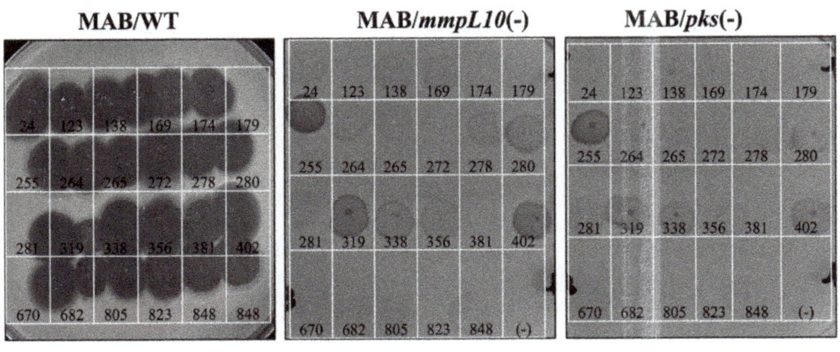

Figure 8. Phage resistance patterns in MmpL10 efflux pump and Pks mutants of MAB. The phage susceptibility assay, utilizing the phage plaque formation method, was conducted to evaluate the MAB 0937c/mmpL10 and MAB_0939/pks mutants in comparison to the wild-type MAB control.

4. Discussion

The National Animal Health Monitoring System dairy study in 2007 revealed that 68% of U.S. dairy herds harbor *Mycobacterium avium* subspecies *paratuberculosis* infection in at least one cow, with nearly 100% prevalence in large herds [34]. It is also estimated that 8% of beef cattle herds are infected with MAP. However, because entire beef herds are not routinely tested unless an individual animal tests positive for MAP, Johne's disease is likely more widespread in beef cattle herds than is currently recognized. The current approach to deal with the disease is very limited and of unsatisfactory results. Given the unpreventable and incurable nature of Johne's disease, there is a critical need for effective strategies for its control, prevention, and treatment. In the absence of an effective vaccine, bacteriophages offer a promising alternative to combat MAP infections. Also, traditional antimicrobial agents for large animals are limited, and long-term drug administration for mycobacterial infections is impractical both physically and economically. In contrast, low-cost MAP-specific phages offer numerous advantages [10], including potential applications for treating both the early stages of Johne's disease and later-stage infections in cows to prevent the spread of infection from diseased animals. Beyond treatment, phages can also be used as a 'probiotic-like' preparation and serve as a prophylactic intervention, administered periodically to healthy animals through water or feed additives to reduce the frequency of infections in dairy cows. This approach has the potential to make a significant impact on the industry. Due to the silent nature of MAP infections, which can take years to manifest as disease, administering phages to calves from birth via colostrum and milk could be the most effective strategy for preventing infections and reducing disease occurrence later in life. Additionally, one of the attractive characteristics of phages is their specificity, which could have a significant impact on controlling this environmental bacterium. Active phage preparations can be employed as biocontrol agents to reduce MAP infections by reducing bacterial loads in potentially contaminated dairy farm environments and lowering the overall risk of disease transmission.

The research on MAP phage identification and characterization is currently very limited [31] and, thus, their potential applications in livestock and veterinary medicine remain largely unexplored. To highlight, the majority of fully characterized bacteriophages against mycobacteria have been discovered in *Mycobacterium smegmatis* [29,30]. The isolation of MAP-specific phages is challenging, primarily due to the complexity of working with slow-growing bacteria and their tendency to clump. The primary goal of this research was to establish a robust phage bank targeting MAP and develop phage cocktails as cost-effective therapeutic and prophylactic interventions for Johne's disease.

Phages that specifically infect MAP require a living host to replicate in natural environments, and MAP's ability to survive long-term in these environments is well documented [35]. We hypothesized that environmental samples from dairy farms with a history of MAP infection would harbor lytic phages specifically active against this pathogen. To test this, we focused on manure-exposed sites, such as water sources and decaying soil, which are likely to yield phages with lytic potential against MAP. Samples were collected from diverse geographic regions and microclimates to maximize the likelihood of isolating effective phages. We also recognized that MAP-specific phages may exist at low titers due to the pathogen's slow growth. To address this, we screened all purified environmental samples against *M. smegmatis*, a faster-growing surrogate host, to obtain phages at higher titers. These phages were then re-tested against MAP to enhance the chances of isolating specific phages. Rapid MAP-specific phage propagation and efficient production of therapeutic cocktails require the use of a fast-growing host. Leveraging *M. smegmatis* ensures the quick isolation and turnover of mycobacteriophages, crucial in developing effective phage-based treatments. *M. smegmatis* is a widely used model organism for studying various mycobacterial pathogens, including *M. tuberculosis* and the *M. avium* complex [36–38], making it ideal for phage isolation and propagation. Importantly, global collections of mycobacterial species-specific phages have been isolated using *M. smegmatis*, as all mycobacteriophages can infect this clade for reasons that remain unclear. This model organism replicates rapidly (every 1–3 h), enabling high-titer phage propagation in a cost-effective manner. Purified phage preparations from *M. smegmatis* are safe for use in humans and animals. In contrast, propagating phages in pathogenic strains like MAP would be slow, uneconomical, and unsafe due to the need for extensive purification to remove pathogenic contaminants, which could provoke adverse immune reactions. Additionally, *M. smegmatis* is a non-pathogenic commensal in humans and has been shown to act as a powerful cellular immune adjuvant [39–43].

The direct screening of sample filtrates on the MAP K10 strain identified eight MAP-specific phages that formed clear plaques on this slow-growing pathogen. Re-testing the *M. smegmatis* phage library on MAP K10 revealed 15 additional phages, which, while not all forming plaques, either cleared MAP K10 cultures or significantly reduced bacterial numbers. To develop an effective intervention for Johne's disease, it is essential to identify MAP phages with broad-spectrum activity. In total, 23 active phages found in MAP K10 were tested on five additional MAP isolates. The results show that eight phages originally isolated on MAP K10 retained plaque-forming ability, with seven additional phages exhibiting lytic activity across other MAP strains, as demonstrated by clear plaque formation and/or bacterial clearance observed using a liquid culturing method. Notably, phages capable of forming plaques suppressed bacterial growth for over two months throughout the experiment. Of the 23 phages discovered in the MAP K10 strain, most exhibited activity across various MAP clinical isolates, significantly reducing bacterial loads in vitro. Additionally, we tested these phages against a wide range of NTM species and *M. tuberculosis* strains, finding that some MAP phages demonstrated an extensive host range, with lytic activity against the majority of tested mycobacterial species and strains. The broad host range of this phage collection enhances its therapeutic potential, making it a promising antimicrobial tool for use across multiple species.

Orally administered phages face challenges in the bovine digestive tract due to high temperatures and acidity, which can affect their stability. The rumen, with its complex chemical composition and rich anaerobic microbiota, can also impact phage activity. For prophylactic applications in calves, phages must remain stable and virulent in milk and colostrum. In vitro results demonstrate that the majority of phages tested are stable and maintain lytic efficacy under various stress conditions, including those present in the bovine intestine, rumen, milk, and colostrum. Additionally, tissue culture studies identified robust phages that prevent MAP colonization and invasion in bovine epithelial cells by clearing or significantly reducing viable bacteria.

The diversity and versatility of phages are key advantages for therapeutic applications; however, an important aspect of designing effective phage preparations is mitigating the development of phage resistance in bacteria [9,44]. To achieve this, it is essential to create a cocktail of lytic phages that can bind to different host receptors and employ various mechanisms for absorption into bacterial cells. In this study, we sequenced three phages, all identified as Siphoviridae morphotypes through SEM imaging, but belonging to different genetic clusters of AC, F, and R based on comparative analysis. Moreover, building on our previous discovery of DAT/PAT trehalose polyphleates being the critical binding targets for a subset of phages in *M. abscessus* [33] and supported by findings from other studies [44], we explored their role in MAP phage adsorption, as well. Using MAB knockout mutants lacking the MmpL10 efflux pump or polyketide synthase, we assessed susceptibility to MAP phages, distinguishing those reliant on trehalose polyphleates from those with independent lytic activity.

In this study, we successfully isolated and characterized 23 lytic phages targeting MAP. This work lays a strong foundation for expanding the MAP phage research, particularly in designing phage cocktails that could be used both to prevent and treat Johne's disease in infected animals, aiming to reduce the infection transmission to healthy animals. The MAP-specific phage bank represents a valuable resource for cost-effective probiotic-like product development, offering targeted and sustainable solutions for one of the most persistent and costly diseases in dairy farming.

Author Contributions: Conceptualization, L.D.; Methodology, A.C.G., J.C., R.V.V. and L.D.; Validation, A.C.G., J.C., R.V.V. and L.D.; Formal analysis, R.V.V. and L.D.; Investigation, A.C.G., R.V.V. and L.D.; Resources, L.D.; Data curation, A.C.G., J.C., R.V.V. and L.D.; Writing—original draft, A.C.G.; Writing—review and editing, L.D.; Visualization, A.C.G. and L.D.; Project administration, L.D.; Funding acquisition, L.D. All authors have read and agreed to the published version of the manuscript.

Funding: This research was funded by Animal Health and Disease Research program, USDA National Institute of Food and Agriculture (Grant no. HT0350).

Institutional Review Board Statement: Not applicable.

Informed Consent Statement: Not applicable.

Data Availability Statement: Data are contained within the article.

Conflicts of Interest: The authors declare no conflicts of interest.

References

1. Stabel, J.R. Johne's disease: A hidden threat. *J. Dairy Sci.* **1998**, *81*, 283–288. [CrossRef]
2. Ott, S.L.; Wells, S.J.; Wagner, B.A. Herd-level economic losses associated with Johne's disease on US dairy operations. *Prev. Vet. Med.* **1999**, *40*, 179–192. [CrossRef]
3. Harris, N.B.; Barletta, R.G. Mycobacterium avium subsp. paratuberculosis in Veterinary Medicine. *Clin. Microbiol. Rev.* **2001**, *14*, 489–512. [CrossRef]
4. National Advisory Committee on Microbiological Criteria for Foods. Assessment of food as a source of exposure to Mycobacterium avium subspecies paratuberculosis (MAP). *J. Food Prot.* **2010**, *73*, 1357–1397. [CrossRef]
5. Over, K.; Crandall, P.G.; O'Bryan, C.A.; Ricke, S.C. Current perspectives on Mycobacterium avium subsp. paratuberculosis, Johne's disease, and Crohn's disease: A review. *Crit. Rev. Microbiol.* **2011**, *37*, 141–156. [CrossRef]
6. Honap, S.; Johnston, E.; Agrawal, G.; Al-Hakim, B.; Hermon-Taylor, J.; Sanderson, J. Anti-Mycobacterium paratuberculosis (MAP) therapy for Crohn's disease: An overview and update. *Frontline Gastroenterol.* **2021**, *12*, 397–403. [CrossRef]
7. Aitken, J.M.; Aitken, J.E.; Agrawal, G. Mycobacterium avium ssp. paratuberculosis and Crohn's Disease-Diagnostic Microbiological Investigations Can Inform New Therapeutic Approaches. *Antibiotics* **2024**, *13*, 158. [CrossRef]
8. Gill, C.O.; Saucier, L.; Meadus, W.J. Mycobacterium avium subsp. paratuberculosis in dairy products, meat, and drinking water. *J. Food Prot.* **2011**, *74*, 480–499. [CrossRef]
9. Strathdee, S.A.; Hatfull, G.F.; Mutalik, V.K.; Schooley, R.T. Phage therapy: From biological mechanisms to future directions. *Cell* **2023**, *186*, 17–31. [CrossRef]
10. O'Connell, L.M.; Coffey, A.; O'Mahony, J.M. Alternatives to antibiotics in veterinary medicine: Considerations for the management of Johne's disease. *Anim. Health Res. Rev.* **2023**, *24*, 12–27. [CrossRef]
11. Rappo, U.; Cohen, A.; Kario, E.; Gold, J.; Nevenzal, H.T.; Levy-Saar, I.; Cohen, T.; Weiner, I.; Livnat, H.S.; Slutskin, I.V.; et al. 2891. Nebulized Phage Therapy for Patients with Cystic Fibrosis with Chronic Pseudomonas aeruginosa Pulmonary Infection: A Phase 1b/2a Randomized, Double-Blind, Placebo-Controlled, Multicenter Study. *Open Forum Infect. Dis.* **2023**, *10*, ofad500.2474. [CrossRef]
12. Dedrick, R.M.; Smith, B.E.; Cristinziano, M.; Freeman, K.G.; Jacobs-Sera, D.; Belessis, Y.; Whitney Brown, A.; Cohen, K.A.; Davidson, R.M.; van Duin, D.; et al. Phage Therapy of Mycobacterium Infections: Compassionate Use of Phages in 20 Patients With Drug-Resistant Mycobacterial Disease. *Clin. Infect. Dis.* **2023**, *76*, 103–112. [CrossRef]
13. Oechslin, F.; Piccardi, P.; Mancini, S.; Gabard, J.; Moreillon, P.; Entenza, J.M.; Resch, G.; Que, Y.A. Synergistic Interaction Between Phage Therapy and Antibiotics Clears Pseudomonas Aeruginosa Infection in Endocarditis and Reduces Virulence. *J. Infect. Dis.* **2017**, *215*, 703–712. [CrossRef]
14. Comeau, A.M.; Tetart, F.; Trojet, S.N.; Prere, M.F.; Krisch, H.M. Phage-Antibiotic Synergy (PAS): Beta-lactam and quinolone antibiotics stimulate virulent phage growth. *PLoS ONE* **2007**, *2*, e799. [CrossRef]
15. Kamal, F.; Dennis, J.J. Burkholderia cepacia complex Phage-Antibiotic Synergy (PAS): Antibiotics stimulate lytic phage activity. *Appl. Environ. Microbiol.* **2015**, *81*, 1132–1138. [CrossRef]
16. Torres-Barcelo, C.; Hochberg, M.E. Evolutionary Rationale for Phages as Complements of Antibiotics. *Trends Microbiol.* **2016**, *24*, 249–256. [CrossRef]
17. Squires, R.A. Bacteriophage therapy for management of bacterial infections in veterinary practice: What was once old is new again. *N. Z. Vet. J.* **2018**, *66*, 229–235. [CrossRef]
18. Colavecchio, A.; Goodridge, L.D. Phage Therapy Approaches to Reducing Pathogen Persistence and Transmission in Animal Production Environments: Opportunities and Challenges. *Microbiol. Spectr.* **2017**, *5*. [CrossRef]
19. Smith, H.W.; Huggins, M.B. Successful treatment of experimental Escherichia coli infections in mice using phage: Its general superiority over antibiotics. *J. Gen. Microbiol.* **1982**, *128*, 307–318. [CrossRef]
20. Smith, H.W.; Huggins, M.B. Effectiveness of phages in treating experimental Escherichia coli diarrhoea in calves, piglets and lambs. *J. Gen. Microbiol.* **1983**, *129*, 2659–2675. [CrossRef]

21. Smith, H.W.; Huggins, M.B.; Shaw, K.M. Factors influencing the survival and multiplication of bacteriophages in calves and in their environment. *J. Gen. Microbiol.* **1987**, *133*, 1127–1135. [CrossRef]
22. Smith, H.W.; Huggins, M.B.; Shaw, K.M. The control of experimental Escherichia coli diarrhoea in calves by means of bacteriophages. *J. Gen. Microbiol.* **1987**, *133*, 1111–1126. [CrossRef]
23. Firlieyanti, A.S.; Connerton, P.L.; Connerton, I.F. Campylobacters and their bacteriophages from chicken liver: The prospect for phage biocontrol. *Int. J. Food Microbiol.* **2016**, *237*, 121–127. [CrossRef]
24. Zhang, J.; Li, Z.; Cao, Z.; Wang, L.; Li, X.; Li, S.; Xu, Y. Bacteriophages as antimicrobial agents against major pathogens in swine: A review. *J. Anim. Sci. Biotechnol.* **2015**, *6*, 39. [CrossRef]
25. Wall, S.K.; Zhang, J.; Rostagno, M.H.; Ebner, P.D. Phage therapy to reduce preprocessing Salmonella infections in market-weight swine. *Appl. Environ. Microbiol.* **2010**, *76*, 48–53. [CrossRef]
26. Moye, Z.D.; Woolston, J.; Sulakvelidze, A. Bacteriophage Applications for Food Production and Processing. *Viruses* **2018**, *10*, 205. [CrossRef]
27. FDA. GRAS Notice Inventory 2018. 2018. Available online: https://www.fda.gov/food/generally-recognized-safe-gras/gras-notice-inventory (accessed on 25 September 2024).
28. Wernicki, A.; Nowaczek, A.; Urban-Chmiel, R. Bacteriophage therapy to combat bacterial infections in poultry. *Virol. J.* **2017**, *14*, 179. [CrossRef]
29. Hatfull, G.F. Mycobacteriophages. *Microbiol. Spectr.* **2018**, *6*. [CrossRef]
30. Hatfull, G.F.; Science Education Alliance Phage Hunters Advancing Genomics and Evolutionary Science Program; KwaZulu-Natal Research Institute for Tuberculosis and HIV Mycobacterial Genetics Course Students; Phage Hunters Integrating Research and Education Program. Complete genome sequences of 138 mycobacteriophages. *J. Virol.* **2012**, *86*, 2382–2384. [CrossRef]
31. Harman-McKenna, V.K.; De Buck, J. Effective Isolation and Characterization of Mycobacteriophages with the Ability to Lyse Mycobacterium avium subsp. paratuberculosis. *Viruses* **2023**, *16*, 20. [CrossRef]
32. Stothard, P.; Wishart, D.S. Circular genome visualization and exploration using CGView. *Bioinformatics* **2005**, *21*, 537–539. [CrossRef]
33. Gorzynski, M.; De Ville, K.; Week, T.; Jaramillo, T.; Danelishvili, L. Understanding the Phage-Host Interaction Mechanism toward Improving the Efficacy of Current Antibiotics in Mycobacterium abscessus. *Biomedicines* **2023**, *11*, 1379. [CrossRef]
34. NAHMS. USDA:APHIS:VS, C., National Animal Health Monitoring System. Fort Collins, CO: c2008. Johne's Disease on US Dairies, 1991–2007. 2008, #N521.0408. Available online: https://johnes.org/wp-content/uploads/2020/09/USDA-NAHMS-CEAH_Dairy2007_Johnes.pdf (accessed on 25 September 2024).
35. Whittington, R.J.; Marshall, D.J.; Nicholls, P.J.; Marsh, I.B.; Reddacliff, L.A. Survival and dormancy of Mycobacterium avium subsp. paratuberculosis in the environment. *Appl. Environ. Microbiol.* **2004**, *70*, 2989–3004. [CrossRef]
36. Shiloh, M.U.; Champion, P.A. To catch a killer. What can mycobacterial models teach us about Mycobacterium tuberculosis pathogenesis? *Curr. Opin. Microbiol.* **2010**, *13*, 86–92. [CrossRef]
37. Namouchi, A.; Cimino, M.; Favre-Rochex, S.; Charles, P.; Gicquel, B. Phenotypic and genomic comparison of Mycobacterium aurum and surrogate model species to Mycobacterium tuberculosis: Implications for drug discovery. *BMC Genom.* **2017**, *18*, 530. [CrossRef]
38. Hatfull, G.F. The secret lives of mycobacteriophages. *Adv. Virus Res.* **2012**, *82*, 179–288. [CrossRef]
39. Zhao, S.; Zhao, Y.; Mao, F.; Zhang, C.; Bai, B.; Zhang, H.; Shi, C.; Xu, Z. Protective and therapeutic efficacy of Mycobacterium smegmatis expressing HBHA-hIL12 fusion protein against Mycobacterium tuberculosis in mice. *PLoS ONE* **2012**, *7*, e31908. [CrossRef]
40. Taylor, N.; Bahunde, F.; Thompson, A.; Yu, J.S.; Jacobs, W.R., Jr.; Letvin, N.L.; Haynes, B.F.; Lee, S. Enhanced priming of adaptive immunity by Mycobacterium smegmatis mutants with high-level protein secretion. *Clin. Vaccine Immunol.* **2012**, *19*, 1416–1425. [CrossRef]
41. Zhang, H.; Peng, P.; Miao, S.; Zhao, Y.; Mao, F.; Wang, L.; Bai, Y.; Xu, Z.; Wei, S.; Shi, C. Recombinant Mycobacterium smegmatis expressing an ESAT6-CFP10 fusion protein induces anti-mycobacterial immune responses and protects against Mycobacterium tuberculosis challenge in mice. *Scand. J. Immunol.* **2010**, *72*, 349–357. [CrossRef]
42. Adam, A.; Ciorbaru, R.; Petit, J.F.; Lederer, E.; Chedid, L.; Lamensans, A.; Parant, F.; Parant, M.; Rosselet, J.P.; Berger, F.M. Preparation and biological properties of water-soluble adjuvant fractions from delipidated cells of Mycobacterium smegmatis and Nocardia opaca. *Infect. Immun.* **1973**, *7*, 855–861. [CrossRef]
43. Junqueira-Kipnis, A.P.; de Oliveira, F.M.; Trentini, M.M.; Tiwari, S.; Chen, B.; Resende, D.P.; Silva, B.D.; Chen, M.; Tesfa, L.; Jacobs, W.R., Jr.; et al. Prime-boost with Mycobacterium smegmatis recombinant vaccine improves protection in mice infected with Mycobacterium tuberculosis. *PLoS ONE* **2013**, *8*, e78639. [CrossRef] [PubMed]
44. Oechslin, F. Resistance Development to Bacteriophages Occurring during Bacteriophage Therapy. *Viruses* **2018**, *10*, 351. [CrossRef] [PubMed]

pharmaceuticals

MDPI

Article

Antimicrobial, Anticancer, and Antioxidant Activities of Maize and Clover Pollen Grains Extracts: A Comparative Study with Phytochemical Characterizations

Heba Barnawi [1], Husam Qanash [1,2], Abdu Aldarhami [3], Ghaida Alsaif [1], Bandar Alharbi [1], Majed N. Almashjary [4,5], Alhomidi Almotiri [6] and Abdulrahman S. Bazaid [1,2,*]

1 Department of Medical Laboratory Science, College of Applied Medical Sciences, University of Ha'il, Hail 55476, Saudi Arabia; h.barnawi@uoh.edu.sa (H.B.); h.qanash@uoh.edu.sa (H.Q.); g.alsaif@uoh.edu.sa (G.A.); b.alharbi@uoh.edu.sa (B.A.)

2 Medical and Diagnostic Research Center, University of Ha'il, Hail 55473, Saudi Arabia

3 Department of Medical Microbiology, Qunfudah Faculty of Medicine, Umm Al-Qura University, Al-Qunfudah 21961, Saudi Arabia; ahdarhami@uqu.edu.sa

4 Department of Medical Laboratory Sciences, Faculty of Applied Medical Sciences, King Abdulaziz University, Jeddah 22254, Saudi Arabia; malmashjary@kau.edu.sa

5 Hematology Research Unit, King Fahd Medical Research Center, King Abdulaziz University, Jeddah 22254, Saudi Arabia

6 Department of Clinical Laboratory Sciences, College of Applied Medical Sciences-Dawadmi, Shaqra University, Dawadmi 17464, Saudi Arabia; hsalmutiri@su.edu.sa

* Correspondence: ar.bazaid@uoh.edu.sa

Abstract: The failure to treat infectious diseases due to the continual emergence of drug-resistant microbes poses a huge and serious challenge for human health globally. Currently, the discovery and development of natural therapeutic compounds are attracting considerable attention from researchers worldwide. In this project, two types of pollen grains (maize and clover) were evaluated for potential antimicrobial activities. Extracts of both pollen grains were purified using HPLC, which has been shown to have numerous phenolic and flavonoid compounds. Pyro catechol and methyl gallate were detected in high concentrations (1145.56 and 1056.57 µg/mL, respectively) in the maize extract, while caffeic acid, quercetin, and kaempferol (464.73, 393.05, and 390.93 µg/mL, respectively) were among the compounds observed at high concentrations in the clover pollen grains extract. Staphylococcus aureus, Escherichia coli, Salmonella typhi, and Candida albicans were more sensitive to the clover pollen grains extract with inhibition zones of 22 ± 0.2, 18 ± 0.1, 29 ± 0.3, and 42 ± 0.4 mm compared to the size of the inhibitory zones caused by the maize pollen grains extract (19 ± 0.3, 15 ± 0.4, 27 ± 0.1, and 22 ± 0.4 mm, respectively). Moreover, lower MIC values for the clover pollen grains extract were recorded against C. albicans (1.97 ± 0.04 µg/mL), S. aureus (62.5 ± 1.00 µg/mL), and E. coli (62.5 ± 0.07 µg/mL) than the MICs caused by the maize pollen grains extract. The use of a transmission electron microscope revealed that the E. coli that had been treated with the clover pollen grains extract showed changes in its cell walls compared to that treated with the maize pollen grains extract. The clover pollen grains extract exhibited a stronger antioxidant potential, with an IC50 value of 22.18 µg/mL, compared to an IC50 value of 54.85 µg/mL for the maize pollen grains extract, via a DPPH scavenging assay. Regarding anticancer activity, the maize pollen grains extract was revealed to be more effective in terms of inhibiting the human colon cancer cell line HCT-116, with an IC50 value of 67.02 ± 1.37 µg/mL, compared with the observed toxicity caused by the clover extract, with an IC50 value of 75.03 ± 1.02 µg/mL. Overall, the clover pollen grains extract demonstrated potent antibacterial and antioxidant activities, but not anticancer activity, when compared to the maize grains extract. Thus, the current findings related to both types of pollen grains (clover and maize) highlight their potential therapeutic applications for the treatment of certain infectious diseases and malignancies.

Keywords: biological activities; pollen grains; flavonoids; phenols; cytotoxicity; antibacterial mechanism

1. Introduction

Pollen grains are well known as traditional medicines in certain countries due to their numerous effective phytoconstituents, including steroids carotenoids, flavonoids, terpenoids, and phenolics [1]. Pollen grains' antimicrobial and antioxidant efficiency, as well as their phenolic composition, have been investigated in several studies; for instance, Avşar et al. [2] mentioned that pollen grains of *Castanea sativa* have displayed antimicrobial activity toward certain Gram-positive and Gram-negative bacteria but were less potent against yeasts, although a strong antioxidant activity was reported. Date palm pollen grains have been shown to exhibit high flavonoid and phenolic contents with promising antibacterial and antioxidant activities [3]. Previously, Almaraz-Abarca et al. [4] reported that the antioxidant potential of pollen grains is mainly due to the presence of flavonoid and phenolic compounds. Moreover, anti-inflammatory, anti-toxicant, and hepato-protective properties have been attributed to pollen grains [5–7].

The correlation between the origin of plants' chemical and biological activities, such as antibacterial and antioxidant activities, is fundamental for further studies [8–10]. Velásquez et al. [11] studied the relationship among a plant's origin, its phytoconstituents, and the observed antibacterial properties of its pollen grains. In addition, the same group of researchers have reported that the pollen grains of *Galega officinalis* L. and *Brassica* species have inhibited the growth of *Escherichia coli, Pseudomonas aeruginosa, Streptococcus pyogenes,* and *Staphylococcus aureus,* with higher efficacy than conventional antibiotics. The anticancer activity of pollen grains was reported by Kaur et al. [12], but it was linked with a certain type of plant, including *Cassia glauca,* followed by *Bauhinia variegate, Cassia siamea,* and *Cassia biflora*. Several phenolic and flavonoid compounds have been detected in plant pollen grains, such as naringenin protocatechuic acid, gallic acid, hesperidin, kaempferol, isorhamnetin 3-orutinoside, quercetin, isorhamnetin, rutin, vanillic acid, quercetin 3-*o*-neohesperidoside, luteolin, 3-*o*-rutinoside, *p*-coumaric acid, apigenin, and rhamnetin 3-*o*-neohesperidoride [13]. Chantarudee et al. [14] reported the presence of 0.567 μg/g biotin, 0.199 μg/g invert sugar, and 0.0153 mg β carotene and vitamin A in the pollen grains of corn. Moniruzzaman et al. [15] have linked the antioxidant activity of pollen grains to its high contents of phenolics and flavonoids.

Developments in the utilization of drugs for cancer management are often correlated with unfavorable side effects or recurrence. Mărgăoan et al. [16] revealed the application of pollen grains for cancer treatment. One study in the literature claimed that the polyphenols of pollen grains are one of the main reasons behind their anticancer activity and apoptosis stimulation [17]. Pollen grains have shown antibacterial activity toward certain bacteria, including *E. coli, Staphylococcus aureus,* and *Bacillus subtilis,* but have not shown antifungal activity [18]. In the same study, anticancer activity was observed against HepG-2 cancer, Caco-2 cells, and PC3 cell lines with different IC_{50} values. Recently, Bakour et al. [19] studied the antimicrobial and antioxidant potential of pollen grains for six different plants, including *Quercus ilex, Punica granatum, Centaurium erythraea, Ruta graveolens, Citrus aurantium,* and *Coriandrum sativum.* Pollen grains of *P. granatum* and *Q. ilex* demonstrated a significant inhibition for the growth of *Pseudomonas aeruginosa, E. coli, Enterobacter cloacae, Klebsiella pneumoniae, Staphylococcus aureus,* and *Acinetobacter baumannii* than *C. sativum, C. erythraea,* and *R. graveolens,* while no antibacterial activity was observed for *C. aurantium.* To the best of our knowledge, the data related to the comparative bioactivities of clover and maize pollen grains, especially those collected prior to honeybees entering their hives, are absolutely absent or very limited. Thus, the main aim of this study was to assess the potential antimicrobial, anticancer, and antioxidant activities of the collected clover and maize pollen grains as well as to provide a characterization of their detected phenolic and flavonoid compounds.

2. Results and Discussion

2.1. Phytoconstituents Characterizations

In the present investigation, two types of pollen grains were investigated to evaluate their phenolic and flavonoid contents and their biological activity (Figure 1). An HPLC analysis of the phenolic and flavonoid contents of maize and clover pollen grains was reported (Supplementary Figures S1 and S2) and recorded (Table 1). The pollen grains from the maize pollen grains extract were characterized by the presence of a high amount of pyro catechol (1145.56 µg/mL), followed by methyl gallate (1056.57 µg/mL), and low amount of ferulic acid (5.88 µg/mL), ellagic acid (6.26 µg/mL), and daidzein (7.65 µg/mL). On the other hand, caffeic acid (464.73 µg/mL), quercetin (393.05 µg/mL), and kaempferol 390.93 µg/mL) had the highest detected concentrations, while naringenin (4.07 µg/mL) and vanillin (4.42 µg/mL) had the lowest concentrations in the clover pollen grains. Ten out of the eighteen detected compounds were recorded at higher concentrations in the clover pollen grains than in the maize pollen grains, including catechin, caffeic acid, ellagic acid, ferulic acid, rosmarinic acid, daidzein, quercetin, cinnamic acid, kaempferol, and hesperetin. Seven out of the eighteen detected compounds were recorded at higher concentrations in the maize pollen grains than in the clover pollen grains, except for rutin (68.41 µg/mL), which was in the maize pollen grains only. Therefore, the yellow color of maize pollen grains is due to the presence of rutin and quercetin [20]. In a previous study, 14 types of pollen grains were investigated to detect the flavonoid and phenolic constituents. Caffeic acid derivatives were shown to be the majority of the examined pollen grains [21]. In contrast, there were several phenolic and flavonoid compounds in two types of pollen grains, but their levels depended on the type of pollen grains. Nevertheless, the level/concentration of phenolic and flavonoid compounds in these pollens was dependent on geographical origin and climatic conditions as well as plant status. According to certain scientific papers on the constituents of pollen grains, kaempferol quercetin, naringenin, and caffeic acid represented the main functional constituents in biological activities [22]. A previous study associated with clover and maize pollen grains [23] has reported *p*-coumaric acid, catechin, daidzin, cinnamic acid, and pyrogallic acid as the main constituents of clover pollen grains, followed by caffeic and *p*-hydroxy benzoic acids, gestein and ferulic acid; however, eugenol and kaempferol were present in trace amounts. In addition, pollen grains of maize contain high amounts of cinnamic acid, p-coumaric acid, daidzin, and *p*-hydroxy benzoic, moderate amounts of caffeic acid, ferulic acid and genistein, and trace amounts of eugenol and kaempferol.

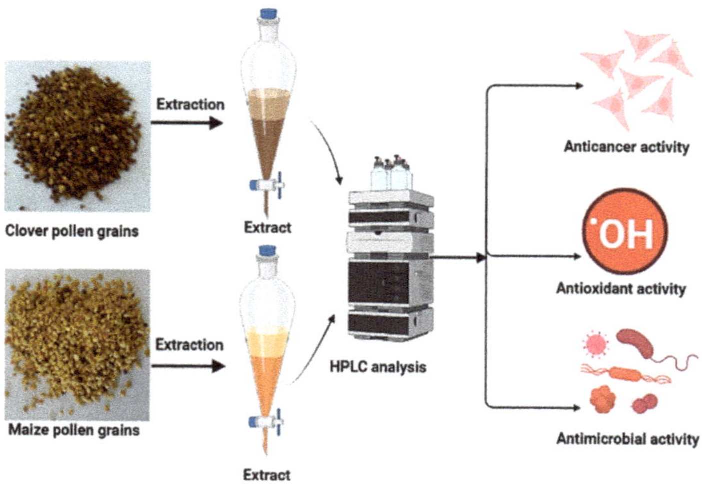

Figure 1. Image summarizing the performed tests for maize and clover pollen grains.

Table 1. The phenolic and flavonoid compounds detected in maize and clover pollen grains.

Compound	Pollen Grains Extract of Maize				Pollen Grains Extract of Clover			
	Retention Time	Area	Area (%)	Conc. (μg/mL)	Retention Time	Area	Area (%)	Conc. (μg/mL)
Gallic acid	3.544	22.87	2.0647	100.21	3.542	22.81	0.7373	99.97
Chlorogenic acid	4.300	15.59	1.4080	107.04	4.301	12.87	0.4159	88.33
Catechin	4.597	3.56	0.3218	41.14	4.714	4.03	0.1304	46.58
Methyl gallate	5.580	404.47	36.5194	1056.57	5.589	38.86	1.2556	101.50
Caffeic acid	6.087	17.13	1.5468	70.61	6.094	112.76	3.6439	464.73
Syringic acid	6.440	0.00	0.00	0.00	6.440	0.00	0.00	0.00
Pyro catechol	6.572	157.19	14.1924	1145.56	6.568	18.73	0.6052	136.48
Rutin	6.917	8.44	0.7624	68.41	6.925	0.00	0.00	0.00
Ellagic acid	7.531	1.45	0.1311	6.26	7.100	8.83	0.2854	38.09
Coumaric acid	8.299	94.09	8.4950	175.21	8.299	7.67	0.2479	14.29
Vanillin	8.942	26.10	2.3567	49.82	8.956	2.32	0.0748	4.42
Ferulic acid	9.802	1.92	0.1734	5.88	9.960	17.63	0.5698	53.98
Naringenin	10.590	12.37	1.1170	59.60	10.580	0.84	0.0273	4.07
Rosmarinic acid	11.723	1.75	0.1583	9.69	12.181	23.82	0.7697	131.66
Daidzein	16.203	2.60	0.2348	7.65	16.025	3.66	0.1183	10.77
Quercetin	17.276	8.63	0.7796	53.90	17.272	62.97	2.0347	393.05
Cinnamic acid	19.541	15.42	1.3923	14.38	19.184	19.26	0.6224	17.97
Kaempferol	20.559	6.59	0.5950	21.84	20.814	117.93	3.8108	390.93
Hesperetin	21.181	5.39	5.38755	13.98	21.181	6.16	0.1990	15.97

2.2. *Antimicrobial Activity of Maize and Clover Pollen Grains Extract*

The pollen grains extract of clover exhibited antibacterial activity with more inhibition zones of 22 ± 0.2, 18 ± 0.1, 29 ± 0.3, and 42 ± 0.4 mm, compared to the maize grains extract which showed inhibition zones of 19 ± 0.3, 15 ± 0.4, 27 ± 0.1 and 22 ± 0.4 mm against some tested microorganisms, including *S. aureus*, *E. coli*, *S. typhi*, and *C. albicans*, respectively. However, *B. subtilis* was more sensitive to the maize pollen grains extract (with an inhibition zone of 42 ± 0.4 mm) than the clover pollen grains extract (with an inhibition zone of 39 ± 0.3 mm) (Figure 2 and Table 2). Furthermore, there was no differences between the MIC of the maize and clover pollen grains extract for *B. subtilis* and *S. typhi*, while a low MIC of the clover pollen grains extract was observed against *C. albicans* (1.97 ± 0.04 μg/mL), *S. aureus* (62.5 ± 1.00 μg/mL), and *E. coli* (62.5 ± 0.07 μg/mL) compared to the MIC of maize pollen grains. In addition, the MBC of the clover pollen grains extract was less than the MBC of the maize pollen grains extract against all tested microorganisms except *S. typhi*. The two pollen grains extracts reflected bactericidal properties against all tested microorganisms. This explanation was documented via the estimation of MBC/MIC index (Table 2). The filamentous fungus (*A. niger*) was resistant to both the maize and clover pollen grains extracts. Results of the current study were in agreement with an earlier report, where ethanolic and aqueous extracts of clover pollen grains exhibited higher inhibition zones (26.70 and 25.40 mm, respectively) than ethanolic and aqueous extracts of maize (20.75 and 22.65 mm, respectively) against *Paenibacillus larvae* [23]. However, Khider et al. [24] reported different results, where the highest antibacterial activity was attributed to the methanolic extract of maize pollen grains, which was followed by clover with MIC values of 320–640 μg/mL using clover pollen grains and of 320–1280 μg/mL using maize pollen grains against tested bacteria. Khider et al. collected the pollen grains from inside the honeybee hives. In this study, however, the pollen grains were collected prior to the honeybees entering their hives. In addition to the antibacterial activity of quercetin and kaempferol (contents of pollen grains extract), they showed anti-yeast activity against *Candida parapsilosis* [25]. *Enterococcus faecalis* was inhibited by the pollen grains extract of *Ailanthus altissima*, *Cupressus arizonica* and *Chenopodium album* with different MIC values of 38.8, 113, and 562 μg/mL, respectively [26]. Concerning antibacterial activity, earlier reports indicated that the natural compounds could possess significant antibacterial activity if their MIC was fewer than 100 μg/mL [27].

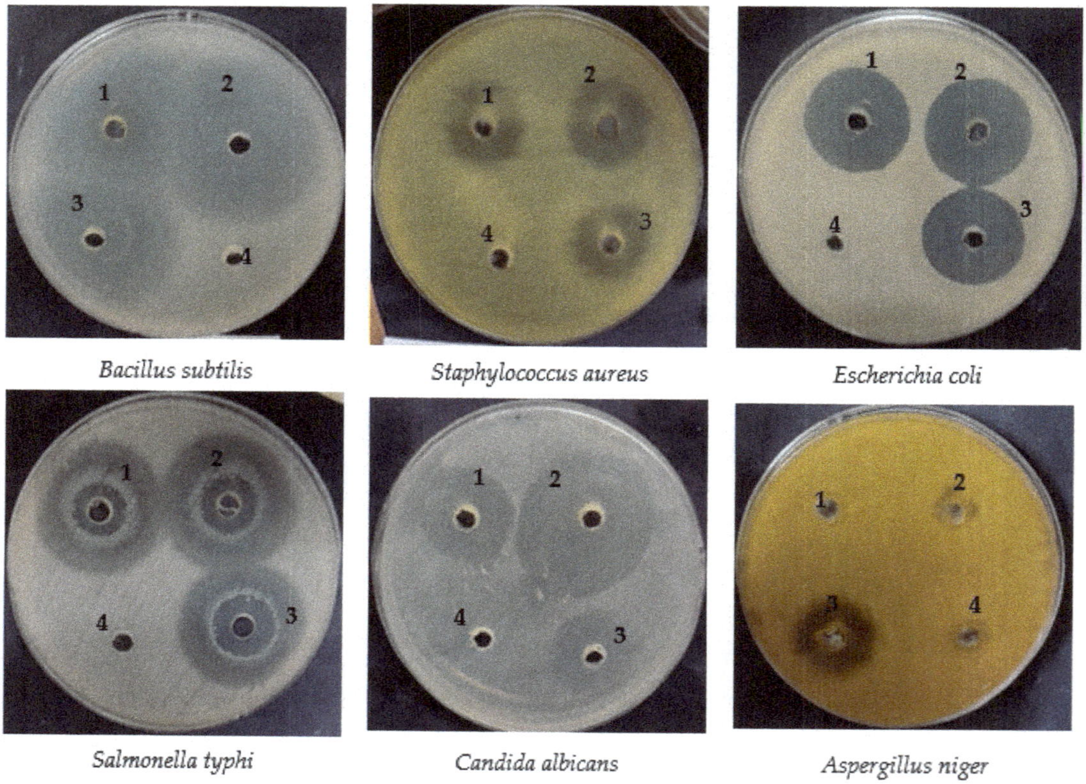

Figure 2. Antimicrobial activity of maize (1) and clover (2) pollen grains extracts compared to the positive (3) and negative (4) controls.

The effect of maize and clover pollen grains on the ultrastructure of *E. coli* was examined and imaged (Figure 3). The impact of two types of pollen grains extract was assessed on the cell structure of *E. coli* compared to unaffected *E. coli*. Clover pollen grains were more effective than maize pollen grains. Cell wall deformation was observed in case of the effect of maize pollen grains (black arrow) with a leakage of cytoplasm because of cell wall rupture (orange arrow). In contrast, control cells of *E. coli* were examined in an identical shape with the presence of perfect and clear healthy cell walls, including membranes and cytoplasm (blue arrows).

2.3. Antioxidant Activity of Maize and Clover Pollen Grains Extract

Extracts of both pollen grains, including maize and clover exhibited antioxidant activity with an increased DPPH scavenging (%) as their concentration was increased. The current results reflected that the clover pollen grains extract had the highest capacity of antioxidant out of all the maize grains at all tested concentrations. DPPH scavenging (%) was 24.1 and 12.1% at 1.95 μg/mL, 68.1 and 58.5% at 125 μg/mL, and 92.1 and 85.6% at 1000 μg/mL, using the clover and maize pollen grains extracts, respectively (Table 3). Moreover, the IC_{50} of the clover pollen grains extract was 22.18 μg/mL, while 54.85 μg/mL was recorded as the IC_{50} for the maize pollen grains extract. All these findings were compared to ascorbic acid as a standard drug which exhibited an IC_{50} value of 2.51

μg/mL. A previous study reported the antioxidant activity of maize pollen grains with an IC_{50} value of 425.4 μg/mL [28]. These differences in the antioxidant activity may depend on several factors, such as the plant origin, phenolic, and flavonoid constituents of pollen grains.

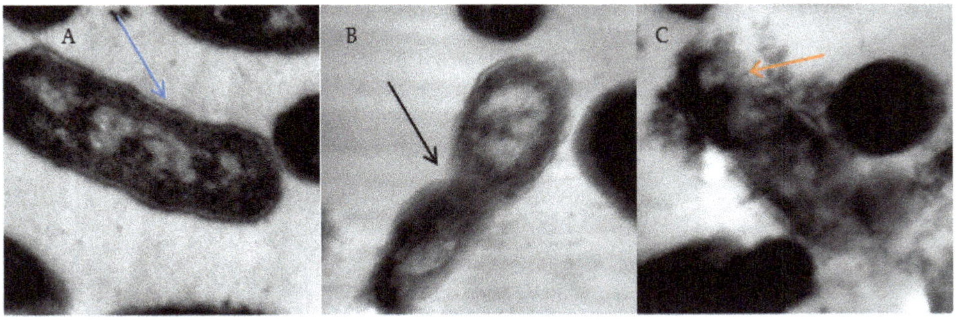

Figure 3. Ultrastructure of *E. coli* (**A**), *E. coli* exposed to maize (**B**) and *E. coli* exposed to clover (**C**) pollen grains extract. Direct magnification 6000×.

In the present study, clover pollen grains possessed a high content of quercetin compared to maize pollen grains. The presence of these flavonoids showed a significant antioxidant activity of clover pollen grains compared to maize pollen grains. Leja et al. [29] reported that the phenolics, flavonoids, and pigments such as β-carotene were responsible for the action mechanism of antioxidants of pollen grains. Marghitaş et al. [30] mentioned that quercetin played a vital role in preventing the oxidative damage of cellular biomolecules, such as nucleic acids, and lipoprotein due to the reactive oxygen species (ROS). Strong antioxidant activity was attributed to maize pollen grains as mentioned by Bujang et al. [31]. Several investigations documented that pollen grains had significant antioxidant potential that was commonly influenced by the presence of phenolic and flavonoid constituents. However, a great difference of the antioxidant activity was observed between pollen grains obtained from various plant species and different geographical areas [32]. In a previous study, the antioxidant activity of the maize pollen grains extract had a lower IC_{50} value than the IC_{50} of clover [24]. According to Avşar et al. [2], the IC_{50} value of pollen grains of *Castanea sativa* via DPPH was 19.5 μg/mL. In the study conducted by Bakour et al. [19], the IC_{50} values of *P. granatum*, *Q. ilex*, and *C. erythraea* pollen grains extracts were 2, 8 and 200 μg/mL, respectively.

Table 2. Efficacy of maize and clover pollen grains against tested microorganisms with MIC, MBC, and MBC/MIC index detection.

Tested Microorganisms	Mean Inhibition Zone (mm)				p Value	MIC (µg/mL)		p Value	MBC (µg/mL)		p Value	MBC/MIC Index		p Value
	Maize	Clover	+ve C *	-ve C *		Maize	Clover		Maize	Clover		Maize	Clover	
Bacillus subtilis	42 ± 0.4 a	39 ± 0.3 b	33 ± 0.1 c	0.0	0.000	3.9 ± 0.33 a	3.9 ± 0.25 a	1.00	7.9 ± 0.35 a	3.9 ± 0.2 b	0.00	2.02 a	1.0 b	0.00
Staphylococcus aureus	19 ± 0.3 b	22 ± 0.2 a	16 ± 0.2 c	0.0	0.001	125 ± 4.0 a	62.5 ± 1.00 b	0.00	125 ± 1.53 a	125 ± 3 a	1.00	1.0 b	2.5 a	0.00
Escherichia coli	15 ± 0.4 b	18 ± 0.1 a	18 ± 0.3 a	0.0	0.016	250 ± 3.0 a	62.5 ± 0.07 b	0.00	500 ± 3.0 a	125 ± 4 b	0.000	2.0 a	2.5 a	0.288
Salmonella typhi	27 ± 0.1 ab	29 ± 0.3 a	25 ± 0.1 b	0.0	0.125	31.25 ± 1.33 a	31.25 ± 0.5 a	1.00	31.25 ± 1.53 b	62.5 ± 1 a	0.00	1.0 b	2.0 a	0.00
Candida albicans	22 ± 0.4 b	42 ± 0.4 a	21 ± 0.3 b	0.0	0.000	15.62 ± 0.35 a	1.97 ± 0.04 b	0.00	15.62 ± 1.32 a	3.97 ± 0.2 b	0.002	1.0 b	2.01 a	0.00
Aspergillus niger	NA	NA	20 ± 0.4	0.0	0.000	-	-		-	-		-	-	

* +ve C, positive control (Gentamycin/Nystatin); -ve C, negative control (solvent of extraction used). Each value is mean of 3 replicates ± standard error of means. Honestly Significant Difference (HSD) at $p \leq 0.05$ by post hoc Tukey's test. The comparison refers to the means of the same row. The same letters mean no significance, while different letters mean significant difference.

2.4. Anticancer Activity of Pollen Grains

The anticancer potential of the two types of pollen grains was examined against human colon cancer cell line (HCT-116) (Table 4). It was clear that the maize pollen grains extract was more effective than clover pollen grains, particularly at concentrations up to 125 µg/mL. There were negligible differences in the cytotoxicity of the two pollen grains extracts at high concentrations (250–1000 µg/mL). A lower IC_{50} value (67.02 ± 1.37 µg/mL) of maize than the IC_{50} value (75.03 ± 1.02 µg/mL) of clover pollen grains was recorded compared with the IC_{50} value of the positive control Adriamycin (43.12 ± 1.25 µg/mL). Wang et al. [33] have reported the anticancer activity of *Rosa rugosa* pollen grains against HCT-116 and HT-29 cell lines, using MTT assay. The pollen grains of other origin, such as *Bauhinia variegate*, *Cassia biflora*, *C. siamea* and *C. glauca* exhibited anticancer activity [12]. The morphological study reported the presence of an alteration in the treated HCT-116 cells with the two pollen grains extracts at a concentration of more than 62.6 µg/mL increasing the damaged and dead cells (Figures 4 and 5). Moreover, the detachment and rounding of HCT-116 cells were observed at 250–1000 µg/mL with a reduction in treated cells and cytoplasm mass. These alterations in the exposed cells to the maize and clover pollen grains extracts have documented some markers of antitumor activity against HCT-116 cells.

Table 3. Antioxidant activity of maize pollen grains extract, clover pollen grains extract and ascorbic acid.

Concentration (µg/mL)	DPPH Scavenging (%)			*p* Value
	Maize	Clover	Ascorbic Acid	
1000	85.6 ± 0.05 [c]	92.1 ± 0.01 [b]	99.2 ± 0.02 [a]	0.000
500	74.1 ± 0.17 [c]	84.4 ± 0.02 [b]	96.1 ± 0.031 [a]	0.002
250	66.2 ± 0.24 [c]	76.2 ± 0.04 [b]	94.6 ± 0.14 [a]	0.000
125	58.5 ± 0.16 [c]	68.1 ± 003 [b]	91.8 ± 0.25 [a]	0.000
62.50	51.3 ± 0.32 [c]	61.4 ± 0.05 [b]	84.3 ± 0.41 [a]	0.001
31.25	43.3 ± 0.21 [c]	53.9 ± 0.15 [b]	76.2 ± 0.08 [a]	0.005
15.63	35.8 ± 0.45 [c]	46.3 ± 0.24 [b]	67.7 ± 0.24 [a]	0004
7.81	28.2 ± 0.21 [c]	38.4 ± 0.18 [b]	60.4 ± 0.21 [a]	0.003
3.90	20.4 ± 0.32 [c]	30.7 ± 0.06 [b]	52.2 ± 0.41 [a]	0.005
1.95	12.1 ± 0.15 [c]	24.1 ± 0.09 [b]	43.6 ± 0.14 [a]	0.000
0.0	0.0	0.0	0.0	
IC_{50}	54.85 ± 0.21 [a] µg/mL	22.18 ± 0.21 [b] µg/mL	2.51 ± 0.24 [c] µg/mL	0.000

Each value is mean of 3 replicates ± standard error of means. Honestly Significant Difference (HSD) at $p \leq 0.05$ by post hoc Tukey's test. The comparison referred to the means of the same row. The same letters mean no significance, while different letters mean significant difference.

Table 4. Anticancer activity of maize and clover pollen grains extract against HCT-116 cells line.

Concentration (µg/mL)	Cytotoxicity (%)		*p* Value
	Maize	Clover	
1000	97.57 ± 0.24 [a]	97.53 ± 0.21 [a]	1.00
500	97.61 ± 0.16 [a]	96.53 ± 0.02 [a]	0.288
250	94.92 ± 0.08 [a]	95.05 ± 0.15 [a]	0.168
125	81.99 ± 0.07 [a]	76.82 ± 0.65 [b]	0.038

Table 4. *Cont.*

Concentration (μg/mL)	Cytotoxicity (%)		*p* Value
	Maize	Clover	
62.50	60.81 ± 0.14 [a]	55.38 ± 0.57 [b]	0.028
31.25	19.84 ± 0.35 [a]	13.06 ± 0.38 [b]	0.021
0.0	0.0	0.0	-
IC_{50} (μg/mL)	67.02 ± 1.37 [b]	75.03 ± 1.02 [a]	0.004

Each value is mean of 3 replicates ± standard error of means. Honestly Significant Difference (HSD) at $p \leq 0.05$ by post hoc Tukey's test. The comparison referred to the means of the same row. The same letters mean no significance, while different letters mean significant difference.

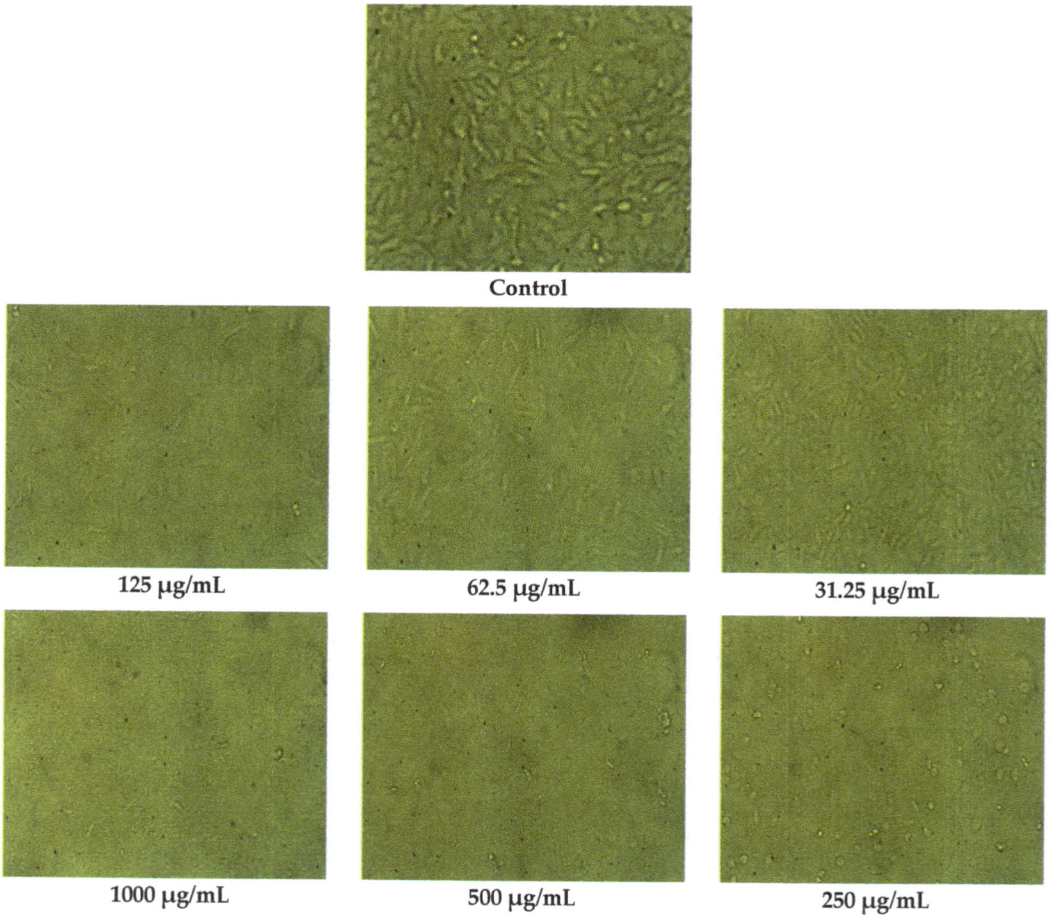

Figure 4. The morphological features of HCT-116 cells line (control) exposed to different concentrations of the maize pollen grains extract. As the concentration of maize pollen grains increased, cancerous cells (HCT-116) become tightly regulated within a cell suicide process.

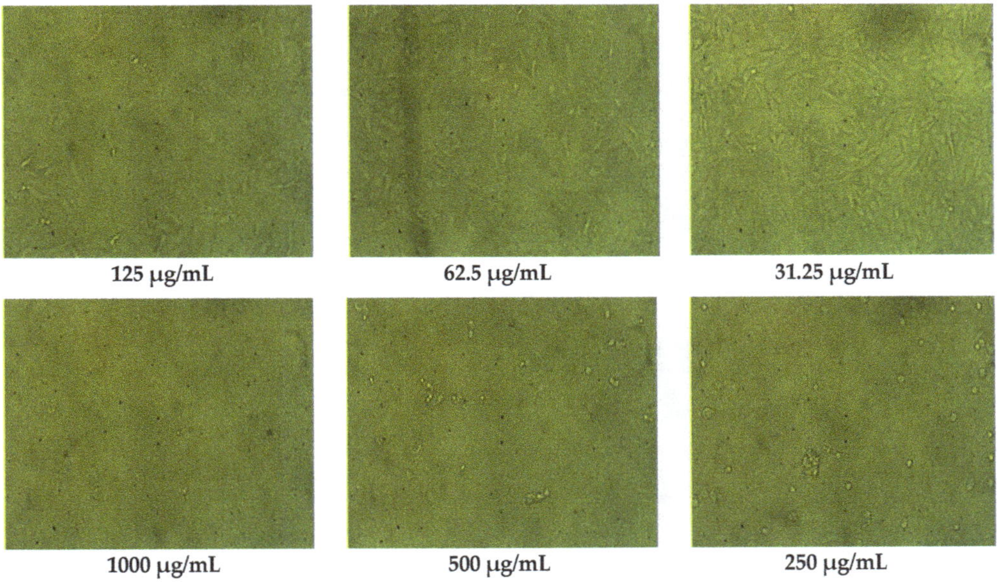

Figure 5. The morphological features of HCT-116 cells line exposed to different concentrations of the clover pollen grains extract. As the concentration of clover pollen grains increased, cancerous cells (HCT-116) become tightly regulated within a programed cell death process.

On the other hand, untreated HCT-116 cells remained in their identical spindle and angular form. Some detected compounds in the current pollen grains extract, such as naringenin, quercetin, kaempferol, and coumaric acid were reported as anticancer agents in vitro [34]. Maize pollen grains were tested against human prostate cancer by Ganash [28], where at 500 μg/mL, it exhibited 77.09% cytotoxicity with an IC_{50} value of 339.81 μg/mL, and the marker changes in the tested cells, such as shrunken and round form, and breakdown of cell DNA were reported. According to Elsayed et al. [18], beebread (mixture of plant pollen and honey fermented with lactic acid) had anticancer potential against liver hepatocellular, colorectal adenocarcinoma, and prostate adenocarcinoma cell lines of human with IC_{50} values of 386, 314 and 262 μg/ mL, respectively.

3. Materials and Methods

3.1. Pollen Grains Source and Extraction

In the current study, the pollen grains were harvested by bees via the fitting of a pollen trap at the hive entrance of the 10 bees for 3 days (Figure 6). The pollen grains of clover were dark yellow and collected during the season of clover (during the period of clover "May" flowering) from a farm that cultivated only clover, while maize pollen grains were yellow and collected during the season of maize from a farm that cultivated only maize (during the period of maize "July" flowering). Moreover, the farms cultivated only clover and maize. In addition to that, the farms were treated with herbicides to prevent other cultivars. Furthermore, the cultivated area was huge, ranging from 6 to 8 hectares. Therefore, further palynological analysis was not required. The harvested pollen grains were dehydrated at 40 °C for 12 h, using an oven. The extracts of the collected pollen grains were prepared by mixing 50 g into 500 mL of methanol. Then, via shaking in a water bath, they were shaken for 24 h, which was followed by filtration. Finally, the extracts of the pollen grains were concentrated to obtain a crude extract under vacuum.

Figure 6. Pollen trap (PT) fitted to the entrance of a hive bee to collect pollen grains (white arrows).

3.2. *Assessment of Phenolic and Flavonoid Constituents of Pollen Grains by HPLC*

HPLC (Agilent 1260 series) was used to determine the phenolic and flavonoid contents in the pollen grain extracts. One Zorbax Eclipse Plus C8 column (4.6 mm × 250 mm i.e., 5 µm) was used for the separation process. The mobile phase (MP) consisted of acetonitrile and water (W) with 0.05% trifluoroacetic acid (A) flowing at a rate of 0.9 mL/min. The MP was automatically programmed in the following flowing order, using a linear gradient: 82% W for 0 min, 75% W for 1 min, 60% W for 11–18 min, and 82% W for 18–24 min. The ultraviolet (UV) detector was used to detect flavonoids at 330 nm and phenolic components at 280 nm. The tested sample solution was injected into a volume of 5 µL, while the column was kept at 40 °C. Utilizing standard molecules of flavonoids and phenolic acid as input data, the extract's chemical composition was semi-quantitatively determined.

3.3. *Tested Microorganisms and Cancer Cell Line*

Escherichia coli ATCC 11293, *Staphylococcus aureus* ATCC 6538, *Bacillus Subtilis* (ATCC 6633), *Salmonella typhi* (ATCC 6539), *Candida albicans* ATCC 14053 and *Aspergillus niger* (Ain Shams University Hospitals and University Mycological Centre, Egypt) were investigated. The tested bacteria/fungi were activated using nutrient/Sabouraud Dextrose agar (SDA) media (Sigma–Aldrich, Steinheim, Germany). Regarding the cancer cell line, the human colon cancer cell line (HCT-116) (Organism: Homo sapiens, human; tissue: colon; cell type: epithelial; disease: colorectal carcinoma) was purchased from the American Type Culture Collection (ATCC, Rockville, MD, USA).

3.4. *Antimicrobial Activity of Pollen Grains Extract*

The well diffusion technique was used to examine the antimicrobial activity of the extracts of pollen grains. For antibacterial activity, 20 mL of medium growth (Mueller–Hinton agar) (Modern Lab Co., Indore, Madhya Pradesh, India) was poured onto plastic petri plates. A bacterial inoculum, containing 0.5 McFarland 2×10^8 CFU/mL was added to each Petri plate containing Mueller–Hinton agar, while Sabouraud agar medium was used for fungi. Through sterile cork borer, 6 mm diameter wells were cut from the medium, which was followed by filling them with 20 µL of the methanolic extract of pollen grains. Solvent served as a negative control, but antibiotic was used as a positive control. The bacteria were incubated at 37 °C for 24 h, while fungi were incubated for at 30 °C 3 days. The zones of inhibition were measured in millimeters [35,36].

3.5. Minimal Bactericidal Concentration (MBC) and Minimum Inhibitory Concentration (MIC) of Pollen Grains Extract

Different dilutions of both extracts of pollen grains were prepared. From each dilution, 10 μL was added to 170 μL of Mueller–Hinton broth into 96 wells of a microplate amended with 20 μL of bacterial inoculum, containing 5×10^5 CFU/mL based on the approach of NCCLS standards [35]. The microtiter plates were then incubated at 37 °C for 24 h. The developed turbidity indicates the growth of the tested microorganism, and the MIC is the lowest concentration of the tested extract where no growth is visually detected. The MBC is detected, where the dilution demonstrates the MIC and at least two of the highly concentrated compound dilutions are plated and counted to estimate the viable CFU/mL of the tested microorganisms. The MBC is the lowest dose that determines a pre-detected reduction (99.9%) in CFU/mL when compared to the MIC dilution.

3.6. Ultrastructure of E. coli

A JEOL-JEM 1010 transmission electron microscope (TEM) (JEOL JEM 1200-, Tokyo, Japan) was used to examine *E. coli* cells exposed to the pollen grains extract, where the treated *E. coli* cells were fixed in 3% glutaraldehyde and rinsed in phosphate buffer. Then, via a solution of potassium permanganate, the cells were post-fixed at 25 °C for 5 min. The treated cells were gradually dehydrated by ethanol (10% to 90% ethanol) at 25 °C. At the end, using the absolute ethanol, the cells were dehydrated for 30 min. The treated *E. coli* cells were infiltrated via epoxy resin and acetone until forming a pure resin. The ultrathin segments of treated cells were loaded on the grids of copper, which was followed by staining them twice with uranyl acetate and lead citrate. At 70 kV, the stained sections were imagined by TEM [37].

3.7. Antioxidant Activity of Pollen Grains Extract

Using the DPPH test, free radical scavenging activity was evaluated in accordance with the technique outlined by Al-Rajhi et al. [38]. Briefly, a PerkinElmer Lambda 40 UV/Vis spectrophotometer was used to measure the absorbance at 517 nm after adding 75 μL of various dilutions of the extracts of each pollen to an 825 μL solution of DPPH (63.4 μM, prepared in ethyl alcohol). Ascorbic acid was used as a positive control. The reaction was carried out in the dark for an hour. Using the following equation, the proportion of the DPPH radical scavenging inhibition was calculated:

$$\text{Inhibition}(\%) = \frac{\text{Absorbanceatcontrol} - \text{Absorbanceattreatement}}{\text{Absorbanceatcontrol}} \times 100$$

The graph of the curve representing the percentage of DPPH inhibition as a function of the pollen grains extract concentration was used to estimate the concentration of each extract needed to scavenge 50% of DPPH (IC_{50}). The values for the IC_{50} were given in μg/mL.

3.8. Anticancer Activity of Pollen Grains Extract and Morphological Characteristics of Cancer Cells

The 3-(4,5-dimethyl-2-thiazolyl)-2,5-diphenyl-2H-tetrazolium bromide (MTT) test was used to determine the cytotoxicity activity of the pollen grains extract against the HCT-116 cell line. Prior to the MTT assay, cells (5×10^4 cells/well) were dispersed into a 96-well sterile microplate and treated for 48 h at 37 °C with a series of varied doses of the pollen grains extract (31.25–1000 μg/mL) in DMSO, and Adriamycin was used as a positive control as well. Following incubation, the media were carefully withdrawn, each well was fortified with 40 μL of MTT (2.5 mg/mL), and incubation was again performed for 4 h at 37 °C at 5% CO_2. Then, 200 μL of DMSO was added to dissolve the purple formazan dye crystals. A SpectraMax® Paradigm® Multi-Mode microplate reader was used to measure the absorbance at 560 nm and subtract the background at 620 nm. Optical density was directly linked with cell mass. In comparison with the untreated control cells, the mean percentage of viable cells was used to express the relative cell viability. Through a phase contrast microscope, the morphological characteristics of HCT-116 cell line cells were examined after 24 h of treatment [39,40].

3.9. Statistical Analysis

Each value is the mean of 3 replicates $\pm$ standard error of means. An Honestly Significant Difference (HSD) at $p \leq 0.05$ by post hoc Tukey's test was applied. IC_{50} was detected by probit analysis, utilizing SPSS 29 software program (SPSS Inc., Chicago, IL, USA).

4. Conclusions

The current findings concluded that the tested extracts of pollen grains (maize and clover) held great potential for therapeutic applications due to the observed potent antimicrobial, antioxidant, and anticancer activities. Clover pollen grains exhibited stronger antimicrobial and antioxidant activities, while lower toxicity toward cancer cells (HCT-116) was monitored when compared to maize pollen grains. Furthermore, the morphological effects of the tested extracts (maize and clover pollen grains) against certain bacterial and cancer cells were represented by the ultrastructure changes in *E. coli* and HCT-116 cells, which confirmed the antibacterial and antitumor activities. Moreover, pollen grains represent a safe substitute in nutritional and pharmacological fields. Thus, maize and clover pollen grains and their derived phenolic and flavonoid compounds can serve as a future food additive for human consumption as well as a source of bioactive and functional constituents in pharmaceutical and nutraceutical industries. Nevertheless, the underlying mechanism of the identified constituents in the pollen grains extracts should be further in vitro examined against a wider range of cancer and bacterial cells.

Supplementary Materials: The following supporting information can be downloaded at https://www.mdpi.com/article/10.3390/ph16121731/s1, Figure S1: Chromatogram of detected phenolic and flavonoid compounds in maize via HPLC; Figure S2: Chromatogram of detected phenolic and flavonoid compounds in clover via HPLC.

Author Contributions: Conceptualization, H.B. and A.S.B.; formal analysis—antimicrobial activity, H.B., A.A. (Abdu Aldarhami) and A.S.B.; formal analysis—anticancer activity, H.Q., B.A. and M.N.A.; formal analysis—antioxidant activity, G.A. and A.A. (Alhomidi Almotiri); investigation, H.B., H.Q., A.A. (Abdu Aldarhami) and A.S.B.; visualization and validation, H.B., H.Q., A.A. (Abdu Aldarhami), G.A., B.A., M.N.A., A.A. (Alhomidi Almotiri) and A.S.B.; supervision, H.Q. and A.S.B.; funding acquisition, A.S.B.; writing—original draft preparation, H.B., H.Q., A.A. (Abdu Aldarhami) and A.S.B.; writing—review and editing; H.B., H.Q., A.A. (Abdu Aldarhami), G.A., B.A., M.N.A., A.A. (Alhomidi Almotiri) and A.S.B. All authors have read and agreed to the published version of the manuscript.

Funding: This research was funded by the Scientific Research Deanship at University of Ha'il- Saudi Arabia via a project number (RG-23152).

Institutional Review Board Statement: Not applicable.

Informed Consent Statement: Not applicable.

Data Availability Statement: Data are contained within the article and Supplementary Materials.

Acknowledgments: All authors thank the Scientific Research Deanship at University of Ha'il for their grant through a project number (RG-23152).

Conflicts of Interest: The authors declare no conflict of interest.

References

1. Kao, Y.T.; Lu, M.J.; Chen, C. Preliminary analyses of phenolic compounds and antioxidant activities in tea pollen extracts. *J. Food Drug Anal.* **2011**, *19*, 470–477. [CrossRef]
2. Avşar, C.; Özler, H.; Berber, I.; Civek, S. Phenolic composition, antimicrobial and antioxidant activity of Castanea sativa Mill. pollen grains from Black Sea region of Turkey. *Int. Food Res. J.* **2016**, *23*, 1711–1716.
3. Daoud, A.; Malika, D.; Bakari, S.; Hfaiedh, N.; Mnafgui, K.; Kadri, A.; Gharsallah, N. Assessment of polyphenol composition, antioxidant and antimicrobial properties of various extracts of Date Palm Pollen (DPP) from two Tunisian cultivars. *Arab. J. Chem.* **2019**, *12*, 3075–3086. [CrossRef]
4. Almaraz-Abarca, N.; da Graça Campos, M.; Ávila-Reyes, J.A.; Naranjo-Jiménez, N.; Herrera Corral, J.; González-Valdez, L.S. Antioxidant activity of polyphenolic extract of monofloral honeybee-collected pollen from mesquite (*Prosopis juliflora*, Leguminosae). *J. Food Compos. Anal.* **2007**, *20*, 119–124. [CrossRef]
5. Uzbekova, D.M.V.; Khvoynitskaya, L.; Slepnev, A. Evaluation of bee collected pollen influence on lipid peroxidation, antioxidant system and liver function in old animals. *Hepatology* **2003**, *38*, 203–208. [CrossRef]
6. Choi, E.M. Antinociceptive and antiinflammatory activities of pine (*Pinus densiflora*) pollen extract. *Phytother. Res.* **2007**, *21*, 471–475. [CrossRef]
7. LeBlanc, B.W.; Davis, O.K.; Boue, S.; DeLucca, A.; Deeby, T. Antioxidant activity of Sonoran Desert bee pollen. *Food Chem.* **2009**, *115*, 1299–1305. [CrossRef]
8. Abdelghany, T.M.; Yahya, R.; Bakri, M.M.; Ganash, M.; Amin, B.H.; Qanash, H. Effect of Thevetia peruviana seeds extract for microbial pathogens and cancer control. *Int. J. Pharmacol.* **2021**, *17*, 643–655. [CrossRef]
9. Al-Rajhi, A.M.H.; Qanash, H.; Almashjary, M.N.; Hazzazi, M.S.; Felemban, H.R.; Abdelghany, T.M. Anti-Helicobacter pylori, Antioxidant, Antidiabetic, and Anti-Alzheimer’s Activities of Laurel Leaf Extract Treated by Moist Heat and Molecular Docking of Its Flavonoid Constituent, Naringenin, against Acetylcholinesterase and Butyrylcholinesterase. *Life* **2023**, *13*, 1512.
10. Qanash, H.; Bazaid, A.S.; Aldarhami, A.; Alharbi, B.; Almashjary, M.N.; Hazzazi, M.S.; Felemban, H.R.; Abdelghany, T.M. Phytochemical Characterization and Efficacy of Artemisia judaica Extract Loaded Chitosan Nanoparticles as Inhibitors of Cancer Proliferation and Microbial Growth. *Polymers* **2023**, *15*, 391. [CrossRef]
11. Velásquez, P.; Rodriguez, K.; Retamal, M.; Giordano Villatoro, A.; Valenzuela Roediger, L.M.; Montenegro Rizzardini, G. Relation between composition, antioxidant and antibacterial activities and botanical origin of multi-floral bee pollen. *J. Appl. Bot. Food Qual.* **2017**, *90*, 306–314. [CrossRef]
12. Kaur, R.; Nagpal, A.; Katnoria, J.K. Exploration of antitumor properties of pollen grains of plant species belonging to fabaceae family. *J. Pharm. Sci. Res.* **2015**, *7*, 127–129.
13. Han, L.; Liu, X.; Yang, N.; Li, J.; Cai, B.; Cheng, S. Simultaneous chromatographic fingerprinting and quantitative analysis of flavonoids in Pollen Typhae by high-performance capillary electrophoresis. *Acta Pharm. Sin. B* **2012**, *2*, 602–609. [CrossRef]
14. Chantarudee, A.; Phuwapraisirisan, P.; Kimura, K.; Okuyama, M.; Mori, H.; Kimura, A.; Chanchao, C. Chemical constituents and free radical scavenging activity of corn pollen collected from Apis mellifera hives compared to floral corn pollen at Nan, Thailand. *BMC Complement. Altern. Med.* **2012**, *12*, 45. [CrossRef] [PubMed]
15. Moniruzzaman, M.; Khalil, M.I.; Sulaiman, S.A.; Gan, S.H. Physicochemical and antioxidant properties of Malaysian honeys produced by Apis cerana, Apis dorsata and Apis mellifera. *BMC Complement. Altern. Med.* **2013**, *13*, 43. [CrossRef] [PubMed]
16. Mărgăoan, R.; Stranț, M.; Varadi, A.; Topal, E.; Yücel, B.; Cornea-Cipcigan, M.; Campos, M.G.; Vodnar, D.C. Bee Collected Pollen and Bee Bread: Bioactive Constituents and Health Benefits. *Antioxidants* **2019**, *8*, 568. [CrossRef] [PubMed]
17. Premratanachai, P.; Chanchao, C. Review of the anticancer activities of bee products. *Asian Pac. J. Trop. Biomed.* **2014**, *4*, 337–344. [CrossRef] [PubMed]
18. Elsayed, N.; El-Din, H.S.; Altemimi, A.B.; Ahmed, H.Y.; Pratap-Singh, A.; Abedelmaksoud, T.G. In Vitro Antimicrobial, Antioxidant and Anticancer Activities of Egyptian Citrus Beebread. *Molecules* **2021**, *26*, 2433. [CrossRef]
19. Bakour, M.; Laaroussi, H.; Ousaaid, D.; Oumokhtar, B.; Lyoussi, B. Antioxidant and Antibacterial Effects of Pollen Extracts on Human Multidrug-Resistant Pathogenic Bacteria. *J. Food Qual.* **2021**, *2021*, 5560182. [CrossRef]
20. Žilić, S.; Vančetović, J.; Janković, M.; Maksimović, V. Chemical composition, bioactive compounds, antioxidant capacity and stability of floral maize (*Zea mays* L.) pollen. *J. Funct. Foods* **2014**, *10*, 65–74. [CrossRef]
21. Bakour, M.; Campos, M.d.G.; Imtara, H.; Lyoussi, B. Antioxidant content and identification of phenolic/flavonoid compounds in the pollen of fourteen plants using HPLC-DAD. *J. Apic. Res.* **2020**, *59*, 35–41. [CrossRef]
22. Sarić, A.; Balog, T.; Sobocanec, S.; Kusić, B.; Sverko, V.; Rusak, G.; Likić, S.; Bubalo, D.; Pinto, B.; Reali, D.; et al. Antioxidant effects of flavonoid from Croatian Cystus incanus L. rich bee pollen. *Food Chem. Toxicol.* **2009**, *47*, 547–554. [CrossRef]
23. Abdel-Rasheed, K.G.; Moustafa, Y.M.; Hassan, E.A.; Abdel-Ati, Y.Y.; Gadel-Hak, S.-E.H. Honeybee Collected Pollen as Antibiotics to Control American Foul Brood Disease, at Giza Governorate. *Egypt. J. Agric. Res.* **2017**, *95*, 1551–1563. [CrossRef]
24. Khider, M.; Elbanna, K.; Mahmoud, A.; Owayss, A.A. Egyptian honeybee pollen as antimicrobial, antioxidant agents, and dietary food supplements. *Food Sci. Biotechnol.* **2013**, *22*, 1–9. [CrossRef]
25. Rocha, M.F.G.; Sales, J.A.; da Rocha, M.G.; Galdino, L.M.; de Aguiar, L.; Pereira-Neto, W.A.; de Aguiar Cordeiro, R.; Castelo-Branco, D.; Sidrim, J.J.C.; Brilhante, R.S.N. Antifungal effects of the flavonoids kaempferol and quercetin: A possible alternative for the control of fungal biofilms. *Biofouling* **2019**, *35*, 320–328. [CrossRef]

26. Bashiri, Z.; Yousefi, M.; Royce, S.G.; Fereidouni, M. Antibacterial activity of aqueous and lipid extracts of five common allergenic pollens. *J. Med. Plants* **2022**, *21*, 11–18. [CrossRef]
27. Kuete, V. Potential of Cameroonian plants and derived products against microbial infections: A review. *Planta Med.* **2010**, *76*, 1479–1491. [CrossRef]
28. Ganash, M. Genotoxic and Antitumor Activity of Pollen Grains against Prostate Cancer Cell Line. *Int. J. Pharma Bio Sci.* **2021**, *11*, 36–46. [CrossRef]
29. Leja, M.; Mareczek, A.; Wyżgolik, G.; Klepacz-Baniak, J.; Czekońska, K. Antioxidative properties of bee pollen in selected plant species. *Food Chem.* **2007**, *100*, 237–240. [CrossRef]
30. Mărghitaş, L.A.; Stanciu, O.G.; Dezmirean, D.S.; Bobiş, O.; Popescu, O.; Bogdanov, S.; Campos, M.G. In vitro antioxidant capacity of honeybee-collected pollen of selected floral origin harvested from Romania. *Food Chem.* **2009**, *115*, 878–883. [CrossRef]
31. Bujang, J.S.; Zakaria, M.H.; Ramaiya, S.D. Chemical constituents and phytochemical properties of floral maize pollen. *PLoS ONE* **2021**, *16*, e0247327. [CrossRef]
32. Gercek, Y.C.; Celik, S.; Bayram, S. Screening of Plant Pollen Sources, Polyphenolic Compounds, Fatty Acids and Antioxidant/Antimicrobial Activity from Bee Pollen. *Molecules* **2021**, *27*, 117. [CrossRef] [PubMed]
33. Wang, B.; Diao, Q.; Zhang, Z.; Liu, Y.; Gao, Q.; Zhou, Y.; Li, S. Antitumor activity of bee pollen polysaccharides from Rosa rugosa. *Mol. Med. Rep.* **2013**, *7*, 1555–1558. [CrossRef] [PubMed]
34. Roszkowski, S. Application of Polyphenols and Flavonoids in Oncological Therapy. *Molecules* **2023**, *28*, 4080. [CrossRef] [PubMed]
35. Peterson, L.R.; Collins, S.M. Bactericidal testing for infectious disease therapy: The why and how to measure if we kill the bugs. *Clin. Microbiol. Newsl.* **2000**, *22*, 153–157. [CrossRef]
36. Bazaid, A.S.; Alsolami, A.; Patel, M.; Khateb, A.M.; Aldarhami, A.; Snoussi, M.; Almusheet, S.M.; Qanash, H. Antibiofilm, Antimicrobial, Anti-Quorum Sensing, and Antioxidant Activities of Saudi Sidr Honey: In Vitro and Molecular Docking Studies. *Pharmaceutics* **2023**, *15*, 2177. [CrossRef] [PubMed]
37. Abdelghany, T.M.; Bakri, M.M.; Al-Rajhi, A.M.H.; Al Abboud, M.A.; Alawlaqi, M.M.; Shater, A.R.M. Impact of copper and its nanoparticles on growth, ultrastructure, and laccase production of Aspergillus niger using corn cobs wastes. *BioResources* **2020**, *15*, 3289–3306. [CrossRef]
38. Al-Rajhi, A.M.H.; Yahya, R.; Abdelghany, T.M.; Fareid, M.A.; Mohamed, A.M.; Amin, B.H.; Masrahi, A.S. Anticancer, Anticoagulant, Antioxidant and Antimicrobial Activities of Thevetia peruviana Latex with Molecular Docking of Antimicrobial and Anticancer Activities. *Molecules* **2022**, *27*, 3165. [CrossRef] [PubMed]
39. Qanash, H.A.; Alotaibi, K.; Bazaid, A.S.; Ganash, M.; Saeedi, N.H.; Ghany, T.A. Effectiveness of Oil-based Nanoemulsions With Molecular Docking of its Antimicrobial Potential. *BioResources* **2023**, *18*, 1554–1576. [CrossRef]
40. Qanash, H.; Bazaid, A.S.; Binsaleh, N.K.; Patel, M.; Althomali, O.W.; Sheeha, B.B. In Vitro Antiproliferative Apoptosis Induction and Cell Cycle Arrest Potential of Saudi Sidr Honey against Colorectal Cancer. *Nutrients* **2023**, *15*, 3448. [CrossRef]

pharmaceuticals

MDPI

Article

Comprehensive Analysis of Bioactive Compounds in Wild *Ganoderma applanatum* Mushroom from Kerala, South India: Insights into Dietary Nutritional, Mineral, Antimicrobial, and Antioxidant Activities

Akbar Rijia [1,†], Raman Krishnamoorthi [2,†], Madhusoodhanan Rasmi [3], Pambayan Ulagan Mahalingam [1,*] and Kwang-sun Kim [4,*]

1 Department of Biology, The Gandhigram Rural Institute (Deemed to be University), Gandhigram, Dindigul 624302, Tamil Nadu, India; rijiazareen@gmail.com
2 Pharmaceutics Laboratory, Graduate Institute of Natural Products, Chang Gung University, Kweishan, Taoyuan 33302, Taiwan; krishmicro3@gmail.com
3 Department of Microbiology, Bharathidasan University, Tiruchirappalli 620024, Tamil Nadu, India; rasmi.m@bdu.ac.in
4 Department of Chemistry and Chemistry Institute for Functional Materials, Pusan National University, Busan 46241, Republic of Korea
* Correspondence: pumahalingam.gri@gmail.com (P.U.M.); kwangsun.kim@pusan.ac.kr (K.-s.K.)
† These authors contributed equally to this work.

Abstract: The present study focused on the mushroom Ganoderma, which has been used in Eastern countries for centuries as a food and medicinal source. Specifically, the fruiting bodies of *Ganoderma applanatum* from the Kerala Forest Research Institute in Thirussur, Kerala, India, were analyzed for their nutritional and medicinal properties. The methanolic extracts of *G. applanatum* were used to examine secondary metabolites and proximate profiles, revealing the presence of various phytochemicals such as terpenoids, phenolics, glycosides, alkaloids, flavonoids, and saponins. Further analysis revealed the presence of significant amounts of calcium, sodium, phosphorus, and manganese. The compounds were characterized using chromatographic analysis, FTIR, and GC-MS, which revealed potential therapeutic compounds with C-H and C-O bonds in the amide group, β-glycosides, and C-C/C-O vibrations of phenolic substances. Mushroom extract at a concentration of 100 μg mL^{-1} exhibited potent antimicrobial activity against various pathogens. This study suggests that *G. applanatum* has a rich biochemical composition and pharmacological potential, making it a promising candidate for drug development and traditional medicine, and contributes valuable insights into its diverse therapeutic applications.

Keywords: *Ganoderma applanatum*; antioxidant activity; bioactive compound; antimicrobials

1. Introduction

Mushrooms are a diverse group of macrofungi that are a significant component of forest ecosystems and occupy a prominent position in the biological world because of their clinical, agricultural, food, environmental, and economic importance. The dual significance of mushrooms as both edible delights and potent medicinal agents has spurred growing interest among researchers and health enthusiasts alike [1]. Culinary mushrooms are excellent sources of nutrition because they contain secondary metabolites, polysaccharides, minerals, unsaturated fatty acids, and proteins. Vitamins, such as riboflavin, thiamin, folic acid, ascorbic acid, and ergosterol, and minerals, such as phosphorus, selenium, magnesium, and copper, are widely distributed throughout the fruiting bodies of many mushrooms. Fruiting bodies are abundant in β-glucans, chitin, and dietary fibers [2–5].

Both wild and farmed mushrooms are now staples of the human dietary regimen. To date, there are several wild mushrooms that can be considered edible, and most of them are consumed worldwide owing to their desirable taste, flavor, texture, specific aroma, and a wide range of organic compounds with therapeutic and dietary benefits [6,7]. Several investigations have presented different mushrooms as an abundant supply of biologically active substances, which could be helpful in reducing the risk of illnesses, such as several types of cancer, stroke, Parkinson's disease, Alzheimer's disease, and hypertension [8–10]. Certain types of mushrooms contain fruiting body extracts that have antimicrobial, immune-boosting, and cholesterol-reducing effects [2,4,11]. Many polysaccharides from a variety of mushroom species belonging to the Agaricaceae, Polyporaceae, and Tricholomataceae families have remarkable in vivo antitumor activity [12–14]. Over the last few decades, Oriental and ancient therapies have been extensively used to track the records of therapeutic mushrooms. Combinations of several therapeutic mushrooms, such as *Auricularia, Flammulina, Ganoderma, Grifola, Hericium, Lentinus, Pleurotus, Schizophyllum, Trametes,* and *Tremella* are used for maximum benefit [15,16]. Edible mushrooms *Flammulina velutipes, Grifola frondosa, Lyophyllum shimeji,* and *Pleurotus eryngii* have all been shown to possess health-promoting and therapeutically active substances [17]. However, some effective, palatable therapeutic mushroom types, such as *Ganoderma lucidum, Inonotus obliquus,* and *Trametes versicolor,* are extremely sour or challenging to consume and are thus utilized as extracts, syrups, drinks, spore products, dietary supplements, coffee, and tea powder [18,19]. Additionally, a range of goods including lotions, soaps, and toothpastes are made from the fruiting bodies of *Ganoderma* and marketed as useful dietary and pharmaceutical additives with health advantages [20].

Ganoderma spp. are among the most widely studied medicinal fungi. Poroid wood-degrading macrofungi belonging to the genus *Ganoderma* are mostly found in tropical and subtropical forests worldwide. These mushrooms often grow as saprophytes or parasites on live trees. However, they can occasionally be observed attached to tree stumps and woodlogs. The genus *Ganoderma* has a very complex taxonomy. There are approximately 80 *Ganoderma* species worldwide, and approximately 117 species have been recorded in the family *Ganodermataceae* of Basidiomycota [20]. Certain *Ganoderma* species can break down cellulose, lignin, and various other associated polysaccharides in both angiosperms and gymnosperm hosts, leading to various rot diseases. Many woody trees of different *Ganoderma* species are declining because of stem and root rot. This is one of the advantages of identifying the *Ganoderma* spp. *Ganoderma* spp. have a diverse variety of phytochemicals, including alkaloids, amino acids, fatty acids, phenolic compounds, proteins and polysaccharides, steroids, and terpenoids, with advantages in food and medicine, including antifungal, antibacterial, antiviral, antidiabetic, antioxidant, immunomodulatory, anti-inflammatory, anti-adipogenic [21], and anticancer properties [22–26]. This has led to a great deal of curiosity regarding in-depth analysis of this mushroom. Therefore, it should be promoted as an effective food supplement for maintaining health.

Wild medicinal mushrooms have been extensively studied by many ethnobotanists and medical researchers in other parts of the world, whereas southern India has been relatively less explored and unexploited. The Western Ghats, one of the hotspots of biological diversity, runs parallel to the west coast of India from Maharashtra, South Gujarat, Goa, Karnataka, Kerala, and Tamil Nadu, covering up to 160,000 km^2 [27]. However, different types of farmed mushrooms are well known in this region because of their edibility, nutritional importance, and therapeutic effects. Numerous edible mushroom species that bloom in the forest have to be eaten by various tribes in the southern Indian region because of their exotic flavors and tastes, without knowing their nutritional and medicinal importance [28–30]. In India, several cases of misidentification of fruiting bodies collected from the wild for consumption were reported in different states between 2010 and 2019 [31,32]. However, the dietary significance, toxicity, and medicinal properties of mushrooms need to be explored. In addition, the identification of the most promising species of mushrooms with high nutritive value, development of agrotechnology, and extension of its cultivation technologies among rural growers, unemployed youth, womenfolk, and weaker populations may increase the economic status of people in the southern Indian region.

The number of well-researched mushrooms needs to be improved, even among the recognized species. However, further research is needed to capitalize on these beneficial medicinal properties. Genus *Ganoderma* is a well-explored bioactive secondary-metabolite-rich mushroom with medicinal effects. Extensive research has been conducted on the laccate species (*G. lucidum*). However, very few efforts have been made to explore the non-laccate species (*G. applanatum*) found in Kerala, South India. Pharmacognostic standards must be set before any crude medicine can be successfully bioprospected into herbal pharmacopeia [33]. Despite this, attempts have been made to study the nutritional, chemical, and therapeutic properties of *G. applanatum* in Kerala, South India.

2. Results

2.1. Collection of Wild Mushroom G. applanatum

In the present study, 31 fruiting body samples were obtained from different locations in the Kerala Forest Research Institute (KFRI), Peechi, Thrissur, and Kerala (Figure 1A–C). Throughout this investigation, most of the mushroom samples were collected from the soil, followed by decaying wood and other natural sources.

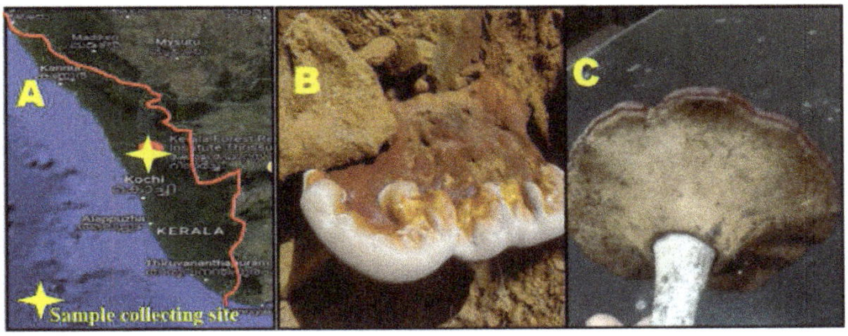

Figure 1. Collection of *G. applanatum*. (**A**) Collection site (source: Google map of India), (**B**) fruiting bodies outer side, and (**C**) fruiting bodies inner side.

2.2. Proximate Composition of G. applanatum Mushroom

In this study, the proximate composition of the collected mushrooms considered for evaluation was moisture, ash, fiber, crude protein, carbohydrate, and fat content. The results of this study showed that the *G. applanatum* mushroom has superabundant nutrients, such as carbohydrates (33.63 ± 0.66%) and proteins (21.01 ± 0.88%) (Figure 2). The total energy value was 262.50 ± 2.12 calories per 100 g of this mushroom. Mushrooms possess a rich amount of crude fibers (16.34 ± 0.33%) and insoluble and soluble (β-glucans and chitosan) fibers. Mushrooms have a low amount of beneficial n-3 fatty acids and a minimal total lipid level (7.91 ± 0.33%). Therefore, they have a limited nutritional contribution and are better for human diets and cardiovascular conditions. In our study, the ash and moisture contents were 10.64 ± 0.66% and 10.47 ± 1.12%, respectively. These findings suggest that *G. applanatum* is an excellent source of ash, carbohydrates, fiber, and protein.

2.3. Qualitative Analysis of Phytochemicals from G. applanatum

The current findings concern the presence of secondary metabolites in *G. applanatum* mushrooms in the Western Ghats of Kerala. Qualitative phytochemical screening of the crude methanol extracts of *G. applanatum* revealed the presence of alkaloids, flavonoids, phenolic substances, saponins, and terpenoids (Table 1).

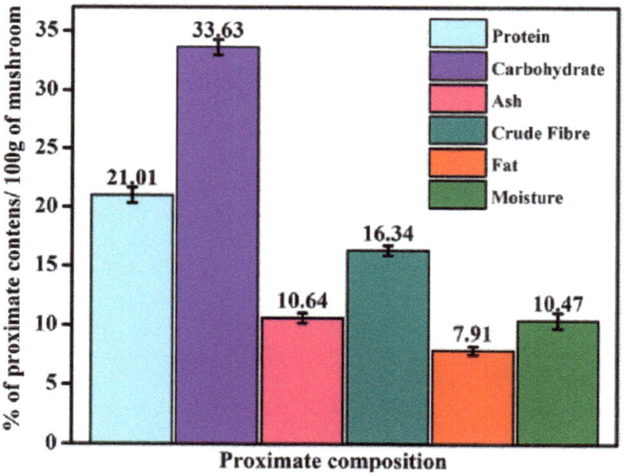

Figure 2. Proximate profile revealing the nutrient constitution of *G. applanatum* mushroom extracts. Values are the mean ± SD from n = 3.

Table 1. Primary phytochemicals of *G. applanatum* using biochemical tests.

Qualitative Analysis of Phyto Compounds			
Sample No.	**Biochemical Test**	**Compound**	***G. applanatum***
1	Salkowski test	Triterpenoids	+
2	$FeCl_3$ test	Tannins	–
3	Mayer's test	Alkaloids	+
4	Alkaline reagent	Flavonoids	+
5	Lead Acetate	Phenol	+
6	Frothing test	Saponins	+
7	Keller–Kiliani test	Glycosides	+

Note: (+) indicates presence; (−) indicates absence.

2.4. Quantitative Analysis of Phytochemicals from G. applanatum

The quantification of phytochemicals revealed that *G. applanatum* mushrooms possess significant amounts of phenolic substances [22.9 mg of gallic acid equivalent (GAE)] and flavonoid substances [15.84 mg of quercetin equivalent (QE)] per gram in the mushroom sample (dry weight). Moreover, terpenoid substances and saponin were present in 0.351 mg and 22.19 µg per gram and milligram, respectively, in dry *G. applanatum* mushrooms (Table 2).

Table 2. Quantitative analysis of phytochemicals from *G. applanatum* mushroom.

Sample No.	Biochemical Test	Amount of Phytochemicals per Dry Weight of Mushroom
1	Phenols	22.9 mg GAE per gram
2	Flavonoid	15.84 mg QE per gram
3	Saponin	22.19 µg per milligram
4	Terpenoid	0.351 mg per gram

Note: GAE, gallic acid equivalents; QE, quercetin equivalents.

2.5. Minerals Constitution in G. applanatum Mushroom

Atomic absorption spectrometry (AAS) was performed using the ash remaining after the whole burning of the dried *G. applanatum* mushroom and found the elemental minerals of the fruiting bodies. As shown in Table 3, calcium (Ca; 283.00 ± 3.12 mg kg^{-1}), sodium (Na; 190.00 ± 2.76 mg kg^{-1}), phosphorus (P; 117.00 ± 1.12 mg kg^{-1}), and a moderate level of potassium (K; 23.9 ± 0.66 mg kg^{-1}) were present in *G. applanatum* mushroom. Additional AAS analysis of four micronutrients including zinc (Zn), copper (Cu), manganese (Mg), and iron (Fe) showed a remarkable elevation of Zn (10.34 ± 1.27 mg kg^{-1}), whereas the other two elements [Fe (4.83 ± 0.33 mg kg^{-1}) and Mg (4.6 ± 0.66 mg kg^{-1})] were present in a low amount, which was in accordance with the results of previous studies [34,35]. A negligible amount of Cu (0.781 ± 0.88 mg kg^{-1}) was also detected.

Table 3. Contents of minerals in *G. applanatum* mushroom.

Sample No.	Minerals	Mineral Contents (mg kg^{-1})
1	Copper (Cu)	0.781 ± 0.88
2	Calcium (Ca)	283 ± 3.12
3	Zinc (Zn)	10.34 ± 1.27
4	Potassium (K)	23.9 ± 0.66
5	Manganese (Mn)	90.73 ± 1.22
6	Magnesium (Mg)	4.6 ± 0.66
7	Iron (Fe)	4.83 ± 0.33
8	Sodium (Na)	190 ± 2.76
9	Phosphorous (P)	117 ± 1.12

Note: The amount of each mineral is expressed as mg per kg of mushroom. Values are the mean ± SD from n = 3.

2.6. *Elucidation and Characterization of Compounds from G. applanatum Mushroom Extracts*

2.6.1. Thin-Layer Chromatography (TLC) Profile

TLC analysis suggested the presence of alkaloid fractions, flavonoids, glycosides, phenols, and terpenoids (R_f values: 0.720, 0.155, 0.560, 0.809, and 0.534, respectively) in the *G. applanatum* mushroom (Figure 3; Table 4). This finding was further supported by qualitative and quantitative Fourier-transform infrared (FTIR) spectroscopy and gas chromatography–mass spectrometry (GC-MS) analyses (Figure 4; Table 5). Remarkably, the corresponding alkaloid, flavonoid, terpenoid, glycoside, and phenol fractions were green, blue, blue, blue, and blue in color, respectively; therefore, the specific conforming band was not noticeable in the TLC plates.

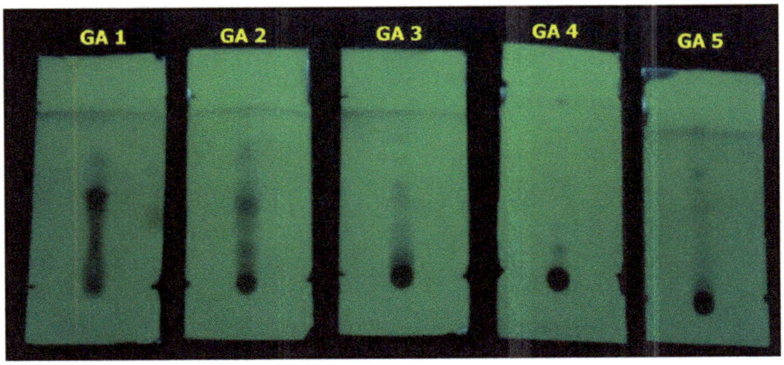

Figure 3. TLC profile of various bioactive compounds in *G. applanatum* mushroom extract.

Table 4. TLC profile of different bioactive compounds in *G. applanatum* mushroom extract.

Lane	Distance Traveled by the Solute (cm)	Distance Traveled by the Solvent (cm)	Band Colour Observed under UV Light	Rf Value	Expected Compound
GA1	2.3	4.3	Green	0.534	Alkaloid
GA2	3.4	4.2	Blue	0.809	Flavonoid
GA3	2.3	4.1	Blue	0.560	Terpenoid
GA4	0.7	4.5	Blue	0.155	Glycoside
GA5	3.1	4.3	Blue	0.720	Phenolic

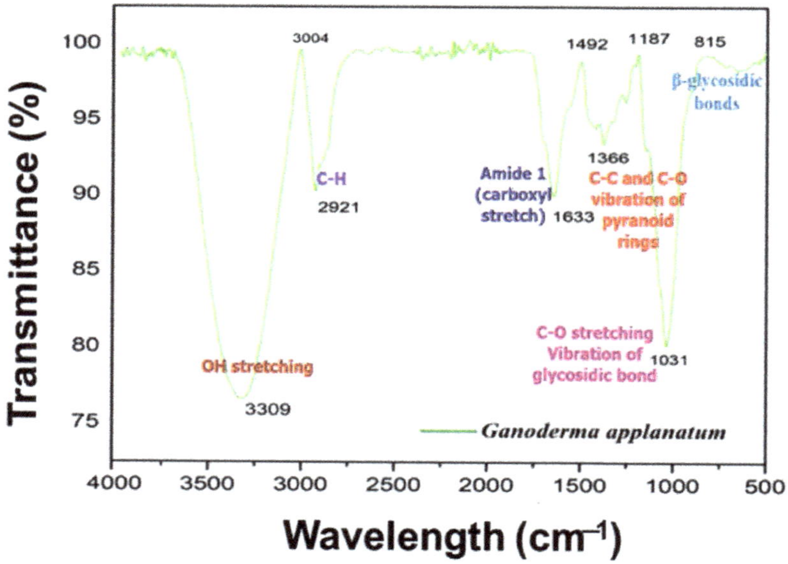

Figure 4. Functional group analysis of *G. applanatum* mushroom extract via FTIR spectroscopy.

Table 5. GC-MS profile of different bioactive compounds in *G. applanatum* mushroom extract.

Peak	Retention Time (RT)	Peak Area (%)	Name	Molecular Weight (g mol^{-1})	Molecular Formula
1	13.91	5.53	2,3,4,4-tretrapropyl-1-(trimethylsilyl)-1-(trimethylsilyloxy)-1,3-diaza-2,4-diborabutane	UN	UN
2	17.527	2.5	2-cyclobuten-1-one, 4-[[(1,1-dimethylethyl)dimethylsilyl]oxy]-2,3-dimethoxy-4-(3-phenyl-1-propynyl)-	UN	UN
3	18.34	5.84	2-tert-butyl-4-(1,1,3,3-tetramethylbutyl)phenol	262.4	$C_{18}H_{30}O$
4	19.942	29.53	1,2-benzenedicarboxylic acid, diethyl ester	222.24	$C_{12}H_{14}O_4$
5	20.766	1.57	1-(3,4-ditrimethylsiloxyphenyl)-2-isopropylaminoethanol	UN	UN
6	22.083	1.04	Isopropenyl dodecanoate	334	$C_{22}H_{38}O_2$
7	24.904	3.04	1,2-benzenedicarboxylic acid, dicyclohexyl ester	330.4	$C_{20}H_{26}O_4$
8	26.493	3.26	Furo[2,3-c]pyridine, 2,3-dihydro-2,7-dimethyl-	149.19	$C_9H_{11}NO$
9	26.59	0.97	2H-3,11c-(epoxymethano)phenanthro[10,1-bc]pyran, picras-3-en-21-oic acid derivative	548.6	$C_{28}H_{36}O_{11}$
10	37.46	23.99	1,4-benzenedicarboxylic acid, bis(2-ethylhexyl)ester	390.6	$C_{24}H_{38}O_4$
11	37.783	0.94	3-pyridinemethanol, 4-[(5H-dibenzo[a,d]cyclohepten-5-ylimino)methyl]-5-hydroxy-6-methyl-,(e)-	UN	UN
12	38.145	0.86	Silane, trimethyl[2-methylene-4,4-bis(phenylsulfonyl)butyl]	UN	UN

Table 5. *Cont.*

Peak	Retention Time (RT)	Peak Area (%)	Name	Molecular Weight (g mol^{-1})	Molecular Formula
13	38.273	3.95	2-propanone, 1,1,1-tris(ethylthio)-3-(4-methoxyphenyl)-3-[(trimethylsilyl)oxy]-	UN	UN
14	38.325	5.13	3-bromo-4-(difluoromethyl)pyridine	208	$C_6H_4BrF_2N$
15	38.48	2.2	7-oxabicyclo [2.2.1]hept-2-ene, 5,6-bis(chloromethyl)-2,3-dimethyl-, (exo,exo)-	UN	UN
16	38.615	4.99	12-azabicyclo(9.2.1)tetradeca-1(14)-ene-13-one	UN	UN
17	38.648	1.46	12-azabicyclo(9.2.1)tetradeca-1(14)-ene-13-one	UN	UN
18	38.86	1.11	12-azabicyclo(9.2.1)tetradeca-1(14)-ene-13-one	UN	UN
19	38.9	1.63	Cyclotrisiloxane, hexamethyl-	222.46	$C_6H_{18}O_3Si_3$
20	40.116	0.48	1H-Furo[3,4-c]pyrrole-4-carboxylic acid, 6-(2-furanyl)hexahydro-1,3-dioxo-4-phenyl-, methyl ester, (3a.alpha.,4.beta.,6.beta.,)	UN	UN

Note: UN, unknown.

2.6.2. FTIR Analysis

Figure 4 shows the FTIR spectrum of the methanol extract of the *G. applanatum* mushrooms. The large absorption peak was the OH stretching peak and C-H was 3309.00 cm^{-1} and 2921.00 cm^{-1}, respectively. The absorption bands at 1633.00 cm^{-1} and 1366 cm^{-1} were attributed to the stretching vibration of the carboxyl bond (C-O) of the amide group and the C-C and C-O vibrations of the phenolic substance or pyranoid rings, respectively. The bands near 1640 cm^{-1} were assigned to C=O stretching vibrations of the carboxyl group, indicating the presence of organic or amino acids [36]. The band between 711.60 cm^{-1} and 912. 16 cm^{-1} indicates the presence of β-glycosidic bonds. The band between 1032.23 cm^{-1} and 1111.76 cm^{-1} arose from the C-O stretching vibration of the glycosidic bond.

2.6.3. GC-MS Chromatogram

The GC-MS chromatograms of the compounds present in the mushroom extracts are listed in Table 5.

Seventeen metabolic compounds were determined from the methanolic extract of *G. applanatum*. Among them, nine compounds were reported in previous studies, whereas eight compounds were newly identified in this study. Among them, 1,2-benzenedicarboxylic acid and diethyl ester were the most abundant compounds (peak area 29.53% and RT of 19.942), followed by 1,4-benzenedicarboxylic acid and bis(2-ethylhexyl) ester (peak area % 23.9 and RT 37.46) (Figure 5).

Some other important compounds, which were much less in abundance, were 2, 3, 4, 4-tretrapropyl-1-(trimethylsilyl)-1-(trimethylsilyloxy)-1,3-diaza-2,4-diborabutane, 2-tert-butyl-4-(1,1,3,3-tetramethylbutyl) phenol, 1,2-benzenedicarboxylic acid, dicyclohexyl ester, 2-cyclobuten-1-one, 4-[[(1,1-dimethylethyl)dimethylsilyl]oxy]-2,3-dimethoxy-4-(3-phenyl-1-propynyl), Furo [2,3-c]pyridine, 2,3-dihydro-2,7-dimethyl, 2H-3,11c-(epoxymethano) phenanthro[10,1-bc]pyran, picras-3-en-21-oic acid deriv, 1-(3,4-ditrimethylsiloxyphenyl)-2-isopropylaminoethanol, 12-azabicyclo(9.2.1)tetradeca-1(14)-ene-13-one, 7-oxabicyclo[2.2.1]hept-2-ene, 5,6-bis(chloromethyl)-2,3-dimethyl-, (exo, exo), isopropenyl dodecanoate, 12-azabicyclo (9.2.1)tetradeca-1(14)-ene-13-one, silane, trimethyl [2-methylene-4,4-bis(phenylsulfonyl)butyl], 2-propanone, 1,1,1-tris(ethylthio)-3-(4-methoxyphenyl)-3-[(trimethylsilyl)oxy], 3-bromo-4-(difluoromethyl)pyridine, cyclotrisiloxane, hexamethyl, 1H-Furo[3,4-c]pyrrole-4-carboxylic acid, 6-(2-furanyl)hexahydro-1,3-dioxo-4-phenyl, methyl ester, (3a.alpha.,4.beta.,6.beta.,6), and 3-pyridinemethanol, and 4-[(5H-dibenzo[a,d] cyclohepten-5-ylimino)methyl]-5-hydroxy-6-methyl-(e)-. The percentage peak area of each metabolite is shown in the GC-MS chromatogram (Figure 5).

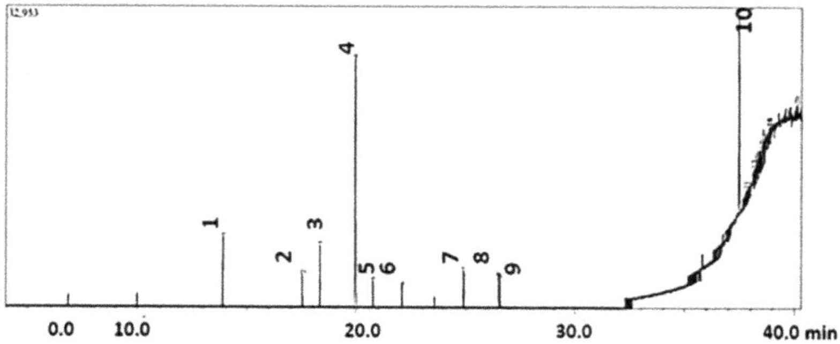

Figure 5. GC-MS chromatogram of *G. applanatum* methanolic extract. Numbers on the peaks indicate individual bioactive compounds listed in Table 5.

2.7. Antimicrobial Potential of G. applanatum Mushroom

In this study, the *G. applanatum* mushroom extract at the dosage of 100 μg mL^{-1} exhibited the highest antimicrobial efficacy against Gram-negative bacterial pathogens by zone of inhibition size (*Shigella flexneri* 19.98 ± 0.88 mm; *Klebsiella pneumoniae* 18.29 ± 1.12 mm) compared with Gram-positive pathogens (*Streptococcus pyogenes* 17.32 ± 0.33 mm; *Enterococcus faecalis* 11.21 ± 0.88 mm) (Figure 6; Table 6).

This is due to structural dissimilarities in the bacterial cell wall, the presence or absence of outer membranes, and the presence or absence of specialized machinery for antibiotic resistance [37,38]. Mushroom extracts showed significantly better antifungal activity against both fungal pathogens (*Candida albicans* 16.54 ± 1.12 mm; *Aspergillus fumigatus* 11.27 ± 0.33 mm) (Table 6). The current study showed that *G. applanatum* mushrooms were capable of inhibiting one or the other of Gram-negative or Gram-positive bacteria, the yeast *C. albicans*, and the mold *A. fumigatus*.

2.8. Antioxidant Activity of G. applanatum Mushroom Extract

2.8.1. Radical Scavenging Activity

In this study, the methanolic extract of *G. applanatum* scavenged 1,1-diphenyl-2-picrylhydrazyl (DPPH) radicals in a dose-dependent manner (Figure 7). The maximum amount of DPPH radical scavenging was observed at 100 μg mL^{-1} (88.38 ± 1.12%). Based on the IC_{50} value of *G. applanatum* at 50 μg mL^{-1}, the DPPH radical scavenging potential was 62.55 ± 0.87%. Compared to a previous report [39], our sample had greater DPPH radical scavenging potential, suggesting that ~500 μg mL^{-1} of mushroom extract inhibits DPPH radicals by up to 40–60%

2.8.2. Reducing Power Activity

The reducing power of the extracts from the wild mushroom *G. applanatum* was determined by measuring the absorbance at 700 nm. The reducing power of the mushroom samples increased in a concentration-dependent manner (Figure 8). By expressing the IC_{50} value, the concentration of the extract showed an absorbance of 0.5 and the highest reducing power was recorded at 100 μg mL^{-1}. The results also showed a positive correlation between antioxidant potential and total phenolic content. This may be due to the reaction specificity of the phenolic compounds present in the samples.

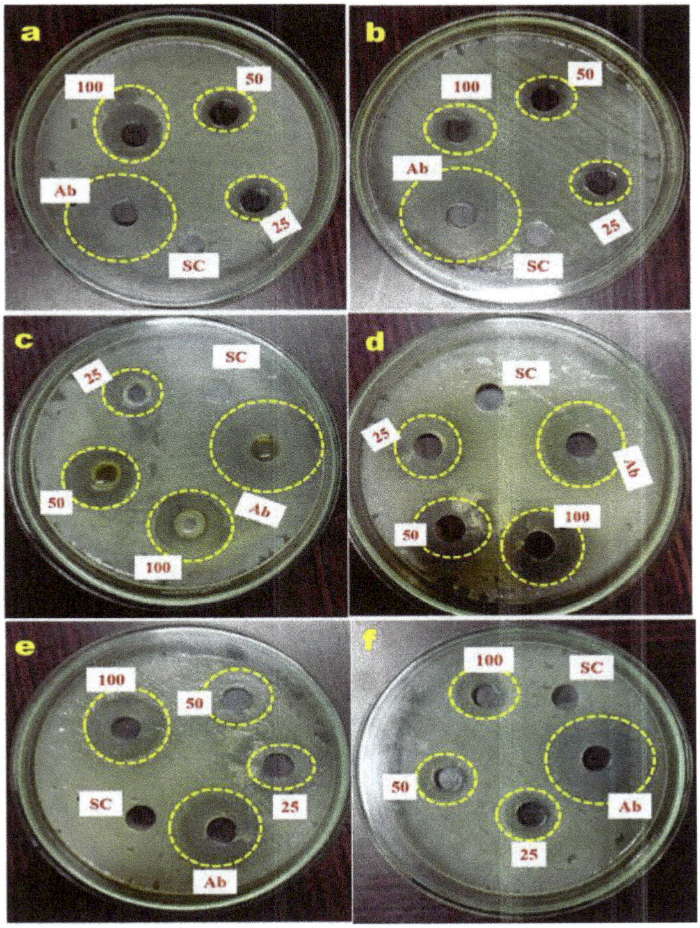

Figure 6. Agar well diffusion technique exhibiting the antimicrobial effects of *G. applanatum* mushroom extract against selected human pathogens: (**a**) *S.* pyogenes, (**b**) *E. faecalis*, (**c**) *K. pneumoniae*, (**d**) *S. flexneri*, (**e**) *C. albicans*, and (**f**) *A. fumigatus*. Various dosages (25, 50, and 100 μg mL^{-1}) of *G. applanatum* mushroom extract were analyzed along with positive control (streptomycin for bacteria and griseofulvin for fungi) and solvent control (methanol; 50 μL) against six selected human pathogens. Note: Ab, antibiotics (streptomycin for bacteria and griseofulvin for fungi); SC, solvent control.

Table 6. Antimicrobial potential of *G. applanatum* mushroom extract showing the zone of inhibition (well size 6 mm) against six selected human pathogens. Values are mean ± SD from n = 3. NA indicates no activity.

Pathogens	Zone of Inhibition (mm) (Well Size 6 mm)				
	Mushroom Extract (μg mL^{-1})			Antibiotic Control	Solvent Control
	25	50	100	(30 μg mL^{-1})	50 μL
Gram-positive bacterial pathogens					
S. pyogenes	10.22 ± 0.33	11.33 ± 0.88	17.32 ± 0.33	26.12 ± 0.33	NA
E. faecalis	9.98 ± 0.88	10.64 ± 0.66	11.21 ± 0.88	25.9 ± 0.12	NA

Table 6. *Cont.*

Pathogens	Zone of Inhibition (mm) (Well Size 6 mm)				
	Mushroom Extract (µg mL^{-1})			Antibiotic Control	Solvent Control
	25	50	100	(30 µg mL^{-1})	50 µL
Gram-negative bacterial pathogens					
K. pneumoniae	9.78 ± 0.88	13.58 ± 0.33	18.29 ± 1.12	26.4 ± 0.33	NA
S. flexneri	11.56 ± 0.33	14.27 ± 0.66	19.98 ± 0.88	21.28 ± 0.33	NA
Fungal pathogens					
C. albicans	9.1 ± 0.33	12.75 ± 0.88	16.54 ± 1.12	18.21 ± 0.66	NA
A. fumigatus	8.97 ± 0.33	9.58 ± 0.66	11.27 ± 0.33	23.9 ± 0.88	NA

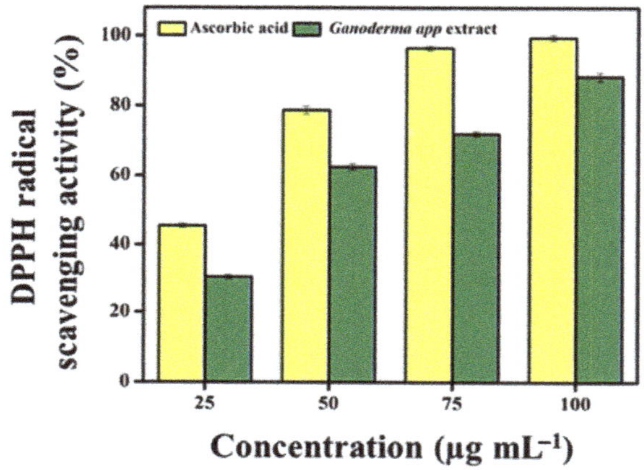

Figure 7. DPPH free radical scavenging activity of *G. applanatum* mushroom extract. Values are expressed as mean ± SD from $n = 3$.

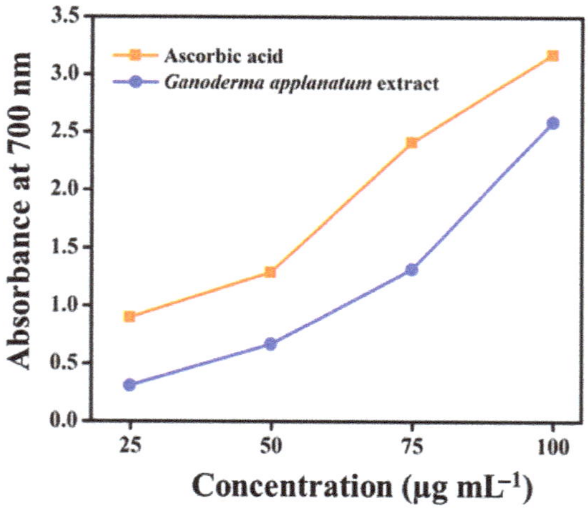

Figure 8. Reducing power efficacy of *G. applanatum* mushroom extract. Values are expressed as mean ± SD from $n = 3$.

3. Discussion

Mushrooms are essential for the treatment of various degenerative disorders; therefore, they have long been highly regarded and valued. Moreover, their distinct flavor, taste, and high protein content have increased the demand for nutritious foods. Further details on their therapeutic potential, dietary significance, and chemical constituents will be useful for exploring their value as nutraceuticals. There is a wide variety of physiological efficacies governed by mushrooms. The present study investigated the nutritional profile, elemental analysis, proximate content estimation, and phytochemical analysis of *G. applanatum* mushrooms for their utilization as functional foods and medicines.

A random distribution of samples at the selected sites was considered for sample collection. The opportunistic sampling method involved walking over collection sites and collecting samples from almost 80% of the target area. The climatic and edaphic conditions of the Western Ghats region of Kerala favor the occurrence of diverse mushroom species with seasonal variations. Frequent rainfall, high humidity, optimum temperature, and acidic soil pH are crucial factors that support mycelial development and fruiting body production [40]. The availability of dense forest areas in the entire KFRI region also helps maintain the diversity of wood-rotting fungi (Basidiomycota). Many previous studies support this finding, as different mushrooms grow in different habitats [41,42].

Proximate and qualitative analyses of the extract of *G. applanatum* growing from Kerala were evaluated in this study, and the results were reported by other authors in different countries where different values were observed in the present study; in some cases, these samples had higher values than in other places. However, very few studies have reported the nutritional profile of *G. applanatum*. Malnutrition is a major health issue in emerging nations. Owing to their high per-unit yield, nutritional content, texture, and flavor, mushrooms are an excellent food source for reducing starvation in underdeveloped nations. Carbohydrate and protein contents of 33–38% and 20–30%, respectively, are presented in various types of edible mushrooms, as documented in earlier research [3, 4,7]. One of the most vital elements in nourishment is protein, which is crucial for the development of bodily tissues. Owing to their high protein content, mushrooms are a valuable supplement to a person's diet when animal sources are scarce [43,44]. Insoluble fibers aid in digestion, whereas soluble fibers combat cardiovascular diseases by lowering cholesterol levels. In addition, mushrooms are rich in dietary fiber, which is also known for its anti-tumorigenic and hypocholesterolemic properties. Nagaraj et al. reported moisture content (42% w/w) results relatively higher than those in the present study [45]. The moisture content should be kept at a minimum to prevent microbial development and unfavorable enzymatic efficacy, which can cause spoiling. The carbohydrate (33.63 $\pm$ 0.66%) and crude fiber (16.34 $\pm$ 0.33%) contents were quite similar, but the protein (21.01 $\pm$ 0.88%) and ash (10.64 $\pm$ 0.66%) contents were higher than those reported in previous studies [4,34].

Mushroom extracts contain phenolic substances, which are considered one of the most vital natural forms of antioxidants and are frequently present in mushrooms and plants [46]. Consequently, since phenols and flavonoids have been shown to possess various antioxidant properties, mushrooms can be used to treat diseases caused by oxidative stress. These compounds have been shown to exhibit antimicrobial properties through microbial adhesion, the suppression of protein synthesis, proteolytic enzymes, and the rupture of cell membranes [3,4,47].

Mineral content plays a key role in the development of mushroom fruiting bodies [48]. Singh et al. documented that the elemental analysis of *G. lucidum* shows that the Ca, Na, Zn, and P are significantly similar to our present finding, but the K (23.9 $\pm$ 0.66 mg kg^{-1}) and Fe (4.83 $\pm$ 0.33 mg kg^{-1}) are much lower than in *G. lucidum* [40] and the amount of Mn (90.73 $\pm$ 1.22 mg kg^{-1}) present is higher in *G. applanatum*. This is comparable to other previously studied results that also recommended Mn as the most abundant mineral among samples of different species. Therefore, *G. applanatum* is the best source of Mn and protects the cardiovascular and cerebrovascular systems [14]. As indicated by earlier research, mineral concentrations of mushrooms are significantly influenced by developing media, habitats, and several other internal and external elements, such as growth circumstances and genetic variation [48]. The consumption of mushrooms with high K content maintains the salt balance in the body and regulates blood pressure; it also maintains cardiovascular health. Earlier studies have documented that human uptake reduces the bioavailability of some metals, particularly Cu, owing to inadequate absorption from the small intestine [49].

The different proportions of the solvent systems used in column chromatography resulted in fractions obtained from 100% ethyl acetate and hexane/ethyl acetate (5:5 and 8.5:1.5, respectively), which showed the presence of flavonoids. Five different fractions were eluted by column chromatography using hexane/ethyl acetate (5:5 and 8.5:1.5) and 100% ethyl acetate as the solvents. A few earlier TLC reports stated that other linked species of *A. bisporus* and *Pleurotus florida* also contain alkaloids and terpenoids that possess antimicrobial efficacy against both Gram-negative and Gram-positive bacterial strains [2,3]. Therefore, alkaloids and terpenoids may play key roles in suppressing the growth of Gram-negative and Gram-positive pathogenic bacterial strains. To confirm this, the TLC purified fractions of the crude extracts were individually tested against the bacterial strains. Previous reports have stated that *A. bisporus* mushrooms also contain phenolic substances with antioxidant activities [4,36].

A comparison of the R_f values of the compounds in the sample with the reference substance revealed several important observations. Notably, the R_f values obtained for the tested compounds generally aligned with the R_f values of the expected compounds, indicating successful separation and identification. Caffeine is the reference substance for the alkaloid group, exhibiting an R_f value of 0.53 [50]. Terpineol is the reference substance for terpenoids with an R_f value of 0.56 [51]. This correspondence strengthens the evidence for the presence of flavonoids in the mushroom samples of this study. Furthermore, TLC analysis revealed a compound with an R_f value of 0.81, which matches the expected R_f value of rutine in this alignment supports the identification of flavonoids in the sample, contributing to our understanding of its chemical composition. Leela and Devi have already documented that glycosides from lichen *Parmelia Perlata* have an R_f value of 0.15, which is similar to our result [52]. Similarly, Sonam et al. reported that a compound with an R_f value of 0.8 from *Reinwardtia indica* is similar to the current result [53].

FTIR spectroscopy was used to identify the functional groups in the samples. In this study, amide, aldehyde, and carboxylic acid groups were obtained from heterocyclic substances, such as proteins found in *P. florida* mushroom extracts [3]. All of the samples had characteristic peaks that indicated the presence of alkaloids, polysaccharides, proteins, and phenolics. The findings of the present study, including C-O stretching in flavonoids and phenols, were consistent with the quantitative examination of total phenolic and flavonoid compounds. The phenolic and amide-specific peaks found in the FTIR spectrum of the *A. bisporus* mushroom extract strongly correlated with the previously documented antioxidant and antimicrobial potential [4,36].

GC-MS was used to identify the various compounds present in mushrooms. Comparing the mass spectra, retention times, areas, peak heights, and the sample data's area-to-height ratio of the sample data with the NIST GC-MS library helped identify the substances included in the extract. The compounds identified through GC-MS have different biopotential properties, including anti-inflammatory, antioxidant, anticancer, and antimicrobial effects [54,55]. Most Ganoderma spp. have been reported to contain many biologically active compounds such as phenolics, alkaloids, saponins, flavonoids, fatty acids, and saponins [46,56]. According to Pal et al., compound 1,2 benzene dicarboxylic acid (19.35%) is found in *Astraeus hygrometricus*, which helps in neurotoxicity, anticancer, antiviral, and antibacterial activities [57,58]. Isopropenyl dodecanoate is an odorous compound first reported in *G. applanatum*. Meanwhile, *Ganoderma* sp. contains 1,2-benzenedicarboxylic acid and dicyclohexyl ester, showing potent antibacterial activity [59].

Wild mushrooms possess antibacterial and antifungal properties that allow them to thrive in their native environment. Thus, wild mushrooms have abundant supplies of naturally occurring phytochemical agents. Initial screening for the antimicrobial activity of methanolic *G. applanatum* extracts of wild mushrooms was carried out using the agar well diffusion technique. Extracts containing bioactive compounds diffused out of the well through the porous agar matrix and inhibited the growth of the target bacteria and fungi around the well, as shown in Figure 6. Menaga et al. reported that the control of Gram-negative bacteria, especially cell-wall-suppressing antimicrobial agents [2], is tougher than Gram-positive bacteria. Therefore, the wild mushrooms studied possessed broad-spectrum antibiotics. Although inhibition by bacteria and yeast was not up to the standard antibiotics, these wild mushrooms were capable of inhibiting a broad range of pathogenic microorganisms. Krishnamoorthi et al. showed that the phenolic and tannin contents of *A. bisporus* have antimicrobial properties through microbial adhesion, the suppression of protein synthesis, proteolytic enzymes, and the rupture of cell membranes [4,60].

It has been noted by many researchers conducting different studies that mushrooms are an excellent source of antioxidant compounds. Many researchers have recommended that antioxidant compounds extracted from different mushrooms exhibit protective properties at different phases of the oxidation process via several mechanisms [61]. However, high concentrations, such as 5–20 mg mL^{-1}, can inhibit 40–60% of DPPH free radicals in a few edible mushroom species, as reported previously [62]. The solvents used in the extraction process play a vital role in determining the antioxidant properties of biological samples. Previous studies have reported the use of methanol and ethyl acetate as extraction solvents to obtain antioxidant compounds with high antioxidant activity [63]. In this study, *G. applanatum* mushroom extract had a wider number of antioxidant compounds such as flavonoids, phenols, and alkaloids, and scavenged free radicals and the IC_{50} value of *G. applanatum* at 50 µg mL^{-1}, especially 2-tert-butyl-4-(1,1,3,3-tetramethylbutyl) phenol compound, had a higher radical scavenging potential. According to Herawati et al. [64], different wild mushrooms such as *Trametes versicolor*, *Microporus xantophus*, *Schizophyllum commune*, and *Auricularia auricula* reported a lower IC_{50} value (494.455, 251.200, 121.046, and 499.249 µg mL^{-1}, respectively) than the present study. They also showed that bioactive compounds affect antioxidant capacity. This is strongly associated with the influence of the habitat where they grow [64]. The capacity of a compound to transport electrons is related to its reducing power; therefore, it is considered a significant measure of its antioxidant activity. The presence of a chemical substance with antioxidant capability causes the reduction of the ferric cyanide complex to its ferrous form, resulting in a color change in the reaction mixture from yellow to greenish-blue, based on the ability to reduce the power of each reaction mixture [65]. This is demonstrated by the fact that antioxidant activity increases with increasing environmental stress [66]. A comparison of *G. applanatum* with other wild mushrooms showed greater nutritional value and mildly strong antioxidant activity. Additional research on mushrooms cultivated using diverse substrates must be conducted to assess their medicinal use.

4. Materials and Methods

4.1. Sample Collection

The fruiting bodies of *G. applanatum* grown on dead and unknown stems were obtained from the Kerala Forest Research Institute, Thirussur, Kerala, India. Samples were collected using a sterile knife, digging apparatus, and pluckers. The collected samples were authenticated by Dr. Shambhu Kumar, Senior Scientist, Department of Forest Pathology, Forest Health Division, Kerala, India. The collected fungal fruiting bodies were kept clean, dry (40–50 °C for 2 days), and powdered, with labeled polythene bags. The collected fungal fruiting bodies were subjected to lyophilization for dehydration and were preserved for long-term storage. Parts of the samples were pulverized using a mixer grinder and stored for further study.

4.2. Methanol Extraction from G. applanatum

Methanol extraction from *G. applanatum* mushrooms was performed using a previously reported protocol [67]. Briefly, 30 g of finely ground mushroom powder was extracted with 70% methanol for 8–10 h using Soxhlet extraction apparatus. The extract was then filtered through a Whatman No.1 filter paper. The resulting filtrate was subjected to rotary vacuum evaporation at 40 °C to eliminate the solvent, resulting in a concentrated extract of *G. applanatum*. The concentrated extract was lyophilized using a lyophilizer (Labconco Corporation, Kansas City, MO, USA) and stored at −20 °C.

4.3. Proximate Analysis of G. applanatum

4.3.1. Estimation of Carbohydrate

The carbohydrate content of dried *G. applanatum* mushrooms was measured using a standard procedure [68]. Dried mushrooms (0.1 g) with 5 mL of 2.5 N hydrochloric acid (HCl) were placed in a test tube and boiled in a water bath for 3 h. After cooling to room temperature, the reaction mixture was maintained at pH 7 using solid sodium carbonate. This was conducted gradually until fizzing (effervescence). The reaction mixture was spun at 4000 rpm for 5 min and the resulting supernatant (0.5 and 1.0 mL) was used for further carbohydrate analysis. Standard carbohydrate solutions were prepared in a range of concentrations (0–1.0 mL) to create a calibration curve. The volume of each standard and sample tube was adjusted to 1 mL of distilled water (DW), and 4 mL of anthrone reagent was added to each tube, which reacted with carbohydrates to produce a green to dark green color. The tubes were then heated in a boiling water bath for 8 min and quickly cooled, and the intensity of the developed color was measured at 630 nm using a spectrophotometer.

$$\text{Carbohydrate (\%)} = [\text{Mg of glucose}/\text{volume of the sample}]/100$$

4.3.2. Estimation of Protein

DW (10 mL) was used to dissolve 5 mg of bovine serum albumin (BSA), and ammonium sulfate was gradually added, equaling 50–100% of its saturation level to the BSA solution, ensuring that the mixture was stirred continuously. Once ammonium sulfate was fully incorporated, the mixture was left to stand undisturbed for 30 min. The sample was then spun at 11,000 rpm for 5 min. The supernatant was discarded, and the pellet was resuspended in DW. This reaction was repeated twice and the pellet was resuspended in 1 mL of DW. From this final suspension, 200 μL of the sample was used for protein quantification using the Lowry method.

$$\text{Recovery (\%)} = (\text{Ending protein content}/\text{Starting protein content}) \times 100$$

4.3.3. Estimation of Moisture

The moisture concentration of powdered *G. applanatum* mushrooms was assessed following the standard procedure outlined in the AOAC [68]. Initially, 10 g of sample was dried at 105 °C for 6 h. The process continued until the sample reached a consistent weight, indicating the complete removal of moisture. The moisture contents of the samples were determined using the following formula:

$$\text{Moisture level (\%)} = [(\text{Starting weight} - \text{Ending weight})/\text{weight of sample}] \times 100$$

4.3.4. Estimation of Fat

The fat content of dried *G. applanatum* mushrooms was assessed following the standard procedure outlined by the AOCC [68]. Dried mushroom samples (5 g) were placed inside a thimble in an extraction tube in a Soxhlet apparatus. The heater temperature was adjusted to ensure continuous drip of ether on the sample within the extraction tube. The extraction process used petroleum ether with a boiling point range of 40–60 °C and lasted for 16 h. After completion of the extraction, the sample was removed and the solvent was allowed to evaporate under a fume hood. The resulting extract was thoroughly dried in a hot air oven for 30 min at 105 °C. After cooling in a desiccator, the weight of the extracts was recorded. The determination of crude fat content was subsequently carried out using the following formula:

$$\text{Crude Fat (\%)} = [\text{Weight of sample}/\text{Weight of Fat in sample}] \times 100$$

4.3.5. Estimation of Ash Content

The procedure for estimating the ash content of *G. applanatum* mushrooms was performed in compliance with the methodology specified by the AOAC [68]. Initially, the crucible with its lid was heated in a furnace at 550 °C overnight and then cooled in a desiccator for 30 min. The cooled crucible, still with its lid, was precisely weighed to three decimals. Approximately 5 g of mushrooms was added to the crucible. The sample was then gently heated using a Bunsen burner with a lid partially covering the crucible, until no additional smoke was produced. The crucible, which was fully covered with a lid, was placed back in the furnace and heated overnight at 550 °C. After heating, the crucible was covered with a lid and cooled in a desiccator. This process helps to remove all organic material through combustion, leaving behind inorganic ash for measurement. The ash level was determined using the following formula:

$$\text{Ash content (\%)} = [\text{Weight of sample}/\text{Weight of ash sample}] \times 100$$

4.3.6. Estimation of Crude Fiber

Crude fiber content was assessed using a method involving the sequential extraction of *G. applanatum* as outlined in the AOAC [68]. The extraction process involved the use of 1.25% sulfuric acid (H_2SO_4) and 1.25% sodium hydroxide (NaOH) solutions, utilizing a fiber bag as a container for the samples. Following the extraction, the samples were dried and ashed for further analysis. The samples were dried in a crucible for 5 h at 105 °C, and subsequently ashed in a muffle furnace overnight at 525 °C.

$$\text{Crude Fiber (\%)} = [\text{Weight of sample}/\text{loss of weight after ignition}] \times 100$$

4.3.7. Estimation of Energy Values

The energy values were investigated using the following equation:

$$\text{Energy values (Cal/100 g)} = (4.2 \times \%\ \text{carbohydrate}) + (8.37 \times \%\ \text{fat}) + (2.62 \times \%\ \text{proteins})$$

4.4. *Preliminary Phytochemical Analysis Using Biochemical Methods*

G. applanatum mushroom extract was subjected to phytochemical screening using various known qualitative biochemical tests, as used in a previous report, with minor modifications [69].

4.4.1. Phenol Test (Lead Acetate Test)

A few drops of a 1% lead acetate solution was mixed with 1 mL of the methanolic mushroom extract. The production of a bluish-black cloud showed the existence of phenolic compounds.

4.4.2. Flavonoid Test (Alkaline Reagent Test)

A few drops of NaOH solution were added to the mushroom extract. The emergence of an intense yellow color, transitioning to colorlessness upon adding a few drops of diluted acid, proved that flavonoids were present.

4.4.3. Terpenoid Test (Salkowski Test)

The mushroom extract underwent treatment with a few drops of concentrated H_2SO_4, followed by thorough shaking and standing for a specific duration. The appearance of a yellow-colored lower layer was indicative of the presence of triterpenoids.

4.4.4. Steroid Test

A few drops of strong H_2SO_4 were added to the mushroom extract, which was then thoroughly mixed and let to stand for a while. The emergence of red color in the bottom layer indicated the existence of steroids.

4.4.5. Alkaloid Test (Mayer's Test)

A few drops of Mayer's reagent were added to 5 mL of mushroom extract. The reactive solution exhibited a indicated a white, creamy precipitate, indicating a positive result.

4.4.6. Test for Tannins (Ferric Chloride Test)

A few drops of ferric chloride were mixed to 1 mL of the mushroom extract. The emergence of a deep bluish-black color suggested the existence of tannins.

4.4.7. Saponin Test (Frothing Test)

Here, 1 mL of mushroom extract was added to 4 mL of distilled water. The mixture was shaken and maintained for at least 15 min, and the presence of continuous foam indicated the presence of saponin.

4.5. *Quantitative Estimation of Phytochemicals from G. applanatum Mushroom Extracts*

4.5.1. Total Amount of Flavonoid

The colorimetric technique using aluminum chloride was used to determine the total amount of flavonoid in the *G. applanatum* mushroom extract [70]. Briefly, 0.5 mL of mushroom extract was mixed with 2.8 mL of DW, 0.1 mL of 1 M potassium acetate, 0.1 mL of 10% $AlCl_3$, and 1.5 mL of methanol for 30 min at room temperature. After incubation, the intensity of the resulting mixture was calculated using the absorbance values at 415 nm. Quercetin was used as a standard. The extracts' total flavonoid content (TFC) was determined from the quercetin calibration graph using the formula $T = C \times (V/M)$. Here, T stands for TFC (mg g^{-1}), C is the quercetin concentration (mg mL^{-1}) obtained from the calibration graph, V is the extract volume (mL), and M is the extract weight (g). The milligrams of quercetin equivalent (QE) per gram of sample dry weight indicates how the TFC is stated.

4.5.2. Total Amount of Terpenoid

The total terpenoid content of phenolics in the *G. applanatum* mushroom extract was assessed using the standard technique, as previously described [71]. Briefly, 100 mg of mushroom extract was immersed in 9 mL of ethanol for 24 h and then filtered through a Whatman No.1 filter paper. The resulting solution was subsequently extracted using a separating funnel containing 10 mL petroleum ether. The resulting ether extract was then divided into pre-weighed glass vials and allowed to undergo thorough drying until completion. The dried ether extract was considered the total terpenoid content.

4.5.3. Total Amount of Phenolics

The total amount of phenolics in the *G. applanatum* mushroom extract was determined using Folin–Ciocalteu reagent [72]. Briefly, Folin–Ciocalteu reagent (2.5 mL) was thoroughly mixed with a 0.5 mL aliquot of the extract. After a 5 min interval, 2 mL of 7.5% sodium carbonate (Na_2CO_3) solution was added to the reaction mixture and incubated at 45 °C for 15 min. The reaction mixture was developed in blue and the absorbance was measured at 765 nm. Gallic acid served as a standard, and the total phenolic content (TPC) was determined from the validation plot of gallic acid. The TPC of the extract was calculated using the formula T = C × (V/M), where T represents the TPC (mg g^{-1}) of the mushroom extract, C shows the dose of gallic acid (mg mL^{-1}) obtained from the validation plot, V represents the volume of the mushroom extract obtained (mL), and M is the weight of the mushroom extract (g). The results are represented as mg of gallic acid equivalents (GAE) per gram of sample dry weight.

4.5.4. Total Saponin Content

To determine the total saponin content of *G. applanatum* mushrooms, a vanillin–sulfuric acid assay was performed using a standard procedure. Briefly, 2.5 mL of 72% (*v*/*v*) H_2SO_4 in water, reagent blank or standards with 0.25 mL of 0.8% (*w*/*v*) vanillin in ethanol and 1 mg ml^{-1} of mushroom extracts were kept in a shaking water bath at 60 °C for 15 min. Diosgenin was used as the standard, and methanol as the reagent blank. The content of the mushroom extracts and standard was determined by the absorbance at 544 nm using UV-Vis spectroscopy [71].

4.6. Elemental Analysis

One gram of methanolic *G. applanatum* mushroom extract was subjected to ash in an oven at 650 °C for a duration—5–6 h. Subsequently, it was dissolved in HNO_3, followed by digestion with H_2O_2 and HCl. The solution was allowed to cool and then centrifuged. After centrifugation, the solution was diluted to 10 mL using distilled H_2O_2. Elemental analysis was performed using an Atomic Absorption Spectrophotometer (AAS) [73].

4.7. Compound Elucidation and Characterization

4.7.1. Thin-Layer Chromatography (TLC)

TLC was performed to identify the different compounds present in the *G. applanatum* mushroom extract as per the standard procedure [74]. A suitable developing solvent (mobile phase) was prepared based on the compounds used for analysis. The mobile phase was prepared by dissolving different solvents for different bioactive compounds, such as alkaloids [chloroform/methanol/toluene (4:2:2)], flavonoids [chloroform/methanol/toluene (33:7:10)], terpenoids [toluene/chloroform/ethanol (40:40:10)], glycosides [toluene/chloroform (9:1)] and phenols [petroleum ether/ethyl acetate (7:3)]. After the chromatogram was developed, the retention factor (Rf) was calculated using the formula. The target compounds were purified using column chromatography.

4.7.2. Column Chromatography

This technique was performed using silica gel (Nice—60–120 Mesh), as per the protocol [74]. The methanol extract was then chromated using a silica gel column. Initially, the column was eluted with hexane (100%), followed by hexane/ethyl acetate (10:1), hexane/ethyl acetate (1:5), hexane/ethyl acetate (1:10), ethyl acetate (100%), ethyl acetate/methanol (10:5), ethyl acetate/methanol (1:5), and methanol (100%).

4.7.3. Fourier-Transform Infrared Spectroscopy (FTIR)

FTIR spectra were employed for the characterization of *G. applanatum* mushroom using a methanol extract sample through Attenuated Total Reflectance (ATR) in the area of 400–4000 cm^{-1}. Functional groups were identified using an Alpha-II FTIR-ATR instrument provided by Bruker International [4].

4.7.4. Gas Chromatography–Mass Spectrometry (GC-MS)

The secondary bioactive compounds presented in the methanol extract of *G. applanatum* were screened using GC-MS and recorded on a SHIMADZU/QP2020 equipped with a column model of SH-RXi-5Sil MS as per the standard procedure [67]. Helium was used as the flow gas. The column temperature mode was 50 °C at the initial point, and then changed to 280 °C at a rate of 6.00 °C for a holding time of 2 min. By comparing the retention times, areas of proportion, and mass spectra fragmentation patterns of the chemicals in the mushroom extracts to those kept in computer libraries and in the published literature, the identities of the constituents were determined. The chemicals that were determined were matched using NIST08s.LIB and WILEY8.LIB library sources.

4.8. Antimicrobial Activity of Methanolic G. applanatum Mushroom Extract

The antimicrobial activity of the methanolic crude *G. applanatum* mushroom extracts was determined against six selected human pathogenic strains [two fungal pathogens: *A. fumigatus* (MTCC 2550) and *C. albicans* (MTCC 3017), and four bacterial pathogens: *E. faecalis* (MTCC 439), *S. flexneri* (MTCC 1457), *K. pneumoniae* (MTCC 432), and *S. pyogenes* (MTCC 442)] using the well diffusion method [60]. The pathogen was procured from the Microbial Type Culture Collection (MTCC; Chandigarh, India). The bacterial and fungal strains were pre-grown in Nutrient Broth (NB) and potato dextrose broth (PDB). To assess antimicrobial activity, Mueller–Hinton agar (MHA) plates were prepared and spread on pathogenic strains on the surface and the wells were made (6 mm diameter) using a sterile well borer. Following this, the different dosages (100, 50, and 25 μg mL^{-1}) of the methanolic mushroom extracts were loaded into the respective wells of the plates along with 50 μg mL^{-1} of positive control (streptomycin for bacteria and griseofulvin for fungi) and negative control methanol (50 μL) against each tested pathogenic strain, and the plates were incubated for 24 h at 37 °C. A zone of inhibition encircling the wells suggested that positive antibacterial activity was observed. Three replicates were carried out for this examination.

4.9. Antioxidant Activity of Methanolic G. applanatum Mushroom Extract

4.9.1. Free Radical Scavenging Assay

The methanolic *G. applanatum* mushroom extract was used to assess DPPH free radical scavenging potential using previously described methods [4]. Different dosages (25, 50, 75, and 100 μg mL^{-1}) of the methanolic mushroom extract combined with 1 mL of DPPH solution were kept at 28 °C for 30 min in dark. Methanol served as the blank, the control consisted of DPPH in methanol without the sample, ascorbic acid was utilized as the standard reference, and absorbance at 517 nm (Ab_{517}) was recorded. The DPPH radical scavenging capacity was determined using the following formula:

% DPPH scavenging potential = [(Ab517 (control) - Ab_{517} (sample)/Ab_{517} (control)) × 100, where Ab_{517} (control) and Ab_{517} (sample) is the absorbance of DPPH.

4.9.2. Reducing Power Assay

The reduction power was assessed in accordance with the previous method [65]. Various dosages (25, 50, 75, and 100 μg mL^{-1}) of the methanolic *G. applanatum* mushroom extract were combined with 1.5 mL of 1% potassium ferricyanide and 1.5 mL of 0.2 M sodium phosphate buffer (pH 6.6). The reaction mixture was mixed with 5 mL of 10% trichloroacetic acid and incubated for 20 min at 50 °C. After 20 min, the reaction mixture was centrifuged at 8000 rpm at 4 °C for 7 min. Then, 1.5 mL of the supernatant was taken from the top layer, 0.1% ferric chloride (300 μL) and DW (1.5 mL) were added, and the absorbance was measured at 700 nm. Ascorbic acid was used as a reference material. An increase in the absorbance within the resultant mixture signified a higher level of reducing power. The concentration at 0.5 absorbance was determined to be the benchmark for evaluating the scavenging capacity.

4.10. Statistical Analysis

Statistical computations were performed using Origin 8.1 (Origin Lab Corporation, Northampton, USA), and the mean of three replicates ± standard deviation was displayed as the outcome. The Student's *t*-test for dataset was used to determine the *p*-value, and $p < 0.05$ was regarded as statistically significant.

5. Conclusions

In conclusion, this study provides a comprehensive analysis of *G. applanatum* By focusing on proximate composition, elemental content, chemical composition, and bioactivity. Proximate analysis revealed that *G. applanatum* is rich in carbohydrate (33.63 ± 0.66%) and protein (21.01 ± 0.88%), making it a nutritious source. The presence of K (283 ± 3.12 mg kg^{-1}), Na (190 ± 2.76 mg kg^{-1}), P (117 ± 1.12 mg kg^{-1}), and Mn (90 ± 1.22 mg kg^{-1}) was confirmed in *G. applanatum* through AAS analysis. Qualitative and quantitative analyses of the bioactive components were performed. This study reported 20 compounds using GC-MS analysis, and of these, 8 unidentified compounds were identified. This presents a great opportunity for further research in this field. The methanolic extract of *G. applanatum* has significant antimicrobial potential against human pathogens such as *S. flexneri* (19.98 ± 0.88 mm), *K. pneumoniae* (18.29 ± 1.12 mm), *S. pyogenes* (17.32 ± 0.33 mm), *C. albicans* (16.54 ± 1.12 mm), *A. fumigatus* (11.27 ± 0.33 mm), and *E. faecalis* (11.21 ± 0.88 mm). *G. applanatum* has significant enrichment with 2-tert-butyl-4-(1,1,3,3-tetramethylbutyl) phenol and may act as a potential therapeutic agent by protecting the biological system from the generation of free radicals, thereby reducing the illness triggered by oxidative stress. The identified compounds, along with their diverse bioactivities, warrant further studies to elucidate their mechanisms and potential health benefits prior to clinical trials.

Author Contributions: A.R.: Conceptualization, methodology, software, data curation, and writing—original draft preparation. R.K.: Methodology, software, data curation, and writing—review and editing. M.R.: Formal analysis and data curation. P.U.M.: Conceptualization, validation, supervision, and writing—review and editing. K.-s.K.: Writing—review and editing, and funding. All authors have read and agreed to the published version of the manuscript.

Funding: This research received no external funding.

Institutional Review Board Statement: Not applicable.

Informed Consent Statement: Not applicable.

Data Availability Statement: Data available on request.

Acknowledgments: We sincerely thank the Department of Biology, The Gandhigram Rural Institute (Deemed to be University), Gandhigram, Dindigul, for the encouragement and research support.

Conflicts of Interest: The authors declare no conflicts of interest.

References

1. Singdevsachan, S.K.; Patra, J.K.; Tayung, K.; Thatoi, H. Chemical Constituents, Antioxidative and Antibacterial Properties of Medicinal Mushrooms Collected from Similipal Biosphere Reserve, Odisha, India. *Proc. Natl. Acad. Sci. India Sect. B-Biol. Sci.* **2017**, *87*, 559–570. [CrossRef]
2. Menaga, D.; Mahalingam, P.U.; Rajakumar, S.; Ayyasamy, P.M. Evaluation of Phytochemical Characteristics and Antimicrobial Activity of Pleurotus Florida Mushroom. *Asian J. Pharm. Clin. Res.* **2012**, *5*, 102–106.
3. Chandran Priyadarshni, K.; Krishnamoorthi, R.; Mumtha, C.; Ulagan Mahalingam, P. Biochemical Analysis of Cultivated Mushroom, Pleurotus Florida and Synthesis of Silver Nanoparticles for Enhanced Antimicrobial Effects on Clinically Important Human Pathogens. *Inorg. Chem. Commun.* **2022**, *142*, 109673. [CrossRef]
4. Krishnamoorthi, R.; Srinivash, M.; Mahalingam, P.U.; Malaikozhundan, B. Dietary Nutrients in Edible Mushroom, Agaricus Bisporus and Their Radical Scavenging, Antibacterial, and Antifungal Effects. *Process Biochem.* **2022**, *121*, 10–17. [CrossRef]
5. Antunes, F.; Marçal, S.; Taofiq, O.; Morais, A.M.M.B.; Freitas, A.C.; Ferreira, I.C.F.R.; Pintado, M. Value-Added Compounds and Potential Applications. *Molecules* **2020**, *1*, 1–40.
6. Ugbogu, E.A.; Emmanuel, O.; Salem, A.Z.M.; Elghandour, M.M.M.Y. Nutritional Composition of Termitomyces Robustus (Agaricomycetes) and Lentinus Squarrosulus (Mont.) Singer in South East Nigeria. *Agrofor. Syst.* **2020**, *94*, 1291–1300. [CrossRef]
7. Kalač, P. A Review of Chemical Composition and Nutritional Value of Wild-Growing and Cultivated Mushrooms. *J. Sci. Food Agric.* **2013**, *93*, 209–218. [CrossRef]
8. Rahman, M.A.; Abdullah, N.; Aminudin, N. Interpretation of Mushroom as a Common Therapeutic Agent for Alzheimer's Disease and Cardiovascular Diseases. *Crit. Rev. Biotechnol.* **2016**, *36*, 1131–1142. [CrossRef] [PubMed]
9. García-Sánchez, A.; Miranda-Díaz, A.G.; Cardona-Muñoz, E.G. The Role of Oxidative Stress in Physiopathology and Pharmacological Treatment with Pro- and Antioxidant Properties in Chronic Diseases. *Oxid. Med. Cell. Longev.* **2020**, *2020*, 2082145. [CrossRef]

10. Tang, C.; Hoo, P.C.X.; Tan, L.T.H.; Pusparajah, P.; Khan, T.M.; Lee, L.H.; Goh, B.H.; Chan, K.G. Golden Needle Mushroom: A Culinary Medicine with Evidenced-Based Biological Activities and Health Promoting Properties. *Front. Pharmacol.* **2016**, *7*, 474. [CrossRef]
11. Erdoğan Eliuz, E.A. Antibacterial Activity and Antibacterial Mechanism of Ethanol Extracts of Lentinula Edodes (Shiitake) and Agaricus Bisporus (Button Mushroom). *Int. J. Environ. Health Res.* **2021**, *32*, 1828–1841. [CrossRef] [PubMed]
12. Khatun, M.A.; Sato, S.; Konishi, T. Obesity Preventive Function of Novel Edible Mushroom, Basidiomycetes-X (Echigoshirayukidake): Manipulations of Insulin Resistance and Lipid Metabolism. *J. Tradit. Complement. Med.* **2020**, *10*, 245–251. [CrossRef] [PubMed]
13. Khinsar, K.H.; Abdul, S.; Hussain, A.; Ud Din, R.; Lei, L.; Cao, J.; Abbasi, M.; Ur Rehman, A.; Farooqui, N.; Yi, X.; et al. Anti-Tumor Effect of Polysaccharide from Pleurotus Ostreatus on H22 Mouse Hepatoma Ascites in-Vivo and Hepatocellular Carcinoma in-Vitro Model. *AMB Express* **2021**, *11*, 1–15. [CrossRef] [PubMed]
14. Zhang, Y.; Zhang, Z.; Liu, H.; Wang, D.; Wang, J.; Liu, M.; Yang, Y.; Zhong, S. A Natural Selenium Polysaccharide from Pleurotus Ostreatus: Structural Elucidation, Anti-Gastric Cancer and Anti-Colon Cancer Activity in Vitro. *Int. J. Biol. Macromol.* **2022**, *201*, 630–640. [CrossRef] [PubMed]
15. Lee, K.H.; Morris-Natschke, S.; Yang, X.; Huang, R.; Zhou, T.; Wu, S.F.; Shi, Q.; Itokawa, H. Recent Progress of Research on Medicinal Mushrooms, Foods, and Other Herbal Products Used in Traditional Chinese Medicine. *J. Tradit. Complement. Med.* **2012**, *2*, 1–12. [CrossRef]
16. Wu, Q.; Dai, T.; Song, J.; Liu, X.; Song, S.; Li, L.; Liu, J.; Pugazhendhi, A.; Jacob, J.A. Effects of Herbal and Mushroom Formulations Used in Traditional Chinese Medicine on in Vitro Human Cancer Cell Lines at the Preclinical Level: An Empirical Review of the Cell Killing Mechanisms. *Process Biochem.* **2020**, *94*, 136–142. [CrossRef]
17. Rathore, H.; Prasad, S.; Sharma, S. Mushroom Nutraceuticals for Improved Nutrition and Better Human Health: A Review. *PharmaNutrition* **2017**, *5*, 35–46. [CrossRef]
18. Ajith, T.A.; Janardhanan, K.K. Indian Medicinal Mushrooms as a Source of Antioxidant and Antitumor Agents. *J. Clin. Biochem. Nutr.* **2007**, *40*, 157–162. [CrossRef] [PubMed]
19. Selvi, S.; Umadevi, P.; Murugan, S.; Giftson Senapathy, J. Anticancer Potential Evoked by Pleurotus Florida and Calocybe Indica Using T 24 Urinary Bladder Cancer Cell Line. *African J. Biotechnol.* **2011**, *10*, 7279–7285. [CrossRef]
20. Bulam, S.; Üstün, N.Ş.; Pekşen, A. Health Benefits of Ganoderma Lucidum as a Medicinal Mushroom. *Turkish J. Agric.-Food Sci. Technol.* **2019**, *7*, 84–93. [CrossRef]
21. Peng, X.R.; Wang, Q.; Su, H.G.; Zhou, L.; Xiong, W.Y.; Qiu, M.H. Anti-Adipogenic Lanostane-Type Triterpenoids from the Edible and Medicinal Mushroom Ganoderma Applanatum. *J. Fungi* **2022**, *8*, 331. [CrossRef]
22. Martínez-Montemayor, M.M.; Ling, T.; Suárez-Arroyo, I.J.; Ortiz-Soto, G.; Santiago-Negrón, C.L.; Lacourt-Ventura, M.Y.; Valentín-Acevedo, A.; Lang, W.H.; Rivas, F. Identification of Biologically Active Ganoderma Lucidum Compounds and Synthesis of Improved Derivatives That Confer Anti-Cancer Activities in Vitro. *Front. Pharmacol.* **2019**, *10*, 115. [CrossRef]
23. Lau, M.F.; Chua, K.H.; Sabaratnam, V.; Kuppusamy, U.R. In Vitro and in Silico Anticancer Evaluation of a Medicinal Mushroom, Ganoderma Neo-Japonicum Imazeki, against Human Colonic Carcinoma Cells. *Biotechnol. Appl. Biochem.* **2021**, *68*, 902–917. [CrossRef]
24. Ahmad, M.F.; Alsayegh, A.A.; Ahmad, F.A.; Akhtar, M.S.; Alavudeen, S.S.; Bantun, F.; Wahab, S.; Ahmed, A.; Ali, M.; Elbendary, E.Y.; et al. Ganoderma Lucidum: Insight into Antimicrobial and Antioxidant Properties with Development of Secondary Metabolites. *Heliyon* **2024**, *10*, e25607. [CrossRef] [PubMed]
25. Karadeniz, M.; Bakır, T.K.; Ünal, S. Investigation of The Antioxidant Activity And Total Phenolic Substance of Fomes Fomentarius And Ganoderma Applanatum Mushrooms Showing Therapeutic Properties. *Bilge Int. J. Sci. Technol. Res.* **2024**, *8*, 14–18. [CrossRef]
26. Kiddane, A.T.; Kang, M.J.; Ho, T.C.; Getachew, A.T.; Patil, M.P.; Chun, B.S.; Kim, G. Do Anticancer and Apoptotic Activity in Cervical Adenocarcinoma HeLa Using Crude Extract of Ganoderma Applanatum. *Curr. Issues Mol. Biol.* **2022**, *44*, 1012–1026. [CrossRef]
27. Prabhu, N.; Karunakaran, S.; Devi, S.K.; Ravya, R.; Subashree, P.; Senthil Vadivu, M. Biogenic Synthesis of Myconanoparticles From Mushroom Extracts and Its Medical Applications: A Review. *Int. J. Pharm. Sci. Res.* **2019**, *10*, 2108–2118. [CrossRef]
28. Ao, T.; Seb, J.; Ajungla, T.; Deb, C.R. Diversity of Wild Mushrooms in Nagaland, India. *Open J. For.* **2016**, *06*, 404–419. [CrossRef]
29. Ao, T.; Deb, C.R. Nutritional and Antioxidant Potential of Some Wild Edible Mushrooms of Nagaland, India. *J. Food Sci. Technol.* **2019**, *56*, 1084–1089. [CrossRef]
30. Manoharachary, C.; Sridhar, K.; Singh, R.; Adholeya, A.; Suryanarayanan, T.S.; Rawat, S.; Johri, B.N. Fungal Biodiversity: Distribution, Conservation and Prospecting of Fungi from India. *Curr. Sci.* **2005**, *89*, 58–71.
31. Verma, N.; Bhalla, A.; Kumar, S.; Dhiman, R.K.; Chawla, Y.K. Wild Mushroom Poisoning in North India: Case Series with Review of Literature. *J. Clin. Exp. Hepatol.* **2014**, *4*, 361–365. [CrossRef]
32. Upadhyay, M.K.; Shukla, A.; Yadav, P.; Srivastava, S. A Review of Arsenic in Crops, Vegetables, Animals and Food Products. *Food Chem.* **2019**, *276*, 608–618. [CrossRef]
33. Chumbhale, D.S.; Upasani, C.D. Pharmacognostic Standardization of Stems of Thespesia Lampas (Cav.) Dalz & Gibs. *Asian Pac. J. Trop. Biomed.* **2012**, *2*, 357–363. [CrossRef] [PubMed]

34. Nakalembe, I.; Kabasa, J.D.; Olila, D. Comparative Nutrient Composition of Selected Wild Edible Mushrooms from Two Agro-Ecological Zones, Uganda. *Springerplus* **2015**, *4*, 1–15. [CrossRef] [PubMed]
35. Mleczek, M.; Budka, A.; Siwulski, M.; Mleczek, P.; Gąsecka, M.; Jasińska, A.; Kalač, P.; Sobieralski, K.; Niedzielski, P.; Proch, J.; et al. Investigation of Differentiation of Metal Contents of Agaricus Bisporus, Lentinula Edodes and Pleurotus Ostreatus Sold Commercially in Poland between 2009 and 2017. *J. Food Compos. Anal.* **2020**, *90*, 103488. [CrossRef]
36. Krishnamoorthi, R.; Bharathakumar, S.; Malaikozhundan, B.; Mahalingam, P.U. Mycofabrication of Gold Nanoparticles: Optimization, Characterization, Stabilization and Evaluation of Its Antimicrobial Potential on Selected Human Pathogens. *Biocatal. Agric. Biotechnol.* **2021**, *35*, 102107. [CrossRef]
37. Kombrink, A.; Tayyrov, A.; Essig, A.; Stöckli, M.; Micheller, S.; Hintze, J.; van Heuvel, Y.; Dürig, N.; Lin, C.W.; Kallio, P.T.; et al. Induction of Antibacterial Proteins and Peptides in the Coprophilous Mushroom Coprinopsis Cinerea in Response to Bacteria. *ISME J.* **2019**, *13*, 588–602. [CrossRef] [PubMed]
38. Meikle, T.G.; Dyett, B.P.; Strachan, J.B.; White, J.; Drummond, C.J.; Conn, C.E. Preparation, Characterization, and Antimicrobial Activity of Cubosome Encapsulated Metal Nanocrystals. *ACS Appl. Mater. Interfaces* **2020**, *12*, 6944–6954. [CrossRef]
39. Boonsong, S.; Klaypradit, W.; Wilaipun, P. Antioxidant Activities of Extracts from Five Edible Mushrooms Using Different Extractants. *Agric. Nat. Resour.* **2016**, *50*, 89–97. [CrossRef]
40. Moore, D.; Gange, A.C.; Gange, E.G.; Boddy, L. Chapter 5 Fruit Bodies: Their Production and Development in Relation to Environment. *Br. Mycol. Soc. Symp. Ser.* **2008**, *28*, 79–103. [CrossRef]
41. Gbolagade, J.; Ajayi, A.; Oku, I.; Wankasi, D. Nutritive Value of Common Wild Edible Mushrooms from Southern Nigeria. *Glob. J. Biotechnol. Biochem.* **2006**, *1*, 16–21.
42. Adeniyi, M.; Odeyemi, Y.; Odeyemi, O. Ecology, Diversity and Seasonal Distribution of Wild Mushrooms in a Nigerian Tropical Forest Reserve. *Biodiversitas* **2018**, *19*, 285–295. [CrossRef]
43. Singh, R.; Kaur, N.; Shri, R.; Singh, A.P.; Dhingra, G.S. Proximate Composition and Element Contents of Selected Species of Ganoderma with Reference to Dietary Intakes. *Environ. Monit. Assess.* **2020**, *192*, 1–15. [CrossRef] [PubMed]
44. Singh, R.; Shri, R.; Singh, A.P.; Dhingra, G.S. Valorization of Ganoderma Species: Chemical Characterization and Antidepressant-Like Activity. *Waste and Biomass Valorization* **2021**, *12*, 2025–2036. [CrossRef]
45. Nagaraj, K.; Mallikarjun, N.; Naika, R.; Venugopal, T.M. Antioxdative Activities of Wild Macro Fungi Ganoderma Applanatum (PERS.) PAT. *Asian J. Pharm. Clin. Res.* **2014**, *7*, 166–171.
46. Kolayli, S.; Sahin, H.; Aliyazicioglu, R.; Sesli, E. Phenolic Components and Antioxidant Activity of Three Edible Wild Mushrooms from Trabzon, Turkey. *Chem. Nat. Compd.* **2012**, *48*, 137–140. [CrossRef]
47. Ibiam, U.A.; Uti, D.E.; Ejeogo, C.C.; Orji, O.U.; Aja, P.M.; Nwamaka, E.N.; Alum, E.U.; Chukwu, C.; Aloke, C.; Chinedum, K.E.; et al. In Vivo and in Silico Assessment of Ameliorative Effects of Xylopia Aethiopica on Testosterone Propionate-Induced Benign Prostatic Hyperplasia. *Pharm. Front.* **2023**, *5*, e64–e76. [CrossRef]
48. Mallikarjuna, S.E.; Ranjini, A.; Haware, D.J.; Vijayalakshmi, M.R.; Shashirekha, M.N.; Rajarathnam, S. Mineral Composition of Four Edible Mushrooms. *Hindawi* **2012**, *2013*, 805284.
49. Matute, R.G.; Figlas, D.; Curvetto, N. Agaricus Blazei Production on Non-Composted Substrates Based on Sunflower Seed Hulls and Spent Oyster Mushroom Substrate. *World J. Microbiol. Biotechnol.* **2011**, *27*, 1331–1339. [CrossRef]
50. Pradhan, D.; Biswasroy, P.; Kapil; Kajol; Jatin; Pradhan, R. Qualitative and Quantitative Analysis of Caffeine in Some Commercial Brands of Tea Consumed in India. *J. Ayurvedic Herb. Med.* **2017**, *3*, 200–204. [CrossRef]
51. Kristanti, H.; Tunjung, W.A.S. Detection of alkaloid, flavonoid, and terpenoid compounds in bread (Artocarpus Communis Forst.) leaves and pulps. *KnE Life Sci.* **2015**, *2*, 129. [CrossRef]
52. Leela, K.; Anchana Devi, C. Isolation, Purification and Application of Secondary Metabolites From Lichen Parmelia Perlata. *Biosci. Biotechnol. Res. Asia* **2017**, *14*, 1413–1428. [CrossRef]
53. Sonam, M.; Singh, R.P.; Saklani, P. Phytochemical Screening and TLC Profiling of Various Extracts of Reinwardtia Indica. *Int. J. Pharmacogn. Phytochem. Res.* **2017**, *9*, 8125. [CrossRef]
54. Barman, S.; Chakraborty, U.; Chakraborty, B. GC-MS Analyses of Biochemical Constituents in Fruit Body of Agaricus Bisporus. *J. Mycopathol. Res.* **2021**, *59*, 47–61.
55. Kaseleht, K.; Leitner, E.; Paalme, T. Determining Aroma-Active Compounds in Kama Flour Using SPME-GC/MS and GC-Olfactometry. *Flavour Fragr. J.* **2011**, *26*, 122–128. [CrossRef]
56. Udu-Ibiam, O.E.; Ogbu, O.; Ibiam, U.A.; Nnachi, A.U.; Agah, M.V.; Ukaegbu, C.O.; Chukwu, O.S.; Agumah, N.B.; Ogbu, K.I. Phytochemical and Antioxidant Analyses of Selected Edible Mushrooms, Ginger and Garlic from Ebonyi State, Nigeria. *IOSR J. Pharm. Biol. Sci.* **2014**, *9*, 86–91. [CrossRef]
57. Pal, A.; Ray, R.; Acharya, K.; Paul, S. Assessment of the Anti-Leukemic and Antioxidant Potential of the Methanol Extract of a Wild, Edible, and Novel Mushroom, Astraeus Hygrometricus, and Unraveling Its Metabolomic Profile. *J. Adv. Biotechnol. Exp. Ther.* **2021**, *4*, 388–404. [CrossRef]
58. Pal, A.; Ray, R.; Chouni, A.; Hajra, S.; Paul, S. Novel Wild Edible Mushroom Astraeus Hygrometricus Induces Robust Apoptosis on Human Acute Lymphoblastic Leukemia Cells through a RONS-Subsisted Mitochondria-Dependent Pathway. *J. Tradit. Chinese Med. Sci.* **2023**, *11*, 67–77. [CrossRef]
59. Arora, A. In-Vitro Antibacterial Activity of Some Ganoderma Species: A Review. *Qeios* **2023**, 1–18. [CrossRef]

60. Krishnamoorthi, R.; Ulagan Mahalingam, P.; Malaikozhundan, B. Edible Mushroom Extract Engineered Ag NPs as Safe Antimicrobial and Antioxidant Agents with No Significant Cytotoxicity on Human Dermal Fibroblast Cell. *Inorg. Chem. Commun.* **2022**, *139*, 109362. [CrossRef]
61. Kozarski, M.; Klaus, A.; Vunduk, J.; Zizak, Z.; Niksic, M.; Jakovljevic, D.; Vrvic, M.M.; Van Griensven, L.J.L.D. Nutraceutical Properties of the Methanolic Extract of Edible Mushroom Cantharellus Cibarius (Fries): Primary Mechanisms. *Food Funct.* **2015**, *6*, 1875–1886. [CrossRef] [PubMed]
62. Jeena, G.S.; Punetha, H.; Prakash, O.; Chandra, M.; Kushwaha, K.P.S. Study on in Vitro Antioxidant Potential of Some Cultivated Pleurotus Species (Oyster Mushroom). *Indian J. Nat. Prod. Resour.* **2014**, *5*, 56–61.
63. Akata, I.; Zengin, G.; Picot, C.M.N.; Mahomoodally, M.F. Enzyme Inhibitory and Antioxidant Properties of Six Mushroom Species from the Agaricaceae Family. *South African J. Bot.* **2019**, *120*, 95–99. [CrossRef]
64. Herawati, E.; Ramadhan, R.; Ariyani, F.; Marjenah; Kusuma, I.W.; Suwinarti, W.; Mardji, D.; Amirta, R.; Arung, E.T. Phytochemical Screening and Antioxidant Activity of Wild Mushrooms Growing in Tropical Regions. *Biodiversitas* **2021**, *22*, 4716–4721. [CrossRef]
65. Campos, C.; Guzmán, R.; López-Fernández, E.; Casado, Á. Evaluation of the Copper(II) Reduction Assay Using Bathocuproinedisulfonic Acid Disodium Salt for the Total Antioxidant Capacity Assessment: The CUPRAC-BCS Assay. *Anal. Biochem.* **2009**, *392*, 37–44. [CrossRef] [PubMed]
66. Michalak, A. Phenolic Compounds and Their Antioxidant Activity in Plants Growing under Heavy Metal Stress. *Polish J. Environ. Stud.* **2006**, *15*, 523–530.
67. Suseem, S.R.; Saral, M.; Reddy, N.; Gregory, M. Evaluation of the Analgesic Activity of Ethyl Acetate, Methanol and Aqueous Extracts of Pleurotus Eous Mushroom. *Asian Pac. J. Trop. Med.* **2011**, *4*, 117–120. [CrossRef] [PubMed]
68. AOAC. *Official Method of Analysis: Association of Analytical Chemists*; AOAC: Washington, DC, USA, 2012.
69. Shabbir, M.; Khan, M.R.; Saeed, N. Assessment of Phytochemicals, Antioxidant, Anti-Lipid Peroxidation and Anti-Hemolytic Activity of Extract and Various Fractions of Maytenus Royleanus Leaves. *BMC Complement. Altern. Med.* **2013**, *13*, 1–13. [CrossRef] [PubMed]
70. Shraim, A.M.; Ahmed, T.A.; Rahman, M.M.; Hijji, Y.M. Determination of Total Flavonoid Content by Aluminum Chloride Assay: A Critical Evaluation. *LWT* **2021**, *150*, 111932. [CrossRef]
71. Ogidi, O.I.; Oguoma, L.M.O.; Adigwe, P.C.; Anthony, B.B. Phytochemical Properties and In-Vitro Antimicrobial Potency of Wild Edible Mushrooms (Pleurotus Ostreatus) Obtained from Yenagoa, Nigeria. *J. Phytopharm.* **2021**, *10*, 180–184. [CrossRef]
72. Ainsworth, E.A.; Gillespie, K.M. Estimation of Total Phenolic Content and Other Oxidation Substrates in Plant Tissues Using Folin-Ciocalteu Reagent. *Nat. Protoc.* **2007**, *2*, 875–877. [CrossRef] [PubMed]
73. Altaf, U.; Lalotra, P.; Sharma, Y.P. Nutritional and Mineral Composition of Four Wild Edible Mushrooms from Jammu and Kashmir, India. *Indian Phytopathol.* **2020**, *73*, 313–320. [CrossRef]
74. Cai, H.; Liu, X.; Chen, Z.; Liao, S.; Zou, Y. Isolation, Purification and Identification of Nine Chemical Compounds from Flammulina Velutipes Fruiting Bodies. *Food Chem.* **2013**, *141*, 2873–2879. [CrossRef] [PubMed]

pharmaceuticals

MDPI

Review

Phage Encapsulation and Delivery Technology: A Strategy for Treating Drug-Resistant Pathogenic Microorganisms

Yang Yue [1], Zhenbo Xu [1,2,*], Thanapop Soteyome [3], Mahesh Premarathna [1,4], Xiaomao Yin [5] and Junyan Liu [4,6,7]

1 School of Food Science and Engineering, South China University of Technology, Guangzhou 510640, China; mahesh.pr@nifs.ac.lk (M.P.)
2 Department of Laboratory Medicine, the Second Affiliated Hospital of Shantou University Medical College, Shantou 515051, China
3 Home Economics Technology, Rajamangala University of Technology Phra Nakhon, Bangkok 10300, Thailand; thanapop.s@rmutp.ac.th
4 Microbial Biotechnology Unit, National Institute of Fundamental Studies, Kandy 20000, Sri Lanka; yaner0722@hotmail.com
5 Department of Laboratory Medicine, Guangzhou Red Cross Hospital, Jinan University, Guangzhou 510632, China; yinxiaomaoandy@sina.com
6 College of Light Industry and Food Science, Guangdong Provincial Key Laboratory of Lingnan Specialty Food Science and Technology, Academy of Contemporary Agricultural Engineering Innovations, Zhongkai University of Agriculture and Engineering, Guangzhou 510225, China
7 Key Laboratory of Green Processing and Intelligent Manufacturing of Lingnan Specialty Food, Ministry of Agriculture, Guangzhou 510225, China
* Correspondence: zhenbo.xu@hotmail.com

Abstract: Antimicrobial resistance (AMR) is one of the most critical challenges to global public health in the 21st century, posing a significant threat to healthcare systems and human health due to treatment failure and high mortality. The World Health Organization (WHO) estimates that, without effective interventions, AMR-associated infections could cause 10 million deaths annually and economic losses of up to 100 trillion US dollars by 2050. The rapid spread of drug-resistant strains, especially in hospital and community settings, has significantly reduced the efficacy of traditional antibiotics. With the continuous advancements in relevant research, bacteriophage (Phage) therapy is constantly innovating in the antimicrobial field. The application of frontier technologies, such as phage cocktails and engineered phages, has significantly enhanced the broad spectrum and high efficiency of phage therapy, which is gradually becoming a new generation of tools to replace antibiotics and effectively combat pathogenic bacteria. However, phage therapy is facing several challenges, including phage inactivation by gastric acid, enzymes, ultraviolet light, and mechanical stress, as well as the potential risk of bacterial phage resistance. Advanced encapsulation technologies such as electrospun fibers, liposomes, chitosan nanoparticles, and electrospray provide solutions to these problems by protecting phage activity and enabling controlled release and targeted delivery. This review addresses phage therapeutic studies of *Salmonella*, *Pseudomonas aeruginosa*, *Staphylococcus aureus*, *Escherichia coli*, and *Listeria monocytogenes*, summarizes the recent advances in phage research, and details the current development and applications of encapsulated phage technologies across various delivery modes.

Keywords: phage; delivery; encapsulation; CRISPR-CAS; medicine

1. Introduction

1.1. Pathogenic Microorganisms

Pathogenic microorganisms widely exist on the surfaces of objects frequently touched by human beings or in inadequately treated foodstuffs, becoming a potential health threat. These microorganisms can induce a wide range of diseases, such as skin inflammation, digestive disorders, or respiratory infections, thus posing a significant impact on individual health and public health. Their characteristics of spreading in the environment through direct contact, diet, or airborne transmission threaten human societies. *Salmonella* spp. is a widespread foodborne pathogen, a group of Gram-negative bacteria belonging to the Enterobacteriaceae family, and is one of the most important foodborne pathogens globally [1].

Salmonella is widespread in the natural environment, including soil, water, and the intestinal tracts of animals, and has become a common cause of food contamination and human infections because of its environmental adaptability and ability to spread [2]. The use of antibiotics in livestock may select for antibiotic-resistant bacteria, posing particular concern for the safety of our food supply chain. Antibiotic resistance, especially multidrug-resistant (MDR) Salmonella, is a major issue causing approximately 100,000 illnesses annually and is commonly associated with food from all possible contamination sources [3,4]. Antibiotic resistance in *Salmonella* has emerged as a major challenge in global food safety, with rising resistance rates closely linked to the overuse of antimicrobial drugs in livestock farming. As a key virulence mechanism, *Salmonella*'s ability to form biofilms enables the bacteria to evade host immune clearance and impede antibiotic penetration, significantly enhancing its environmental adaptability and pathogenicity. Furthermore, the presence of virulence genes (such as *invA*, *spvC*, *spvB1*, and *agfA*) further elevates the pathogenic potential of these strains. In a study of 38 *Salmonella* isolates, high resistance rates were observed against common antimicrobials, specifically: cefepime (60.5%), chloramphenicol (57.9%), trimethoprim-sulfamethoxazole (50.0%), amoxicillin-clavulanic acid (39.5%), azithromycin (28.9%), ofloxacin (26.3%), and cefuroxime (13.2%). Notably, 47.4% of these strains were identified as multidrug resistant (MDR), underscoring the severe risks posed by *Salmonella* resistance to public health and food safety [5–8]. However, it has now been shown that *Salmonella* is able to rely on the resistance nodulation cell division (RND) family of efflux pumps to efficiently efflux a wide range of antibiotics, such as chloramphenicol, using a proton motive force, which reduces the intracellular concentration of the antibiotics and diminishes their bactericidal efficacy. Together, these mechanisms allow *Salmonella* to survive antibiotic pressure and make treatment more difficult. Currently, regarding biofilms, *Salmonella* in mixed biofilms increases the risk of Salmonella contamination by enhancing biomass formation, resistance to disinfectants and food processing stresses, antibiotic tolerance, and human cytotoxicity [9]. This ability to form biofilms—involving bacterial aggregation and embedding within self-produced matrices—constitutes a key persistence mechanism for *Salmonella*, enabling its sustained survival in both environmental and host settings. For instance, non-typhoidal *Salmonella* serotypes can aggregate and embed within self-produced extracellular matrices to form so-called biofilms, posing significant challenges for microbial clearance [10,11].

Staphylococcus aureus is a Gram-positive bacterium which is frequently isolated from a variety of foods, such as egg products, vegetables, raw milk and dairy products, meat and seafood, and poses a significant threat to human health [12]. Methicillin-resistant *S. aureus* (MRSA) poses significant clinical and economic challenges due to increased incidence, prolonged hospital stays, and rising healthcare costs. For instance, a nationwide study in Japan estimated that MRSA infections result in over $2 billion in annual healthcare overspending and more than 14,000 deaths [13,14]. MRSA's resistance is encoded by multiple genes it carries, which collectively confer a spectrum of resistance to β-lactam antibiotics, lincosamides, macrolides, clindamycin, and aminoglycosides [15]. The virulence of *S. aureus* is controlled by a complex regulatory system, including SigB (alternative σ-factor), Sar-family DNA-binding proteins, and the Agr group sensing system, which triggers severe infections mainly through a variety of toxins (e.g., haemolysins, PVLs, PSMs, EDINs, ETs, SEs, and TSST-1) [16,17]. *S. aureus* has been reported to be resistant to antibiotics mediated by *mecA* genes, VanA manipulators, *erm* genes, and modifying enzymes. Furthermore, colonization on implanted biomaterial surfaces begins with microbial surface component recognition and adhesion matrix molecules (MSCRAMMs). Subsequently, the biofilm matures through the synthesis of an extracellular polymeric matrix composed of host factors, polysaccharide intercellular adhesin (PIA), proteins, extracellular DNA, and lipids. This matrix effectively shields against host immune responses and antimicrobial agents, making infections difficult to eradicate. It also further enhances antibiotic resistance, rendering treatment increasingly challenging [18,19].

Escherichia coli O157:H7 is an important foodborne pathogen capable of causing severe gastrointestinal diseases [20], and such cases are reported from contaminated food, undercooked beef, vegetables, dairy products, and tap water, all of which can contribute to the spread of the organism [21]. For *E. coli*, the extracellular matrix primarily consists of amyloid fibrils formed by amyloid proteins and polysaccharide cellulose fibers. The biofilm microenvironment of *E. coli* significantly promotes the evolution of gentamicin resistance, though adaptation pathways vary among different strains. Food-isolated strains achieve high-efficiency resistance through point mutations in key genes such as *aceE* and *atpG*, whereas reference strains rely on large-scale deletions in genes like *sbmA*, significantly altering cellular adaptability. These unique biofilm-specific characteristics serve as powerful drivers of bacterial evolution, promoting the emergence of antibiotic-resistant clones even in the absence of external pressure [22,23].

Pseudomonas aeruginosa is a Gram-negative bacterium widely distributed in the natural environment. Carbapenems are critical drugs for treating severe infections caused by multidrug-resistant pathogens. However, infections caused by carbapenem-resistant *P. aeruginosa* (CRPA) often respond poorly to most antimicrobial agents, with some strains exhibiting resistance to all available antibiotics. Of particular concern, certain CRPA strains can produce carbapenemases—enzymes that efficiently hydrolyze carbapenems and other β-lactam antibiotics—leading to treatment failure [24–26]. *P. aeruginosa* can be planktonic or sessile, and biofilms are typical of the sessile state, consisting of an extracellular matrix [27]. The pathogenicity of *P. aeruginosa* is primarily mediated by multiple virulence factors acting in concert, including exotoxins (such as ExoS and ToxA), elastase (LasB), and pyroxins (PhzM). These factors synergistically enhance the bacterium's ability to invade host tissues and promote immune evasion. Furthermore, the bacterium can form biofilms on various surfaces, not only enhancing its environmental survival but also significantly increasing its resistance to conventional disinfection methods [28]. *P. aeruginosa* also expresses a variety of efflux pumps, such as MexAB-OprM, MexXY-OprM, etc., which excrete β-lactam, fluoroquinolone, and aminoglycoside antibiotics [29]. In addition, β-lactam resistance in *P. aeruginosa* can arise through a combination of mechanisms such as altered PBP3 target proteins, reduced pore protein uptake, increased antibiotic efflux, β-lactamase degradation, horizontal gene transfer, and metabolic changes [30]. *P. aeruginosa* forms biofilms by adhering to surfaces and secreting extracellular matrix, undergoing multiple stages including adhesion, proliferation, maturation, and dispersion. This structured microbial community effectively resists antibiotics, oxidative stress, and host immune attacks, leading to persistent chronic infections and inflammation in medical devices and the lungs of cystic fibrosis patients. The final dispersal step of biofilms can also lead to systemic dissemination of infection, posing a major challenge for clinical treatment [31,32].

Listeria monocytogenes is a deadly foodborne pathogen commonly found in food processing facilities and can cause a zoonotic disease. Milk and dairy products are considered excellent substrates for the development of *L. monocytogenes* [33]. *L. monocytogenes* is a severe foodborne illness with a mortality rate ranging from 20% to 30% [34]. Its cold tolerance and ubiquitous nature make it prone to contaminating food, particularly ready-to-eat (RTE) products [35]. Numerous *L. monocytogenes* outbreaks have also been linked to RTE salads, fruits, and vegetables, whose consumption is increasingly high and globally widespread. Antibiotic resistance in bacteria has become a growing concern due to excessive antibiotic use in clinical settings over the past decades [36]. *L. monocytogenes* has been reported to exhibit resistance rates of 66.7% to clindamycin, 70% to penicillin G, 50% to amoxicillin, and 55% to vancomycin [37]. In the United States, its outbreaks cause an economic burden of more than $2 billion per year [38]. *L. monocytogenes* significantly enhances its survival and persistence in food processing environments by forming biofilms. Biofilms provide physical protection, enabling bacteria to withstand routine cleaning and disinfection protocols. This structure also confers substantial adaptive antibiotic resistance, with biofilm-associated infections often requiring 10 to 1000 times higher antibiotic doses than planktonic bacteria. This poses severe challenges for food safety and clinical treatment [39,40].

1.2. Antimicrobial Resistance

With increase in antimicrobial resistance, antimicrobial resistance (AMR) is now a chronic public health problem worldwide, with 10 million deaths projected annually by 2050. The increase in AMR is mainly due to the overuse and inappropriate application of antimicrobials, especially antibiotics, which accelerates the evolution and spread of drug-resistant strains of bacteria [41]. A study investigated the cases of infections in Ethiopia from 2015 to 2024 and found that *Acinetobacter baumannii* infections accounted for 9.13% of bacterial infections, with multi-drug resistant (MDR) strains as high as 88.22%, and resistance to antibiotics, such as cefotaxime, was 83.18%, with a higher prevalence especially in cases of sepsis [42]. Another study showed that carbapenem-resistant *Klebsiella pneumoniae* (CRKP) accounted for 14.2% of bloodstream infections (BSIs) and A. baumannii resistance was as high as 87.8% of BSIs in Brazil during the period 2022–2023, and that mortality rates of patients infected with resistant strains were significantly higher than those of non-resistant strains [43]. In the classification of AMR, MDR, and extensive drug resistance (XDR) represent the severity of resistance; MDR refers to bacteria that are resistant to at least three classes of antibiotics, whereas XDR refers to bacteria that are highly resistant to almost all classes of antibiotics, retaining susceptibility to a very small number of drugs, with a range of resistance that extends far beyond MDR [44]. In the case of *P. aeruginosa*, for example, the prevalence of its MDR strain exceeds 40% in Latin America and reaches 28.4% in Europe, significantly increasing the complexity of infection treatment [45]. An analysis of mortality rates and disability-adjusted life years (DALYs) for all age groups and specific age groups from 1990 to 2021 shows that drug-resistant bacteria impose a significant burden on people of all ages, so action must be taken to find alternative treatments to antibiotics [46,47].

1.3. Materials and Methods

Initially, we retrieved a substantial number of publications—approximately 50 of the most relevant documents per keyword category, amounting to a total of around 200 documents. We then reviewed the abstracts of these articles and conducted a preliminary screening to identify literature aligned with the thematic focus of our review, resulting in roughly 100 selected papers. Subsequently, we examined the methodology and conclusions of the research articles and performed a comprehensive overview of the review papers. By integrating these findings with the current state of research, we identified emerging research hotspots and gaps, which we synthesized into a coherent logical structure to form the preliminary framework of the review.

We anchored the framework around three core keywords: 'phage cocktail', 'CRISPR-Cas', and 'phage encapsulation and delivery'. For each keyword, we conducted a more detailed literature search covering the past six years (2019–2025), retrieving over 100 articles per category and accumulating approximately 400 publications in total. After a preliminary assessment of the abstracts and conclusions, we closely examined around 100 papers focusing on current research hotspots. More than half of these were selected to develop the framework, supplemented with critical reflections, leading to the initial draft of the manuscript.

Nevertheless, we recognized that the coverage remained somewhat limited. To enhance the depth of the review, we further refined the search strategy using additional keywords such as "phage oral delivery", "phage skin delivery", "phage lung delivery", and "cocktail of more than two phages". Concurrently, we restructured the overall organization of the article and incorporated critical appraisal to strengthen its reflective and logical qualities. The final version of the review cites 160 references. Details are shown in Figure 1.

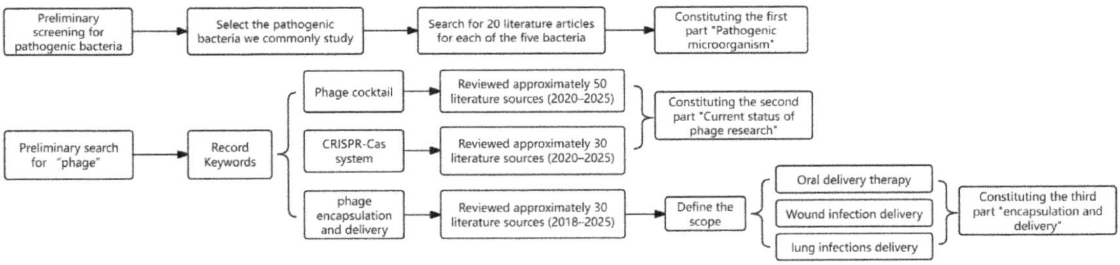

Figure 1. Flowchart of the article screening process, including keywords, number of articles read, and number of final citations.

2. Current Status of Phage Research

Bacteriophages (Phages) are viruses that can specifically infect and replicate within a target bacterium, eliminating 4–50% of bacteria [48]. Phage therapy has re-entered the scientific horizon as a precise and efficient biological approach. Phage therapy is flexible in that it can be personalized to different patients, greatly increasing the efficiency of treatment [49]. Currently, phage therapy programs have been established in the United States, Belgium, France, and Sweden, and there are long-term programs in the Republic of Georgia and Poland [50]. The phage lytic cycle is the process by which phages replicate within bacterial cells and eventually cause cell lysis. The cycle begins with phage attachment to the host cell surface and is divided into two phases, reversible and irreversible. Initially, the phage establishes transient contact with the bacterial cell surface through reversible attachment. Subsequently, the phage enters the irreversible attachment phase by securely binding to a specific receptor on the host cell surface. Following this binding, the phage releases lytic enzymes that degrade the host cell wall, thereby injecting the phage's genetic material into the host cell. The phage genome takes over the host cell's metabolic system and redirects it towards phage DNA replication and protein synthesis to produce new phage particles. At the same time, the host cell's genome is degraded, and phage components are assembled within the cell. Eventually, the host cell undergoes lysis, releasing new phage particles that can infect other cells as shown in Figure 2. The key to the lytic cycle is the action of phage proteins and endolysins, which disrupt the host cell membrane and degrade the cell wall, respectively, leading to cell lysis in response to osmotic imbalance [51,52].

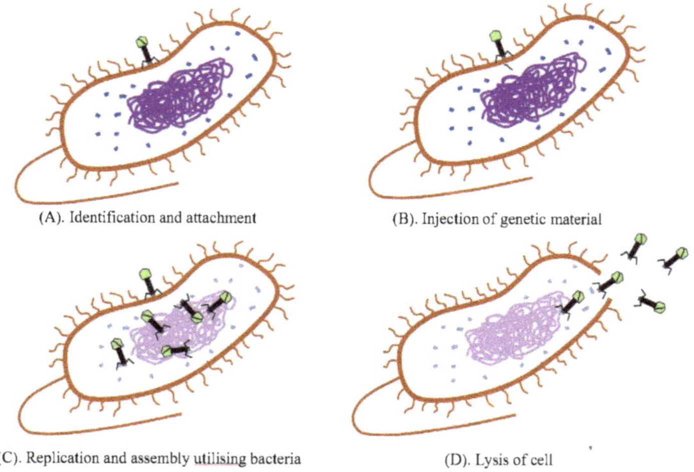

Figure 2. The principle of bacteriophage action. The figure depicts the four stages of phage infection of bacteria. When phages reach sufficient numbers, they trigger lysis, leading to bacterial death. Throughout the entire figure, the purple lines represent the genetic material DNA, the blue lines depict ribosomes, and the orange lines indicate the bacterial cell wall, fimbriae, and flagella.

Phages, with their unique bacterial targeting and self-replicating ability, are able to efficiently lyse drug-resistant strains without disturbing the host microbiota. It has been found that phage SEP4 has a significant inhibitory effect on multidrug-resistant *Salmonella* and its biofilm formation on food matrices, especially at 25 °C. In addition, phage SEP4 was able to completely inhibit the mature biofilm on the surface of lettuce leaves and chicken meat within 4–6 h [53]. However, prolonged use of phages can likewise cause strains of bacteria to become resistant to phages, so it is important to keep developing phages in conjunction with some of the currently popular technologies to meet the human need for phage therapy.

2.1. Phage Cocktails

Phage cocktails, which are mixtures of multiple phages, can target a wide range of bacterial strains (Table 1). There are three types of phage cocktails: multi-species phage cocktail, single-species phage cocktail, and single-strain targeted phage cocktail [54].

Table 1. Application of phage cocktails.

Types	Phages	Targets
multi-species phage cocktail	S19cd, S143_2, N2, C6, C2	*Salmonella*, *E. coli*
	Ph(S), Ph(E), Ph(P)	*Salmonella*, *E. coli*, *P. aeruginosa*
	VB_ECoM_SYGD1, VB ECoP SYGE1, VB ECOM SYGMH, vB_SauP_SLPW, vB SauM JDYN, vB_SauM JDF86, vB PaeS-PAJD-1	*E. coli*, *S. aureus*, *P. aeruginosa*
single-strain and single-species phage cocktail	JDYN, JDF86, SLPW	*S. aureus*
	vB_SauM_JDYN, vB_SauM_JDF86 and vB_SauP_SLPW	*S. aureus*
	HEMP1, HEMP4, HEMP5	*E. coli*

2.1.1. Application of Phage Cocktails

Pelyuntha designed a single-strain phage cocktail, a phage cocktail specifically targeting the species *Salmonella.* Due to the diversity of *Salmonella*, it is difficult for a single phage to fully eradicate this microorganism. By combining three *Salmonella* phages, the cocktail broadens the lysis range, enhances coverage of Salmonella's diversity, and overcomes the narrow host range limitation of a single phage [55]. As an alternative to antibiotic therapy, a mixture of three phages was significantly more effective than one phage alone in lysing *E. coli* against *E. coli* resistant to the antibiotic-resistance gene *mcr-1* [56]. Guo developed a novel phage, PHC-1, comprising three phages, JDYN, JDF86, and SLPW, in a 1:1:1 ratio, which expands the host range compared to a single phage in the fight against drug-resistant *S. aureus* in bovine mastitis, and provides a safe and efficient alternative to antibiotics in the treatment of bovine mastitis [57]. The intestinal flora is highly complex. Liu found that administering a phage cocktail via oral gavage to piglets' intestines actively forms a protective barrier, altering the microbial community composition in a way that promotes intestinal health [58]. Shiue et al. developed a phage cocktail hydrogel dressing incorporating three distinct phages, each specifically lysing its target bacterial host without affecting non-target species, thereby constituting a multi-species phage cocktail [59]. Similarly, Xue et al. selected three pathogenic bacteriophages and formulated them into a phage cocktail for the treatment of bovine mastitis. This multi-species cocktail broadens the treatment spectrum, enhancing the prevention and control of polymicrobial infections on dairy farms [60].

2.1.2. Progress in Design Methodology Research

The design of the phage cocktail is also being innovated. Królikowska developed a systematic phage cocktail production method using anti-mastitis treatment as an example, integrating bioinformatics, including phage genome and receptor protein analyses, lytic activity testing, and resistance assessment, which markedly improved the effectiveness and relevance of the phage cocktail [61]. Hegarty proposes an advanced phage cocktail design approach through integrated modeling, genomics, and experimental validation. The core lies in predicting phage-host interactions, optimizing phage combinations, and assessing environmental impact. The method is more systematic than traditional trial-and-error methods and is a more efficient solution for designing phage cocktails [62]. Kunisch screened two efficient phages out of four and trained them in a targeted manner to develop a phage cocktail production method with enhanced antibacterial and anti-biofilm effects [63]. Li investigated a new method for the rapid construction of an efficient phage cocktail using phage-resistant mutant strains, combining the effective new phage Va3 with the initially used Va1 phage to synthesize a phage cocktail that greatly improved phage lysis [64]. Artificial Intelligence (AI) has grown rapidly in recent years, and its great advantage is the availability of rich, high-quality data for building machine learning models [65]. For example, predicting phage-host relationships and predicting host range design phage cocktails [66–68]. Smug predicted a scenario in which four phages infested *Klebsiella pneumoniae*, with KP34 and KP15/KP27 phages mixed in a 1:1 ratio, were more effective than a single phage [69]. Designing more efficient phage cocktails by combining different phage combinations with different effects on targeted elimination of microorganisms has become a powerful tool for phage cocktails [70].

2.2. Engineered Phages Based on the CRISPR-Cas System

Engineered phages are chemically modified and genetically altered to enhance phage host range, effectiveness, and safety and to impart new functions, including detection and antimicrobial therapy. There are four main types of phage engineering: chemical modification, in vivo genetic engineering, in vitro synthesis of phage genome, and CRISPR-Cas-based phage engineering [71]. Currently, the most prominent research is engineering phages using the CRISPR-Cas system. The CRISPR-Cas system is an adaptive immune system for prokaryotes that identifies and eliminates invading phages or plasmids by storing exogenous DNA fragments [72]. The mechanism of action of the CRISPR-Cas system is divided into three processes: adaptation, expression, and interference. The first process involves recognizing the exogenous DNA fragments and integrating them into CRISPR sequences to form new spacers. These sequences undergo transcription to generate pre-crRNA, which is further processed into mature crRNA. The crRNA then guides the Cas nuclease to recognize and cleave exogenous nucleic acids complementary to the spacer sequence, thus resisting their invasion [73,74]. Although the CRISPR-Cas system is said to be a defense system for prokaryotes, it can also be used to attack prokaryotes. There are different types of CRISPR-Cas systems, which are divided into two broad categories. Class 1 CRISPR-Cas systems (types I, III, and IV) encode multisubunit effector complexes. In contrast, class 2 CRISPR-Cas systems (types II, V, and VI) use only a single-subunit effector protein [75]. Engineered phage based on the CRISPR-Cas system operates primarily from two approaches: one is to modify the tail gene of phage through CRISPR-Cas9 to broaden the host range of phage; the other is to deliver the CRISPR-Cas system that can knock out the drug-resistant genes of the host through phage or directly deliver the phage to kill certain genes of the host. The CRISPR-Cas system is used to kill certain genes of bacteria, so that the bacteria die directly. The most common is the use of type I and type II CRISPR-Cas systems [76,77].

2.2.1. Engineered Phages Based on Type I CRISPR-Cas System

Among CRISPR-Cas systems, type I systems are characterized by the presence of the distinctive nuclease Cas3 and are the most abundant and diverse CRISPR-Cas systems in nature. Type I systems display a complex structure, comprising conserved adaptation modules (Cas1 and Cas2), as well as multiprotein effector complexes containing effector modules of the characteristic nuclease Cas3 and constituting cascades [78]. Gencay utilized *E. coli*'s type I-E CRISPR-Cas system to design a new CRISPR-Cas system combined with an engineered phage. The phage's tail was altered through gene editing, and CRISPR sequences with Cas nuclease targeting *E. coli* were incorporated into the phage's genetic material, forming CRISPR-Cas-armed phages (CAPs). Four optimal CAPs were obtained through validation screening, and they were integrated into a phage cocktail named SNIPR001, which combines the most advanced phage screening with phage tail fiber engineering and the CRISPR-Cas system and opens up a new avenue for phage therapeutics [79].

2.2.2. Engineered Phages Based on Type II CRISPR-Cas System

Type II CRISPR-Cas is considered to be the most compact CRISPR-Cas system and the most widely used CRISPR-Cas gene editing system, which can perform targeting and cutting functions with only a single effector protein Cas9, and has applications in gene editing, knocking out drug-resistant genes, and genetic modification, and it is a pair of scissors for genetic engineering [80–83]. Using a type II CRISPR-Cas9 system, Park developed an antibacterial approach targeting the *nuc* gene of *S. aureus*, which was able to recognize the PAM sequence next to the *nuc* gene, cleave the genomic DNA, and trigger bacterial death. Additionally, 10 superantigens and 11 cytotoxic genes of the host were eliminated to improve safety after lysing the bacteria, and expanding the host by tail fibronectin modification scope [84].

However, the therapeutic efficacy of phages is often compromised by environmental vulnerability during administration. This critical limitation has driven the rapid development of advanced encapsulation and delivery technologies, which we will explore in the following sections.

3. Recent Research Advances in Phage-Based Encapsulation and Delivery

Encapsulation entails the use of polymers, proteins, or other materials to create a continuous layer of membranes, gels, microcapsules, films, or fibrous webs that encapsulate the active substance internally or embed it in a matrix by physical or chemical means [85]. Phages exhibit very low environmental tolerance and can be rendered non-functional by denaturation, internal DNA damage, and internal structural changes caused by harsh environments. Therefore, it is necessary to use some strategies to protect the phage from performing its functions when necessary. Membrane embedding technology is very effective in protecting phages, such as sodium alginate membrane, cellulose acetate membrane, chitosan membrane, PCL membrane, electro discharge fiber embedding and microcapsule protection, which have been used in food packaging to reduce the number of E. coli, *Salmonella*, *P. aeruginosa*, *L. monocytogenes*, and *Vibrio parahaemolyticus* in foods to a large extent, reducing the chances of foodborne pathogens attacking the body at the source of infection (Table 2) [86]. In medical phage therapy, encapsulation protects the phage, preserves its biological activity, and enhances therapeutic efficacy [87]. Currently, encapsulation technology is constantly being optimized, and mainstream research is divided into three main areas: improving the encapsulation rate, enhancing the performance of encapsulated phage, and developing new encapsulation materials [88]. Currently, there are three main routes of drug delivery, as shown in Figure 3.

Table 2. Encapsulation materials to cope with different modes of delivery.

Delivery	Encapsulation Methods	Advantages	Disadvantages	Encapsulation Materials and Techniques	Target
Gastrointestinal	Particle encapsulation	Acid resistance Highly antibacterial Multifunctionality	Complex process Imprecise targeting	Cs-nps	*S. aureus*
	PH-responsive material	Acid resistance Precise targeting	Complex process Poor stability	Pectin	
				Methacrylic acid-ethyl acrylate copolymer	
				Alginate	
				Alginate-carrageenan	*Salmonella*
Wound	Hydrogel encapsulation	Biocompatibility Sustained effect	High cost Poor stability	Binary temperature-sensitive hydrogel consisting of Pf-127 and HPMC	*A. baumannii*
				Hydrogel consisting of sa, cmc and ha	*E.faecalis*
	Fiber encapsulation	Sustained effect Good stability	Complex process Material constraints	Electrospun fiber encapsulated materials	*P. aeruginosa*
Lung	Powder inhalation	Good stability long-term storage	Equipment requirements Complex process	Liposome encapsulation based on the thin-film hydration method	*P. aeruginosa*
				Inhalation pmps-pis	*S. aureus*
	Nebulized inhalation	Simple process High flexibility	Equipment requirements Poor stability Short shelf life	Liposome atomization encapsulation based on vibrating mesh nebulizer	
				Electrospun fiber encapsulated materials	*P. aeruginosa*

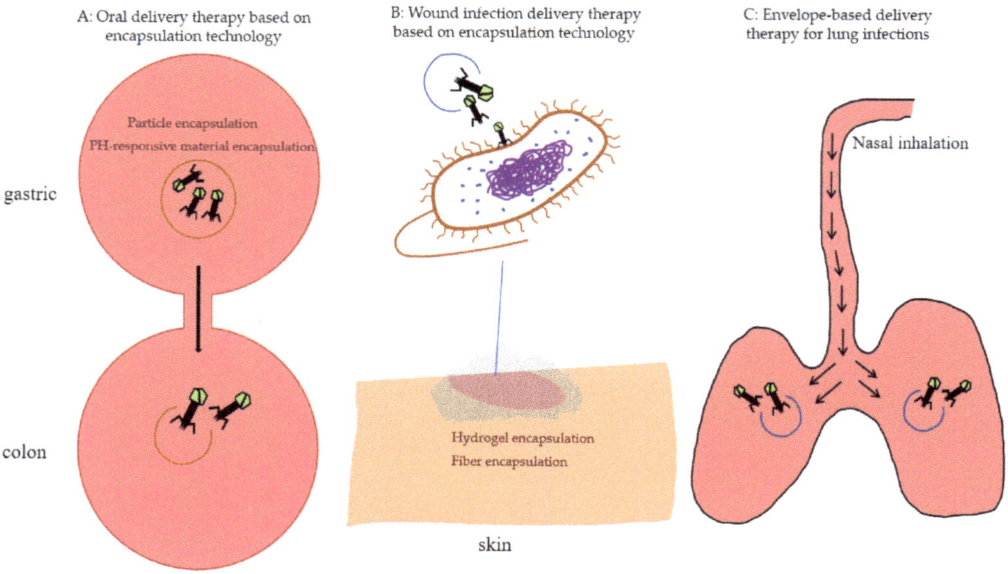

Figure 3. Phage encapsulation and delivery. The figure illustrates three phage delivery pathways: As shown in section (**A**), encapsulated phages successfully traverse the gastric environment to reach the intestinal tract for release. As shown in section (**B**), encapsulated phages enable sustained release at wound sites. As shown in section (**C**), encapsulated phages enter the lungs via inhalation, where release achieves pathogen eradication. In (**A**,**C**), the circles enveloping the bacteriophages represent the encapsulation material. In (**A**), the entire pink outer circle represents different sections of the gastrointestinal tract. (**B**) depicts the skin wound and the bacteriophage dressing applied to its surface. (**C**) represents the lungs, with the arrows indicating the delivery pathway of the bacteriophages during the inhalation process.

3.1. *Optimization of Traditional Encapsulation Techniques*

Traditional encapsulation techniques have significantly improved the efficiency and stability of phage encapsulation by optimizing the material ratio and preparation process. In the context of the ongoing research and development of encapsulation materials, Koçer carbonated the amount of sodium alginate for encapsulation of *Salmonella* phage on the basis of sodium alginate membranes, and found that the highest encapsulation rate of *Salmonella* phage was achieved when the ratio of encapsulating material to phage was 2:1 [89]. Another classical strategy, liposome encapsulation, is favored because it mimics the structure of the cell membrane [90]. Choi's research also encompassed the investigation of a co-encapsulation technique, wherein the optimal ratio of a mixed system comprising maltodextrin and alginate was determined for the material of this technique. It was observed that this ratio not only improved the encapsulation rate after encapsulation of the T4 phage, but also resulted in the phage having a high lysogenic activity and strong antimicrobial activity [91]. Guo optimized the conditions for the preparation of phage SaTp-04 CA-SA membranes using casein (CA) and sodium alginate (SA) as substrates and improved the encapsulation rate of SaTp-04 [92]. The encapsulation rate of phage has been enhanced through the utilization of conventional encapsulation technology. This development has been instrumental in achieving stable encapsulation of phage, thereby markedly enhancing its efficiency in various delivery modalities, including oral, topical, and inhalation methods. This advancement serves as a crucial foundation for the effective implementation of medical phage therapy.

3.2. Oral Delivery

Gastrointestinal infections represent the most prevalent infectious diseases, with an estimated 1.9 million deaths worldwide each year due to gastrointestinal diseases. Oral administration represents a highly convenient method for drug delivery, allowing patients to administer the medication themselves. However, a significant challenge in orally delivering therapeutics targeting the intestinal tract is the necessity to traverse the harsh gastric environment, characterized by hydrochloric acid and digestive enzymes, such as pepsin, which facilitates food digestion and eliminates pathogenic bacteria. The gastric pH typically ranges from 1.5 to 3.5 [93]. To ensure effective oral delivery, one of the problems that needs to be solved is the need for acid-resistant encapsulation materials to encapsulate the phage, so that it can successfully pass through the strong acidic environment of gastric acid to reach the intestinal tract and perform its function.

3.2.1. Particle Encapsulation

Particle encapsulation refers to the physical or chemical encapsulation of a phage in the interior of tiny particles to protect the phage [94,95]. The encapsulated phage exhibits enhanced acid resistance, enabling effective delivery to the gastrointestinal tract for therapeutic purposes. Elsayed developed chitosan nanoparticles (CS-NPs) for encapsulation of MRSA phage, a material that protects the phage from acidic environments and enzymatic damage [96]. They prepared chitosan nanoparticles (CS-NPs) of phage HK6 by cross-linking chitosan with sodium tripolyphosphate (TPP) using an ionic gel method, which can achieve a high encapsulation efficiency of 97%. Temsaah further demonstrated that CS-NPs encapsulated phage HK6 exhibited stronger antibacterial activity compared to conventional chitosan-encapsulated phage, especially in the MRSA infection model [97]. Phages encapsulated in CS-NPs benefit from a more resilient protective coating than those encapsulated in agarose, alginate, or latex. Karthika introduced a phage mixture encapsulated in a chitosan-based matrix to address the limitation of single phage encapsulation, which also allows effective delivery to the gut for the treatment of gastrointestinal disorders beyond the strong acidic environment [98]. The advantages of this approach are the introduction of a phage cocktail combinatorial strategy to broaden the antimicrobial spectrum and reduce the risk of resistance, the bead-like structure suitable for gastrointestinal delivery, and the high efficiency verified by in vitro biofilm removal experiments. This particle packaging creates a protective barrier for the phage, not only in the gastric acid environment, but also for other applications such as topical administration and cryopreservation, offering versatile solutions for phage delivery. However, existing studies are subject to several key limitations. Although the encapsulation efficiency (EE%) is widely reported, neither the phage release kinetics nor the underlying release mechanisms have been quantitatively characterized. For instance, Karthika's study asserted effective penetration in a strongly acidic environment; however, it did not provide cumulative release profiles at intestinal pH levels, which are essential to determine whether burst release or premature leakage occurred. Current encapsulation strategies primarily rely on the passive dissolution and diffusion of chitosan-based systems, lacking deliberate design and investigation into active, stimuli-responsive controlled release. Future research should therefore prioritize a mechanistic understanding of particle encapsulation, with emphasis on release rates and trigger mechanisms.

3.2.2. PH-Responsive Material Encapsulation

The disadvantages of particle encapsulation can be effectively avoided with pH-responsive materials, which are chosen to dissolve or swell under colonic pH conditions as the ideal material for achieving colonic-targeted delivery [99–101]. In contrast to basic phage protection, pH-responsive materials enable targeted release of phages in specific pH environments, optimizing their therapeutic efficacy. Pectin is not digested in the stomach, but can be degraded by pectinase secreted by intestinal bacteria to produce short-chain fatty acids, while releasing encapsulated material in the colon, making it a good pH-responsive encapsulation material [102]. A groundbreaking "infect first, encapsulate later" strategy was proposed. Rather than directly encapsulating the fragile bacteriophages, they employed pH-responsive polymers (methacrylic acid-ethyl acrylate copolymer) to encapsulate bacteria that had already been successfully infected by the phages [103]. Alginate also serves as a robust pH-responsive material; the pKa values of β-D-mannuronic acid and α-L-guluronic acid on the glycosyl chain of sodium alginate are higher than the pH of the gastric juice, and when it reaches the colon, the pKa values are lower than the pH, which results in the dissolution of the sodium alginate gel. Encapsulation of phage by dropwise addition of sodium alginate solution to calcium chloride solution to form microspheres to provide acid-resistant protection is a simple process, but a single material may limit protection in more extreme environments [104]. Dlamini found that the addition of calcium carbonate to alginate-carrageenan encapsulation prolonged the active duration of phage encapsulated in alginate-carrageenan microcapsules in acidic conditions. This technology forms a physical barrier that protects phage from pepsin degradation, allowing for a large amount of phage to be released in the intestinal environment, making it suitable for oral treatment of intestinal infections such as *Salmonella* enteritis [105].

The considerable pH variation between the colon and the stomach, while somewhat improving release rates compared to particle encapsulation systems, introduces a series of challenges. For pectin-based encapsulation materials, their relatively low mechanical strength makes them prone to rupture in the acidic gastric environment, leading to premature phage release [106]. Similarly, certain copolymers can cause abrupt phage release rather than sustained delivery, thereby failing to achieve continuous therapeutic effects. Moreover, the degradation of these polymers may cause intestinal damage—an aspect overlooked in the cited study [107]. Similarly, certain copolymers can cause abrupt phage release rather than sustained delivery, thereby failing to achieve continuous therapeutic effects. Moreover, the degradation of these polymers may cause intestinal damage—an aspect overlooked in the cited study [108]. An even more critical issue with pH-responsive materials lies in the similar pH ranges of the small intestine (pH 6–8) and the colon (pH 5–7.5). Since many responsive materials initiate release at pH 6.8, a substantial proportion of phages may be released prematurely in the small intestine, resulting in insufficient dosage reaching the colon [109]. Therefore, more refined material designs are necessary to overcome this limitation and enable precise site-specific release.

Different encapsulation methods emphasize different aspects depending on the intended purpose. For broad-spectrum inhibition of pathogens throughout the gastrointestinal tract, robust particulate encapsulation is more suitable. For targeted treatment of colonic diseases, further optimization of pH-responsive materials is required, or the development of materials with multiple response mechanisms such as enzyme-responsive or time-responsive properties.

In contrast to the systemic challenge of gastrointestinal delivery, topical management of localized wound infections demands a different set of characteristics from the delivery system, primarily focusing on biocompatibility and sustained release at the site of infection.

3.3. Wound Infection Delivery

Wound infections are frequently caused by drug-resistant organisms such as *P. aeruginosa* or MRSA. Phage encapsulation improves therapeutic efficacy through topical delivery, and wound therapy is the most common application of phage therapy. In contrast to oral delivery, wound delivery generally places a higher value on the adherence and biocompatibility of encapsulated materials.

3.3.1. Hydrogel Encapsulation

As wound dressings, hydrogels are a favorite material for scientists to consider, and various hydrogel materials have been employed for phage encapsulation. Mukhopadhyay developed a binary temperature-sensitive hydrogel consisting of Planctin F-127 (PF-127) and hydroxypropyl methylcellulose (HPMC), which stabilizes phages and mucins. PF-127 and HPMC binary temperature-sensitive hydrogel achieves stable encapsulation and high encapsulation efficiency of phage and mucin through the temperature-sensitivity of PF-127 and hydrophilic network of HPMC, which is suitable for wound site administration. This encapsulation technology. This formulation resulted in a phage-activated scaffold (PAS) wound dressing that effectively treatment of wound infections caused by MDR-*Acinetobacter baumannii*, which significantly improves the antimicrobial efficiency, inhibits biofilms, and promotes wound healing [110]. Abed engineered a novel hydrogel based on SA, CMC, and HA, integrating for the first time the antimicrobial effect of SAM-E.f 12 phage and the pro-healing properties of HA to achieve efficient antimicrobial and rapid healing of Enterococcus faecalis infected wounds. It resolves the challenge of mutual constraints between antimicrobial effects and tissue regeneration in traditional therapies, representing the integrated future direction of wound treatment [111]. Khazani formulated a composite hydrogel consisting of SA, CMC, and HA for stable encapsulation and sustained release of phage through optimized ratios and Ca^{2+} cross-linking for enhanced wound infection therapy. The use of Ca^{2+} to form an ionic cross-linking network with SA increases the mechanical strength and elasticity of the hydrogel, allowing it to withstand the mechanical stresses of the wound site and the wet environment [112].

As wound dressings, the most critical performance parameter is adhesive strength. Although all three evaluated dressings exhibit sufficient mechanical strength to withstand skin stretching and folding without damage, their adhesive properties have not been systematically investigated. For instance, clinically available hyaluronic acid (HA) hydrogels are known for their poor bioadhesion, which often leads to inadequate fixation and thus significantly compromises therapeutic outcomes [113]. Additionally, hydrogels are generally costly to produce and typically require large-scale manufacturing. To address these challenges, researchers could explore alternative strategies such as single-step cross-linking or 3D printing technologies to streamline the fabrication process [114,115].

3.3.2. Fiber Encapsulation

Electrospun fiber encapsulation is a method of encapsulating active substances in nanofibers by electrospinning technology. It is divided into uniaxial electrospinning and coaxial electrospinning. Uniaxial electrospinning involves blending a polymer solution with phages and extruding them into fine fibers under a high-voltage electric field, ensuring uniform phage distribution throughout the fiber matrix. In contrast, coaxial electrospinning utilizes coaxial needles to spray two kinds of solutions at the same time, one is the mix of polymer solution and phage, and the other is the shell layer solution, which forms the fiber of the core–shell layer structure [116,117]. Liu developed a novel electrospun fiber encapsulation material that embeds anti-*P. aeruginosa* phage (Neko) into nanofibers by uniaxial and coaxial electrostatic spinning, which can be applied directly to infected wounds with sustained release of the active phage, providing a dressing solution for the treatment of infected wounds [118]. Using coaxial electrospinning technology, phages were encapsulated within the core of the fiber while protected by a PVP shell layer, and nanofibres with a core–shell structure were successfully prepared. The rapid dissolution, biocompatibility, and mechanical properties of PVP make it an ideal material for wound dressing. Electrospun fibers offer high flexibility in local delivery, but the preparation process is complex and costly. Coaxial electrospinning provides better shell layer protection than uniaxial, but the rapid solubility of PVP may lead to excessive release.

Moving from external wounds to internal respiratory infections, the delivery paradigm shifts again, requiring technologies that can navigate the complex architecture of the lungs to achieve deep tissue deposition.

3.4. Pulmonary Delivery

Respiratory infections, especially infections caused by drug-resistant bacteria, have always been a challenge to treat due to limitations in the way treatments are administered. In 2019, an estimated 1.5 million deaths globally were attributed to respiratory infections caused by drug-resistant pathogens [119]. The key challenge in treating lung biofilm infections is drug delivery; intravesical phage administration is commonly used but predicts poor biodistribution [120]. Whereas phage encapsulation for respiratory therapy is currently of two main types: inhalation therapies utilizing dry powder formulations and those employing nebulized liquid phage formulations.

3.4.1. Powder Inhalation Therapy

The liquid mixture of phages and encapsulated material is atomized into small droplets by spray drying and spray freeze drying, followed by solvent evaporation through hot air or lyophilization to produce dry powder particles. These particles are delivered deep into the lungs using some techniques, such as a dry powder inhaler, to achieve therapeutic effects [121]. Sawant developed a liposome encapsulation system for *P. aeruginosa* phage, using thin film hydration to encapsulate the phage in liposomes, showing the promising potential of inhaled phage liposomes in the treatment of lung infections caused by *P. aeruginosa* [122]. Liu investigated an inhaled PLGA porous microsphere dry powder formulation (PMPs-PI), which achieved efficient synergistic treatment of MRSA pneumonia by combining heat-resistant phage NKU-1 with ICG photothermal therapy [123]. This design achieves the integration of physical encapsulation protection with physicochemical synergistic therapy. The microspheres themselves shield the phage and ensure its efficient pulmonary deposition, while photothermal therapy and phage therapy generate a powerful synergistic effect, offering a novel non-invasive combined treatment approach for deep-tissue MRSA pneumonia. Liposomes exhibit high deposition efficiency in pulmonary delivery and can be stored for long periods of time, but they are expensive to prepare and suitable for mass production.

3.4.2. Nebulized Inhalation Therapy

The design of liquid nebulization is much simpler compared to dry powder production, which requires further drying after liquid nebulization and is more demanding in terms of encapsulation materials. Nebulized liquid formulations bypass the drying phase, making them a more convenient inhalation therapy modality. Li innovated an inhaled liposome-phage K nanocomplex in which he nebulized a solution of liposome-phage K liquid mixture into an aerosol via a vibrating mesh nebulizer for pulmonary drug delivery to target Staphylococcus aureus infections. The EPC-phage K formulation exhibited superior nebulization stability, lung deposition, and intracellular antimicrobial efficacy [124]. Cao optimized the liposome formulation by applying liposome encapsulation technology to the nebulized inhalation delivery of phage. By studying both jet and vibrating mesh nebulizers, he highlighted the critical role of liposomes in protecting phage from the mechanical stresses of nebulization and found that vibrating mesh nebulizers were better at achieving efficient lung deposition [125]. These studies have elevated liposomes from simple protective carriers to highly efficient delivery platforms, overcoming the longstanding technical challenges of poor stability and inadequate pulmonary distribution in liquid phage aerosol formulations. This breakthrough paves the way for the clinical translation of inhalation therapies. The electrospinning technology mentioned in the previous section can also be nebulized and designed for the treatment of respiratory tract infections. Liquid atomized encapsulated products have poor storage stability and should not be kept for long periods of time, but because of their simple process, it is more suitable for small-scale rapid development.

For commercial products requiring long-term storage and distribution, powder inhalation is the inevitable direction, but its core lies in developing mild drying protection technology. For individualized treatments prepared on-site in hospitals, optimizing nebulized formulations and protective agents represents a more practical and flexible approach.

3.5. Novel Encapsulation Material Breakthrough

As encapsulation technology continues to evolve, innovative materials have been explored to enhance phage delivery. Liu identified a natural material that can withstand UV, developing sporopollenin exine capsules (SEC) derived from plant pollen grains. This novel encapsulation material, for the first time, effectively shielded phages from UV-induced damage [126]. Sporopollenin is one of the most durable materials known in nature. This study demonstrates for the first time that SEC provides exceptional UV protection for phages, thereby resolving the core challenge of rapid inactivation caused by sunlight exposure in outdoor applications such as agriculture and environmental disinfection. This breakthrough paves the way for a new "nature-inspired" approach to phage-based biopesticides [127]. Nano-emulsion, a colloidal dispersion system within the realm of nanotechnology, has also been utilized. Kaur developed a novel nano-emulsion encapsulated material to disperse phage in a system of ultra-small droplets, which facilitates wound washing and enhances phage-bacteria contact efficiency through atomization design [128]. Natural products have also been used in this field. Wdowiak employed green tea extract as a reducing and stabilizing agent to synthesize environmentally friendly nanoparticles (G-TeaNPs), suitable for complex wound environments [129]. Easwaran combines liposome technology with phage mixtures to form L-PCT, providing a novel treatment for postoperative wound infections caused by multidrug-resistant bacteria MDR [130].

3.6. Application of Simulation and Prediction Techniques to Phage Encapsulation

Currently, more researchers in this field are using simulation and prediction techniques to simulate and predict the stability and release behavior of phage in different environments (e.g., the gastric acid environment). Elsayed's study developed chitosan nanoparticles (CS-NPs) to encapsulate phage and tested their stability in simulated gastric fluid (SGF, pH 3), and the results showed that CS-NPs encapsulated phage could effectively resist enzymatic destruction by pepsin and significantly improve the ability to tolerate gastric acid [96]. Dlamini's study evaluated the protective effect of sodium alginate with κ-carrageenan (ALG-CG) and a composite containing calcium carbonate (ALG-CG-CA) by simulating the gastrointestinal environment, and found that the recovery rate of phage encapsulated by ALG-CG-CA was significantly higher than that in an acidic environment. ALG-CG or unencapsulated phage, and could extend the duration of phage activity and provide more uniform protection [105]. Zhang et al. conducted simulated gastric environment tests on composite microcapsules of xanthan gum (XG), sodium alginate (SA), calcium chloride ($CaCl_2$), and chitooligosaccharides (COS) to optimize ratios to enhance the stability and release efficiency of phage in an acidic environment [131]. Jamaledin et al. developed a predictive model using computational fluid dynamics (CFD) technology to optimize the phage encapsulation of polylactic acid-hydroxyacetic acid copolymer (PLGA) microspheres, comparing two methods: direct encapsulation by double emulsification technology (W/O/W) and preparation of empty microspheres followed by encapsulation by adsorption or osmotic encapsulation (post-encapsulation), and found that post-encapsulation was more optimal for in vivo release, showing higher [132]. Tao's team designed a bioresponsive hydrogel (Gel-ADH and DF-PEG matrix) and validated its encapsulated phage cocktail (FPR-30) against *P. aeruginosa* infection in an ex vivo human skin model to simulate the wound environment, demonstrating that the hydrogel can release phage on-demand when triggered by bacterial secretion enzymes or pH changes, significantly prolonging the duration and improving the stability for localized wounds. This demonstrates that the hydrogel can release phage on demand, triggered by bacterial secretory enzymes or pH changes, significantly prolonging the duration of action and improving stability, providing an innovative solution for the treatment of localized wound infection. Bacteriophages are released only in the presence of bacteria, significantly enhancing drug delivery efficiency and preventing indiscriminate medication use. This approach holds promise for extending therapeutic duration and reducing dosing frequency, representing a crucial step toward precision medicine [133]. The simulation prediction technique is a good approach for phage therapy research, as it provides a key support to ensure efficient and precise phage delivery to the target area by verifying phage release behavior through indirect experiments.

4. Limitations

Despite the remarkable progress in phage encapsulation technology, there are still several critical research gaps and challenges that need to be addressed before it can be widely applied in clinical and industrial settings.

4.1. Research on Encapsulation Technology

Alginate, chitosan, and liposomes provide good protection, but retention of phage infectivity after encapsulation tends to vary widely and lacks systematic quantification [134]. Further optimization of material composition and process parameters is needed in the future to minimize phage loss and extend shelf life. Moreover, the current research on the release rate is very lacking, which should be complemented in the study of encapsulation materials. Although encapsulation can enhance phage stability, achieving targeted delivery to specific infection sites remains an unsolved challenge. The development of smart controlled-release vectors capable of responding more precisely to environmental signals is still at an early stage and requires further in-depth research. The complexity of bacterial-phage interactions in the infection environment and the difficulty of simulating biofilms or polymicrobial environments in the laboratory lead to the fact that drug delivery strategies studied in vitro often fail in vivo. Researchers should conduct more in vivo experiments to ensure the usefulness of the results of research and development [134].

4.2. Scale Production

Current encapsulation technologies are mostly limited to the laboratory scale. It remains a challenge to translate these methods into a stable process that meets the requirements of industrial production and ensures consistency in particle size, encapsulation efficiency, and phage activity [135]. Problems such as batch-to-batch variation, inefficient encapsulation, and high production costs must be overcome through engineering innovation and process standardization. At the same time, production standards are difficult to control, and problems such as unclear Good Manufacturing Practice (GMP) guidelines, inconsistent endotoxin standards, and concentration limit contenders have slowed production development [136].

4.3. Security

The interaction between encapsulated phage and the host immune system is currently less studied [137]. Encapsulation may reduce immunogenicity by avoiding rapid phage clearance, but some materials may themselves cause an immune response [138]. In addition, there are potential risks associated with carrier accumulation or off-target release, such as the release of endotoxin from lysing Gram-negative bacteria, which may cause inflammatory reactions or even endotoxemia. This also needs to be systematically evaluated in relevant animal models and future clinical studies [138].

5. Discussion and Perspective

The emergence of phage therapy has significantly boosted the treatment of multidrug-resistant bacterial infections, providing a promising alternative to conventional antibiotics. However, the full potential of phage therapy has yet to be realized, particularly in terms of knowledge gaps in pharmacokinetics and pharmacodynamics (PK/PD), which have not kept pace with the escalating demand for this treatment modality [139]. Sustained research efforts are imperative to address these deficiencies and optimize treatment outcomes.

The development of phage cocktails requires the exploration of new phage combinations that minimize phage interferences, broaden the host range, and achieve precise and comprehensive bacterial killing. These advances are critical not only for clinical efficacy but also for applications such as food safety and environmental decontamination. In addition, phage engineering through CRISPR/Cas genome editing, homologous recombination, and synthetic phage DNA assembly needs to be explored more extensively. Engineered phage is the basis for the expansion of phage therapy and determines the upside of this therapy. Currently, there is limited research on CRISPR/Cas-based phage engineering, and increasing research in this area is critical to the potential of phage therapy. On the other hand, combination with other antimicrobial agents has also been found to potentially enhance the activity of phages. Research indicates that the combined use of bacteriophages and antibiotics has emerged as a novel strategy against biofilms. For instance, pretreatment with plasma-activated water effectively enhances the antibacterial activity of bacteriophages. In complex infection models such as wound biofilms and dual-bacterial communities, this combination therapy demonstrates synergistic effects, significantly improving overall antimicrobial efficacy [140–142].

The evolution of phage encapsulation technology clearly demonstrates a stepwise development path from fundamental optimization to functional integration. At the foundational level, traditional techniques such as those using alginate, chitosan nanoparticles, and liposomes were optimized primarily by adjusting material ratios and processes to enhance encapsulation efficiency and stability, laying the groundwork for subsequent applications. However, as the technology advances into the functional design phase for specific clinical applications, the strengths and weaknesses of different delivery routes become apparent: In oral delivery, physical particle encapsulation provides a robust barrier and excellent resistance to gastric acid, but its release mechanism is predominantly passive diffusion, lacking targeting and controllability [143]. In contrast, pH-responsive materials enable colon-targeted release, yet their insufficient mechanical strength may cause premature leakage in the stomach [144,145]. Furthermore, the similar pH levels in the small intestine and colon make precise control of the release site challenging. For wound healing, hydrogels are favored for their excellent biocompatibility and ability to maintain a moist environment, but their poor adhesion and inadequate mechanical strength may compromise efficacy. Electrospun fibers offer superior mechanical properties and controllable sustained release characteristics, yet their preparation is complex and costly [146]. For pulmonary delivery, dry powder inhalers (DPIs) excel at preserving phage activity and prolonging lung residence time, but drying stresses during preparation can inactivate phages; nebulized liquid formulations are simpler to prepare but suffer from poor stability and lung deposition efficiency [147,148]. These technical limitations have driven researchers to explore more advanced solutions. One promising direction involves developing composite encapsulation systems—such as combining chitosan nanoparticles with pH-responsive polymers, which may enhance system stability while enabling more precise targeted delivery—or integrating hydrogels with electrospun fibers to create novel dressings that combine excellent mechanical properties, biocompatibility, and stability. For dry powder inhalers, a "pre-encapsulation" strategy can be employed, where phages are first encapsulated in emulsions or liposomes before processing into dry powder, thereby providing effective protection during the drying process. Furthermore, cutting-edge research has advanced toward developing novel materials and intelligent responsive systems: innovative materials like spore wall capsules and green tea nanoparticles demonstrate the advantages of natural substances in providing unique protection and biocompatibility, while mimetic technologies have enabled a leap from in vitro validation to predictive design. This has ultimately led to the emergence of bio-responsive hydrogels capable of "on-demand release" triggered by bacterial activity itself. These advancements mark the transition of phage delivery from passive sustained release to an era of active, intelligent responsiveness. However, developing these composite materials is far from a simple accumulation of functions; it requires comprehensive consideration of the intricate interplay between material functionality, production processes, and cost. Future research should focus on deeper mechanistic analysis and efficacy validation of synergistic effects within these composite systems. By skillfully integrating the strengths of different materials and leveraging advanced simulation and prediction tools, the ultimate goal is to construct next-generation delivery systems capable of intelligently sensing and precisely responding to complex infection environments. This will propel phage therapeutics toward the critical leap from laboratory research to clinical application.

Future research should proactively shift toward technology pathways compatible with continuous production, such as utilizing microfluidic technology for precise control of nanoparticle synthesis or developing continuous electrospinning processes for efficient preparation of fiber dressings. When selecting materials and methods, it is imperative to establish a comprehensive evaluation system that transcends the singular metric of "encapsulation efficiency," treating scalability, production costs, and batch-to-batch reproducibility as core indicators equally important to protective efficacy. Concurrently, the biosafety assessment of encapsulation systems must evolve from simple in vitro cytotoxicity testing toward more comprehensive immunopharmacological studies. This necessitates systematically analyzing the innate and adaptive immune responses potentially triggered by the carrier material itself during the preclinical phase. It also requires in-depth investigation into how encapsulation affects the distribution, metabolism, and clearance kinetics of phages within the body—data critical for predicting real immune responses in humans and avoiding potential inflammatory risks or carrier accumulation toxicity. Only by deeply integrating the design philosophy of "scalable production" with the research paradigm of "end-to-end immunological safety monitoring" can phage therapeutics be robustly advanced from innovative laboratory concepts to widely applied clinical products.

As these technological and safety hurdles are addressed, the scope of phage-based solutions is also expanding beyond their traditional antimicrobial role. Moreover, the potential of phage therapy extends far beyond treating human infectious diseases. Its high specificity and programmability are driving a range of interdisciplinary applications [149,150]. In human health, phages are being explored for precisely regulating the gut microbiome [151]. By targeting and eliminating specific pathogenic or opportunistic pathogens to restore ecological balance, they offer novel approaches for treating metabolic disorders, autoimmune diseases, and even neurological conditions [152]. In oncology, the excellent biocompatibility and modifiability of phage capsids make them ideal carriers for targeted delivery of anticancer drugs or imaging agents [153–155]. They can also be engineered for tumor diagnostic imaging and therapy [156]. In industrial and environmental applications, phage-based biofilm removal strategies are deployed as surface coatings for food processing pipelines, water treatment systems, and medical devices to prevent biofilm contamination, enhance safety, and extend service life [157]. In agriculture, phage preparations represent a green, precision biological control technology demonstrating significant potential in combating plant bacterial diseases caused by pathogens such as Xanthomonas and Erwinia [158–160]. This approach holds promise for reducing reliance on chemical pesticides, safeguarding food security, and promoting ecological sustainability. These emerging applications significantly expand the boundaries of phage technology, signaling its potential to play a pivotal role across multiple sectors.

Innovation: This review highlights composite material applications and explores a broader spectrum of delivery pathways, aiming to construct a more comprehensive knowledge framework for the reader. To strengthen the novelty of our work, we have incorporated a range of recent research advances to enrich the content. Furthermore, the article identifies current research gaps and challenges within phage technology, providing valuable insights and future directions for researchers.

Funding: This work was supported by National Natural Science Foundation of China (grant number: 32472465).

Data Availability Statement: No new data were created or analyzed in this study. Data sharing is not applicable to this article.

Conflicts of Interest: The authors declare no conflict of interest.

References

1. Zhao, X.; Van Overbeek, L.; Burgess, C.M.; Holden, N.; Brennan, F.; Johannessen, G.S.; Allende, A.; Höfte, M.; Cottyn, B.; Pothier, J.F.; et al. Human Pathogenic Microorganisms in Fresh Produce Production: Lessons Learned When Plant Science Meets Food Safety. *J. Food Prot.* **2025**, *88*, 100551. [CrossRef]
2. Chen, B.; Yang, X.; Yang, C.; Chan, E.W.-C.; Wang, X.; Chen, S. Ultrasensitive monoclonal antibody-based Salmonella detection through fluorescence single bacterial cell imaging selectively enhanced by tyramide signal amplification. *Innov. Food Sci. Emerg. Technol.* **2025**, *105*, 104229. [CrossRef]
3. Sripaurya, B.; Pelyuntha, W.; Ngasaman, R.; Ching, C.L.; Guyonnet, V.; Vongkamjan, K. Antibiotic resistance, sequence typing, and virulence gene profiles of Salmonella enterica isolated from the broiler production chain in southern Thailand. *Res. Vet. Sci.* **2025**, *196*, 105891. [CrossRef] [PubMed]
4. Liu, Z.; Liu, Y.; He, J.; Pei, X.; Lin, Y.; Magnani, M.; Liu, D.; Ding, T.; Feng, J. Milkfat influences thermal tolerance and biofilm formation of Salmonella Typhimurium during pasteurization. *Food Res. Int.* **2025**, *221*, 117538. [CrossRef] [PubMed]
5. Polat, İ.; Güngör, İ.; Şen, B. Prevalence of Salmonella enterica Serotypes Isolated From Broiler Liver and Their Antibiotic Resistance Profiles. *J. Food Prot.* **2025**, *88*, 100535. [CrossRef]
6. Khanal, S.; Luitel, H.; Adhikari, S.; Marasini, A.; Bhattarai, R.K. Molecular characterization, antimicrobial resistance profiling, and biofilm analysis of Salmonella isolates from dead-in-shell embryonated eggs. *Poult. Sci.* **2025**, *104*, 105773. [CrossRef]
7. Woh, P.Y.; Chen, Y.; Nurjadi, D. Bridging the phylodynamic antimicrobial resistance of Salmonella in Southeast Asia and Hong Kong through a One Health Lens. *Int. J. Food Microbiol.* **2025**, *443*, 111431. [CrossRef]
8. Zhan, Z.; Zhou, Z.; Hu, M.; Tai, C.; Shi, X. Characterization of colistin resistance mechanisms development in Salmonella isolates from retail meat in Shanghai, China. *LWT* **2025**, *232*, 118424. [CrossRef]
9. Xu, J.; Wang, J.; Kang, Z.; Xu, H.; Tang, C.; Chen, J.; Hu, H. Mixed-species biofilm with Salmonella in food industry: Persistence, interspecies interaction, and control. *Food Res. Int.* **2025**, *221*, 117348. [CrossRef]
10. Thiers, I.; Lissens, M.; Langie, H.; Lories, B.; Steenackers, H. Salmonella biofilm formation diminishes bacterial proliferation in the C. elegans intestine. *Biofilm* **2024**, *8*, 100225. [CrossRef]

11. Tulin, G.; Méndez, A.A.E.; Figueroa, N.R.; Smith, C.; Folmer, M.P.; Serra, D.; Wade, J.T.; Checa, S.K.; Soncini, F.C. Integration of BrfS into the biofilm-controlling cascade promotes sessile Salmonella growth at low temperatures. *Biofilm* **2025**, *9*, 100254. [CrossRef] [PubMed]
12. Liang, T.; Liang, Z.; Wu, S.; Ding, Y.; Wu, Q.; Gu, B. Global prevalence of Staphylococcus aureus in food products and its relationship with the occurrence and development of diabetes mellitus. *Med. Adv.* **2023**, *1*, 53–78. [CrossRef]
13. Apiratwarrasakul, S.; Sresuwadjarey, P.; Phumthanakorn, N.; Withatanung, P.; Thongdee, M.; Lerdsittikul, V. The broad-spectrum kayvirus phage disrupts biofilms formed by methicillin-resistant Staphylococcus aureus and methicillin-resistant Staphylococcus pseudintermedius. *Virology* **2025**, *611*, 110657. [CrossRef] [PubMed]
14. Konatam, S.; Roy, D.N. GC-MS unveils monosaccharide D-Allose in Terminalia arjuna bark extract acting against biofilm regulatory proteins (SrtA and SarA) of Staphylococcus aureus: A drug discovery approach. *Int. J. Mass Spectrom.* **2025**, *518*, 117527. [CrossRef]
15. Rana, E.A.; Nur, S.; Fazal, M.A.; Alam, M.; Ahasan, A.S.M.L.; Rocky, M.J.H.; Rakib, T.M. Genome sequences of three methicillin-resistant Staphylococcus aureus strains isolated from subclinical mastitic cows. *Microbiol. Resour. Announc.* **2025**, *14*, e00397-25. [CrossRef]
16. Cheung, G.Y.C.; Bae, J.S.; Otto, M. Pathogenicity and virulence of Staphylococcus aureus. *Virulence* **2021**, *12*, 547–569. [CrossRef]
17. Ahmad-Mansour, N.; Loubet, P.; Pouget, C.; Dunyach-Remy, C.; Sotto, A.; Lavigne, J.P.; Molle, V. Staphylococcus aureus Toxins: An Update on Their Pathogenic Properties and Potential Treatments. *Toxins* **2021**, *13*, 677. [CrossRef]
18. Sivori, F.; Cavallo, I.; Truglio, M.; De Maio, F.; Sanguinetti, M.; Fabrizio, G.; Licursi, V.; Francalancia, M.; Fraticelli, F.; La Greca, I.; et al. Staphylococcus aureus colonizing the skin microbiota of adults with severe atopic dermatitis exhibits genomic diversity and convergence in biofilm traits. *Biofilm* **2024**, *8*, 100222. [CrossRef]
19. Turner, A.B.; Zermeño-Pérez, D.; Mysior, M.M.; Giraldo-Osorno, P.M.; García, B.; O'Gorman, E.; Oubihi, S.; Simpson, J.C.; Lasa, I.; Cróinín, T.Ó; et al. Biofilm morphology and antibiotic susceptibility of methicillin-resistant *Staphylococcus aureus* (MRSA) on poly-D,L-lactide-co-poly(ethylene glycol) (PDLLA-PEG) coated titanium. *Biofilm* **2024**, *8*, 100228. [CrossRef]
20. Rahal, E.A.; Kazzi, N.; Nassar, F.J.; Matar, G. Escherichia coli O157: H7—Clinical aspects and novel treatment approaches. *Front. Cell. Infect. Microbiol.* **2012**, *2*, 138. [CrossRef]
21. Rangel, J.M.; Sparling, P.H.; Crowe, C.; Griffin, P.M.; Swerdlow, D. Epidemiology of Escherichia coli O157: H7 outbreaks, united states, 1982–2002. *Emerg. Infect. Dis.* **2005**, *11*, 603. [CrossRef]
22. Charron, R.; Lemée, P.; Huguet, A.; Minlong, O.; Boulanger, M.; Houée, P.; Soumet, C.; Briandet, R.; Bridier, A. Strain-dependent emergence of aminoglycoside resistance in Escherichia coli biofilms. *Biofilm* **2025**, *9*, 100273. [CrossRef]
23. Zhang, T.; Ray, S.; Melican, K.; Richter-Dahlfors, A. The maturation of native uropathogenic Escherichia coli biofilms seen through a non-interventional lens. *Biofilm* **2024**, *8*, 100212. [CrossRef]
24. Lin, Z.; Duan, X.; Wan, X.; Han, S.; Ji, L.; Zeng, Y.; Shen, J.; Lu, H. Coevolutionary phage training expands phage host range: Driven by tail fiber mutations in Pseudomonas aeruginosa lytic phage? *Microbiol. Res.* **2025**, *301*, 128315. [CrossRef]
25. Horcajada, J.P.; Gales, A.; Isler, B.; Kaye, K.S.; Kwa, A.L.; Landersdorfer, C.B.; Montero, M.M.; Oliver, A.; Pogue, J.M.; Shields, R.K.; et al. How do I manage difficult-to-treat Pseudomonas aeruginosa infections? Key questions for today's clinicians. *Clin. Microbiol. Infect.* **2025**, *31*, 1797–1806. [CrossRef] [PubMed]
26. Alvarez, C.; Bello, J.; Vay, C.; Famiglietti, Á.; Nastro, M.; Rodríguez, H. Antimicrobial resistance profiles in multidrug-resistant Pseudomonas aeruginosa isolates in a university hospital in Argentina. Activity of new antibiotic combinations. *Rev. Argent. Microbiol.* 2025, in press. [CrossRef] [PubMed]
27. Tuon, F.F.; Dantas, L.R.; Suss, P.H.; Tasca Ribeiro, V.S. Pathogenesis of the Pseudomonas aeruginosa Biofilm: A Review. *Pathogens* **2022**, *11*, 300. [CrossRef] [PubMed]
28. Silva, B.S.; Furtado, M.M.; Freire, L.; Lee, S.; Jorge, G.P.; Daldosso, G.; Silva, C.E.; Vieira-Junior, L.; Lemos Junior, W.J.F.; Oliveira, C.A.F.; et al. A large sampling study on the occurrence and characteristics of Pseudomonas aeruginosa and heterotrophic bacteria in mineral water over seasons and in different containers. *Int. J. Food Microbiol.* **2025**, *443*, 111427. [CrossRef]
29. Fernández-Billón, M.; Llambías-Cabot, A.E.; Jordana-Lluch, E.; Oliver, A.; Macià, M.D. Mechanisms of antibiotic resistance in Pseudomonas aeruginosa biofilms. *Biofilm* **2023**, *5*, 100129. [CrossRef]
30. Glen, K.A.; Lamont, I.L. β-lactam Resistance in Pseudomonas aeruginosa: Current Status, Future Prospects. *Pathogens* **2021**, *10*, 1638. [CrossRef]
31. Schlichter Kadosh, Y.; Muthuraman, S.; Nisaa, K.; Ben-Zvi, A.; Karsagi Byron, D.L.; Shagan, M.; Brandis, A.; Mehlman, T.; Gopas, J.; Saravana Kumar, R.; et al. Pseudomonas aeruginosa quorum sensing and biofilm attenuation by a di-hydroxy derivative of piperlongumine (PL-18). *Biofilm* **2024**, *8*, 100215. [CrossRef] [PubMed]
32. Rima, M.; Villeneuve-Faure, C.; Pilloux, L.; Roques, C.; El Garah, F.; Makasheva, K. From adhesion to biofilms formation and resilience: Exploring the impact of silver nanoparticles-based biomaterials on Pseudomonas aeruginosa. *Biofilm* **2025**, *9*, 100267. [CrossRef] [PubMed]

33. Steinle, J.S.; Bertolini, A.B.; Pereira, J.G.; Paes, A.C.; da Silva, L.P.; dos Santos Leite, B.; Ribeiro, T.M.; Aires, I.N.; Rangel, M.V.; Curci, V.C.M.; et al. Molecular detection of Listeria monocytogenes in refrigerated raw milk and geospatial distribution in Brazilian small farms. *Food Sci. Technol.* **2025**, *45*. [CrossRef]
34. Fan, C.; Wang, Y.; Zheng, T.; Li, W.; Li, X.; Wu, P. Rapid detection of Listeria monocytogenes in ready-to-eat foods using a one-tube recombinase polymerase amplification and photosensitization colorimetric assay. *Sens. Actuators B Chem.* **2025**, *444*, 138489. [CrossRef]
35. França, L.; Rauecker, U.N.; Santos, D.L.S.d.; Almeida, N.A.; Gebara, C.; Prado, C.S.; Cavicchioli, V.Q.; Lage, M.E.; Duarte, F.O.; Rezende, C.S.M.E.; et al. Occurrence of *Listeria* spp. and *Listeria monocytogenes* in half-carcasses, meat cuts, equipment, and the environment of bovine slaughterhouses in Brazil. *Res. Vet. Sci.* **2025**, *196*, 105869. [CrossRef]
36. Li, Z.; Shao, Y.; Liu, X.; Wan, X.; Xiong, P.; Wang, L.; Yuan, J. Steap3 is a key node in regulating the phagosome escape of Listeria monocytogenes. *Mol. Immunol.* **2025**, *182*, 96–106. [CrossRef]
37. Upham, J.P.; Markell, J.A. Challenge Testing to Determine Growth of Listeria monocytogenes on Fresh Enoki Mushrooms at 4 °C and 10 °C Storage Temperatures. *J. Food Prot.* **2025**, *88*, 100577. [CrossRef]
38. Bechtel, T.D.; Hershelman, J.; Ghoshal, M.; McLandsborough, L.; Gibbons, J. Chemical mutagenesis of Listeria monocytogenes for increased tolerance to benzalkonium chloride shows independent genetic underpinnings and off-target antibiotic resistance. *PLoS ONE* **2024**, *19*, e0305663. [CrossRef]
39. Rolon, M.L.; Voloshchuk, O.; Bartlett, K.V.; LaBorde, L.F.; Kovac, J. Multi-species biofilms of environmental microbiota isolated from fruit packing facilities promoted tolerance of Listeria monocytogenes to benzalkonium chloride. *Biofilm* **2024**, *7*, 100177. [CrossRef]
40. Liu, X.; Liu, S.; Wang, Y.; Shi, Y.; Chen, Q. New insights into the antibiofilm activity and mechanism of Mannosylerythritol Lipid-A against Listeria monocytogenes EGD-e. *Biofilm* **2024**, *7*, 100201. [CrossRef]
41. Tang, K.W.K.; Millar, B.C.; Moore, J.E. Antimicrobial resistance (AMR). *Br. J. Biomed. Sci.* **2023**, *80*, 11387. [CrossRef]
42. Asmare, Z.; Tamrat, E.; Erkihun, M.; Endalamaw, K.; Alelign, D.; Getie, M.; Sisay, A.; Gashaw, Y.; Reta, M.A. Antimicrobial resistance pattern of Acinetobacter baumannii clinical isolate in Ethiopia. A systematic review and meta-analysis. *Syst. Rev. Meta-Anal.* **2025**, *25*, 518. [CrossRef]
43. Antochevis, L.C.; Wilhelm, C.M.; Arns, B.; Sganzerla, D.; Sudbrack, L.O.; Nogueira, T.C.; Guzman, R.D.; Martins, A.S.; Cappa, D.S.; Dos Santos, Â.C.; et al. World Health Organization priority antimicrobial resistance in Enterobacterales, Acinetobacter baumannii, Pseudomonas aeruginosa, Staphylococcus aureus and Enterococcus faecium healthcare-associated bloodstream infections in Brazil (ASCENSION): A prospective, multicentre, observational study. *Lancet Reg. Health–Am.* **2025**, *43*, 101004. [PubMed]
44. Pallett, S.J.; Morkowska, A.; Woolley, S.D.; Potochilova, V.V.; Rudnieva, K.L.; Iungin, O.S.; Sgro, V.; Boyd, S.E.; Reece, N.; Lambert, Z.L.; et al. Evolving antimicrobial resistance of extensively drug-resistant Gram-negative severe infections associated with conflict wounds in Ukraine: An observational study. *Lancet Reg. Health–Eur.* **2025**, *52*, 101274. [CrossRef] [PubMed]
45. Yin, C.; Alam, M.Z.; Fallon, J.T.; Huang, W. Advances in development of novel therapeutic strategies against multi-drug resistant Pseudomonas aeruginosa. *Antibiotics* **2024**, *13*, 119. [CrossRef] [PubMed]
46. Naghavi, M.; Vollset, S.E.; Ikuta, K.S.; Swetschinski, L.R.; Gray, A.P.; Wool, E.E.; Aguilar, G.R.; Mestrovic, T.; Smith, G.; Han, C. Global burden of bacterial antimicrobial resistance 1990–2021: A systematic analysis with forecasts to 2050. *Arch. Virol.* **2024**, *404*, 1199–1226. [CrossRef]
47. Villalpando-Aguilar, J.L.; Matos-Pech, G.; López-Rosas, I.; Castelán-Sánchez, H.G.; Alatorre-Cobos, F. Phage therapy for crops: Concepts, experimental and bioinformatics approaches to direct Its application. *Int. J. Mol. Sci.* **2023**, *24*, 32. [CrossRef]
48. Ackermann, H.-W. 5500 Phages examined in the electron microscope. *Arch. Virol.* **2007**, *152*, 227–243. [CrossRef]
49. Strathdee, S.A.; Hatfull, G.F.; Mutalik, V.K.; Schooley, R.T. Phage therapy: From biological mechanisms to future directions. *Cell* **2023**, *186*, 17–31. [CrossRef]
50. Munteanu, D.I.; Dunn, J.; Apjok, G.; Kintses, B.; Griselain, J.; Steurs, G.; Cochez, C.; Djebara, S.; Merabishvili, M.; Pirnay, J.P.; et al. Phage Therapy for Orthopaedic Infections: The First Three Cases from the United Kingdom. *Antibiotics* **2025**, *14*, 114. [CrossRef]
51. Teklemariam, A.D.; Al-Hindi, R.R.; Qadri, I.; Alharbi, M.G.; Ramadan, W.S.; Ayubu, J.; Al-Hejin, A.M.; Hakim, R.F.; Hakim, F.F.; Hakim, R.F.; et al. The Battle between Bacteria and Bacteriophages: A Conundrum to Their Immune System. *Antibiotics* **2023**, *12*, 381. [CrossRef]
52. Helmy, Y.A.; Taha-Abdelaziz, K.; Hawwas, H.A.E.-H.; Ghosh, S.; AlKafaas, S.S.; Moawad, M.M.; Saied, E.M.; Kassem, I.I.; Mawad, A. Antimicrobial resistance and recent alternatives to antibiotics for the control of bacterial pathogens with an emphasis on foodborne pathogens. *Antibiotics* **2023**, *12*, 274. [CrossRef]
53. Taj, M.I.; Ding, Y.; Bao, Y.; Huang, Y.; Guan, P.; Zhang, Y.; Liu, X.; Ming, Z.; Wang, X. Biocontrol characteristics and application of phage SEP4 against multidrug-resistant Salmonella biofilm on food matrix. *Food Biosci.* **2024**, *60*, 104251. [CrossRef]

54. Teklemariam, A.D.; Al Hindi, R.; Qadri, I.; Alharbi, M.G.; Hashem, A.M.; Alrefaei, A.A.; Basamad, N.A.; Haque, S.; Alamri, T.; Harakeh, S.; et al. Phage cocktails–an emerging approach for the control of bacterial infection with major emphasis on foodborne pathogens. *Biotechnol. Genet. Eng. Rev.* **2024**, *40*, 36–64. [CrossRef]
55. Pelyuntha, W.; Ngasaman, R.; Yingkajorn, M.; Chukiatsiri, K.; Guyonnet, V.; Vongkamjan, K. Phage cocktail administration to reduce Salmonella load in broilers. *Res. Vet. Sci.* **2024**, *169*, 105163. [CrossRef]
56. Shin, H.; Ahn, E.; Kim, J.; Ryu, S. Development of phage cocktail for controlling colistin-resistant Escherichia coli harboring mcr-1 and its biofilm formation in retail raw chicken. *LWT* **2024**, *200*, 116153. [CrossRef]
57. Guo, M.; Zhang, Y.; Wu, L.; Xiong, Y.; Xia, L.; Cheng, Y.; Ma, J.; Wang, H.; Sun, J.; Wang, Z.; et al. Development and mouse model evaluation of a new phage cocktail intended as an alternative to antibiotics for treatment of Staphylococcus aureus-induced bovine mastitis. *J. Dairy Sci.* **2024**, *107*, 5974–5987. [CrossRef]
58. Liu, Y.; Lin, Y.; Zhu, W. Systemic effects of a phage cocktail on healthy weaned piglets. *Biology* **2024**, *13*, 271. [CrossRef] [PubMed]
59. Shiue, S.J.; Wu, M.S.; Chiang, Y.H.; Lin, H.Y. Bacteriophage—Cocktail hydrogel dressing to prevent multiple bacterial infections and heal diabetic ulcers in mice. *J. Biomed. Mater. Res. Part A* **2024**, *112*, 1846–1859. [CrossRef] [PubMed]
60. Xue, Y.; Gao, Y.; Guo, M.; Zhang, Y.; Zhao, G.; Xia, L.; Ma, J.; Cheng, Y.; Wang, H.; Sun, J. Phage cocktail superimposed disinfection: A ecological strategy for preventing pathogenic bacterial infections in dairy farms. *Environ. Res.* **2024**, *252*, 118720. [CrossRef] [PubMed]
61. Królikowska, D.; Szymańska, M.; Krzyżaniak, M.; Guziński, A.; Matusiak, R.; Kajdanek, A.; Kaczorek-Łukowska, E.; Maszewska, A.; Wójcik, E.A.; Dastych, J. A New Approach for Phage Cocktail Design in the Example of Anti-mastitis Solution. *Pathogens* **2024**, *13*, 839. [CrossRef]
62. Hegarty, B. Making waves: Intelligent phage cocktail design, a pathway to precise microbial control in water systems. *Water Res.* **2025**, *268*, 122594. [CrossRef]
63. Kunisch, F.; Campobasso, C.; Wagemans, J.; Yildirim, S.; Chan, B.K.; Schaudinn, C.; Lavigne, R.; Turner, P.E.; Raschke, M.J.; Trampuz, A. Targeting Pseudomonas aeruginosa biofilm with an evolutionary trained bacteriophage cocktail exploiting phage resistance trade-offs. *Nat. Commun.* **2024**, *15*, 8572. [CrossRef]
64. Li, C.; Shi, T.; Sun, Y.; Zhang, Y. A novel method to create efficient phage cocktails via use of phage-resistant bacteria. *Appl. Environ. Microbiol.* **2022**, *88*, e02323-21. [CrossRef] [PubMed]
65. Yuan, X.; Fan, L.; Jin, H.; Wu, Q.; Ding, Y. Phage engineering using synthetic biology and artificial intelligence to enhance phage applications in food industry. *Curr. Opin. Food Sci.* **2025**, *62*, 101274. [CrossRef]
66. Mahony, J. Biological and bioinformatic tools for the discovery of unknown phage–host combinations. *Curr. Opin. Microbiol.* **2024**, *77*, 102426. [CrossRef] [PubMed]
67. Keith, M.; Park de la Torriente, A.; Chalka, A.; Vallejo-Trujillo, A.; McAteer, S.P.; Paterson, G.K.; Low, A.S.; Gally, D.L. Predictive phage therapy for Escherichia coli urinary tract infections: Cocktail selection for therapy based on machine learning models. *Proc. Natl. Acad. Sci. USA* **2024**, *121*, e2313574121. [CrossRef]
68. Lood, C.; Boeckaerts, D.; Stock, M.; De Baets, B.; Lavigne, R.; van Noort, V.; Briers, Y.J.C.O.i.V. Digital phagograms: Predicting phage infectivity through a multilayer machine learning approach. *Curr. Opin. Virol.* **2022**, *52*, 174–181. [CrossRef]
69. Smug, B.J.; Majkowska-Skrobek, G.; Drulis-Kawa, Z. PhREEPred: Phage Resistance Emergence Prediction Web Tool to Foresee Encapsulated Bacterial Escape from Phage Cocktail Treatment. *J. Mol. Biol.* **2022**, *434*, 167670. [CrossRef]
70. Seguritan, V.; Alves Jr, N.; Arnoult, M.; Raymond, A.; Lorimer, D.; Burgin, A.B.; Salamon, P.; Segall, A.M. Artificial neural networks trained to detect viral and phage structural proteins. *PLoS Comput. Biol.* **2012**, *8*, e1002657. [CrossRef]
71. Peng, H.; Chen, I.A.; Qimron, U. Engineering phages to fight multidrug-resistant bacteria. *Chem. Rev.* **2024**, *125*, 933–971. [CrossRef]
72. Nidhi, S.; Anand, U.; Oleksak, P.; Tripathi, P.; Lal, J.A.; Thomas, G.; Kuca, K.; Tripathi, V. Novel CRISPR–Cas systems: An updated review of the current achievements, applications, and future research perspectives. *Int. J. Mol. Sci.* **2021**, *22*, 3327. [CrossRef]
73. Stoltzfus, M.J.; Workman, R.E.; Keith, N.C.; Modell, J.W. A dynamic subpopulation of CRISPR–Cas overexpressers allows Streptococcus pyogenes to rapidly respond to phage. *Nat. Microbiol.* **2024**, *9*, 2410–2421. [CrossRef] [PubMed]
74. Workman, R.E.; Stoltzfus, M.J.; Keith, N.C.; Euler, C.W.; Bondy-Denomy, J.; Modell, J.W. Anti-CRISPR proteins trigger a burst of CRISPR-Cas9 expression that enhances phage defense. *Cell Rep.* **2024**, *26*, 43. [CrossRef] [PubMed]
75. Koonin, E.V.; Makarova, K.S.; Zhang, F. Diversity, classification and evolution of CRISPR-Cas systems. *Curr. Opin. Microbiol.* **2017**, *37*, 67–78. [CrossRef] [PubMed]
76. Ran, F.A.; Hsu, P.D.; Wright, J.; Agarwala, V.; Scott, D.A.; Zhang, F. Genome engineering using the CRISPR-Cas9 system. *Nat. Protoc.* **2013**, *8*, 2281–2308. [CrossRef]
77. Jia, H.-J.; Jia, P.-P.; Yin, S.; Bu, L.-K.; Yang, G.; Pei, D.-S. Engineering bacteriophages for enhanced host range and efficacy: Insights from bacteriophage-bacteria interactions. *Front. Microbiol.* **2023**, *14*, 1172635. [CrossRef]
78. Hidalgo-Cantabrana, C.; Barrangou, R. Characterization and applications of Type I CRISPR-Cas systems. *Biochem. Soc. Trans.* **2020**, *48*, 15–23. [CrossRef]

79. Gencay, Y.E.; Jasinskytė, D.; Robert, C.; Semsey, S.; Martínez, V.; Petersen, A.Ø.; Brunner, K.; de Santiago Torio, A.; Salazar, A.; Turcu, I.C. Engineered phage with antibacterial CRISPR–Cas selectively reduce E. coli burden in mice. *Nat. Biotechnol.* **2024**, *42*, 265–274. [CrossRef]
80. Chylinski, K.; Makarova, K.S.; Charpentier, E.; Koonin, E.V. Classification and evolution of type II CRISPR-Cas systems. *Nucleic Acids Res.* **2014**, *42*, 6091–6105. [CrossRef]
81. Agha, A.S.A.; Al-Samydai, A.; Aburjai, T. New frontiers in CRISPR: Addressing antimicrobial resistance with Cas9, Cas12, Cas13, and Cas14. *Heliyon* **2025**, *11*, e42013. [CrossRef]
82. Hosen, A.; Nishat, M.N.H.; Soaib, M.M.H.; Sharkar, O.S.; Sahabuddin, M.; Sharif, I.H.; Bhajan, S.K. A Review: CRISPR Cas System and the Mechanism With an Inhibition of Binding of CRISPR Cas—9. *Nano Sel.* **2025**, *6*, e202400009. [CrossRef]
83. Tsubota, T.; Takasu, Y.; Yonemura, N.; Sezutsu, H. Enhancements of the CRISPR-Cas System in the Silkworm Bombyx mori. *Cris. J.* **2025**, *8*, 155–164. [CrossRef]
84. Park, J.Y.; Moon, B.Y.; Park, J.W.; Thornton, J.A.; Park, Y.H.; Seo, K.S. Genetic engineering of a temperate phage-based delivery system for CRISPR/Cas9 antimicrobials against Staphylococcus aureus. *Sci. Rep.* **2017**, *7*, 44929. [CrossRef]
85. Ranveer, S.A.; Dasriya, V.; Ahmad, M.F.; Dhillon, H.S.; Samtiya, M.; Shama, E.; Anand, T.; Dhewa, T.; Chaudhary, V.; Chaudhary, P. Positive and negative aspects of bacteriophages and their immense role in the food chain. *NPJ Sci. Food* **2024**, *8*, 1. [CrossRef]
86. Wagh, R.V.; Priyadarshi, R.; Rhim, J.-W. Novel bacteriophage-based food packaging: An innovative food safety approach. *Coatings* **2023**, *13*, 609. [CrossRef]
87. Colom, J.; Cano-Sarabia, M.; Otero, J.; Cortés, P.; Maspoch, D.; Llagostera, M. Liposome-encapsulated bacteriophages for enhanced oral phage therapy against *Salmonella* spp. *Appl. Environ. Microbiol.* **2015**, *81*, 4841–4849. [CrossRef] [PubMed]
88. Yang, Y.; Du, H.; Zou, G.; Song, Z.; Zhou, Y.; Li, H.; Tan, C.; Chen, H.; Fischetti, V.A.; Li, J. Encapsulation and delivery of phage as a novel method for gut flora manipulation in situ: A review. *J. Control. Release* **2023**, *353*, 634–649. [CrossRef] [PubMed]
89. Koçer Alaşalvar, G.; Yıldırım, Z.J. Encapsulation of SE—P47 phage specific to Salmonella Enteritidis and evaluation of its stability. *J. Food Saf.* **2024**, *44*, e13103. [CrossRef]
90. Selim, H.M.R.M.; Gomaa, F.A.M.; Alshahrani, M.Y.; Morgan, R.N.; Aboshanab, K.M. Phage therapeutic delivery methods and clinical trials for combating clinically relevant pathogens. *Ther. Deliv.* **2025**, *16*, 247–269. [CrossRef]
91. Choi, I.; Lee, J.-S.; Han, J. Maltodextrin-trehalose miscible system-based bacteriophage encapsulation: Studies of plasticizing effect on encapsulated phage activity and food application as an antimicrobial agent. *Food Control* **2023**, *146*, 109550. [CrossRef]
92. Guo, P.; Zhang, L.; Ning, M.; Cai, T.; Long, F.; Yuan, Y.; Yue, T. Characterization and release of casein—Sodium alginate embedding phage edible film and the application in controlling of Salmonella contamination in food. *Int. J. Food Microbiol.* **2025**, *434*, 111137. [CrossRef]
93. Simadibrata, D.M.; Syam, A.F.; Lee, Y.Y. A comparison of efficacy and safety of potassium—Competitive acid blocker and proton pump inhibitor in gastric acid—Related diseases: A systematic review and meta—Analysis. *J. Gastroenterol. Hepatol.* **2022**, *37*, 2217–2228. [CrossRef]
94. Javeedvali, S.; Gopalakrishnan, C.; Kannan, R.; Manonmani, S.; Prasanthrajan, M.; Varanavasiappan, S. Strategies for sustainable rice bacterial leaf blight management: A holistic approach through phage biocontrol and nanoparticle encapsulation. *J. Plant Pathol.* **2024**, 1–12. [CrossRef]
95. Xu, Y.; Yang, T.; Miao, Y.; Zhang, Q.; Yang, M.; Mao, C. Injectable Phage-Loaded Microparticles Effectively Release Phages to Kill Methicillin-Resistant Staphylococcus aureus. *ACS Appl. Mater. Interfaces* **2024**, *16*, 17232–17241. [CrossRef] [PubMed]
96. Elsayed, M.M.; Elkenany, R.M.; El-Khateeb, A.Y.; Nabil, N.M.; Tawakol, M.M.; Hassan, H.M. Isolation and encapsulation of bacteriophage with chitosan nanoparticles for biocontrol of multidrug-resistant methicillin-resistant Staphylococcus aureus isolated from broiler poultry farms. *Sci. Rep.* **2024**, *14*, 4702. [CrossRef] [PubMed]
97. Temsaah, H.R.; Abdelkader, K.; Ahmed, A.E.; Elgiddawy, N.; Eldin, Z.E.; Elshebrawy, H.A.; Kasem, N.G.; El-Gohary, F.A.; Azmy, A.F. Chitosan nano-formulation enhances stability and bactericidal activity of the lytic phage HK6. *BMC Biotechnol.* **2025**, *25*, 3. [CrossRef] [PubMed]
98. Karthika, C.; Malligarjunan, N.; Pandian, S.K.; Gowrishankar, S. Chitosan-encapsulated bacteriophage cocktail as promising oral delivery system to surpass gastrointestinal infection caused by Klebsiella aerogenes. *Int. J. Biol. Macromol.* **2025**, *292*, 139236. [CrossRef]
99. Salatin, S.; Yari Khosroushahi, A. Overviews on the cellular uptake mechanism of polysaccharide colloidal nanoparticles. *J. Cell Mol. Med.* **2017**, *21*, 1668–1686. [CrossRef]
100. Li, J.; Ma, L.; Wang, C.; Jiang, P.; Cui, P.; Wang, J. Rationally designed oral DOX gels for colon-specific administration. *Gels* **2022**, *8*, 759. [CrossRef]
101. Guan, Z.; Feng, Q. Chitosan and chitooligosaccharide: The promising non-plant-derived prebiotics with multiple biological activities. *Int. J. Mol. Sci.* **2022**, *23*, 6761. [CrossRef]
102. Li, Y.; Li, X.-M.; Duan, H.-Y.; Yang, K.-D.; Ye, J.-F. Advances and optimization strategies in bacteriophage therapy for treating inflammatory bowel disease. *Front. Immunol.* **2024**, *15*, 1398652. [CrossRef] [PubMed]

103. Lin, S.; Xie, G.; He, J.; Meng, L.; Pang, Y.; Liu, J. Enhancing phage therapy by coating single bacteriophage-infected bacteria with polymer to preserve phage vitality. *Nat. Biomed. Eng.* **2025**, *9*, 1–17. [CrossRef] [PubMed]
104. Chen, H.; Zhi, W.; Bai, B.; Anjum, F.R.; Jia, Z.; Kong, R.; Liu, Q.; Wang, B.; Ma, C.; Ma, D. Impact of sodium alginate hydrogel containing bacteriophage peptides that specifically bind to the EtCab protein on the inhibition of Eimeria tenella infection. *Vet. Res.* **2025**, *56*, 18. [CrossRef] [PubMed]
105. Dlamini, S.B.; Gigante, A.M.; Hooton, S.P.; Atterbury, R.J. Efficacy of different encapsulation techniques on the viability and stability of diverse phage under simulated gastric conditions. *Microorganisms* **2023**, *11*, 2389. [CrossRef]
106. Truong-Le, Q.A.; Lee, S.O.; Ubeyitogullari, A. Encapsulation of Bifidobacterium bifidum into a pH-sensitive alginate-pectin gel system using 3D food printing: Enhanced viability and targeted release. *Int. J. Biol. Macromol.* **2025**, *318*, 145134. [CrossRef]
107. Guadarrama-Escobar, O.R.; Sánchez-Vázquez, I.; Serrano-Castañeda, P.; Chamorro-Cevallos, G.A.; Rodríguez-Cruz, I.M.; Sánchez-Padrón, A.Y.; Anguiano-Almazán, E.; Peña-Juárez, M.C.; Méndez-Albores, A.; Domínguez-Delgado, C.L.; et al. Development, Characterization, Optimization, and In Vivo Evaluation of Methacrylic Acid-Ethyl Acrylate Copolymer Nanoparticles Loaded with Glibenclamide in Diabetic Rats for Oral Administration. *Pharmaceutics* **2021**, *13*, 2023. [CrossRef]
108. Li, N.; Zhang, G.; Zhang, X.; Liu, Y.; Kong, Y.; Wang, M.; Ren, X. A rapid-floating natural polysaccharide gel-raft with double-effect for the treatment of gastroesophageal reflux disease. *Int. J. Biol. Macromol.* **2024**, *261*, 129667. [CrossRef]
109. Brinck, J.E.; Sinha, A.K.; Laursen, M.F.; Dragsted, L.O.; Raes, J.; Uribe, R.V.; Walter, J.; Roager, H.M.; Licht, T.R. Intestinal pH: A major driver of human gut microbiota composition and metabolism. *Nat. Rev. Gastroenterol. Hepatol.* **2025**, *22*, 639–656. [CrossRef]
110. Mukhopadhyay, S.; To, K.K.; Liu, Y.; Bai, C.; Leung, S.S. A thermosensitive hydrogel formulation of phage and colistin combination for the management of multidrug-resistant Acinetobacter baumannii wound infections. *Biomater. Sci.* **2024**, *12*, 151–163. [CrossRef]
111. Abed, S.; Beig, M.; Barzi, S.M.; Shafiei, M.; Hashemi Shahraki, A.; Sadeghi, S.; Sohrabi, A. Development of phage-containing hydrogel for treating Enterococcus faecalis-infected wounds. *PLoS ONE* **2024**, *19*, e0312469.
112. Khazani Asforooshani, M.; Elikaei, A.; Abed, S.; Shafiei, M.; Barzi, S.M.; Solgi, H.; Badmasti, F.; Sohrabi, A. A novel Enterococcus faecium phage EF-M80: Unveiling the effects of hydrogel-encapsulated phage on wound infection healing. *Front. Microbiol.* **2024**, *15*, 1416971. [CrossRef]
113. Peng, X.; Wang, T.; Dai, B.; Zhu, Y.; Ji, M.; Yang, P.; Zhang, J.; Liu, W.; Miao, Y.; Liu, Y.; et al. Gene Therapy for Inflammatory Cascade in Intrauterine Injury with Engineered Extracellular Vesicles Hybrid Snail Mucus-enhanced Adhesive Hydrogels. *Adv. Sci.* **2025**, *12*, e2410769. [CrossRef]
114. An, D.; Wang, Z.; Ning, Y.; Yue, Y.; Xuan, H.; Hu, Y.; Yang, M.; Zhou, H.; Liu, Q.; Wang, X.; et al. One-Step Physical and Chemical Dual-Reinforcement with Hydrophobic Drug Delivery in Gelatin Hydrogels for Antibacterial Wound Healing. *ACS Omega* **2024**, *9*, 34413–34427. [CrossRef]
115. Shakeri, A.; Najm, L.; Khan, S.; Tian, L.; Ladouceur, L.; Sidhu, H.; Al-Jabouri, N.; Hosseinidoust, Z.; Didar, T.F. Noncontact 3D Bioprinting of Proteinaceous Microarrays for Highly Sensitive Immunofluorescence Detection within Clinical Samples. *ACS Nano* **2024**, *18*, 31506–31523. [CrossRef]
116. Dehnad, D.; Emadzadeh, B.; Ghorani, B.; Rajabzadeh, G.; Tucker, N.; Jafari, S.M. Bioactive-loaded nanovesicles embedded within electrospun plant protein nanofibers; a double encapsulation technique. *Food Hydrocoll.* **2023**, *141*, 108683. [CrossRef]
117. Xu, C.; Ma, J.; Liu, Z.; Wang, W.; Liu, X.; Qian, S.; Chen, L.; Gu, L.; Sun, C.; Hou, J. Preparation of shell-core fiber-encapsulated Lactobacillus rhamnosus 1.0320 using coaxial electrospinning. *Food Chem.* **2023**, *402*, 134253. [CrossRef]
118. Ju, T.; Li, J.; Weston, A.; Satta, G.; Bolognini, S.; Di Luca, M.; Gaisford, S.; Williams, G.R. Anti—Pseudomonas Aeruginosa Bacteriophage Loaded Electrospun Fibers for Antibacterial Wound Dressings. *Macromol. Rapid Commun.* **2025**, *13*, 2400744. [CrossRef] [PubMed]
119. Murray, C.J.; Ikuta, K.S.; Sharara, F.; Swetschinski, L.; Aguilar, G.R.; Gray, A.; Han, C.; Bisignano, C.; Rao, P.; Wool, E. Global burden of bacterial antimicrobial resistance in 2019: A systematic analysis. *Lancet* **2022**, *399*, 629–655. [CrossRef] [PubMed]
120. Khosravi, A.; Chen, Q.; Echterhof, A.; Koff, J.L.; Bollyky, P.L. Phage therapy for respiratory infections: Opportunities and challenges. *Lung* **2024**, *202*, 223–232. [CrossRef] [PubMed]
121. Baldelli, A.; Liang, M. Design of respirable sprayed microparticles of encapsulated bacteriophages. *Front. Drug Deliv.* **2023**, *3*, 1209534. [CrossRef]
122. Sawant, S.S.; Ahmed, M.U.; Gantala, N.-G.; Chiu, C.; Qu, L.; Zhou, Q. Development of Inhalable Bacteriophage Liposomes Against Pseudomonas aeruginosa. *Pharmaceutics* **2025**, *17*, 405. [CrossRef]
123. Liu, M.-Y.; Liu, X.; Wang, C.-Y.; Wan, Q.-Q.; Tian, Y.-F.; Liu, S.-L.; Pang, D.-W.; Wang, Z.-G. Inhalable polymeric microparticles for phage and photothermal synergistic therapy of Methicillin-Resistant Staphylococcus aureus pneumonia. *Nano Lett.* **2024**, *24*, 8752–8762. [CrossRef]
124. Li, J.; Zheng, H.; Leung, S.S.Y. Investigating the effectiveness of liposome-bacteriophage nanocomplex in killing Staphylococcus aureus using epithelial cell coculture models. *Int. J. Pharm.* **2024**, *657*, 124146. [CrossRef] [PubMed]
125. Cao, Y.; Wang, Y.; Li, M.; Khanal, D.; Chan, H.K. Can bacteriophage be stabilised by lipid encapsulation when nebulised for inhalation delivery against Pseudomonas aeruginosa? *Int. J. Pharm.* **2025**, *678*, 125670. [CrossRef] [PubMed]

126. Liu, S.; Quek, S.Y.; Huang, K. An Ecofriendly Nature—Inspired Microcarrier for Enhancing Delivery, Stability, and Biocidal Efficacy of Phage—Based Biopesticides. *Small* **2024**, *20*, 2403465. [CrossRef] [PubMed]
127. Yaacob, S.; Jamil, R.Z.R.; Suah, F.B.M. Sporopollenin based materials as a versatile choice for the detoxification of environmental pollutants—A review. *Int. J. Biol. Macromol.* **2022**, *207*, 990–1004. [CrossRef]
128. Singh, L.S. Nano-emulsion encapsulation for the efficient delivery of bacteriophage therapeutics. *Biologicals* **2024**, *85*, 101725.
129. Wdowiak, M.; Raza, S.; Grotek, M.; Zbonikowski, R.; Nowakowska, J.; Doligalska, M.; Cai, N.; Luo, Z.; Paczesny, J. Phage-Nanoparticle Cocktails as a Novel Antibacterial Approach: Synergistic Effects of Bacteriophages and Green-Synthesized Silver Nanoparticles. *BioRxiv* **2025**. [CrossRef]
130. Easwaran, M.; Raja, N.; Saravanan, M.; Belete, M.A. Liposome-loaded phage cocktail: A promising therapeutic option against postsurgical wound infections–A critical appraisal. *Ann. Med. Surg.* **2024**, *86*, 4319–4321. [CrossRef]
131. Zhang, B.; Wang, Y.; Wang, F.; Zhang, Y.; Hao, H.; Lv, X.; Hao, L.; Shi, Y. Microencapsulated phage composites with increased gastrointestinal stability for the oral treatment of Salmonella colonization in chicken. *Front. Vet. Sci.* **2022**, *9*, 1101872. [CrossRef]
132. Jamaledin, R.; Sartorius, R.; Di Natale, C.; Onesto, V.; Manco, R.; Mollo, V.; Vecchione, R.; De Berardinis, P.; Netti, P.A. PLGA microparticle formulations for tunable delivery of a nano-engineered filamentous bacteriophage-based vaccine: In vitro and in silico-supported approach. *J. Nanostruct. Chem.* **2023**, *14*, 307–322. [CrossRef]
133. Tao, S.; Hu, A.; Bavel, E.; Yu, C.H.; Zhang, S.; Kissling, V.M.; Li, Z.; Moriaty, T.F.; Maniura-Weber, K.; Ren, Q. Bio—Responsive Hydrogel for Targeted on—Demand Release of a Phage Cocktail for Treatment of Pseudomonas aeruginosa Infection. *Adv. Funct. Mater.* **2025**, e09360. [CrossRef]
134. Youssef, R.A.; Sakr, M.M.; Shebl, R.I.; Aboshanab, K.M. Recent insights on challenges encountered with phage therapy against gastrointestinal-associated infections. *Gut Pathog.* **2025**, *17*, 60. [CrossRef] [PubMed]
135. Taati Moghadam, M.; Mohebi, S.; Sheikhi, R.; Hasannejad-Bibalan, M.; Shahbazi, S.; Nemati, S. Phage and Endolysin Therapy Against Antibiotics Resistant Bacteria: From Bench to Bedside. *MedComm* **2025**, *6*, e70280. [CrossRef] [PubMed]
136. Mohammadi, E.; Rahimian, M.; Panahi, B. Bridging the gap: Phage manufacturing processes from laboratory to agri-food industry. *Virus Res.* **2025**, *353*, 199537. [CrossRef]
137. Fatima, R.; Hynes, A.P. Phage-Antibiotic Combinations for Pseudomonas: Successes in the Clinic and In Vitro Tenuously Connected. *Microb. Biotechnol.* **2025**, *18*, e70193. [CrossRef]
138. Banicod, R.J.S.; Javaid, A.; Tabassum, N.; Jo, D.M.; Hassan, M.I.; Kim, Y.M.; Khan, F. Marine Bacteriophages as Next-Generation Therapeutics: Insights into Antimicrobial Potential and Application. *Viruses* **2025**, *17*, 971. [CrossRef]
139. Nang, S.C.; Lin, Y.-W.; Fabijan, A.P.; Chang, R.Y.; Rao, G.G.; Iredell, J.; Chan, H.-K.; Li, J. Pharmacokinetics/pharmacodynamics of phage therapy: A major hurdle to clinical translation. *Clin. Microbiol. Infect.* **2023**, *29*, 702–709. [CrossRef]
140. Akturk, E.; Pinto, G.; Ostyn, L.; Crabbé, A.; Melo, L.D.R.; Azeredo, J.; Coenye, T. Combination of phages and antibiotics with enhanced killing efficacy against dual-species bacterial communities in a three-dimensional lung epithelial model. *Biofilm* **2025**, *9*, 100245. [CrossRef]
141. Akturk, E.; Melo, L.D.R.; Oliveira, H.; Crabbé, A.; Coenye, T.; Azeredo, J. Combining phages and antibiotic to enhance antibiofilm efficacy against an in vitro dual species wound biofilm. *Biofilm* **2023**, *6*, 100147. [CrossRef]
142. Shambharkar, A.; Thompson, T.P.; McClenaghan, L.A.; Bourke, P.; Gilmore, B.F.; Skvortsov, T. Plasma activated water pre-treatment substantially enhances phage activity against Proteus mirabilis biofilms. *Biofilm* **2024**, *8*, 100230. [CrossRef] [PubMed]
143. Sokolov, I. Ultrabright fluorescent particles via physical encapsulation of fluorescent dyes in mesoporous silica: A mini-review. *Nanoscale* **2024**, *16*, 10994–11004. [CrossRef] [PubMed]
144. Zhang, X.; Liu, Y.; Wang, Z.; Zhao, H.; Zhan, L.; Gui, H.; Xu, X.; Ma, X.; Ma, B. pH-responsive and self-adaptive injectable sodium alginate/carboxymethyl chitosan hydrogel accelerates infected wound healing by bacteriostasis and immunomodulation. *Carbohydr. Polym.* **2025**, *354*, 123322. [CrossRef] [PubMed]
145. Xue, J.; Yao, Y.; Wang, M.; Wang, Z.; Xue, Y.; Li, B.; Ma, Y.; Shen, Y.; Wu, H. Recent studies on proteins and polysaccharides-based pH-responsive fluorescent materials. *Int. J. Biol. Macromol.* **2024**, *260*, 129534. [CrossRef]
146. Yadav, P.; Gupta, M.; Singh, S.P.; Parashar, P. Biocompatible Electrospun Hydrogel Fibers for Advanced Wound Healing Therapies. *Curr. Pharm. Des.* **2024**, *30*, 3240–3254. [CrossRef]
147. Gordon, A.; Li, B.; Witten, J.; Nguyen, H.; Anderson, D.G. Inhalable Dry Powders for Lung mRNA Delivery. *Adv. Healthc. Mater.* **2024**, *13*, e2400509. [CrossRef]
148. Naureen, F.; Shah, Y.; Rehman, M.U.; Fazli Nasir, F.N.; Pirzada, A.S.; Al-Otaibi, J.S.; Daglia, M.; Khan, H. Innovative inhalable dry powder: Nanoparticles loaded with Crizotinib for targeted lung cancer therapy. *BMC Cancer* **2025**, *25*, 1526. [CrossRef]
149. Rodriguez-Gonzalez, R.A.; Leung, C.Y.; Chan, B.K.; Turner, P.E.; Weitz, J.S. Quantitative Models of Phage-Antibiotic Combination Therapy. *MSystems* **2020**, *5*, 10–1128. [CrossRef]
150. Qureshi, K.A.; Parvez, A.; Fahmy, N.A.; Abdel Hady, B.H.; Kumar, S.; Ganguly, A.; Atiya, A.; Elhassan, G.O.; Alfadly, S.O.; Parkkila, S.; et al. Brucellosis: Epidemiology, pathogenesis, diagnosis and treatment-a comprehensive review. *Ann. Med.* **2023**, *55*, 2295398. [CrossRef]

151. Shanahan, F.; Ghosh, T.S.; O'Toole, P.W. The Healthy Microbiome-What Is the Definition of a Healthy Gut Microbiome? *Gastroenterology* **2021**, *160*, 483–494. [CrossRef]
152. Cao, Z.; Sugimura, N.; Burgermeister, E.; Ebert, M.P.; Zuo, T.; Lan, P. The gut virome: A new microbiome component in health and disease. *EBioMedicine* **2022**, *81*, 104113. [CrossRef]
153. Ruzzi, F.; Semprini, M.S.; Scalambra, L.; Palladini, A.; Angelicola, S.; Cappello, C.; Pittino, O.M.; Nanni, P.; Lollini, P.L. Virus-like Particle (VLP) Vaccines for Cancer Immunotherapy. *Int. J. Mol. Sci.* **2023**, *24*, 12963. [CrossRef]
154. Yue, H.; Li, Y.; Yang, T.; Wang, Y.; Bao, Q.; Xu, Y.; Liu, X.; Miao, Y.; Yang, M.; Mao, C. Filamentous phages as tumour-targeting immunotherapeutic bionanofibres. *Nat. Nanotechnol.* **2025**, *20*, 167–176. [CrossRef]
155. Petrov, G.; Dymova, M.; Richter, V. Bacteriophage-Mediated Cancer Gene Therapy. *Int. J. Mol. Sci.* **2022**, *23*, 14245. [CrossRef] [PubMed]
156. Wang, S.W.; Gao, C.; Zheng, Y.M.; Yi, L.; Lu, J.C.; Huang, X.Y.; Cai, J.B.; Zhang, P.F.; Cui, Y.H.; Ke, A.W. Current applications and future perspective of CRISPR/Cas9 gene editing in cancer. *Mol. Cancer* **2022**, *21*, 57. [CrossRef] [PubMed]
157. Suh, G.A.; Lodise, T.P.; Tamma, P.D.; Knisely, J.M.; Alexander, J.; Aslam, S.; Barton, K.D.; Bizzell, E.; Totten, K.M.C.; Campbell, J.L.; et al. Considerations for the Use of Phage Therapy in Clinical Practice. *Antimicrob. Agents Chemother.* **2022**, *66*, e0207121. [CrossRef] [PubMed]
158. Dy, R.L.; Rigano, L.A.; Fineran, P.C. Phage-based biocontrol strategies and their application in agriculture and aquaculture. *Biochem. Soc. Trans.* **2018**, *46*, 1605–1613. [CrossRef]
159. Miller, M.; Deiulio, A.; Holland, C.; Douthitt, C.; McMahon, J.; Wiersma-Koch, H.; Turechek, W.W.; D'Elia, T. Complete genome sequence of Xanthomonas phage RiverRider, a novel N4-like bacteriophage that infects the strawberry pathogen Xanthomonas fragariae. *Arch. Virol.* **2020**, *165*, 1481–1484. [CrossRef]
160. Vique, G.; Mendoza-Barberá, E.; Ramos-Barbero, M.D.; Blanco-Picazo, P.; Sala-Comorera, L.; Quirós, P.; Atares, S.; Salaet, I.; Muniesa, M.; Rodríguez-Rubio, L. Efficacy of Erwinia amylovora and Xanthomonas campestris pv campestris phages to control fire blight and black rot in vivo. *Microbiol. Spectr.* **2025**, *13*, e0028025. [CrossRef]

Journal of
Clinical Medicine

MDPI

Article

Linezolid-Associated Thrombocytopenia: Assessment of Risk Factors in Patients without Hemato-Oncologic Diseases

Abrar K. Thabit [1,*], Arwa A. Alghamdi [2], Afnan K. Alsaeed [2], Nesereen M. Magbool [2], Yazed S. Alsowaida [3], Ahmad J. Mahrous [4], Alya Alruwaili [5], Ziyad K. Albakistani [6], Basem O. Albangali [6], Anas M. Alghumuy [6], Sara A. Youssef [7], Reem M. Alodayli [7] and Masaad Saeed Almutairi [8]

1 Pharmacy Practice Department, Faculty of Pharmacy, King Abdulaziz University, Jeddah 22254-2265, Saudi Arabia
2 Faculty of Pharmacy, King Abdulaziz University, Jeddah 22254-2265, Saudi Arabia
3 Department of Clinical Pharmacy, College of Pharmacy, University of Ha'il, Hail 55473, Saudi Arabia
4 Clinical Pharmacy Department, College of Pharmacy, Umm Al Qura University, Makkah 21955, Saudi Arabia
5 Department of Pharmaceutical Care, King Fahad Medical City, Riyadh 12231, Saudi Arabia
6 College of Pharmacy, Umm Al Qura University, Makkah 21955, Saudi Arabia
7 Department of Pharmaceutical Care, Saudi German Hospital, Hail 55481, Saudi Arabia
8 Department of Pharmacy Practice, College of Pharmacy, Qassim University, Qassim 51452, Saudi Arabia
* Correspondence: akthabit@kau.edu.sa

Abstract: Background: Linezolid is used for Gram-positive bacterial infections. Thrombocytopenia is one of its main adverse effects resulting from myelosuppression. Several studies have assessed risk factors that may increase the risk of this adverse effect. However, most studies included patients with hemato-oncologic diseases, which may confound such assessments. This study aimed to investigate risk factors for linezolid-associated thrombocytopenia in patients without hemato-oncologic diseases. **Methods**: This was a multicenter retrospective case-control study of adult patients treated with linezolid twice daily for ≥3 days. Patients with hemato-oncologic diseases, active dengue fever, active COVID-19, baseline platelet count <100 × 10^3/mm^3, concurrent therapy with trimethoprim/sulfamethoxazole or valproic acid, and a recent platelet transfusion within 7 days were excluded. Thrombocytopenia was defined as a drop in platelet count below 100 × 10^3/mm^3. **Results**: Out of 158 evaluated patients, 33 developed thrombocytopenia, indicating an incidence rate of 20.9%. Of all the risk factors assessed, creatinine clearance of <60 mL/min and bacteremia/infective endocarditis were significantly associated with linezolid-associated thrombocytopenia (adjusted odds ratios, 3.25 and 5.95; 95% CI 1.12–9.45 and 1.23–28.66; $p = 0.031$ and 0.026, respectively). End of therapy platelet counts were significantly lower in the cases than in the controls (79 vs. 243 × 10^3/mm^3; $p < 0.001$). Similarly, the percentage of platelet count change was significantly different (−55.1% vs. −10.2%; $p < 0.001$). **Conclusions**: In our study, the incidence rate of linezolid-associated thrombocytopenia was 20.9%, and we found that patients with renal impairment and bacteremia may need close monitoring of platelet counts. Prospective studies are warranted to evaluate the potential need for renal dose adjustment.

Keywords: linezolid; thrombocytopenia; bacteremia; creatinine; creatinine clearance; risk factor; renal insufficiency

1. Introduction

Linezolid is the first member of the oxazolidinone class of antibiotics [1]. It exhibits its activity by inhibiting the protein synthesis of Gram-positive cocci, including *Staphylococcus aureus* (both methicillin-resistant and susceptible), *Streptococci*, *Enterococci* (including vancomycin-resistant strains), and *Listeria monocytogenes* [2]. Common adverse effects associated with linezolid use are thrombocytopenia, peripheral and ocular neuropathy, anemia, and lactic acidosis [1]. Thrombocytopenia is a consequence of myelosuppression, usually occurring after long-term linezolid treatment [3]. In the literature, the incidence of thrombocytopenia varied between 16.7% and 60.5% [3–8]. Such adverse effects can lead to treatment discontinuation due to the increased risk of bleeding [9]. The mechanisms of linezolid-associated thrombocytopenia are not clearly identified.

A considerable body of literature has studied the incidence and risk factors of linezolid-associated thrombocytopenia. Most of these studies were limited to patients with hemato-oncologic diseases, which may confound the assessment of thrombocytopenia. Receiving linezolid therapy for more than 14 days, low pretreatment platelet counts, renal insufficiency, chronic liver disorder, respiratory tract infections, and low body weight have all been reported as potential risk factors in these patients [3,10–13]. Moreover, intensive care unit patients who developed thrombocytopenia after treatment with linezolid had remarkably higher mortality [14]. One study evaluated the incidence and clinical characteristics of linezolid-associated thrombocytopenia in 19 patients treated with linezolid. Overall, thrombocytopenia (platelet count < 100,000 platelets/mm^3) was observed in 32% of patients who received linezolid for more than 10 days [5]. Other studies found that linezolid treatment duration, renal function, and hepatic function play a key role in linezolid-associated thrombocytopenia [6,8,11].

To date, only a few studies have investigated factors that may increase the risk of linezolid-associated thrombocytopenia in patients without hemato-oncologic diseases. These studies have established a variety of risk factors, such as linezolid therapy for ≥7 days, baseline platelet count < 150 × 10^3/mm^3, creatinine clearance (CrCl) < 30 mL/min, concurrent low-dose aspirin therapy, a higher daily per kg dose, and a high white blood cell (WBC) count [15,16]. Nonetheless, no studies have assessed the potential effects of artificial heart valve surgeries and blood vessel grafts, and only one study reported the risk of concomitant aspirin therapy. It is important to ensure the safe use of linezolid by preventing its adverse effects through early identification of risk factors that may increase the incidence of linezolid-associated thrombocytopenia. Therefore, this study aimed to add to the current body of evidence and investigate risk factors for linezolid-associated thrombocytopenia in patients without hemato-oncologic diseases who received linezolid.

2. Methods

2.1. Study Design and Definitions

This was a multicenter retrospective case-control study of patients receiving linezolid therapy in four tertiary hospitals during the period from January 2013 to December 2022.

The primary endpoint was thrombocytopenia, which was defined as a decrease in platelet count to <100 × 10^3/mm^3 (i.e., less than approximately 75% of the lower limit of normal of 150 × 10^3/mm^3) [17]. Cases were defined as patients who developed thrombocytopenia during therapy with linezolid and up to 10 days post therapy, whereas controls were patients who did not develop thrombocytopenia despite exposure to linezolid. Several patient- and treatment-related factors were assessed for the potential augmentation of linezolid-associated thrombocytopenia.

2.2. Inclusion and Exclusion Criteria for Patients

Included patients were hospitalized adults (≥18 years) who received oral or intravenous (IV) linezolid at a dose of 1200 mg daily (600 mg twice daily) for at least 72 h. Patients having hemato-oncologic disease (including iron deficiency anemia, hemophilia, sickle cell disease, thalassemia, bone marrow transplant, leukemia, and lymphoma as well as cancer of other organs), active dengue fever, active coronavirus infection 2019 (COVID-19), baseline platelet count < 100 × 10^3/mm^3, concomitant therapy with trimethoprim/sulfamethoxazole and valproic acid, received platelet transfusion within 7 days prior to linezolid therapy, had no baseline or end of therapy laboratory platelet data (at least within 3 days of each time point) were excluded from the study. Patients with active dengue fever and COVID-19 were excluded due to the high rate of thrombocytopenia associated with these conditions [18,19].

2.3. Ethical Approval

The study protocol was approved by the Biomedical Ethics Committee of the Faculty of Medicine (reference no. 464-20), which waived the requirement for patient consent given the retrospective nature of the study.

2.4. Statistical Analysis

Categorical data were presented as numbers and percentages and compared using the chi-square test, whereas continuous data were presented as the median [interquartile range, IQR] and compared using the Mann–Whitney test. The Shapiro–Wilk test of normality showed the lack of a normal distribution for all continuous variables. Multivariable regression was performed for variables with a $p < 0.1$ in the univariate analysis to generate adjusted odds ratios (AOR). The omnibus test of model coefficients was used to determine the significance of the model, whereas the Hosmer–Lemeshow test was used to establish the goodness of fit of the model. A log-rank test was performed to assess the effect of these factors on the platelet count trend over time. A p value of <0.05 was used to determine statistical significance. SPSS version 24.0 software (IBM, Corp., Armonk, NY, USA) was used for analysis.

2.5. Sample Size Calculation

Based on two previous similar studies on patients without hemato-oncologic diseases that found a thrombocytopenia rate of 29.2% and 42%, respectively, a total of 145 patients were needed to meet a study power of 95% considering a difference in the incidence of thrombocytopenia (i.e., effect size) of 30% (rounded down from a mean of 35.6% to maximize the sample size) and an α-error probability of 5% [15]. This effect size was selected based on this study since it was the only study that was found in the literature that excluded patients who had hemato-oncologic diseases.

3. Results

3.1. Baseline Characteristics

A total of 158 patients met the eligibility criteria and were included in the study. Table 1 lists the characteristics of patients. The median age of the whole cohort was 62 years, with the majority being males (57.6%). Median [IQR] durations of linezolid therapy were 8 [6.5–10.5] vs. 7 [5–9.5] days (p = 0.344) in the cases and control groups, respectively. More than half of the patients were in the intensive care unit (58.2%), and many (70.3%) received linezolid intravenously. Notably, more than half of the patients (61.1%) had a creatinine clearance (CrCl) < 60 mL/min, with the majority being from the cases group (81.8% vs. 55.6%; p = 0.006).

Table 1. Baseline characteristics of the patients.

Characteristic	All Patients (n = 158)	Cases (n = 33)	Controls (n = 125)	p Value
Age, years	62 [47.5–74]	62 [57–76]	62 [42–73]	0.165
Sex, male	91 (57.6)	20 (60.6)	71 (56.8)	0.694
Body mass index (n = 156), kg/m^2	26 [23–30]	26 [24–31]	26 [23–30]	0.639
Location				0.479
Inpatient medical ward	66 (41.8)	12 (36.4)	54 (43.2)	
Intensive care unit	92 (58.2)	21 (63.6)	71 (56.8)	

Table 1. *Cont.*

Characteristic	All Patients (n = 158)	Cases (n = 33)	Controls (n = 125)	p Value
Route of linezolid administration				0.437
Oral	47 (29.7)	8 (24.2)	39 (31.2)	
Intravenous	111 (70.3)	25 (75.8)	86 (68.8)	
Duration of therapy, days	7.5 [5–10]	8 [6.5–10.5]	7 [5–9.5]	0.344
Duration of therapy				0.328
≤7 days	79 (50)	14 (42.4)	65 (52)	
>7 days	79 (50)	19 (57.6)	60 (48)	
History of cardiovascular disease	89 (56.3)	23 (69.7)	66 (52.8)	0.082
Having more than one cardiovascular disease	45 (50.6)	13 (56.5)	32 (48.5)	0.507
History of cardiovascular conditions				0.321
Cerebrovascular accident	40 (25.3)	8 (24.2)	32 (25.6)	
Myocardial infarction	36 (22.8)	10 (30.3)	26 (20.8)	
Arrhythmia	22 (13.9)	7 (21.2)	15 (12)	
Heart failure	16 (10.1)	5 (15.2)	11 (8.8)	
Deep venous thrombosis	12 (7.6)	4 (12.1)	8 (6.4)	
Angina	9 (5.7)	3 (9.1)	6 (4.8)	
Cardiomyopathy	4 (2.5)	1 (3)	3 (2.4)	
Pulmonary embolism	4 (2.5)	1 (3)	3 (2.4)	
Valvular heart disease	2 (1.3)	1 (3)	1 (0.8)	
Rheumatic heart disease	2 (1.3)	1 (3)	1 (0.8)	
History of cardiovascular surgery	13 (9.2)	3 (10)	10 (9)	0.868
Type of cardiovascular surgery				0.276
CABG	9 (5.7)	1 (3)	8 (6.4)	
Heart valve	4 (2.5)	2 (6.1)	2 (1.6)	
Creatinine clearance				0.006
≥60 mL/min	61 (38.9)	6 (18.2)	55 (44.4)	
<60 mL/min	96 (61.1)	27 (81.8)	69 (55.6)	
Indication for linezolid				
Pneumonia	53 (33.5)	5 (15.2)	48 (38.4)	0.012
Empiric	47 (39.7)	13 (39.4)	34 (27.2)	0.173
Soft and skin tissue infections	16 (10.1)	2 (6.1)	14 (11.2)	0.384
Urinary tract infection	14 (8.9)	4 (12.1)	10 (8)	0.459
Bacteremia/infective endocarditis	11 (7)	6 (18.2)	5 (4)	0.004
Intra-abdominal infection	7 (4.4)	1 (3)	6 (4.8)	0.660
Tuberculosis	5 (3.2)	2 (6.1)	3 (2.4)	0.285

Table 1. *Cont.*

Characteristic	All Patients (n = 158)	Cases (n = 33)	Controls (n = 125)	p Value
Central nervous system	3 (1.9)	0 (0)	3 (2.4)	0.369
Bone and joint infection	2 (1.3)	0 (0)	2 (1.6)	0.465
Use of antiplatelet drugs	60 (38)	16 (48.5)	44 (35.2)	0.162
Antiplatelet drug used *				0.244
Aspirin	42 (26.6)	10 (30.3)	12 (9.6)	
Clopidogrel	39 (24.7)	12 (36.4)	27 (21.6)	
Ticlopidine	1 (0.6)	0 (0)	1 (1.6)	
Previous vancomycin use within 14 days	46 (29.1)	14 (42.4)	32 (25.6)	0.058
Baseline blood laboratory data				
Platelets < 150 × $10^3/mm^3$	19 (12)	8 (24.2)	11 (8.8)	0.015
Low hemoglobin **	118 (77.1)	29 (93.5)	89 (73)	0.015
Leukocytosis	93 (58.9)	14 (42.4)	73 (63.2)	0.031

Data are presented as n (%) or median [interquartile range]; CABG, coronary artery bypass graft; * 18 patients received both aspirin and clopidogrel; ** The lower limit of normal is 14 g/dL for males and 12 g/dL for females.

3.2. Platelet Outcomes

Table 2 lists the platelet count outcomes in both groups. Overall, thrombocytopenia occurred in 33 patients, indicating an incidence rate of 20.9%. The median duration of thrombocytopenia was 8 days post-linezolid initiation. Figure 1 shows the trend of platelet counts over 14 days in the cases and the controls.

Table 2. Outcomes of platelet count in patients who developed and did not develop thrombocytopenia.

Outcome	All Patients (n = 158)	Cases (n = 33)	Controls (n = 125)	p Value
Time to thrombocytopenia, days	N/A	8 [5.5–11]	N/A	N/A
EOT platelet count, ×$10^3/mm^3$	213 [132–298]	79 [63.5–114]	243 [168.5–328.5]	<0.001
EOT change in platelet count, %	−15.4 [−46.6–17.2]	−55.1 [−76.2–−37.85]	−10.2 [−30–21.3]	<0.001

Data are presented as median [interquartile range]; EOT, end of therapy; N/A, not applicable.

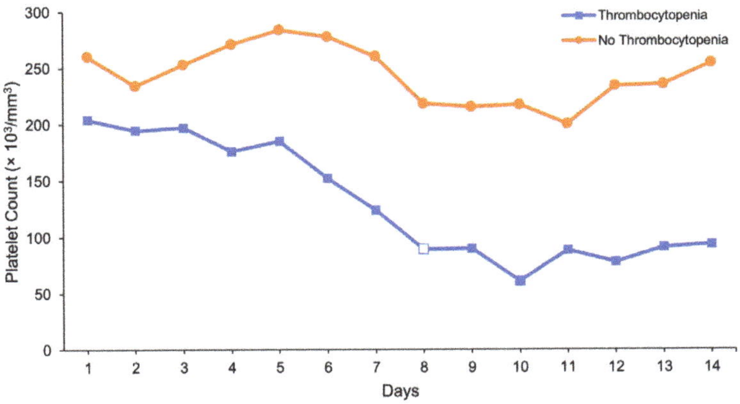

Figure 1. Trend of platelet counts over 14 days in the cases (thrombocytopenia; n = 33) and the controls (no thrombocytopenia; n = 125). The white square indicates the median duration of thrombocytopenia (platelet count < 100 × $10^3/mm^3$).

3.3. Factors Associated with Thrombocytopenia

In the univariate analysis, several factors were significantly associated with linezolid-associated thrombocytopenia, including low CrCl, previous vancomycin use, bacteremia, leukocytosis, and low hemoglobin (Table 3). However, only low CrCl (AOR, 3.25; 95% CI, 1.12–9.45, $p = 0.031$), bacteremia with or without infective endocarditis as linezolid indication (AOR, 5.95; 95% CI, 1.23–28.66, $p = 0.026$), and low baseline platelet count (AOR, 3.22; 95% CI, 1.01–10.26; $p = 0.048$) remained significantly associated with linezolid-associated thrombocytopenia in the multivariable regression analysis. Additionally, results of log-rank tests assessing the effect of all these factors on the platelet count trend over time showed no significant difference between patients who had the factors and those who did not.

Table 3. Factors associated with thrombocytopenia in patients on linezolid therapy.

Factor	Univariate Analysis			Multivariable Analysis		
	OR	95% CI	*p* Value	AOR	95% CI	*p* Value
Previous vancomycin use	0.47	0.21–1.04	0.058	1.07	0.40–2.88	0.893
History of cardiovascular disease	2.06	0.90–4.67	0.082	1.52	0.59–3.88	0.386
Bacteremia or infective endocarditis	0.19	0.05–0.66	0.004	5.95	1.23–28.66	0.026
Pneumonia	3.50	1.26–9.62	0.012	0.39	0.13–1.21	0.104
CrCl < 60 mL/min	3.59	1.38–9.3	0.006	3.25	1.12–9.45	0.031
Baseline platelets < 150 × 10^3/mm^3	0.30	0.11–0.83	0.015	3.22	1.01–10.26	0.048
Leukocytosis	2.33	1.07–5.08	0.031	0.66	0.26–1.64	0.369
Low hemoglobin	0.19	0.04–0.82	0.015	3.30	0.64–17.01	0.154

AOR, adjusted odds ratio; CrCl, creatinine clearance; OR, odds ratio.

4. Discussion

Linezolid is an important antibiotic for several infections, and it is widely used in clinical practice. In our study, we evaluated the incidence rate of linezolid-associated thrombocytopenia and identified the risk factors that may potentially augment this risk. In our study, 33 of 158 patients (20.9%) developed thrombocytopenia with linezolid therapy. We found that impaired kidney function, the presence of bacteremia with or without infective endocarditis, and a low baseline platelet count were significantly associated with thrombocytopenia in patients receiving linezolid for therapy.

The incidence rate of thrombocytopenia with linezolid therapy reported in the package insert is 2.4% [20]. Several studies have evaluated factors that may increase the risk of linezolid-associated thrombocytopenia. Choi et al. evaluated risk factors for linezolid-associated thrombocytopenia in 264 patients without a history of hemato-oncologic conditions [15]. They found a thrombocytopenia incidence of 29.2%, where a duration of linezolid therapy of 7 days or longer was significantly associated with thrombocytopenia (AORs were 7.25, 19.51, and 28.80; 95% CI, 1.92–27.38, 4.76–79.95, and 6.48–127.92 for 7–13 days, 14–20 days, and ≥21 days, respectively; $p < 0.01$ for all durations). Furthermore, they found that a baseline CrCl of <30 mL/min (AOR, 4.19; 95% CI, 1.59–11.06; $p = 0.004$), concomitant low-dose aspirin therapy (≤160 mg/day) (AOR 2.99; 95% CI, 1.26–7.08; $p = 0.013$), and a baseline platelet count of <150 × 10^3/mm^3 (AOR, 5.08; 95% CI, 2.06–12.55; $p < 0.001$) were significantly associated with thrombocytopenia associated with linezolid therapy. While our study also showed a correlation between low CrCl and low baseline platelet counts and the increased risk of linezolid-associated thrombocytopenia, our findings did not suggest a correlation with the use of any antiplatelet drug used by our patient population. Another retrospective study by Han, et al. evaluated 320 patients on linezolid therapy, where they found that thrombocytopenia (platelet count < 100 × 10^3/mm^3) was significantly associated with a low baseline platelet count (OR, 0.99; 95% CI, 0.98–0.99; $p < 0.001$), a long duration of linezolid therapy (OR, 3.46; 95% CI, 1.80–6.68; $p < 0.001$ for 7–13 days and OR, 2.51; 95% CI, 1.05–6.00; $p = 0.039$ for 14 days or longer), and shock (OR, 2.09; 95% CI, 1.13–3.88; $p = 0.019$) [6]. However, only a long duration of linezolid therapy (OR, 3.77; 95% CI, 2.21–6.41; $p < 0.001$ for 7–13 days and OR, 2.62; 95% CI, 1.26–5.43; $p = 0.010$ for 14 days or longer) and shock (OR, 1.91; 95% CI, 1.13–3.21; $p = 0.015$) were associated with a ≥25% drop in platelet count from baseline. The impact of a low baseline platelet count on linezolid-associated thrombocytopenia was also seen in our study; though, the duration of therapy (≤7 days vs. >7 days) did not impact such an incidence. Interestingly, low baseline platelet counts of $\leq 200 \times 10^3/\text{mm}^3$ also significantly increased the risk of linezolid-associated thrombocytopenia in a small study of 66 patients (AOR, 5.66; 95% CI, 1.15–27.9; $p = 0.045$) [7]. Similar to our findings, the study also found that a baseline CrCl < 60 mL/min and a baseline serum creatinine of >1.5 mg/dL were associated with a higher risk of linezolid-associated thrombocytopenia (AOR, 9.41; 95% CI, 1.09–80.54; $p = 0.043$ and AOR, 4.57; 95% CI, 1.26–16.5; $p = 0.035$, respectively). A therapy duration of >10 days was also seen as a factor associated with linezolid-associated thrombocytopenia in a case series of 19 patients, where 6 patients (32%) developed this adverse effect [5].

A potential explanation behind the increased risk of linezolid-associated thrombocytopenia in patients with renal impairment is that linezolid and its two major metabolites (aminoethoxyacetic acid metabolite and hydroxyethyl glycine) were found to accumulate in such patients, particularly in patients with end-stage renal disease [21]. However, the mechanism of thrombocytopenia associated with linezolid, including the potential effects of its two metabolites, is not yet fully understood.

In our study, we noticed that patients with bacteremia with or without infective endocarditis have an increased risk of thrombocytopenia in patients receiving linezolid for therapy (AOR, 5.95; 95% CI, 1.23–28.66; $p = 0.026$). This could be attributed to the bacterial effect on platelets. For instance, *S. aureus* is known to activate and aggregate platelets, as observed in previous in vitro studies [22,23]. Clinically, multiple studies found thrombocytopenia rates of 20–47.6% in patients with bacteremia due to various pathogens [24–26].

The median time to thrombocytopenia in our study was 8 days (IQR, 5.5–11). Several studies have addressed the time to thrombocytopenia after the initiation of linezolid. Giunio-Zorkin et al. assessed the new onset of thrombocytopenia caused by linezolid in a small study of 18 patients [27]. They found that the first thrombocytopenic values occurred within 14 days in 9 of 18 (50%) of the cases, resulting in a discontinuation rate of 61% (11/18), while one patient received a platelet transfusion. While this study did not exclude cancer patients as in the case of our study, they excluded patients with hematological disorders associated with platelet reduction, as well as patients who received bone marrow suppressive therapy within two weeks before linezolid initiation. Although the incidence of thrombocytopenia occurred later than in our study (8 days), newer and larger studies found that the incidence of thrombocytopenia occurred sooner, as in the study by Choi, et al., who found it happening after a median of 11.2 days [15]. In a small prospective study of 31 patients, Nukui, et al. found that the median time to thrombocytopenia was 11 days after the initiation of linezolid treatment, which occurred in 17 patients, indicating an incidence rate of 56.7% [28]. In their study, the authors measured the plasma concentration of linezolid and assessed the kidney function of their patients. They found that patients who developed thrombocytopenia had significantly higher trough concentrations of linezolid, particularly in patients with baseline CrCl < 60 mL/min. Regardless of kidney function, median linezolid trough concentrations in patients who developed thrombocytopenia vs. those who did not develop it were 13.4 mg/L vs. 4.3 mg/L on day 3, 15.3 mg/L vs. 3.8 mg/L, and 15.2 mg/L vs. 5 mg/L on days 3, 7, and 14, respectively ($p < 0.0001$). When divided based on CrCl value, median linezolid trough concentrations were 14.7 mg/L vs. 4.8 mg/L on day 3, 16.3 mg/L vs. 4.9 mg/L, and 13.6 mg/L vs. 7.3 mg/L on days 3, 7, and 14, respectively, in patients with CrCl < 60 mL/min vs. those with higher CrCl values

($p < 0.0001$). As such, the authors found that the strongest predictors of linezolid-associated thrombocytopenia were a CrCl < 60 mL/min (OR, 39; 95% CI, 3.8–399.8; $p = 0.0002$) and a linezolid trough concentration > 7.5 mg/L on day 3 (OR, 90; 95% CI, 7.3–1115.9; $p < 0.0001$). Findings from this study coincide with our findings that renal impairment at baseline can be a predictor of thrombocytopenia with linezolid, which warrants closer monitoring of platelet counts in patients with such a condition. When compared with its first alternative, vancomycin, linezolid was found to be associated with thrombocytopenia four times more than vancomycin (OR 4.39; 95% CI 2.38–8.08), as reported in a retrospective study of 453 patients by Al-Harbi, et al. [29]. In this study, the onset of severe thrombocytopenia (platelet count < $50 \times 10^3/\text{mm}^3$) occurred within an average of 14 ± 9.3 days of linezolid initiation. Although all these studies had a later onset of thrombocytopenia than in our study (8 days), our findings indicate that platelet monitoring should be started within a few days after treatment initiation.

Studies conducted on patients with hemato-oncologic diseases found that the most common factors associated with increased risk of linezolid-associated thrombocytopenia were poor kidney function (CrCl < 50 mL/min) ($p < 0.001$), duration of linezolid therapy of ≥14 days ($p < 0.001$), concomitant use of fluoroquinolones ($p = 0.009$), and respiratory tract infections ($p = 0.03$) [3]. In patients with a CrCl of < 50 mL/min, the time to the onset of thrombocytopenia was significantly shorter than in patients with a CrCl of >50 mL/min ($p = 0.04$) [3]. In our study, we excluded patients with hemato-oncologic conditions because these conditions are significantly associated with thrombocytopenia even without the presence of linezolid, which may confound our results. We also excluded patients with active dengue fever (which has been reported in our country) given the high rate of reported thrombocytopenia, reaching 90.1% [30]. Although one study found that parenteral administration of linezolid resulted in a higher rate of thrombocytopenia ($p = 0.04$) [3], no such difference was found in our study between oral and parenteral formulations, perhaps due to the high oral bioavailability of linezolid, which can reach approximately 100% [20].

In our study, we excluded patients with COVID-19 to match the inclusion criteria of a previous similar study by Han, et al. [6]. This was also because COVID-19 has been linked to thrombocytopenia of varying degrees, as reported in previous studies, case reports, and case series [19,31–38]. Thus, we wanted to exclude this condition as a potential confounding factor. Moreover, since our study spanned from before COVID-19 (2013–2019) until 3 years during/after COVID-19 (2020–2022), patients with COVID-19 would have been only included from the last 3 years of the study, which may have created an imbalance in the distribution of patients across the study period. Nonetheless, a study by Alkhalifa, et al. found that exposure to linezolid increased the likelihood of thrombocytopenia in patients with COVID-19 (chi-square $p < 0.05$); though, the authors did not mention the number of patients who received linezolid among the 256 included patients [39]. Therefore, it is suggested to assess the impact of a current or previous episode of COVID-19 on the incidence of thrombocytopenia in patients receiving linezolid.

Dengue fever is a viral infectious disease that impacts total blood cellular counts, including platelets. The World Health Organization guidelines included thrombocytopenia as one of the manifestations that correlate with the clinical severity of the disease [18]. A study by Castilho, et al. found a thrombocytopenia incidence rate of 40.3% in patients with laboratory-confirmed dengue fever [40]. The study did not assess the potential impact of concomitant drugs, including linezolid, on such incidence. However, the already high rate of thrombocytopenia associated with dengue fever may have made it a significant confounding factor if patients with the disease were to be included in the current study; therefore, we opted to exclude them. Nevertheless, evaluating the impact of both factors combined (i.e., linezolid and dengue fever) on the incidence of thrombocytopenia is worthy of investigation in a future study.

This study was perhaps limited by its sample size, despite the involvement of four medical centers. This can be attributed to the fact that linezolid was not available in our country until 2012 and that it is usually reserved for cases where other cheaper therapies (for example, vancomycin or trimethoprim/sulfamethoxazole) cannot be used, such as in cases of vancomycin resistance or drug-resistant tuberculosis. We also excluded several patients who matched the exclusion criteria to eliminate a potential confounding effect. As such, and given the observed incidence of linezolid-associated thrombocytopenia in our study, patients were included in a 1:3 ratio (i.e., one case for every three controls). Despite being a multicenter study, the limited sample size may limit the generalizability of the findings, especially for patients with conditions that were excluded from the study. Additionally, the retrospective design of the study may have introduced selection bias. Furthermore, to overcome information bias, we excluded patients with insufficient platelet data in their electronic medical records.

5. Conclusions

Linezolid represents an important therapeutic option for many kinds of infections, particularly those caused by drug-resistant Gram-positive bacteria or tuberculosis. The incidence rate of linezolid-associated thrombocytopenia in our study was 20.9%. Clinicians should exercise caution when using it in patients with impaired kidney function, bacteremia and/or infective endocarditis, and a low pre-treatment platelet count by close and frequent monitoring of the platelet count and the potential risk of bleeding. Alternatively, seeking another antibiotic to which the pathogen is susceptible may be more prudent for patients at risk while weighing the benefit vs. risk ratio. Future research is needed to address gaps in the knowledge of the mechanism behind such an adverse effect of linezolid and its metabolites, as well as the potential compounding effect of the excluded conditions, such as dengue fever and COVID-19.

Author Contributions: Conceptualization, methodology, formal analysis, supervision, writing—review and editing, A.K.T.; data curation, investigation, writing—original draft, A.A.A., A.K.A., N.M.M., Z.K.A., B.O.A., A.M.A., S.A.Y. and R.M.A.; supervision, writing—review and editing, A.J.M. and Y.S.A.; data curation, investigation, writing—review and editing, A.A.; writing—original draft, writing—review and editing, M.S.A. All authors have read and agreed to the published version of the manuscript.

Funding: This research received no external funding.

Institutional Review Board Statement: This study was approved by the Biomedical Ethics Committee of the Faculty of Medicine (reference no. 464-20; 8 September 2020).

Informed Consent Statement: Patient consent was waived due to the retrospective study design.

Data Availability Statement: The datasets generated and analyzed in the current study are available from the corresponding author at reasonable request.

Acknowledgments: This study was presented as a poster at the Infectious Diseases Society of America annual meeting, IDWeek, in Boston, MA, 2023 (abstract number 384).

Conflicts of Interest: The authors declare no conflicts of interest.

References

1. Hashemian, S.M.R.; Farhadi, T.; Ganjparvar, M. Linezolid: A review of its properties, function, and use in critical care. *Drug Des. Dev. Ther.* **2018**, *12*, 1759–1767. [CrossRef]
2. Clemett, D.; Markham, A. Linezolid. *Drugs* **2000**, *59*, 815–827; discussion 828. [CrossRef]
3. Takahashi, Y.; Takesue, Y.; Nakajima, K.; Ichiki, K.; Tsuchida, T.; Tatsumi, S.; Ishihara, M.; Ikeuchi, H.; Uchino, M. Risk factors associated with the development of thrombocytopenia in patients who received linezolid therapy. *J. Infect. Chemother.* **2011**, *17*, 382–387. [CrossRef]
4. Ikuta, S.; Tanimura, K.; Yasui, C.; Aihara, T.; Yoshie, H.; Iida, H.; Beppu, N.; Kurimoto, A.; Yanagi, H.; Mitsunobu, M.; et al. Chronic liver disease increases the risk of linezolid-related thrombocytopenia in methicillin-resistant Staphylococcus aureus-infected patients after digestive surgery. *J. Infect. Chemother.* **2011**, *17*, 388–391. [CrossRef]
5. Attassi, K.; Hershberger, E.; Alam, R.; Zervos, M.J. Thrombocytopenia associated with linezolid therapy. *Clin. Infect. Dis.* **2002**, *34*, 695–698. [CrossRef]
6. Han, X.; Wang, J.; Zan, X.; Peng, L.; Nie, X. Risk factors for linezolid-induced thrombocytopenia in adult inpatients. *Int. J. Clin. Pharm.* **2022**, *44*, 330–338. [CrossRef]

7. Lima, L.S.; Brito, E.; Mattos, K.; Parisotto, E.B.; Perdomo, R.T.; Weber, S.S. A retrospective cohort study to screen linezolid-induced thrombocytopenia in adult patients hospitalized in the Midwestern Region of Brazil. *Hematol. Transfus. Cell Ther.* **2020**, *42*, 230–237. [CrossRef]
8. Ichie, T.; Suzuki, D.; Yasui, K.; Takahashi, H.; Matsuda, M.; Hayashi, H.; Sugiura, Y.; Sugiyama, T. The association between risk factors and time of onset for thrombocytopenia in Japanese patients receiving linezolid therapy: A retrospective analysis. *J. Clin. Pharm. Ther.* **2015**, *40*, 279–284. [CrossRef]
9. Arnold, D.M. Bleeding complications in immune thrombocytopenia. *Hematol. Am. Soc. Hematol. Educ. Program* **2015**, *2015*, 237–242. [CrossRef]
10. Hirano, R.; Sakamoto, Y.; Tachibana, N.; Ohnishi, M. Retrospective analysis of the risk factors for linezolid-induced thrombocytopenia in adult Japanese patients. *Int. J. Clin. Pharm.* **2014**, *36*, 795–799. [CrossRef]
11. Hanai, Y.; Matsuo, K.; Ogawa, M.; Higashi, A.; Kimura, I.; Hirayama, S.; Kosugi, T.; Nishizawa, K.; Yoshio, T. A retrospective study of the risk factors for linezolid-induced thrombocytopenia and anemia. *J. Infect. Chemother.* **2016**, *22*, 536–542. [CrossRef]
12. Rabon, A.D.; Fisher, J.P.; MacVane, S.H. Incidence and Risk Factors for Development of Thrombocytopenia in Patients Treated with Linezolid for 7 Days or Greater. *Ann. Pharmacother.* **2018**, *52*, 1162–1164. [CrossRef]
13. Chen, C.; Guo, D.H.; Cao, X.; Cai, Y.; Xu, Y.; Zhu, M.; Ma, L. Risk factors for thrombocytopenia in adult chinese patients receiving linezolid therapy. *Curr. Ther. Res. Clin. Exp.* **2012**, *73*, 195–206. [CrossRef]
14. Kim, H.S.; Lee, E.; Cho, Y.J.; Lee, Y.J.; Rhie, S.J. Linezolid-induced thrombocytopenia increases mortality risk in intensive care unit patients, a 10 year retrospective study. *J. Clin. Pharm. Ther.* **2019**, *44*, 84–90. [CrossRef]
15. Choi, G.W.; Lee, J.Y.; Chang, M.J.; Kim, Y.K.; Cho, Y.; Yu, Y.M.; Lee, E. Risk factors for linezolid-induced thrombocytopenia in patients without haemato-oncologic diseases. *Basic Clin. Pharmacol. Toxicol.* **2019**, *124*, 228–234. [CrossRef]
16. Natsumoto, B.; Yokota, K.; Omata, F.; Furukawa, K. Risk factors for linezolid-associated thrombocytopenia in adult patients. *Infection* **2014**, *42*, 1007–1012. [CrossRef]
17. National Institute of Allergy and Infectious Diseases. Division of AIDS (DAIDS) Table for Grading the Severity of Adult and Pediatric Adverse Events. Available online: https://rsc.niaid.nih.gov/sites/default/files/daidsgradingcorrectedv21.pdf (accessed on 24 August 2023).
18. World Health Organization. Dengue Guidelines, for Diagnosis, Treatment, Prevention and Control. Available online: https://iris.who.int/bitstream/handle/10665/44188/9789241547871_eng.pdf (accessed on 16 April 2024).
19. Choudhury, S.; Dubey, S.; Zumu, M.; Mustafa, P.P.M.; Burma, A.; Kumar, S. The Role of Platelet Parameters in Predicting the Disease Severity of COVID-19 Patients: A Hospital-Based Study. *Cureus* **2024**, *16*, e51523. [CrossRef]
20. *Product Information: ZYVOX®, Linezolid Intravenous Injection, Tablets, and Oral Suspension*; Pfizer, Inc.: Philadelphia, PA, USA, 2023.
21. Brier, M.E.; Stalker, D.J.; Aronoff, G.R.; Batts, D.H.; Ryan, K.K.; O'Grady, M.; Hopkins, N.K.; Jungbluth, G.L. Pharmacokinetics of linezolid in subjects with renal dysfunction. *Antimicrob. Agents Chemother.* **2003**, *47*, 2775–2780. [CrossRef]
22. Bayer, A.S.; Sullam, P.M.; Ramos, M.; Li, C.; Cheung, A.L.; Yeaman, M.R. *Staphylococcus aureus* induces platelet aggregation via a fibrinogen-dependent mechanism which is independent of principal platelet glycoprotein IIb/IIIa fibrinogen-binding domains. *Infect. Immun.* **1995**, *63*, 3634–3641. [CrossRef]
23. O'Brien, L.; Kerrigan, S.W.; Kaw, G.; Hogan, M.; Penades, J.; Litt, D.; Fitzgerald, D.J.; Foster, T.J.; Cox, D. Multiple mechanisms for the activation of human platelet aggregation by *Staphylococcus aureus*: Roles for the clumping factors ClfA and ClfB, the serine-aspartate repeat protein SdrE and protein A. *Mol. Microbiol.* **2002**, *44*, 1033–1044. [CrossRef] [PubMed]
24. Johansson, D.; Rasmussen, M.; Inghammar, M. Thrombocytopenia in bacteraemia and association with bacterial species. *Epidemiol. Infect.* **2018**, *146*, 1312–1317. [CrossRef]
25. Gafter-Gvili, A.; Mansur, N.; Bivas, A.; Zemer-Wassercug, N.; Bishara, J.; Leibovici, L.; Paul, M. Thrombocytopenia in *Staphylococcus aureus* bacteremia: Risk factors and prognostic importance. *Mayo Clin. Proc.* **2011**, *86*, 389–396. [CrossRef]
26. Venkata, C.; Kashyap, R.; Farmer, J.C.; Afessa, B. Thrombocytopenia in adult patients with sepsis: Incidence, risk factors, and its association with clinical outcome. *J. Intensive Care* **2013**, *1*, 9. [CrossRef]
27. Giunio-Zorkin, N.; Brown, G. Real-Life Frequency of New-Onset Thrombocytopenia during Linezolid Treatment. *Can. J. Hosp. Pharm.* **2019**, *72*, 133–138. [CrossRef]
28. Nukui, Y.; Hatakeyama, S.; Okamoto, K.; Yamamoto, T.; Hisaka, A.; Suzuki, H.; Yata, N.; Yotsuyanagi, H.; Moriya, K. High plasma linezolid concentration and impaired renal function affect development of linezolid-induced thrombocytopenia. *J. Antimicrob. Chemother.* **2013**, *68*, 2128–2133. [CrossRef]
29. Al-Harbi, D.; Alturaiki, A.; Alshngeetee, A.; Aldabas, H.; AlBreacan, L.; Aljohani, R.; Alshahrani, E.H.; Althemery, A.; Esba, L.C.A. Linezolid vs Vancomycin in Induced Thrombocytopenia. *Infect. Dis. Ther.* **2022**, *11*, 1649–1660. [CrossRef]
30. Giang, H.T.N.; Banno, K.; Minh, L.H.N.; Trinh, L.T.; Loc, L.T.; Eltobgy, A.; Tai, L.L.T.; Khan, A.; Tuan, N.H.; Reda, Y.; et al. Dengue hemophagocytic syndrome: A systematic review and meta-analysis on epidemiology, clinical signs, outcomes, and risk factors. *Rev. Med. Virol.* **2018**, *28*, e2005. [CrossRef]
31. Sadeghi-Nodoushan, F.; Zare-Khormizi, M.R.; Hekmatimoghaddam, S.; Pourrajab, F. Blood Features Associated with Viral Infection Severity: An Experience from COVID-19-Pandemic Patients Hospitalized in the Center of Iran, Yazd. *Int. J. Clin. Pract.* **2024**, *2024*, 7484645. [CrossRef]

32. Izhari, M.A.; Hadadi, M.A.A.; Alharbi, R.A.; Gosady, A.R.A.; Sindi, A.A.A.; Dardari, D.M.M.; Alotaibi, F.E.; Klufah, F.; Albanghali, M.A.; Alharbi, T.H. Association of Coagulopathy and Inflammatory Biomarkers with Severity in SARS-CoV-2-Infected Individuals of the Al-Qunfudhah Region of Saudi Arabia. *Healthcare* **2024**, *12*, 729. [CrossRef] [PubMed]
33. Asrie, F.; Tekle, E.; Gelaw, Y.; Dagnew, M.; Gelaw, A.; Negash, M.; Kassa, E.; Bizuneh, S.; Wudineh, D. Baseline Thrombocytopenia and Disease Severity Among COVID-19 Patients, Tibebe Ghion Specialized Hospital COVID-19 Treatment Center, Northwest Ethiopia. *J. Blood Med.* **2022**, *13*, 315–325. [CrossRef] [PubMed]
34. Nagori, E.K.; Ghantarchyan, H.; Qadir, A.; Arabian, S. COVID-19-Induced Thrombocytopenia: A Brief Literature Review and Case Report. *Cureus* **2022**, *14*, e30993. [CrossRef]
35. Nguyen, H.; Nguyen, M.; Olenik, A. Immune Thrombocytopenic Purpura Following COVID-19 Infection: A Case Report and Literature Review. *Cureus* **2023**, *15*, e39342. [CrossRef] [PubMed]
36. Shinno, K.; Banno, Y.; Kamimaki, I. Severe immune thrombocytopenia that developed immediately after COVID-19 in a school-aged patient: A case report. *Front. Pediatr.* **2023**, *11*, 1120093. [CrossRef] [PubMed]
37. Deruelle, E.; Ben Hadj Salem, O.; Sep Hieng, S.; Pichereau, C.; Outin, H.; Jamme, M. Immune thrombocytopenia in a patient with COVID-19. *Int. J. Hematol.* **2020**, *112*, 883–888. [CrossRef]
38. Mocan, M.; Chiorescu, R.M.; Tirnovan, A.; Buksa, B.S.; Farcas, A.D. Severe Thrombocytopenia as a Manifestation of COVID-19 Infection. *J. Clin. Med.* **2022**, *11*, 1088. [CrossRef]
39. Alkhalifa, H.; Alsalman, Z.; Alfaraj, A.; Algaraash, M.; Alsalman, M. Thrombocytopenia and clinical outcomes among patients with COVID-19 disease: A cohort study. *Health Sci. Rep.* **2023**, *6*, e1111. [CrossRef]
40. Castilho, B.M.; Silva, M.T.; Freitas, A.R.R.; Fulone, I.; Lopes, L.C. Factors associated with thrombocytopenia in patients with dengue fever: A retrospective cohort study. *BMJ Open* **2020**, *10*, e035120. [CrossRef]

Journal of
Clinical Medicine

MDPI

Opinion

Platelet Pathogen Reduction Technology—Should We Stay or Should We Go…?

Andrea Piccin [1,2,3,*], Allameddine Allameddine [1], Gilbert Spizzo [4], Katrina M. Lappin [5] and Daniele Prati [6]

1 Northern Ireland Blood Transfusion Service (NIBTS), Belfast BT9 7TS, UK; allameddine.allameddine@nibts.hscni.net
2 Department of Internal Medicine V, Medical University of Innsbruck, 6020 Innsbruck, Austria
3 Department of Industrial Engineering, University of Trento, 38122 Trento, Italy
4 Department of Oncology, Brixen Hospital, 39042 Bolzano, Italy; gilbert.spizzo@gmail.com
5 Patrick G Johnston Centre for Cancer Research, Queen's University Belfast, Belfast BT7 1NN, UK; k.lappin@qub.ac.uk
6 Servizio Trasfusionale, Ospedale Ca' Granda, 20122 Milano, Italy; daniele.prati@policlinico.mi.it
* Correspondence: andrea.piccin@nibts.hscni.net

Abstract: The recent COVID-19 pandemic has significantly challenged blood transfusion services (BTS) for providing blood products and for keeping blood supplies available. The possibility that a similar pandemic event may occur again has induced researchers and transfusionists to investigate the adoption of new tools to prevent and reduce these risks. Similarly, increased donor travelling and globalization, with consequent donor deferral and donor pool reduction, have contributed to raising awareness on this topic. Although recent studies have validated the use of pathogen reduction technology (PRT) for the control of transfusion-transmitted infections (TTI) this method is not a standard of care despite increasing adoption. We present a critical commentary on the role of PRT for platelets and on associated problems for blood transfusion services (BTS). The balance of the cost effectiveness of adopting PRT is also discussed.

Keywords: platelets; INTERCEPT®; THERAFLEX®; MIRASOL®; amotosalen; riboflavin; bacterial contaminations; irradiation; TA-GvHD

1. Introduction

Overview of Platelet Transfusions and Transfusion-Transmitted Infections Risks

The risk of transfusion-transmitted infections (TTI) in resource-rich countries has been significantly reduced due to the continuous implementation of donor selection policies, enhanced serological testing, the introduction of nucleic acid testing (NAT), and other crucial measures such as precise skin disinfection and the use of diversion bags during blood collection. Current risks vary according to local epidemiology but are deemed acceptable by most national transfusion systems. According to recent data from the UK and the US, the risks are as follows: HBV 1 in 3 million donations, HIV 1 in 5 million donations, and HCV 1 in 1 million donations [1].

Nevertheless, the possibility of acquiring infections from emerging pathogens (e.g., the West Nile virus and Dengue virus), as well as from bacteria and protozoa, remains a concern in transfusion medicine. Bacterial contamination of stored blood products, particularly platelets (PLTs), is still a threat for the safety of blood supply. PLT bags are kept on agitators at room temperature (20–24 °C) to preserve PLT quality and function, but this storage condition can facilitate bacterial growth, leading to transfusion reactions and sepsis. This risk is not negligible; a meta-analysis estimates it to be approximately 1 in 1900 for buffy coat (BC) PLTs and 1 in 4348 for apheresis-collected PLTs [2]. Furthermore, one-quarter of contaminated PLT bags result in sepsis [3].

Blood transfusion services may choose to mitigate this risk using bacterial screening approaches. The BacT/ALERT© system, is an efficient tool that promptly detects bacterial growth by measuring CO_2 production in samples from PLT bags. Samples are incubated in dedicated culture bottles, and any increase in CO_2 triggers an alarm, prompting the operator to remove the implicated unit from the blood product inventory. The unit remains quarantined until further investigation. However, BacT/ALERT© cannot always detect bacteria such as *Staphylococcus epidermidis, Staphylococcus aureus*, and other species [4]. Additionally, BacT/ALERT© has a false positivity rate of 0.2–3.2%, leading to unnecessary product discards [5].

In addition, a further reduction in the possibility of acquiring TTIs is crucial to ensure patient safety and maintain product supply. Globalization and international travel have increased TTI risks, particularly from zoonoses. This was highlighted at the onset of the COVID-19 pandemic, with concerns about potential blood transmission. Another challenge for BTS is emerging pathogens. When a new pathogen, such as the Chikungunya or Zika virus, emerges, specific measures like donor deferral, screening, and recall are adopted, significantly impacting donor pools. Implementing these measures requires substantial time, staff, and resources, and their efficacy in reducing TTI risk is only partial [6]. To prevent TTIs and ensure adequate blood stocks, BTS often import blood products from other regions or countries, leading to service disruptions and high costs.

Evaluating and implementing new pathogen reduction technology (PRT) is thus essential. A recent systematic review by Angel Gimenez-Richarte and colleagues demonstrated that UV-based PRT is particularly effective against arboviruses, significantly reducing the viral load of West Nile virus, Chikungunya virus, Dengue virus, and Zika virus [7]. The efficacy of UV-based PRT varies with the inactivation method used. Importantly, PRT does not neutralize donor antibodies (e.g., convalescent plasma), allowing the recipients of treated blood products to benefit from donor antibodies against new infectious agents.

Recently, the European Directorate for the Quality of Medicines (EDQM) suggested implementing PRT as a part of blood processing in its Blood Supply Contingency and Emergency Plan (B-SCEP) (https://freepub.edqm.eu/publications/PUBSD-174/detail, Accessed on 31 August 2024). Implementing PRT will help harmonize TTI risks across countries. The new PRT types available use UVA, UVB, and UVC light and may include special compounds that crosslink or associate with DNA/RNA after UVA irradiation (e.g., INTERCEPT® and MIRASOL®) or UVC light only (THERAFLEX®). This review aims to revisit existing knowledge on the efficacy, practicality, and challenges of using UV light-mediated PRT systems for PLTs.

2. Understanding Pathogen Reduction Technology (PRT)

2.1. Explanation and Evolution of PRT (Short Analysis of the Three Main Methods)

The need to ensure better blood safety and to prevent TTI became urgent following the blood scandals of the 1980s and early 1990s. In the attempt to reduce plasma contamination, treatment with solvent detergent (SD) was adopted for blood-derived products (not for PLT). This is an excellent first example of PRT for blood-derived products. SD is very effective against lipid-enveloped viruses (also against protozoa), but not against unenveloped viruses such as HAV. Also, SD badly affects alpha-2 antitrypsin and Protein-S (an important naturally occurring anticoagulant) [8]. Another drug, similarly used as a PRT agent, is methylene blue (MB). In 2003, to reduce the risk of transmission of variant Jacob Creutzfeldt Disease (vCJD) for the population born post-1996, fresh-frozen plasma (FFP) was outsourced and imported from the USA (lower vCJD geographical risk). At the time, plasma was also treated with MB to ensure that the attempt towards the reduction in vCJD risk would have not caused an increase in other blood-borne viruses. MB is associated with allergic reactions and hypotension, currently its use is not recommended. In Belgium and in some other countries (e.g., Poland and Spain) it is still used as an accepted PRT method. Recently, UV-based PRT systems became available, and this is the topic of this review.

2.2. Short Analysis of the Three Main Methods

The existing pathogen reduction (PR) technology systems for platelets based on UV light are the following: (i) INTERCEPT® PR system, (ii) MIRASOL® PR system, (iii) THERAFLEX® UVC PR system [9].

The exact mechanism of action of these three methods are illustrated in Figure 1. Key characteristics are also reported in Table 1.

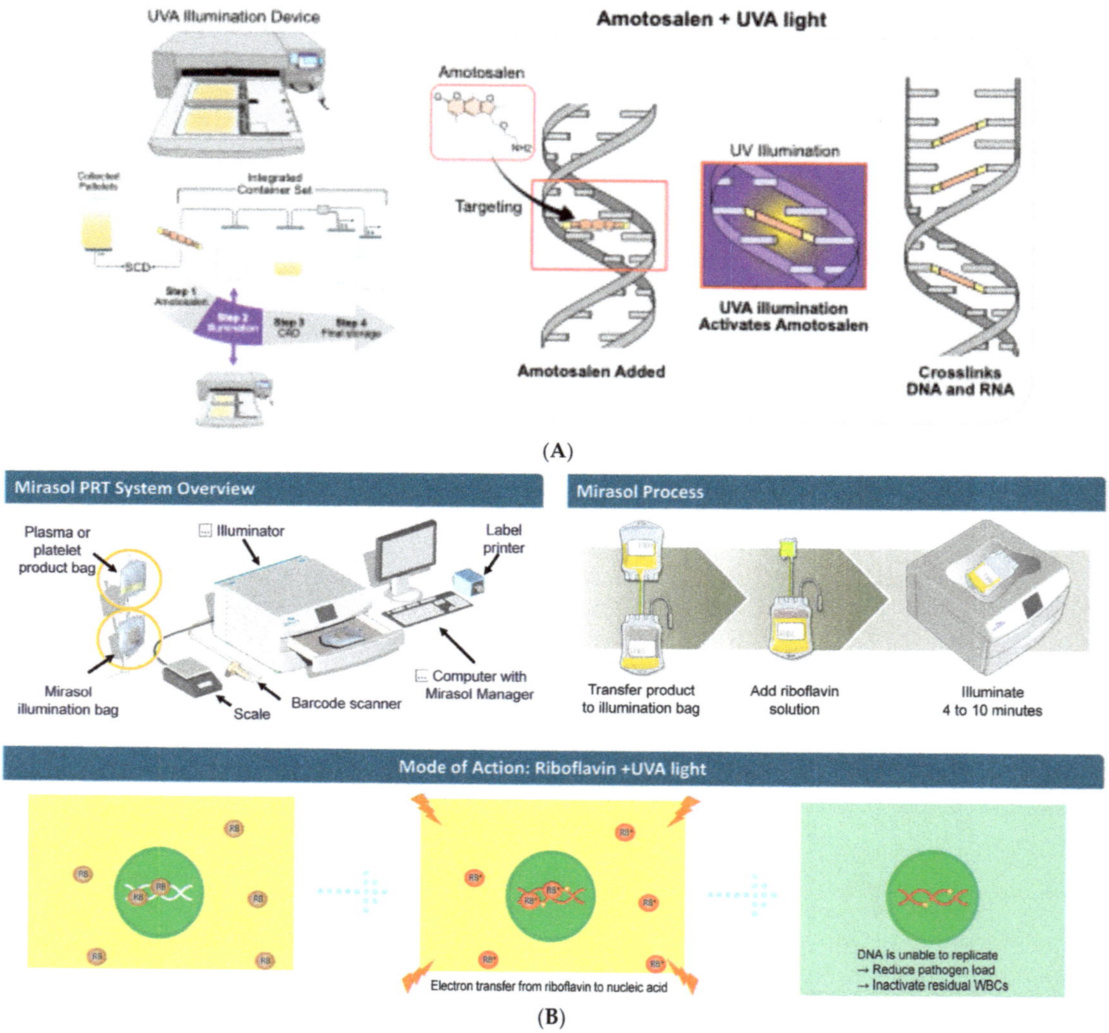

Figure 1. (**A**) The first cartoon shows the mechanism of action of the INTERCEPT® PR system. The inactivation process is based on a direct crosslinking of amotosalen with the DNA chain, following exposure to UVA light (step 2). After the UVA illumination step, the CAD device is used to remove any amotosalen in excess to reduce toxicity. (**B**) The mechanism of action of MIRASOL® PR system. Riboflavin acts as a photosensitizer that is associated with nucleic acids and mediates an oxygen-independent electron transfer process leading to the modification of DNA/RNA upon exposure to UV light. This system utilizes riboflavin (vitamin B2) which is not toxic, therefore a CAD device is not needed.

Table 1. Summary of key characteristics of the three types of PRT.

	Intercept PR System	**Mirasol PR System**	**Theraflex UVC PR System**
UV Type	**UVA**	**UVB**	**UVC**
Mechanism of action	Following exposure to UVA, an amotosalen compound will crosslink to DNA and RNA chains.	Riboflavin (Vitamin B2) associates with nucleic acids and mediates an oxygen-independent electron transfer process leading to the modification of DNA/RNA upon exposure to UV light.	
Adsorption device	The residual amotosalen is removed using a compound adsorption device (CAD).	Not needed.	Not needed.
Toxicity	No toxicologically relevant effects. The safety margins of amotosalen alone in toxicity studies are >1000.	No toxicity.	No toxicity.
Advantages	Excellent microbiological spectrum.	Good microbiological spectrum.	Good microbiological spectrum. Cost contained.
Disadvantages	High cost. Possibility of some saving by treating double dose.	High cost.	Does not protect against a large spectrum of microbiological agents. Large studies are lacking.

We will now analyze all three methods in depth.

I. INTERCEPT® PR system

This PRT method uses amotosalen as the inactivation agent. Amotosalen is a synthetized photoactive psoralen compound. PLTs are resuspended in approximately 35% plasma and 65% PAS, or in 100% plasma. Amotosalen undergoes photoactivation with UVA light at 320–400 nm. A total dose of 3 joules/cm^2 is delivered, causing amotosalen to crosslink nucleic acids (DNA or RNA). Over time, this binding becomes permanent, impairing the replication of bacteria and other microorganisms. The residual amotosalen is subsequently removed using a dedicated compound adsorbtion device (CAD), which results in a certain degree of platelet loss (between 7 and 10%) (Figure 1A). For these reasons, Gathof BS et al. recommend increasing the lower limit of PLT units to 2.5×10^{11} [10]. However, the recent EDQM guidelines have suggested a lower limit of 2.0×10^{11} [11].

The efficacy of the INTERCEPT® PR system has been evaluated in several randomized clinical trials. McCullough reported on a randomized study in bleeding patients (SPRINT trial) with WHO grade 2 degrees (mild blood loss, clinically significant). Patients in the study group were transfused with PLT treated with the INTERCEPT® PR system, while patients in the control group received standard PLT [12]. This study did not show a difference in bleeding events; however, patients transfused with PRT PLT required more products. It also emerged that PRT units had a lower concentration of PLT ($<3.0 \times 10^{11}$). In the same study, post-transfusion CCI at 1 h and at 24 h were lower in the arm photochemically treated (PCT) with PRT-PLT. The transfusion interval was shorter, leading overall to a higher prescription of PLT. Nevertheless, no differences in clinical bleeding were seen. Additionally, fewer transfusion reactions were observed when PRT PLT was administered (3.0% PCT versus 4.4% control; $p = 0.02$) [12].

Another randomized study (PRT-treated PLT unit versus standard PLT unit) on the INTERCEPT® PR system, carried out in over 100 patients with thrombocytopenia (euroSPRITE), showed no statistically significant difference in CCI at 1 h and at 24 h of follow-up. Patients treated with PRT units had slightly lower platelet counts. Hemorrhagic and thrombotic events after transfusion were similar in both groups. However, in 2010, worrisome results were highlighted by the Dutch–Belgian HOVON cooperative group. In a randomized study, Kerkhoffs JLH et al. showed that the use of the INTERCEPT® PR system was associated with decreased CCI and with grade ≥ 2 bleeding. These findings were not confirmed by randomized controlled trials such as the SPRINT trial and the EFFIPAP trial, where the primary endpoint was WHO grade-2 bleeding [13]. However, in the EFFIPAP trial, the results differed depending on the control arm used (PLT in PAS or PLT in plasma). In the HOVON study, amotosalen was added to platelets suspended in approximately 35% plasma/65% PASIII (study group) and compared with PLT resuspended in plasma only and PASIII only (control groups) [14]. This may have in part contributed to the overall negative results. Moreover, a recent review of PRT types and their efficacy showed that INTERCEPT® PR system is certainly more efficacious against arboviruses compared to MIRASOL® PR system [7]. This is also confirmed by a better Log Reduction Factor (LRF) of 4 or more (>4 log10) with the INTERCEPT® PR system. Similarly, the INTERCEPT® PR system inactivates leukocytes, and its implementation could also make irradiation redundant [15]. Moreover, the possibility of a 7 day platelet storage has also been shown with the INTERCEPT® PR system [16].

II. MIRASOL® PR system

This PRT method utilizes riboflavin (vitamin B2), a photoreactive compound that is non-toxic (Figure 1B). Therefore, no further manipulation of the blood components is required after the illumination step, resulting in minimal loss of platelets during the inactivation treatment.

The compound is released in platelets resuspended either in plasma or in plasma with platelet additive solution (PAS). This is followed by photoactivation of riboflavin using UV-B light with a total energy power of 6.2 Joules/cm^2. A compound adsorption device (CAD) is not required because riboflavin is non-toxic. This method is highly effective against enveloped viruses, such as influenza virus, cytomegalovirus (CMV), HIV, respiratory syncytial virus (RSV), and human coronaviruses.

One of the most notable results achieved by the MIRASOL® PR system was demonstrated by Allain JP and colleagues in the AIMS study conducted in Ghana where the risk of blood parasitemia by malaria is >50% [17,18]. In this study, whole blood treated with the MIRASOL® PR system showed a reduction in the incidence of transfusion-transmitted malaria ($p = 0.039$) [18]. Furthermore, the MIRASOL® PR system has been reported to have moderate inactivation of enveloped viruses such as HEV [19,20].

Randomized studies have shown a 25% decrease in corrected count increment (CCI) when platelets inactivated with riboflavin-PRT were infused. This finding was confirmed in another study at 1 h after transfusion. However, no major bleeding or other relevant side effects were observed, although a larger requirement of platelets in some patient categories was reported [21]. A recent study (miPLATE trial) did not show an inferiority of the MIRASOL® PR system compared to conventional platelets when the endpoint was calculated in days $\geq$ Grade 2 bleeding [22].

Additionally, the platelet shelf life was extended to 7 days without any major side effects, except for a higher rate of lactate increase and glucose depletion starting from day 5. The platelet loss with the MIRASOL® PR system was lower than previously reported with the INTERCEPT® PR system, although a loss of the swirling effect and increased apoptosis (assessed by Annexin) were also reported [23]. Furthermore, the MIRASOL® PR system inactivates leukocytes. This could theoretically suggest that leukoreduction/filtration may not be needed, but Norris P and colleagues have found contradictory results [24].

The main side effect noted with the MIRASOL® PR system is ATP consumption, which suggests that glycolytic pathways are activated [25]. A randomized study (MERIT trial) from Uganda University is currently underway to clarify the potential benefits of the MIRASOL® PR system for transfusion-transmitted infection (TTI) prevention [17].

III. THERAFLEX® UVC PR SYSTEM

The THERAFLEX® UVC PR system is based on the use of UVC (wavelength 200–280 nm) [26,27]. This system is highly promising against high concentrations of various bacterial species [28]. However, it is not very effective against HIV inactivation. The strongest efficacy (4 or more log steps) was demonstrated against the vesicular stomatitis virus, porcine parvovirus (a model virus for parvovirus B19), encephalomyocarditis virus (a model virus for hepatitis A), and Sindbis virus (a model virus for hepatitis C). Other viruses, such as the Pseudorabies virus (a model virus for hepatitis B) and West Nile virus, were less effectively inactivated, with reduction factors of approximately 2–3 and 3.5–4 log steps, respectively [27]. Eickmann M and colleagues conducted an in vitro study demonstrating that THERAFLEX® UVC PR system can effectively reduce the infectivity of severe acute respiratory syndrome coronavirus (SARS-CoV), Crimean–Congo hemorrhagic fever virus (CCHFV), and Nipah virus (NiV) in platelets and plasma [29]. THERAFLEX® UVC PR system also induced a mean log reduction greater than 3.5 against HEV [30].

The THERAFLEX® UVC PR system is also effective in reducing transfusion-associated graft-versus-host disease (TA-GvHD) by inactivating leukocytes [31]. It differs from the INTERCEPT® PR system and MIRASOL® PR system as it does not require a photosensitizing agent. A German study involving 171 patients reported no significant side effects associated with the THERAFLEX® UVC PR system [31]. However, this study was underpowered to detect side effects, and the primary efficacy outcome was not achieved, as PRT platelets were found to be less effective than standard platelets [9].

The THERAFLEX® UVC PR system does carry some quality-related side effects, such as increased annexin V binding, lactate accumulation, and higher expression of CD41/61 [32]. Currently, the THERAFLEX® PR system is the only PRT that provides time-to-treatment studies, which are crucial for determining the optimal time to start pathogen reduction treatment. The THERAFLEX® PR system requires no photosensitizer, eliminating additional processing time for the removal of such chemicals. Consequently, it allows for the faster delivery of blood products and enhances safety by eliminating handling and transportation errors [33].

3. Advantages of PRT for Platelet Transfusions

Overall PRT has several effects which seem to be mainly positive. We have summarized and commented on most of those here below:

(i) PRT extends PLT shelf-life.
(ii) PRT and transfusion reactions.
(iii) PRT reduces Ta-GvHD.
(iv) PRT and TTI.
(v) PRT impact on coagulation proteins.
(vi) PRT harmonizes TTI risk between countries.
(vii) PRT does not affect the presence of pre-existing antibodies.
(viii) PRT and sepsis.

3.1. PRT Extends PLT Shelf-Life

The lifespan of platelets (PLT) is currently limited to 5 days. Therefore, any tool that could extend the shelf life of PLT is highly desirable for any blood transfusion service (BTS). The most widely adopted system for monitoring PLT contamination is BacT/ALERT®. This system is particularly prevalent in the UK and Ireland and helps monitor bacterial growth to prevent transfusion-transmitted infections (TTI). BacT/ALERT® can identify contaminated PLT units at a very early stage. Recent studies have shown that BacT/ALERT® can help extend PLT shelf life to 7 days by performing daily sampling for BacT/ALERT cultures and bacterial quantification [34]. Additionally, the INTERCEPT® PR system has demonstrated the ability to preserve PLT quality over 7 days of storage [35]. The potential for extending PLT shelf life should be considered in any cost analysis of PRT implementation. A Spanish study indicated that PRT can help reduce the rate of PLT outdates [36].

3.2. PRT Reduces Transfusion Reactions

PRT has been associated with a reduction in PLT transfusion reactions [10,37]. The exact reason is unknown; however, this may simply be due to the inactivation of white blood cells and the consequent abrogation of cytokine production. Some authors have hypothesized a reduction in reactive stimuli by LPS, anti CD3+ and anti CD28 [38].

3.3. PRT Reduces Transfusion-Associated Graft versus Host Disease

The irradiation of blood products is used to prevent the development of transfusion-associated graft versus host disease (TA-GvHD). Indications for irradiated blood components include patients undergoing intrauterine transfusion, neonatal exchange transfusion, those with congenital cell-mediated immunodeficiencies, patients with Hodgkin lymphoma (for all transfusions during their lifetime), hematopoietic stem cell transplant recipients (both autologous and allogeneic), donors of allogeneic hematopoietic stem cells (in the week prior to and during stem cell harvest), CAR-T therapy recipients (7 days prior to collection and 3 months post-infusion), patients receiving purine analog therapy (e.g., fludarabine, cladribine, deoxycoformycin, bendamustine, clofarabine), those undergoing anti-thymocyte globulin (ATG) or alemtuzumab therapy for hematologic diseases, recipients of donations from biologic relatives, and recipients of donations selected on the basis of HLA matching. Early reports in large cohorts of patients in France and Switzerland have shown that PRT has strong efficacy for the prevention of TA-GvHD [39]. A study by Fast L and colleagues demonstrated that PRT is more effective than irradiation in the inactivation of T-cell lymphocytes and their proliferation [15]; irradiation does not fully impair CD69 expression [40]. Recently, Sim J and colleagues conducted a randomized prospective study to better evaluate the efficacy and safety of PRT. They studied a relatively large allogeneic transplant cohort ($n = 30$) where blood products were treated only with PRT and did not undergo any leukoreduction or irradiation [41]. This study was successful, as none of the patients developed TA-GvHD. This finding is particularly relevant within hematology, highlighting the potential of PRT to make blood product irradiation and leukoreduction redundant. Additionally, from an economic perspective, this could significantly reduce costs [42]. Liebermann and colleagues also commented on the prevention of CMV infection through leukoreduction, noting that PRT can effectively replace leukoreduction for CMV prevention [43].

3.4. PRT Reduces TTI

The capacity of reducing the risk of TTI varies according to PRT type. It also differs if PLT are resuspended in plasma or in platelet additive solutions (PAS). The published data on INTERCEPT® PR system showed a high efficacy against the following: (i) protozoa (Plasmodium falciparum, Trypanosoma cruzi, Leishmania mexicana, Leishmania major Jish, Babesia microti); (ii) Gram-negative bacteria (Escherichia coli, Serratia marcescens, Klebsiella pneumoniae, Pseudomonas aeruginosa, Salmonella choleraesuis, Yersinia enterocolitica, Enterobacter cloacae, Anaplasma phagocytophilum); (iii) Gram-positive bacteria (Staphylococcus epidermis, Staphylococcus aureus, Streptococcus pyogenes, Listeria monocytogenes, Corynebacterium minutissimum, Bacillus cereus (vegetative), Bifidobacterium adolescentis, Propionibacterium acnes, Lactobacillus species, Clostridium perfringens (vegetative form), Spirochete Bacteria such as Treponema pallidum and Borrelia burgdorferi); (iv) enveloped viruses (HIV-1, HBV (strain MS-2), HCV (strain Hutchinson), HTLV-I, HTLV-II, CMV, Bovine Viral Diarrhea Virus, Duck Hepatitis B Virus, Pseudorabies virus, West Nile Virus (WNV), SARS-CoV2, Chikungunya virus, Influenza A H5N1 virus (Avian Influenza, Zika, Dengue, Yellow Fever); (v) non-enveloped viruses (Bluetongue Virus type 11, Calicivirus, Human Adenovirus-5, Parvovirus B19) [44,45]. The technical data sheet available on MIRASOL® PR system is not so informative, but clearly states that even this PRT is efficacious against enveloped and non-enveloped viruses, gram-negative and gram-positive bacteria, and parasites [46].

In the case of the THERAFLEX® UVC PR system, activity against the following bacteria is reported: Enterobacter cloacae, Escherichia coli, Klebsiella Pneumoniae, Morganella morganii, Proteus mirabilis, Pseudomonas fluorescens, Serratia marcescens, Staphylococcus aureus, Staphylococcus epidermidis, Streptococcus bovis, Streptococcus dysgalactiae, Streptococcus pyogenes, Listeria monocytogenes, Acinetobacter baumannii, Streptococcus agalactiae, Streptococcus pneumoniae, Bacillus cereus, Bacillus thuringiensis, Propionibacterium acnes [47]. Also, activity against parasites such as Babesia divergens [48] and African mosquito-borne flaviviruses such as WNV and Usutu virus (USUV) have been shown [49].

The use of PRT is quite promising. At present, there are several reports on the use of PRT. For example, Pitman reported on 1,221,031 platelet transfusions in France over 2010–2020 [37]. Escolar and colleagues reported that >875,000 transfusion treated with INTERCEPT® and >700,000 treated with MIRASOL® have been administered so far, and no case of TTI or sepsis was reported [26].

3.5. PRT Impact on Plasma Coagulation Proteins

Plasma products treated with PRT will have reduced levels of key coagulation proteins. Statistically significant reductions in FII, FV, FVIII, FIX, and FXI have been shown [50]. However, this may not necessarily impair the hemostatic function of the plasma product. For example, a thrombin generation (TG) study conducted on plasma treated with INTERCEPT® PR system showed TG levels similar to those of untreated plasma, suggesting that hemostatic function is preserved [51].

Studies have shown that plasma treated with solvent/detergent or with methylene blue (MB) can affect Protein-S levels [52–54]. This is relevant because Protein-S is a co-factor of Protein-C and is essential for its conversion into activated Protein-C (APC). APC balances the actions of FVa and FVIIIa [55]. MB-treated plasma best preserves FV and FXI levels compared to the INTERCEPT® PR system [56]. This information is relevant because often platelets (PLT) are resuspended in plasma [57].

4. Criticisms and Limitations of PRT

The following points were identified as criticisms and limitations of PRT:

(i) Bleeding risk.
(ii) PLT increment count.
(iii) Biochemical changes.
(iv) Mitochondrial DNA inactivation.
(v) PRT alters microRNA.
(vi) Pathogens not impacted by PRT.
(vii) Costs.

4.1. Bleeding Risk and PLT Count Increment

The possibility that PRT may affect platelet quality and contribute to a higher bleeding risk has been a concern for several investigators and continues to be so for some. An exhaustive Cochrane review on the risk of bleeding associated with PRT was carried out in 2017 [58]. More recently, a systematic review of all published randomized controlled trials was published by the Italian group of Pati I in 2022. In this study, Pati I and colleagues reported no difference in clinically significant bleeding between the INTERCEPT® PR system and the MIRASOL® PR system. However, a higher number of bleeding events was observed in the PRT groups [59].

Among these studies, the most concerning is the one conducted by the Dutch group of Kerkhoffs JLH and colleagues [14]. In this study, a reduced 1 h corrected count increment (CCI) was documented using the INTERCEPT® PR system. Moreover, an increased risk of bleeding was also documented. Interestingly, patients with hematological malignancies such as AML, ALL, and LNH were included. This is relevant since these patients have very low PLT counts and are quite prone to bleeding.

Similarly, the Italian Platelet Technology Assessment Study (IPTAS) showed that the absolute risk difference in WHO Grade $\geq$ 2 bleeding for the INTERCEPT® PR system was 6.1% (with an upper one-sided 97.5% confidence limit of 19.2%), while the absolute risk difference for the MIRASOL® PR system was better at 4.1% (with an upper one-sided 97.5% confidence limit of 18.4%) [60]. However, in neither case was the absolute risk difference statistically significant. In both trials, post-transfusion platelet count increments were significantly lower in treated versus control patients [60].

A randomized French study (EFFIPAP) did not show that PLTs treated with PRT (INTERCEPT® PR system) and resuspended in PAS were associated with any higher rate of WHO Grade $\geq$ 2 bleeding. However, a 12.5% difference margin was considered for this non-inferiority study [61].

Osselaer JC and colleagues studied a large cohort of patients (n = 651) treated with 5106 blood components transfused with the INTERCEPT® PR system. The authors reported no changes in the number of PLT doses administered with and without PRT implementation [62]. Cazenave JP and colleagues also reported similar findings. Interestingly, these authors also found a reduced number of transfusion reactions associated with PRT using the INTERCEPT® PR system [63]. More recently, an Austrian study reported a similar consumption of blood products before and after PRT implementation (INTERCEPT® PR system) over 21 months of follow-up. A similar finding was also reported by Infanti L and colleagues, in a study that also showed the possibility of extending PLT shelf life to 7 days [64].

This suggests that no significant increase in bleeding was seen with PRT implementation [65]. The same group of authors also studied a large cohort of patients treated with massive transfusion [66]. Interestingly, no increased bleeding risk was seen after PRT implementation. Additionally, both the SPRINT [67] and the EFFIPAP [61] studies enrolled patients with severe hematological conditions and low PLT counts, suggesting that these "fragile" patients are not adversely affected when PRT is implemented.

4.2. PLT Increment Count

Several reports have suggested that PRT-treated PLT lead to an inferior post-transfusion PLT count compared to that obtained when PLT without any PRT treatment are transfused [62]. To analyze this aspect, the concept of a corrected count increment (CCI) has been introduced. The CCI was initially developed with the intent to facilitate physicians dealing with differential diagnosis of idiopathic thrombocytopenic purpura (ITP) and is currently used for analyzing platelet refractoriness. This index directly correlates to the PLT dose transfused. The CCI is based on a small calculation on PLT numbers based on a post-transfusion specimen collected 1 h after transfusion (CCI_{1hour}). This index also considers the patient's body surface area (BSA). Usually, after a transfusion of a standard unit of PLT in a patient with not-active bleeding, the PLT count shows an increment between 5 and 10.000/μL.

$$CCI = \frac{[(\text{post} - \text{transfusion PLT count 1 h after transfusion}/\mu L) - (\text{pretransfusion PLT count}/\mu L)] \times \text{BSA}}{(\text{number of PLT transfused}) \times 10^{11}/\mu L}$$

If the CCI does not show an increment, further investigations would be required to identify the cause (e.g., immunization to HLA- or HPA-antigens, ITP, sepsis, etc.). A Canadian simulation study reported on the possible risk of wastages and PLT shortages caused by PRT implementation. The authors described a reduced CCI when the INTERCEPT® PR system was used. This would have led to prescribing more products, and enhanced patient's exposure to more blood donors. The authors concluded that this would impact on supply chain and potentially increase the risk of TTI because of larger exposure to blood products [68]. Different results were shown by the SPRINT randomized study. In particular, the CCIs at 1 h and at 24 h post-transfusion were lower for patients transfused with PRT-treated PLT. Also, the transfusion interval was shorter leading overall to a higher prescription of PLT; however, no differences in clinical bleeding were seen [69]. However, other studies have not confirmed these data and have not reported any increased PLT consumption for the INTERCEPT® PR system [62–66].

4.3. Biochemical Changes

Malvaux N et al. reported on changes in biochemical markers in PLT concentrate treated with either the MIRASOL® PR system or the INTERCEPT® PR system. A higher production of lactate, along with a greater reduction in glucose, was observed for MIRASOL®. Swirling at day 5 was also worse when PLT were treated with the MIRASOL® PR system, suggesting that the INTERCEPT® PR system should be preferred [23]. Conversely, PLT treated with the INTERCEPT® PR system preserved swirling until day 7 [23,35]. However, some biochemical studies have shown problems even when the INTERCEPT® PR system is used. Specifically, Stivala S and colleagues reported that PLT treated with the INTERCEPT® PR system exhibited reduced function in response to activation agonists (i.e., collagen, thrombin, and von Willebrand Factor). The same study showed increased PLT apoptosis through Bak upregulation and a caspase-dependent pathway, enhancing PLT clearance from the circulation, with obvious clinical implications [70].

4.4. Mitochondrial DNA Inactivation

Following ultraviolet (UV) light-based pathogen inactivation, mitochondrial DNA (mitDNA) damage can occur. Specific PCR multiplex assays have been developed for testing mitDNA. These assays have demonstrated that pathogen reduction technology is responsible for this damage [71,72]. Furthermore, Bakkour and colleagues showed that mitDNA damage is primarily UV-dependent and not related to the presence of amotosalen [73].

4.5. PRT Alters microRNA

Diallo I and his colleagues were the first to report alterations in microRNAs contained in platelet-derived microparticles (PMP) [74]. The authors suggested that pathogen reduction technology might influence microRNA loading into PMP, which may indirectly affect PMP bioactivity. Alterations of microRNAs (miRNAs) stored within PMP were observed with the INTERCEPT® PR system. No such alterations were seen with the MIRASOL® PR system. The authors attributed these alterations to the presence of the additive storage solution (amotosalen) used with INTERCEPT®. Arnason and colleagues reported that the INTERCEPT® PR system affects only the expression of miRNA-96-5p, showing downregulation of this particular miRNA out of the 25 miRNAs analyzed [75].

4.6. Pathogens Not Impacted by PRT

PRT can significantly contribute to reducing the pathogen load associated with transfusion, but no type of PRT ensures the full sterility of blood products. Additionally, PRT is clearly ineffective against prions, spores, and some viruses without a lipid envelope (e.g., Hepatitis A). However, the MIRASOL® PR system and the THERAFLEX® UVC PR system have been shown to inhibit HEV [20].

4.7. PRT High Costs

The cost analysis of implementing PRT varies between studies and in different geographical, epidemiological and economical settings. Several factors affect this cost, including the type of PRT procedure, shorter intervals of sustained transfusion, extended shelf-life from 5 to 7 days, the elimination of test procedures (such as CMV testing and bacterial culture), the elimination of gamma irradiation.

Cicchetti A and colleagues conducted a cost analysis on PRT in Italian patients, considering over 50,000 transfused patients in one year. They predicted an overall cost increase of 30% [6]. Additionally, the same authors suggested that while PRT can significantly reduce the risk of transfusion-transmitted infections (TTI), it does not guarantee protection against all TTIs. Furthermore, they indicated that PRT is not applicable to all blood products (e.g., red cells) and may cause alterations in product quality [6].

In a constructive criticism, Farrugia emphasized that although current PRT is not perfect, previous regulatory authorities, such as the US Blood Product Advisory Committee, investigating blood inquiries related to HIV transmission, have clearly stated that "partial solutions should be welcome and where possible implemented by BTS" [76,77]. We concur with this statement. Our perspective is that ensuring and improving patient safety should always remain a priority for any BTS.

CF Bell and colleagues, in an earlier cost analysis, also noted that introducing PRT would render bacterial testing redundant, leading to indirect cost savings [78]. The cost saving associated with the prevention of a potentially emerging virus is very significant when considered in economic studies.

Most studies on cost analysis have been conducted on the INTERCEPT® PR system only. In this respect, a study sponsored by INTERCEPT showed that by adopting PRT the incremental cost effectiveness ratio (ICER) of PLT is certainly higher; however, this may be justified if new emergent pathogens arise with a 10-fold risk of transmission with transfusion [79]. Gregoire Y and colleagues from Québec showed an ICER on quality-adjusted life years (QALY) equal to USD 8.1 million/QALY [80]. However, in the presence of a new blood-borne pathogen the ICER was reduced to USD 123,063. Rosskopf K and his colleagues reported on the advantages of not having gamma irradiation and bacterial screening in place. The analysis concluded that implementing PRT added an additional cost of +7.5% [81]. Mc Cullough and colleagues reported on significant cost saving for each PLT unit treated with PRT, if standard microbiological testing were removed. Furthermore, the same authors stressed that a shelf life extension to 7 days would contribute significantly to cost reduction [12].

5. Future Perspectives

Potential Improvements and Advancements in PRT

The prevention of TTI cannot prescind from implementing PRT or similar systems for red cell concentrate (RCC) and plasma products. Recently, a randomized trial called ReCePI was conducted on the INTERCEPT® PR system for RCC in patients undergoing cardiovascular surgery. The trial demonstrated non-inferiority for the INTERCEPT® PR system-treated RCC compared to conventional RCC as measured by the incidence of acute kidney injury (AKI).

Similarly, studies on fibrinogen compounds like cryoprecipitate are currently ongoing [82].

6. Analysis of Arguments for and against PRT

A critical review of previous randomized control trials (RCT) did not identify relevant clinical complications when PRT was implemented. For example, the worrisome higher bleeding (>WHO grade 2) or lower CCI did not really emerge, even if data have shown that PRT may impact on the post-transfusion PLT count. From the perspective of improving blood safety, there is no clear reason why PRT should not be further implemented. The economic cost associated with PRT implementation may have been an obvious obstacle. However, considering the benefits of PRT (reduced risk of TTI, better TA-GvHD prevention and no need of irradiation), it is not clear to these authors why PRT has not been widely adopted. Perhaps, the large number of RCT designed to prove that PRT does not impact on product quality (reduced CCI count, or increased bleedings), may have contributed to creating the perception that PRT is unsafe. Interestingly, those centers that have already adopted PRT have not moved away from this practice. This indirectly validates PRT. However, the wait and watch approach adopted by other BTS indicates that a large resistance towards implementing PRT still exists.

7. Conclusions

7.1. Summary of Key Points

TTI in transfusion medicine have always been a challenge, particularly because new emerging disease will always be present, even in the future. The recent COVID-19 pandemic is a good example of this. For this reason, PRT in transfusion medicine offers the possibility to reduce and control future events caused by unknown pathogens. The relationship between zoonosis and TTI (West Nile Virus, Zika, MERS, SARS, Malaria, Chikungunya virus, etc.) is large and well reported in the literature. The possibility to implement blood safety by using UV lights with or without intercalating agents able to block DNA and RNA replication signaling, is likely to become a cornerstone of transfusion medicine. Moreover, the introduction of this technology will even potentiate the safety of the existing PCR screening practice, ensuring that any false negative samples that are potentially infective will still undergo inactivation. This transfusion concept has also been recently debated by Farrugia in his description of a "safety tripod" made by (1) appropriate donor selection, (2) blood screening with molecular technology, and (3) the adoption of viral inactivation processes [77].

7.2. Personal Recommendation and Final Remarks

An unquestionable advantage of introducing PRT is the potential to extend the shelf life of platelets (PLTs), which would offer significant inventory management benefits by making more products available for longer periods of time. Although implementing PRT in low-income countries, such as those in sub-Saharan Africa where the risk of transfusion-transmitted infections (TTIs) is high, could theoretically be beneficial, it remains impractical and unrealistic due to the high costs involved [18]. Even in resource-rich countries, however, the widespread adoption of Pathogen Reduction Technology (PRT) encounters resistance, often due to objective concerns such as potential damage to blood products, adverse effects like alloimmunization, logistical challenges, and significant costs. While PRT may offer substantial benefits in specific epidemiological scenarios, it is essential for each national transfusion authority to define the most appropriate strategy based on an evidence-based approach. What is needed now is a shared risk assessment methodology across different countries to support decision-making, ensuring that the process is as objective and reproducible as possible, thereby guaranteeing all patients access to care that is balanced in terms of both efficacy and safety.

Author Contributions: Conceptualization, A.P. and D.P.; methodology, A.P., A.A., G.S., K.M.L. and D.P.; writing—original draft preparation, A.P., A.A., G.S., K.M.L. and D.P.; writing—review and editing, A.P., A.A., G.S., K.M.L. and D.P.; visualization, A.P., A.A., G.S., K.M.L. and D.P.; supervision, D.P.; project administration, A.P.; funding acquisition, A.P. and D.P. All authors have read and agreed to the published version of the manuscript.

Funding: This research received no external funding.

Acknowledgments: We are grateful to Marco Marzola, Piccin Nuova Libraria, Padova, Italy for editing Figure 1A. We are grateful to Marcia Cardoso, Terumo Blood and Cell Technologies for providing us with Figure 1B.

Conflicts of Interest: D.P. has received travel and research grants, speaking fees, and teaching fees from Diasorin, Diamed, Diatech Pharmacogenetics, Grifols, Immucor, Macopharma, Ortho Clinical Diagnostics, Terumo, Cerus. All authors have no conflicts of interest or any potential conflicts of interest.

Abbreviations

AKI	acute kidney injury
ALL	acute lymphoblastic leukaemia
AML	acute myeloid leukaemia
APC	activated Protein C
ATG	anti-thymocyte globulin
BSA	body surface area
B-SCEP	Blood Supply Contingency and Emergency Plan
BTS	Blood Transfusion Service
CAD	compound adsorption device

CCI	corrected count increment
CCHFV	Crimean-Congo hemorrhagic fever virus
CMV	cytomegalovirus
EDQM	European Directorate for the Quality of Medicines
FFP	fresh frozen plasma
GvHD	Graft versus Host Disease
HEV	hepatitis E virus
HIV	human immunodeficiency virus
ICER	incremental cost effectiveness ratio
IPTAS	Italian Platelet Technology Assessment Study
ITP	idiopathic thrombocytopenic purpura
LNH	lymphoma non-Hodgkin
LRF	Log Reduction Factor
MB	methylene blue
mitDNA	mitochondrial DNA
NAT	nucleic acid testing
NiV	Nipah virus
PAS	platelet additive solution
PCT	photochemically treated
PLTs	platelets
PMP	platelet-derived microparticles
PR	pathogen reduction
PRT	Pathogen Reduction Technology
QALY	quality-adjusted life years
RCC	red cell concentrate
RCT	randomized control trials
RSV	respiratory syncytial virus
SARS-CoV	severe acute respiratory syndrome coronavirus
SD	solvent detergent
TA-GvHD	transfusion-associated graft-versus-host disease
TG	Thrombin generation
TTI	Transfusion-transmitted Infections
USUV	Usutu virus
UV	ultraviolet
vCJD	variant Jacob Creutzfeldt Disease
WNV	West Nile virus

References

1. Ainley, L.I.; Hewitt, P.E. Haematology patients and the risk of transfusion transmitted infection. *Br. J. Haematol.* **2018**, *180*, 473–483. [CrossRef] [PubMed]
2. White, S.K.; Schmidt, R.L.; Walker, B.S.; Metcalf, R.A. Bacterial contamination rate of platelet components by primary culture: A systematic review and meta-analysis. *Transfusion* **2020**, *60*, 986–996. [CrossRef] [PubMed]
3. Hong, H.W.X.; Lazarus, H.M.; Good, C.E.; Maitta, R.W.; Jacobs, M.R. Detection of septic transfusion reactions to platelet transfusions by active and passive surveillance. *Blood* **2016**, *127*, 496–502. [CrossRef] [PubMed]
4. Brailsford, S.R.; Tossell, J.; Morrison, R.; McDonald, C.P.; Pitt, T.L. Failure of bacterial screening to detect Staphylococcus aureus: The English experience of donor follow-up. *Vox Sang.* **2018**, *113*, 540–546. [CrossRef] [PubMed]
5. Corean, J.; White, S.K.; Schmidt, R.L.; Walker, B.S.; Fisher, M.A.; Metcalf, R.A. Platelet Component False Positive Detection Rate in Aerobic and Anaerobic Primary Culture: A Systematic Review and Meta-Analysis. *Transfus. Med. Rev.* **2021**, *35*, 44–52. [CrossRef]
6. Cicchetti, A.; Coretti, S.; Sacco, F.; Rebulla, P.; Fiore, A.; Rumi, F.; Di Bidino, R.; Urbina, L.I.; Refolo, P.; Sacchini, D.; et al. Budget impact of implementing platelet pathogen reduction into the Italian blood transfusion system. *Blood Transfus.* **2018**, *16*, 483–489.
7. Giménez-Richarte, A.; Ortiz de Salazar, M.I.; Giménez-Richarte, M.; Larrea, L.; Arbona, C.; Marco, P.; Ramos-Rincón, J.M. Pathogen inactivation methods to prevent transfusion-transmissible arboviruses: A systematic review and meta-analysis. *Trop. Med. Int. Health* **2023**, *28*, 262–274. [CrossRef]
8. Klein, H.G.; Dodd, R.Y.; Dzik, W.H.; Luban, N.L.; Ness, P.M.; Pisciotto, P.; Schiff, P.D.; Snyder, E.L. Current status of solvent/detergent-treated frozen plasma. *Transfusion* **1998**, *38*, 102–107. [CrossRef]
9. Rebulla, P.; Prati, D. Pathogen reduction for platelets-a review of recent implementation strategies. *Pathogens* **2022**, *11*, 142. [CrossRef]

10. Gathof, B.S.; Tauszig, M.E.; Picker, S.M. Pathogen inactivation/reduction of platelet concentrates: Turning theory into practice. *ISBT Sci. Ser.* **2010**, *5*, 114–119. [CrossRef]
11. Guide to the Preparation, Use and Quality Assurance of Blood Components. European Committee EDQM. 21th Edition 2023. Available online: https://freepub.edqm.eu/publications/AUTOPUB_48/detail (accessed on 15 August 2024).
12. McCullough, J.; Goldfinger, D.; Gorlin, J.; Riley, W.J.; Sandhu, H.; Stowell, C.; Ward, D.; Clay, M.; Pulkrabek, S.; Chrebtow, V.; et al. Cost implications of implementation of pathogen-inactivated platelets. *Transfusion* **2015**, *55*, 2312–2320. [CrossRef] [PubMed]
13. Cid, J.; Escolar, G.; Lozano, M. Therapeutic efficacy of platelet components treated with amotosalen and ultraviolet A pathogen inactivation method: Results of a meta-analysis of randomized controlled trials. *Vox Sang.* **2012**, *103*, 322–330. [CrossRef] [PubMed]
14. Kerkhoffs, J.L.H.; van Putten, W.L.J.; Novotny, V.M.J.; Te Boekhorst, P.A.W.; Schipperus, M.R.; Zwaginga, J.J.; van Pampus, L.C.M.; de Greef, G.E.; Luten, M.; Huijgens, P.C.; et al. Dutch-Belgian HOVON cooperative group. Clinical effectiveness of leucoreduced, pooled donor platelet concentrates, stored in plasma or additive solution with and without pathogen reduction. *Br. J. Haematol.* **2010**, *150*, 209–217. [CrossRef]
15. Castro, G.; Merkel, P.A.; Giclas, H.E.; Gibula, A.; Andersen, G.E.; Corash, L.M.; Lin, J.S.; Green, J.; Knight, V.; Stassinopoulos, A. Amotosalen/UVA treatment inactivates T cells more effectively than the recommended gamma dose for prevention of transfusion-associated graft-versus-host disease. *Transfusion* **2018**, *58*, 1506–1515. [CrossRef] [PubMed]
16. Lozano, M.; Knutson, F.; Tardivel, R.; Cid, J.; Maymó, R.M.; Löf, H.; Roddie, H.; Pelly, J.; Docherty, A.; Sherman, C.; et al. A multi-centre study of therapeutic efficacy and safety of platelet components treated with amotosalen and ultraviolet A pathogen inactivation stored for 6 or 7 d prior to transfusion. *Br. J. Haematol.* **2011**, *153*, 393–401. [CrossRef]
17. Kasirye, R.; Hume, H.A.; Bloch, E.M.; Lubega, I.; Kyeyune, D.; Shrestha, R.; Ddungu, H.; Musana, H.W.; Dhabangi, A.; Ouma, J.; et al. The Mirasol Evaluation of Reduction in Infections Trial (MERIT): Study protocol for a randomized controlled clinical trial. *Trials* **2022**, *23*, 257. [CrossRef]
18. Allain, J.P.; Owusu-Ofori, A.K.; Assennato, S.M.; Marschner, S.; Goodrich, R.P.; Owusu-Ofori, S. Effect of Plasmodium inactivation in whole blood on the incidence of blood transfusion-transmitted malaria in endemic regions: The African Investigation of the Mirasol System (AIMS) randomised controlled trial. *Lancet* **2016**, *387*, 1753–1761. [CrossRef]
19. Ojea, A.M.; Seco, C.; Mata, P.; Muñoz, M.D.C.; Álvarez Argüelles, M.E.; Rodríguez-Frías, F.; Quer Sivila, J.; Rando Segura, A.; García-Gala, J.M.; Rodriguez, M. Transfusion-transmission of hepatitis E virus through red blood cell transfusion but not through platelet concentrates: A case report from Spain. *Transfusion* **2023**, *63*, 1767–1772. [CrossRef]
20. Owada, T.; Kaneko, M.; Matsumoto, C.; Sobata, R.; Igarashi, M.; Suzuki, K.; Matsubayashi, K.; Mio, K.; Uchida, S.; Satake, M.; et al. Establishment of culture systems for Genotypes 3 and 4 hepatitis E virus (HEV) obtained from human blood and application of HEV inactivation using a pathogen reduction technology system. *Transfusion* **2014**, *54*, 2820–2827. [CrossRef] [PubMed]
21. Jimenez-Marco, T.; Garcia-Recio, M.; Girona-Llobera, E. Our experience in riboflavin and ultraviolet light pathogen reduction technology for platelets: From platelet production to patient care. *Transfusion* **2018**, *58*, 1881–1889. [CrossRef]
22. Koepsell, S.A.; Stolla, M.; Sedjo, R.L.; Carson, J.; Knudson, M.; Cook, R.; Fasano, R.; Ngamsuntikul, S.G.; Cohn, C.; Gorlin, J.; et al. Results of clinical effectiveness of conventional versus Mirasol-treated Apheresis Platelets in Patients with Hypoproliferative Thrombocytopenia (MiPLATE) trial. *Transfusion* **2024**, *64*, 457–465. [CrossRef] [PubMed]
23. Malvaux, N.; Defraigne, F.; Bartziali, S.; Bellora, C.; Mommaerts, K.; Betsou, F.; Schuhmacher, A. In Vitro Comparative Study of Platelets Treated with Two Pathogen-Inactivation Methods to Extend Shelf Life to 7 Days. *Pathogens* **2022**, *11*, 343. [CrossRef] [PubMed]
24. Norris, P.J.; Kaidarova, Z.; Maiorana, E.; Milani, S.; Lebedeva, M.; Busch, M.P.; Custer, B.; Rebulla, P. Ultraviolet light-based pathogen inactivation and alloimmunization after platelet transfusion: Results from a randomized trial. *Transfusion* **2018**, *58*, 1210–1217. [CrossRef]
25. Li, J.; de Korte, D.; Woolum, M.D.; Ruane, P.H.; Keil, S.D.; Lockerbie, O.; McLean, R.; Goodrich, R.P. Pathogen reduction of buffy coat platelet concentrates using riboflavin and light: Comparisons with pathogen-reduction technology-treated apheresis platelet products. *Vox Sang.* **2004**, *87*, 82–90. [CrossRef]
26. Escolar, G.; Diaz-Ricart, M.; McCullough, J. Impact of different pathogen reduction technologies on the biochemistry, function, and clinical effectiveness of platelet concentrates: An updated view during a pandemic. *Transfusion* **2022**, *62*, 227–246. [CrossRef] [PubMed]
27. Seltsam, A.; Müller, T.H. UVC Irradiation for Pathogen Reduction of Platelet Concentrates and Plasma. *Transfus. Med. Hemother.* **2011**, *38*, 43–54. [CrossRef]
28. Gravemann, U.; Handke, W.; Müller, T.H.; Seltsam, A. Bacterial inactivation of platelet concentrates with the THERAFLEX UV-Platelets pathogen inactivation system. *Transfusion* **2019**, *59*, 1324–1332. [CrossRef]
29. Eickmann, M.; Gravemann, U.; Handke, W.; Tolksdorf, F.; Reichenberg, S.; Müller, T.H.; Seltsam, A. Inactivation of three emerging viruses—Severe acute respiratory syndrome coronavirus, Crimean–Congo haemorrhagic fever virus and Nipah virus—In platelet concentrates by ultraviolet C light and in plasma by methylene blue plus visible light. *Vox Sang.* **2020**, *115*, 146–151. [CrossRef]
30. Praditya, D.; Friesland, M.; Gravemann, U.; Handke, W.; Todt, D.; Behrendt, P.; Müller, T.H.; Steinmann, E.; Seltsam, A. Hepatitis E virus is effectively inactivated in platelet concentrates by ultraviolet C light. *Vox Sang.* **2020**, *115*, 555–561. [CrossRef]

31. Brixner, V.; Bug, G.; Pohler, P.; Krämer, D.; Metzner, B.; Voss, A.; Casper, J.; Ritter, U.; Klein, S.; Alakel, N.; et al. Efficacy of UVC-treated, pathogen-reduced platelets versus untreated platelets: A randomized controlled non-inferiority trial. *Haematologica* **2021**, *106*, 1086–1096. [CrossRef]
32. Bashir, S.; Cookson, P.; Wiltshire, M.; Hawkins, L.; Sonoda, L.; Thomas, S.; Seltsam, A.; Tolksdorf, F.; Williamson, L.M.; Cardigan, R. Pathogen inactivation of platelets using ultraviolet C light: Effect on in vitro function and recovery and survival of platelets. *Transfusion* **2013**, *53*, 990–1000. [CrossRef] [PubMed]
33. Schulze, T.J.; Graveman, U.; Seltsam, A. THERAFLEX ultraviolet C (UVC)-based pathogen reduction technology for bacterial inactivation in blood components: Advantages and limitations. *Ann. Blood* **2002**, *7*, 28. [CrossRef]
34. Ramirez-Arcos, S.; Evans, S.; McIntyre, T.; Pang, C.; Yi, Q.L.; DiFranco, C.; Goldman, M. Extension of platelet shelf life with an improved bacterial testing algorithm. *Transfusion* **2020**, *60*, 2918–2928. [CrossRef] [PubMed]
35. Ohlsson, S.; Diedrich, B.; Uhlin, M.; Sandgren, P. Optimized processing for pathogen inactivation of double-dose buffy-coat platelet concentrates: Maintained in vitro quality over 7-day storage. *Vox Sang.* **2018**, *113*, 611–621. [CrossRef]
36. Gorria, C.; Labata, G.; Lezaun, M.; López, F.J.; Pérez Aliaga, A.I.; Pérez Vaquero, M.Á. Impact of implementing pathogen reduction technologies for platelets on reducing outdates. *Vox Sang.* **2020**, *115*, 167–173. [CrossRef]
37. Pitman, J.P.; Payrat, J.M.; Park, M.S.; Liu, K.; Corash, L.; Benjamin, R.J. Longitudinal analysis of annual national hemovigilance data to assess pathogen reduced platelet transfusion trends during conversion to routine universal clinical use and 7-day storage. *Transfusion* **2023**, *63*, 711–723. [CrossRef]
38. Fast, L.D.; Dileone, G.; Li, J.; Goodrich, R. Functional inactivation of white blood cells by Mirasol treatment. *Transfusion* **2006**, *46*, 642–648. [CrossRef]
39. Corash, L.; Benjamin, R.J. The role of hemovigilance and postmarketing studies when introducing innovation into transfusion medicine practice: The amotosalen-ultraviolet A pathogen reduction treatment model. *Transfusion* **2016**, *56* (Suppl. S1), S29–S38. [CrossRef] [PubMed]
40. Fast, L.D.; DiLeone, G.; Marschner, S. Inactivation of human white blood cells in platelet products after pathogen reduction technology treatment in comparison to gamma irradiation. *Transfusion* **2011**, *51*, 1397–1404. [CrossRef]
41. Sim, J.; Tsoi, W.C.; Lee, C.K.; Leung, R.; Lam, C.C.K.; Koontz, C.; Liu, A.Y.; Huang, N.; Benjamin, R.J.; Vermeij, H.J.; et al. Transfusion of pathogen-reduced platelet components without leukoreduction. *Transfusion* **2019**, *59*, 1953–1961. [CrossRef]
42. Fast, L.D. Developments in the prevention of transfusion-associated graft-versus-host disease. *Br. J. Haematol.* **2012**, *158*, 563–568. [CrossRef] [PubMed]
43. Lieberman, L.; Devine, D.V.; Reesink, H.W.; Panzer, S.; Wong, J.; Raison, T.; Benson, S.; Pink, J.; Leitner, G.C.; Horvath, M.; et al. Prevention of transfusion-transmitted cytomegalovirus (CMV) infection: Standards of care. *Vox Sang.* **2014**, *107*, 276–311. [CrossRef] [PubMed]
44. INTERCEPT Platelets, Technical Data Sheet. Available online: https://www.interceptbloodsystem.com/sites/default/files/resources/prd-tds_00121_v10.0_reduced_file_size_-_secured.pdf (accessed on 15 August 2024).
45. Lanteri, M.C.; Santa-Maria, F.; Laughhunn, A.; Girard, Y.A.; Picard-Maureau, M.; Payrat, J.M.; Irsch, J.; Stassinopoulos, A.; Bringmann, P. Inactivation of a broad spectrum of viruses and parasites by photochemical treatment of plasma and platelets using amotosalen and ultraviolet A light. *Transfusion* **2020**, *60*, 1319–1331. [CrossRef]
46. Available online: https://www.terumobct.com/Public/306690232.pdf (accessed on 25 August 2024).
47. Available online: https://aob.amegroups.org/article/view/6833/html (accessed on 25 August 2024).
48. Castro, E.; González, L.M.; Rubio, J.M.; Ramiro, R.; Gironés, N.; Montero, E. The efficacy of the ultraviolet C pathogen inactivation system in the reduction of Babesia divergens in pooled buffy coat platelets. *Transfusion* **2014**, *54*, 2207–2216. [CrossRef] [PubMed]
49. Gravemann, U.; Boelke, M.; Könenkamp, L.; Söder, L.; Maurer, M.; Ziegler, U.; Schulze, T.J.; Seltsam, A.; Becker, S.C.; Steffen, I. West Nile and Usutu viruses are efficiently inactivated in platelet concentrates by UVC light using the THERAFLEX UV-Platelets system. *Vox Sang.* **2024**, *119*, 827–833. [CrossRef]
50. Coene, J.; Devreese, K.; Sabot, B.; Feys, H.B.; Vandekerckhove, P.; Compernolle, V. Paired analysis of plasma proteins and coagulant capacity after treatment with three methods of pathogen reduction. *Transfusion* **2014**, *54*, 1321–1331. [CrossRef]
51. Ravanat, C.; Dupuis, A.; Marpaux, N.; Naegelen, C.; Mourey, G.; Isola, H.; Laforet, M.; Morel, P.; Gachet, C. Is the thrombin generation test, a useful research tool, suitable to characterize therapeutic plasma? In Proceedings of the ISTH 2017, Berlin, Germany, 8–13 July 2017.
52. Flamholz, R.; Jeon, H.R.; Baron, J.M.; Baron, B.W. Study of three patients with thrombotic thrombocytopenic purpura exchanged with solvent/detergent-treated plasma: Is its decreased protein S activity clinically related to their development of deep venous thromboses? *J. Clin. Apher.* **2000**, *15*, 169–172. [CrossRef]
53. Yarranton, H.; Lawrie, A.S.; Purdy, G.; Mackie, I.J.; Machin, S.J. Comparison of von Willebrand factor antigen, von Willebrand factor-cleaving protease and protein S in blood components used for treatment of thrombotic thrombocytopenic purpura. *Transfus. Med.* **2004**, *14*, 39–44. [CrossRef]
54. Hellstern, P.; Solheimb, B.G. The Use of Solvent/Detergent Treatment in Pathogen Reduction of Plasma. *Transfus. Med. Hemother.* **2011**, *38*, 65–70. [CrossRef]
55. Piccin, A.; O'Connor-Byrne, N.; Daves, M.; Lynch, K.; Farshbaf, A.D.; Martin-Loeches, I. Autoimmune disease and sickle cell anaemia: 'Intersecting pathways and differential diagnosis'. *Br. J. Haematol.* **2022**, *197*, 518–528. [CrossRef]

56. Backholer, L.; Wiltshire, M.; Proffitt, S.; Cookson, P.; Cardigan, R. Paired comparison of methylene blue- and amotosalen-treated plasma and cryoprecipitate. *Vox Sang.* **2016**, *110*, 352–361. [CrossRef] [PubMed]
57. Larrea, L.; Ortiz-de-Salazar, M.I.; Martínez, P.; Roig, R. Quantitative analysis of plasma proteins in whole blood-derived fresh frozen plasma prepared with three pathogen reduction technologies. *Transfus. Apher. Sci.* **2015**, *52*, 305–310. [CrossRef]
58. Estcourt, L.J.; Malouf, R.; Hopewell, S.; Trivella, M.; Doree, C.; Stanworth, S.J.; Murphy, M.F. Pathogen-reduced platelets for the prevention of bleeding. *Cochrane Database Syst. Rev.* **2017**, *7*, CD009072. [CrossRef]
59. Pati, I.; Masiello, F.; Pupella, S.; Cruciani, M.; De Angelis, V. Efficacy and Safety of Pathogen-Reduced Platelets Compared with Standard Apheresis Platelets: A Systematic Review of RCTs. *Pathogens* **2022**, *11*, 639. [CrossRef] [PubMed]
60. Rebulla, P.; Vaglio, S.; Beccaria, F.; Bonfichi, M.; Carella, A.; Chiurazzi, F.; Coluzzi, S.; Cortelezzi, A.; Gandini, G.; Girelli, G.; et al. Clinical effectiveness of platelets in additive solution treated with two commercial pathogen-reduction technologies. *Transfusion* **2017**, *57*, 1171–1183. [CrossRef] [PubMed]
61. Garban, F.; Guyard, A.; Labussière, H.; Bulabois, C.E.; Marchand, T.; Mounier, C.; Caillot, D.; Bay, J.O.; Coiteux, V.; Schmidt-Tanguy, A.; et al. Evaluation of the Efficacy of Platelets Treated with Pathogen Reduction Process (EFFIPAP) Study Group. Comparison of the Hemostatic Efficacy of Pathogen-Reduced Platelets vs Untreated Platelets in Patients with Thrombocytopenia and Malignant Hematologic Diseases: A Randomized Clinical Trial. *JAMA Oncol.* **2018**, *4*, 468–475.
62. Osselaer, J.C.; Doyen, C.; Defoin, L.; Debry, C.; Goffaux, M.; Messe, N.; Van Hooydonk, M.; Bosly, A.; Lin, J.S.; Lin, L.; et al. Universal adoption of pathogen inactivation of platelet components: Impact on platelet and red blood cell component use. *Transfusion* **2009**, *49*, 1412–1422. [CrossRef]
63. Cazenave, J.P.; Isola, H.; Waller, C.; Mendel, I.; Kientz, D.; Laforêt, M.; Raidot, J.P.; Kandel, G.; Wiesel, M.L.; Corash, L. Use of additive solutions and pathogen inactivation treatment of platelet components in a regional blood center: Impact on patient outcomes and component utilization during a 3-year period. *Transfusion* **2011**, *51*, 622–629. [CrossRef]
64. Infanti, L.; Holbro, A.; Passweg, J.; Bolliger, D.; Tsakiris, D.A.; Merki, R.; Plattner, A.; Tappe, D.; Irsch, J.; Lin, J.S.; et al. Clinical impact of amotosalen-ultraviolet A pathogen-inactivated platelets stored for up to 7 days. *Transfusion* **2019**, *59*, 3350–3361. [CrossRef]
65. Amato, M.; Schennach, H.; Astl, M.; Chen, C.Y.; Lin, J.S.; Benjamin, R.J.; Nussbaumer, W. Impact of platelet pathogen inactivation on blood component utilization and patient safety in a large Austrian Regional Medical Centre. *Vox Sang.* **2017**, *112*, 47–55. [CrossRef]
66. Nussbaumer, W.; Amato, M.; Schennach, H.; Astl, M.; Chen, C.Y.; Lin, J.S.; Corash, L.; Benjamin, R.J. Patient outcomes and amotosalen/UVA-treated platelet utilization in massively transfused patients. *Vox Sang.* **2017**, *112*, 249–256. [CrossRef] [PubMed]
67. McCullough, J.; Vesole, D.H.; Benjamin, R.J.; Slichter, S.J.; Pineda, A.; Snyder, E.; Stadtmauer, E.A.; Lopez-Plaza, I.; Coutre, S.; Strauss, R.G.; et al. Therapeutic efficacy and safety of platelets treated with a photochemical process for pathogen inactivation: The SPRINT Trial. *Blood* **2004**, *104*, 1534–1541. [CrossRef] [PubMed]
68. Blake, J.T.; McTaggart, K.; Couture, C. Estimating the impact on the inventory of implementing pathogen-reduced platelets in Canada. *Transfusion* **2021**, *61*, 3150–3160. [CrossRef]
69. Murphy, S.; Snyder, E.; Cable, R.; Slichter, S.J.; Strauss, R.G.; McCullough, J.; Lin, J.; Corash, L.; Conlan, M.G. SPRINT Study Group. Platelet dose consistency and its effect on the number of platelet transfusions for support of thrombocytopenia: An analysis of the SPRINT trial of platelets photochemically treated with amotosalen HCl and ultraviolet A light. *Transfusion* **2006**, *46*, 24–33. [CrossRef]
70. Stivala, S.; Gobbato, S.; Infanti, L.; Reiner, M.F.; Bonetti, N.; Meyer, S.C.; Camici, G.G.; Lüscher, T.F.; Buser, A.; Beer, J.H. Amotosalen/ultraviolet A pathogen inactivation technology reduces platelet activatability, induces apoptosis and accelerates clearance. *Haematologica* **2017**, *102*, 1650–1660. [CrossRef]
71. Kim, S.; Handke, W.; Gravemann, U.; Döscher, A.; Brixner, V.; Müller, T.H.; Seltsam, A. Mitochondrial DNA multiplex real-time polymerase chain reaction inhibition assay for quality control of pathogen inactivation by ultraviolet C light in platelet concentrates. *Transfusion* **2018**, *58*, 758–765. [CrossRef]
72. Bruchmüller, I.; Lösel, R.; Bugert, P.; Corash, L.; Lin, L.; Klüter, H.; Janetzko, K. Effect of the psoralen-based photochemical pathogen inactivation on mitochondrial DNA in platelets. *Platelets* **2005**, *16*, 441–445. [CrossRef] [PubMed]
73. Bakkour, S.; Chafets, D.M.; Wen, L.; Dupuis, K.; Castro, G.; Green, J.M.; Stassinopoulos, A.; Busch, M.P.; Lee, T.H. Assessment of nucleic acid modification induced by amotosalen and ultraviolet A light treatment of platelets and plasma using real-time polymerase chain reaction amplification of variable length fragments of mitochondrial DNA. *Transfusion* **2016**, *56*, 410–420. [CrossRef]
74. Diallo, I.; Benmoussa, A.; Laugier, J.; Osman, A.; Hitzler, W.E.; Provost, P. Platelet Pathogen Reduction Technologies Alter the MicroRNA Profile of Platelet-Derived Microparticles. *Front. Cardiovasc. Med.* **2020**, *7*, 31. [CrossRef]
75. Arnason, N.A.; Johannson, F.; Landrö, R.; Hardarsson, B.; Irsch, J.; Gudmundsson, S.; Rolfsson, O.; Sigurjonsson, O.E. Pathogen inactivation with amotosalen plus UVA illumination minimally impacts microRNA expression in platelets during storage under standard blood banking conditions. *Transfusion* **2019**, *59*, 3727–3735. [CrossRef]
76. Farrugia, A. Cost of pathogen reduction for platelets: Reply to Cicchetti et al. Blood Transfus 2018; 16: 483-9. *Blood Transfus.* **2019**, *17*, 84. [PubMed]
77. Farrugia, A. The Evolution of the Safety of Plasma Products from Pathogen Transmission—A Continuing Narrative. *Pathogens* **2023**, *12*, 318. [CrossRef] [PubMed]

78. Bell, C.E.; Botteman, M.F.; Gao, X.; Weissfeld, J.L.; Postma, M.J.; Pashos, C.L.; Triulzi, D.; Staginnus, U. Cost-effectiveness of transfusion of platelet components prepared with pathogen inactivation treatment in the United States. *Clin. Ther.* **2003**, *25*, 2464–2486. [CrossRef] [PubMed]
79. Davison, N.; Willis, A. Abstracts of the 31st Regional ISBT Congress, ISBT In Focus, Virtual meeting, 2–8 June 2021. *Vox Sang.* **2021**, *116* (Suppl. S1), 5–188. [CrossRef]
80. Grégoire, Y.; Delage, G.; Custer, B.; Rochette, S.; Renaud, C.; Lewin, A.; Germain, M. Cost-effectiveness of pathogen reduction technology for plasma and platelets in Quebec: A focus on potential emerging pathogens. *Transfusion* **2022**, *62*, 1208–1217. [CrossRef] [PubMed]
81. Rosskopf, K.; Helmberg, W.; Schlenke, P. Pathogen reduction of double-dose platelet concentrates from pools of eight buffy coats: Product quality, safety, and economic aspects. *Transfusion* **2020**, *60*, 2058–2066. [CrossRef]
82. Keltner, N.M.; Cushing, M.M.; Haas, T.; Spinella, P.C. Analyzing and modeling massive transfusion strategies and the role of fibrinogen-How much is the patient actually receiving? *Transfusion* **2024**, *64* (Suppl. S2), S136–S145. [CrossRef]

Journal of
Clinical Medicine

MDPI

Article

Aerobic Microbiological Spectrum and Antibiotic Resistance in Children Operated for Anorectal Abscesses

Dzhevdet Chakarov [1,2], Elena Hadzhieva [1,2,*], Yordan Kalchev [3,4] and Dimitar Hadzhiev [1,2]

[1] Section of General Surgery, Department of Propedeutics of Surgical Diseases, Faculty of Medicine, Medical University of Plovdiv, 4001 Plovdiv, Bulgaria; dzhevdet.chakarov@mu-plovdiv.bg (D.C.); dimitar.hadzhiev@mu-plovdiv.bg (D.H.)
[2] First Clinic of Surgery, University Hospital St. George, 4001 Plovdiv, Bulgaria
[3] Department of Medical Microbiology and Immunology "Prof. Dr. Elissay Yanev", Faculty of Medicine, Medical University of Plovdiv, 4002 Plovdiv, Bulgaria; yordan.kalchev@mu-plovdiv.bg
[4] Laboratory of Microbiology, University Hospital St. George, 4002 Plovdiv, Bulgaria
* Correspondence: elena.hadzhieva@mu-plovdiv.bg

Abstract: (1) Background: Anorectal abscesses are a relatively rare pathology in childhood. Most often, male children under 1 year of age are affected. The importance of microbiological examination for the diagnosis and treatment of such patients remains debatable among surgeons, resulting in scarce data being available in the literature. We aimed to identify the aerobic microbiological spectrum and antibiotic resistance of isolates in children undergoing operation to treat anorectal abscesses. **(2) Methods**: We performed a case series of 102 children diagnosed and operated for anorectal abscesses over a period of 10 years (2010–2019). Purulent wound exudate was used for microbiological evaluation, which was subsequently cultured on 5% sheep-blood agar and eosin–methylene blue agar. For microbiological identification, conventional biochemical tests and semi-automated (API 20, bioMerieux, Marcy-l'Étoile, France) tests were used, as well as automated systems (Vitek-2 Compact, bioMerieux, France). Antimicrobial susceptibility testing was performed by the disk diffusion method of Bauer–Kirby and by determining the minimal inhibitory concentrations for glycopeptides. The results were interpreted according to the EUCAST standard for the corresponding year. **(3) Results**: Microbiological testing in children operated for anorectal abscesses mainly identified the gut commensals that normally reside in the rectal mucosa. Monocultures were found in just over half of the cases. *Escherichia coli*, *Klebsiella pneumoniae complex*, and *Proteus mirabilis* were the most frequently isolated. In addition, *Staphylococcus aureus* was found in 7% of patients. In Gram-negative bacteria, antibiotic resistance was most often observed in penicillins, cephalosporins, sulfonamides, and fluoroquinolones. **(4) Conclusions**: The increasing rates of antimicrobial resistance impose the need for the local monitoring of circulating commensal bacteria associated with anorectal abscesses in children to guide antibiotic therapy when indicated.

Keywords: anorectal abscess (ARA); microbial spectrum; aerobes; anorectal fistula

1. Introduction

Anorectal abscesses (ARAs) are among the most common inflammatory diseases in the anorectal area in adult patients and, although less common, the disease is also present in childhood [1]. The lower incidence of the disease in children could be attributed to the anatomical, physiological, and behavioral differences between children and adults [2–4]. The disease affects predominantly male children under 1 year of age but can be found in both sexes up to 18 years of age, possibly in association with Crohn's disease, Hirschsprung's disease, penetrating trauma, and immunosuppression [5–7].

ARAs are characterized by the accumulation of pus in the anorectal region. The disease in healthy children is usually self-limiting and often requires only surgical treatment. However, ARAs pose the risk of serious complications, including fistula formation and systemic infection, especially when risk factors are present, thus making the therapeutic approach more challenging [8,9]. Complications and the development of perianal fistula can reach 24% when associated with colonic pathology. The main trigger for the onset of ARA, in both pediatric and adult patients, is the infection of the pararectal tissues. While the cryptoglandular theory explains the infection in more than 90% of adults, it has inconsistencies and points of controversy, especially in younger children [10–12].

The anorectal region has a unique microbial commensal microbiota, composed of both aerobic and anaerobic bacteria that contribute to the pathogenesis of ARAs' formation. Studies on the bacteria associated with ARA have provided some insights into the microbial landscape of the anorectal area, demonstrating a diverse community of bacteria that could contribute to the development of these abscesses. Among these, organisms such as *Escherichia coli*, *Staphylococcus aureus*, *Streptococcus* spp., *Bacteroides* spp., and *Fusobacterium* spp. are frequently implicated. ARAs are often reported to be polymicrobial in nature, involving complex interactions between multiple microbial species. This polymicrobial aspect complicates the diagnosis and treatment, as different bacteria may have varying susceptibility to antibiotics [13–19].

However, in the medical literature, studies of the microbiological spectrum of aerobic agents of ARA in children are scarce. This demonstrates that ARAs in children still lack clarity in terms of their etiology and diagnosis, and the therapeutic approaches. Understanding the microbial causes of anorectal abscesses is crucial for developing effective treatment strategies and improving patient outcomes. The etiology and antimicrobial resistance of isolates can enhance the effective institutional antibiotic policy and the selection of empiric treatment. The following case series aims to determine the aerobic microbiological spectrum and antibiotic resistance of isolates in children operated for anorectal abscesses.

2. Material and Methods

2.1. Selection of Patients

We performed a case series involving 102 children operated for ARA for a period of 10 years (2010–2019). The study was designed to collect data by reviewing medical records that involved sex, age, microbiological results, and antimicrobial susceptibility reports. All children were hospitalized and treated in the First Surgery Clinic and the Paediatric Surgery Clinic of "St. George" University Hospital. We defined the study group based on the following inclusion and exclusion criteria:

Inclusion criteria:

- All children with ARA aged from 0 to 18 years.

Exclusion criteria:

- Patients with anorectal fistula (ARF);
- Cases with suppurated cusp of pilaris regio sacralis;
- Patients with anorectal form of Crohn's disease;
- Cases with Proctitis ulcerohaemorrhagica chronica;
- Cases with chronic specific diseases (TBC, actinomycosis, etc.), leading to the appearance of ARA;
- Patients with Bartolinitis acuta;
- Cases with Pyodermia fistulosa;
- Cases with pararectal purulent inflammation of non-cryptoglandular origin (perianal skin furuncle, suppurative perianal atheroma, purulent hidradenitis, suppurative teratoma).

2.2. Microbiological Examination

For the microbiological examination, purulent wound exudate was taken with a dry sterile swab in a transport medium (Amies, Biolife Italiana S.r.l., Milan, Italy), during the operative intervention, at the initial incision. Inoculation was carried out using 5% sheep-blood agar (Diachim, Sofia, Bulgaria) and a selective medium for Gram-negative bacteria–eosin-methylene blue agar (Diachim, Sofia, Bulgaria). The agars were incubated for 18–22 h at 36–37 °C. Conventional manual biochemical tests, semi-automated (API 20, bioMerieux, Marcy-l'Étoile, France), and automated systems (Vitek-2 Compact, bioMerieux, Marcy-l'Étoile, France) were used for microbiological identification. All cultured microorganisms were subjected to identification and testing to determine antimicrobial susceptibility to antibiotic agents.

Antimicrobial susceptibility was determined via the Bauer–Kirby disk diffusion method, and an automated system was used to establish minimum inhibitory concentrations (Vitek-2 Compact, bioMerieux, Marcy-l'Étoile, France). When interpreting the results of the antimicrobial susceptibility tests, the current version of the EUCAST standard for the relevant year was used.

We did not perform other follow-up microbiological tests because we did not aim to dynamically follow the changes occurring in the aerobic bacterial microbiota. We also aimed to investigate the aerobic microbiological landscape in the operated children with ARA; therefore, no cultures were created to detect anaerobic bacteria.

The following study endpoints encapsulate our primary objectives:

Demographic Characteristics

- Analysis of age profile and gender distribution among children with ARA.

Microbiological Profile of ARA Cases

- Investigation of the structure of microbial isolates as monoculture and microbial associations.
- Distribution of the etiological structure of ARA in children.
- Data on antimicrobial resistance in the most common isolates.

2.3. Statistical Methods

The statistical package SPSS Statistics v.26 (IBM Corp., Chicago, IL, USA) was used for the statistical processing of the results. Numerical values are stated as mean ± standard deviation and categorical data are presented as frequency (n) and percentage (%).

3. Results

Among the children with ARA, 91 (89.2%) were boys and 11 (10.8%) were girls, as shown in Table 1. The mean age of the children included in the study was 1.6 years (SD ± 3.15). The mean age of boys ($n = 91$) was 1.45 years, SD ± 2.93, while the mean age of girls ($n = 11$) was 2.79 yrs, SD ± 4.43. Of the children that were operated on, the youngest patient was 1 month old and the oldest was 17 years old. The most affected age group among children was neonates and infants, with a total of 65 patients (63.7%), as shown in Figure 1. In infancy, seven children (6.9%) were found to have concomitant diseases, most of them with a congenital genesis.

Table 1. Characteristics of the studied contingent.

	N	Mean Age	Standard Deviation
Boys	91	1.45	2.93
Girls	11	2.79	4.43

The distribution of operated children with ARA according to age and sex is presented in Figure 1.

The results obtained from the bacteriological cultures were divided into three main groups: monocultures, mixed cultures, and sterile cultures (Table 2):

Table 2. Distribution of microbiological results by groups.

	N	%
Monocultures	64	62.7
Mixed cultures	12	11.8
Sterile	26	25.5
Total	102	100

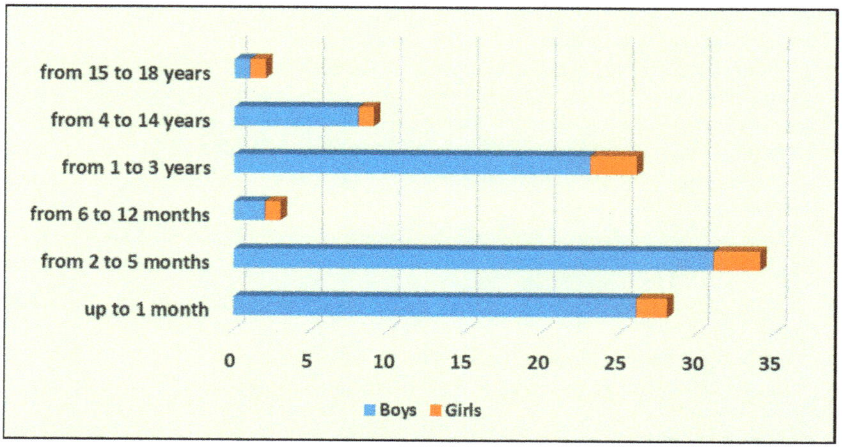

Figure 1. Distribution of the age and sex of operated children with ARA.

Table 3 presents the structure of microbial isolates as a monoculture. The most commonly isolated organism in the monoculture was *E. coli* (56%), followed by *Klebsiella* spp. (20%). *S. aureus* was cultured in 12.5% of patients.

Table 3. Data on the etiological structure of monocultures.

Structure of Established Agents as a Monoculture	Frequency	Percent
E. coli	36	56.3%
Klebsiella spp. (*K. pneumoniae complex*, *K. oxytoca*, *K. aerogenes*)	13	20.2%
S. aureus	8	12.5%
Enterococcus faecalis	2	3.1%
Enterobacter cloacae complex	1	1.6%
Proteus mirabilis	1	1.6%
Citrobacter freundii	1	1.6%
Serratia marcescens	1	1.6%
Streptococcus beta-haem. Group A	1	1.6%
Total	64	100%

The resulting microbial associations are presented in Table 4. The most frequently observed microbial association was between *E. coli* and *P. mirabilis* and between *E. coli* and *K. pneumoniae complex*. In addition, *E. coli* participated most often in microbial associations.

The microorganisms that were most represented, either as a monoculture or in microbial associations, were *E. coli*, *K. pneumoniae complex*, *P. mirabilis*, and *S. aureus*, as shown in Table 5.

Table 6 presents the data on the determination of the antimicrobial sensitivity of the four most frequently presented microorganisms in the etiological structure of ARA in children during the observation period. The highest levels of resistance in *E. coli* isolates were found for ampicillin, amoxicillin/clavulanic acid, and piperacillin. Resistance levels to third-generation cephalosporins reached 28%. Resistance to trimethoprim/sulfamethoxazole was found in 27% of the isolates. There was 100% sensitivity to carbapenems.

Table 4. Data on the etiological structure of microbial associations.

Microbial Association	Frequency	Percent
E. coli + *P. mirabilis*	4	33.3%
E. coli + *K. pneumoniae complex*	2	16.7%
E. coli + *E. faecalis*	1	8.3%
E. coli + *E. coli ESBLs* (+)	1	8.3%
Proteus spp. + *K. aerogenes*	1	8.3%
P. mirabilis + *E. cloacae complex*	1	8.3%
Pseudomonas aeruginosa + *P. mirabilis*	1	8.3%
E. cloacae complex + *P. aeruginosa*	1	8.3%
Total	12	100%

Table 5. Data on the etiological structure of ARA in children.

Structure of Isolates in Children	Frequency	Percent
E. coli	45	39.5%
K. pneumoniae complex	11	9.6%
P. mirabilis	8	7.0%
S. aureus	8	7.0%
E. cloacae complex	3	2.6%
E. faecalis	3	2.6%
K. oxytoca	3	2.6%
K. aerogenes	2	1.8%
P. aeruginosa	2	1.8%
Streptococcus beta-haem. Group A	1	0.9%
S. marcescens	1	0.9%
C. freundii	1	0.9%
Sterile cultures	26	22.8%
Total	114	100%

Table 6. Antimicrobial resistance (%) in the most common ARA isolates in children.

	E. coli	*K. pneumoniaeComplex*	*P. mirabilis*		*S. aureus*
AMP	60	i.r.	29	PEN	88
PIP	51	27	25	FOX	29
TZP	19	0	0	ERY	17
AMC	58	36	29	CLI	0
SCF	6	0	25	AMK	14
FOX	14	9	14	SXT	0
CTX	28	20	14	VAN	0
CRO	28	20	14	TEC	0
FEP	21	20	14	LZD	0
IPM	0	0	0	TGC	0
MEM	0	0	0		
AMK	5	0	0		
CIP	9	9	0		
LVX	11	0	0		
SXT	27	0	0		
TGC	7	-	-		

AMP—ampicillin; PIP—piperacillin; TZP—piperacillin/tazobactam; AMC—amoxicillin/clavulanic acid; CSF—cefoperazone/sulbactam; FOX—cefoxitin; CTX—cefotaxime; CRO—ceftriaxone; FEP—Cefepime; IPM—imipenem; MEM—meropenem; AMK—Amikacin; CIP—ciprofloxacin; LVX—levofloxacin; SXT—trimethoprim/sulfamethoxazole; TGC—tigecycline; PEN—penicillin; ERY—erythromycin; CLI—clindamycin; VAN—vancomycin; TEC—teicoplanin; LZD—linezolid; i.r.—intrinsic resistance.

For *K. pneumoniae*, resistance to amoxicillin/clavulanic acid was recorded in 36% and resistance to piperacillin was recorded in 27%. Resistance to third-generation cephalosporins reached 22%. Of the isolates, 9% were resistant to levofloxacin. Sensitivity to carbapenems, piperacillin/tazobactam, cefoperazone/sulbactam, amikacin, and trimethoprim/sulfamethoxazole was maintained.

In *P. mirabilis*, resistance to ampicillin and amoxicillin/clavulanic acid was demonstrated in 29% and resistance to piperacillin and cefoperazone/sulbactam was demonstrated in 25%. ESBL producers were detected in 14%. Sensitivity to carbapenems, piperacillin/tazobactam, and cefoperazone/sulbactam was maintained. Susceptibility to carbapenems, the fluoroquinolones levofloxacin and ciprofloxacin, as well as to amikacin and trimethoprim/sulfamethoxazole, was observed.

Penicillin resistance levels in *S. aureus* reached 88%. MRSA isolates were found in 29%. Tested microorganisms showed sensitivity to trimethoprim/sulfamethoxazole, clindamycin, glycopeptides, linezolid, and tigecycline.

4. Discussion

The lack of clarity and consensus on the etiopathogenesis and the means of occurrence of ARAs in childhood generates controversial questions about their genesis and therapeutic management. Whenever a clinical analysis of the aerobic microbiological landscape in children with ARA is performed, one must consider the age-specific anatomical differences and age-related changes in the gut microbiome. It is well established that weak immunity in early childhood leads to decreased local protective factors, which is an important prerequisite for the development of a purulent inflammatory process [20,21]. This explains the poor resistance of perianal tissues in childhood to purulent infection. The cryptoglandular origin of ARA explains the penetration of the infection in more than 90% of adult patients very well, but regarding the same theory, there are several points of discussion and inconsistencies in children. According to this theory, the origin of purulent inflammation is from the anal glands, but these are functionally inactive in children and only begin to secrete after puberty. Some authors believe that ARA infections in childhood often result from penetrating rectal wall injuries, local pyoderma, skin injuries, cryptitis, proctitis, rhagades, congenital anorectal malformations, anomalies of the ducts of the anal glands, transient intestinal disorders, etc. [22]. The vulnerability of the epidermis, especially in infancy, with the presence of still-underdeveloped basement membranes, allows for lesions, even from microtrauma, to provide a gateway opportunity for infection.

It should be emphasized that electronic databases of medical publications on the aerobic microbial spectrum in operated children with ARA are very scarce. Logically, a bacteriological examination of operated children with ARA will mainly detect the microbial agents that are normally present in the rectal lumen. In a few clinical cases, microbiological analysis may reveal a mixed infection consisting of microbial associations. The present study found evidence of monocultures to be more frequent among aerobic opportunistic enteric bacteria. According to Abercrombie JF et al., the clinical and microbiological characteristics of pediatric perianal abscesses are like those of adults [23]. According to our data, the most common microorganism with an undisputed leading role as the causative agent of ARA in childhood is *E. coli*, with a relative contribution of about 40%. Other common representatives of the aerobic microbiological landscape are *K. pneumoniae complex* (9.6%) and *P. mirabilis*—7.0%. Of interest is the study by Zhu Y et al., reporting that *K. pneumoniae* is the most common pathogen in perianal abscess in infants aged less than 3 months and is usually resistant to ampicillin and nitrofurantoin. Since perianal abscess in infants younger than 3 months is a disease with a tendency to self-heal, a simple surgical intervention with the synchronous administration of antibiotics is suggested as the optimal treatment [24]. For example, A. Niyogi et al. [3] found the presence of enteric microorganisms in a large proportion of recurrent ARA patients. According to Tan Tanny SP. et al., the presence of enteric microorganisms in microscopic and culture studies in pediatric perianal abscess was not associated with fistula formation nor with abscess recurrence. Their studies also showed a predominance of enteric microorganisms in these patients [25].

Our results show that *S. aureus* (7.0%) is not an uncommon isolate of ARA in childhood, and an intriguing observation is that it occurred only as a monoculture and never in microbiological association. It is important to keep in mind that, in children, it is not uncommon to mix true ARA (infection from the intestinal lumen) with other, similar purulent inflammations that are nearby, such as furuncles, abscesses, and purulent adenitis. (infection from the skin and its appendages) [26,27]. These latter non-cryptoglandular purulent-inflammatory processes are not infrequently assumed to be ARAs, and are significantly induced by microorganisms of the genus Staphylococcus or with the major involvement of *S. aureus*. This suggests the possibility of inaccurate diagnosis in childhood with the presence of infected furuncles, skin abscesses, purulent adenitis, local pyodermas, etc., with *S. aureus* as the main causative agent. These conditions may be mistakenly categorized as ARA.

Questions about the need to establish an etiologic spectrum and antimicrobial resistance in ARA are controversial. There are conflicting statements about the clinical effect of identifying the specific microbiological causative agent and its antibiotic treatment, as some physicians refrain from using antibiotics in patients with ARA. It should be noted that all authoritative surgical opinions of the coloproctology community currently have a reserved and cautious attitude regarding the use of antibacterial therapy in ARA, as this is only recommended in a septic and complicated clinical course [28]. According to the American Society of Coloproctologists (ASCRS) medical standard, antibacterial treatment in ARA is selectively required only in the setting of complications with diffuse pararectal cellulitis, the presence of a systemic septic reaction, and evidence of immunosuppression in patients. According to them, the use of antibiotics for uncomplicated ARA in healthy children has no application, does not improve the healing process, and does not reduce new recurrences [29]. The SICCR and the German S3 guideline have similar views, generally considering antibiotics to have a limited role in trivial ARA, believing they should not be used in uncomplicated ARA, and considering them to be inappropriate [30–33]. For example, Shaughnessy MP et al. concluded that the results of microbiological cultures are of limited utility in the treatment of pediatric pilonidal, ischial, and perianal abscesses, as they do not appear to alter treatment, and omitting culture collection is not associated with failure of surgical treatment [34]. M.S. Brar et al. [35] consider that, in the postoperative period, antibacterial therapy is important to prevent fistula formation, and they recommend it for a 7–10-day course. However, given the short-term follow-up of the study, it is unclear how routine antibiotic administration will reduce ARF formation in the long term. V. Mocanu et al. [36], in a 2019 study, reported that antibiotic therapy after the incision and drainage of the ARA was associated with 36% lower odds of fistula formation. The authors considered that an empirical 5–10-day course of antibiotics after surgical drainage may avoid subsequent fistula formation in healthy children, although the quality of the evidence was low. Further randomized trials are needed to fully clarify the role, duration, and type of antibiotics best suited for postoperative ARF prophylaxis after drainage in perirectal abscesses. According to U. Sozener et al. [37], antibiotic treatment after anorectal abscess drainage has no protective effect on the risk of fistula formation.

However, it should be noted that the supporters of such controversial statements are significantly fewer than those supporting the opposite view.

Analyzing the data from the 10-year microbial monitoring in operated children with ARA, and conditionally dividing it into two periods of 5 years, the following conclusions can be drawn:

- A periodic increase in the frequency of enteric bacteria was found.
- During the first 5-year period (2010–2014), there was an increase in the frequency of staphylococci, particularly Staphylococcus aureus, with a subsequent large decrease during the last period.

- In the second 5-year period (2015–2019), an increase in the number of microbial associations was observed, which indicates an increase in the diversity of microorganisms found.
- The various data regarding the presence of sterile cultures are interesting, with 34% being found in the first period and 14.1% in the last period.

The analysis of the obtained results revealed the most pronounced resistance to the group of penicillins, cephalosporins, sulphonamides, and fluoroquinolones. Most likely, this resistance is a consequence of the excessive administration of certain antibiotic agents and the mechanisms of transfer between bacteria of genes mediating antimicrobial resistance, allowing for rapid dissemination.

We believe that clarifying the microbial sensitivity is important in the application of appropriate empirical therapy when necessary and to initiate optimal antimicrobial treatment in a timely manner after the antibiogram result is established. Therefore, the availability of specific antimicrobial sensitivity data of isolates in operated children with ARA can help to make timely and accurate choices for starting empirical antibacterial therapy and, consequently, avoid the administration of drugs to which the microorganisms are resistant [34,37]. Data on the persistence or increase of resistance levels require optimization, or a re-evaluation of the currently performed antibacterial treatment is required, to build a newer and more advanced antibiotic therapy concept for operated children with ARA. Recommendations to improve the efficacy of antimicrobial therapy and a new approach to monitor and limit resistance manifestations are also needed.

5. Conclusions

The *E. coli* and *K. pneumoniae* complex that are commensals in the gut were most frequently found in children with ARA. A common isolate was *S. aureus*, which plays a controversial role in the occurrence of ARA. The highest levels of resistance in *Enterobacterales* were found to occur against penicillins and cephalosporins. Increasing levels of antimicrobial resistance necessitate monitoring of the etiological spectrum and antimicrobial resistance of isolates in ARA in children to guide the choice of antibiotic therapy when indicated. We suggest that monitoring the etiological spectrum and antimicrobial resistance are important factors for improving the care of children with this rare pathology. Furthermore, we believe that, with this research, we could enrich the literature with important epidemiological data on the current etiological spectrum and antimicrobial resistance of bacterial cultures in children with ARA.

Author Contributions: D.C. was responsible for the conceptualization and the original draft writing; E.H. and D.H. were responsible for the patient selections, treatment, and specimen collection; Y.K. was involved in the microbiological evaluation and interpretation. All authors have read and agreed to the published version of the manuscript.

Funding: This research received no external funding.

Institutional Review Board Statement: The current case study was exempt from ethical review and approval since all procedures and tests were made for diagnostic and treatment purposes that are consistent with the established algorithms for children with anorectal abscesses. No additional testing was performed on the children for research purposes.

Informed Consent Statement: Written informed consent was obtained from the patient's parents. All procedures performed in the present study were in accordance with the ethical standards of the institution and with the 1964 Helsinki Declaration and its later amendments.

Data Availability Statement: Data are available from the corresponding author on a reasonable request.

Conflicts of Interest: The authors declare no conflicts of interest.

References

1. Taylor, G.M.; Erlich, A.H. Perianal abscess in a 2-year-old presenting with a febrile seizure and swelling of the perineum. *Oxf. Med. Case Rep.* **2019**, *2019*, omy116. [CrossRef]

2. Ding, W.; Sun, Y.-R.; Wu, Z.-J. Treatment of Perianal Abscess and Fistula in Infants and Young Children: From Basic Etiology to Clinical Features. *Am. Surg.* **2020**, *87*, 927–932. [CrossRef]
3. Niyogi, A.; Agarwal, T.; Broadhurst, J.; Abel, R.M. Management of perianal abscess and fistula-in-ano in children. *Eur. J. Pediatr. Surg.* **2010**, *20*, 35–39. [CrossRef] [PubMed]
4. Ameh, E. Perianal abscess and fistula in children in Zaria. *Niger. Postgrad. Med. J.* **2003**, *10*, 107–109. [CrossRef]
5. Serour, F.; Somekh, E.; Gorenstein, A. Perianal abscess and fistula-in-ano in infants: A different entity? *Dis. Colon Rectum* **2005**, *48*, 359–364. [CrossRef]
6. Festen, C.; van Harten, H. Perianal abscess and fistula-in-ano in infants. *J. Pediatr. Surg.* **1998**, *33*, 711–713. [CrossRef] [PubMed]
7. Chang, H.K.; Ryu, J.G.; Oh, J.-T. Clinical characteristics and treatment of perianal abscess and fistula-in-ano in infants. *J. Pediatr. Surg.* **2010**, *45*, 1832–1836. [CrossRef]
8. Chae, K.J.; Lee, J.; Hwang, J.-H.; Hwang, J.-H. Invasive Hypervirulent *Klebsiella pneumoniae* Syndrome Originating from an Anorectal Abscess as Opposed to a Pyogenic Liver Abscess. *Medicina* **2022**, *58*, 1450. [CrossRef]
9. Sahnan, K.; Askari, A.; Adegbola, S.O.; Tozer, P.J.; Phillips, R.K.S.; Hart, A.; Faiz, O.D. Natural history of anorectal sepsis. *Br. J. Surg.* **2017**, *104*, 1857–1865. [CrossRef] [PubMed]
10. Roseau, G.; Palazzo, L. Echo-endoscopie ano-rectale. Applications en pédiatrie [Anorectal echo-endoscopy: Applications in pediatrics]. *Presse Med.* **1999**, *28*, 1946–1948. [PubMed]
11. Poenaru, D.; Yazbeck, S. Anal fistula in infants: Etiology, features, management. *J. Pediatr. Surg.* **1993**, *28*, 1194–1195. [CrossRef] [PubMed]
12. Schaarschmidt, K.; Schmid, K.W.; Eiff, C.; von Schleef, J.; Kolberg-Schwerdt, A.; Willital, G.H.; Stratmann, U. Pyogene periproktitische Infektionen des Kindesalters [Pyogenic periproctitis infections in childhood]. *Langenbecks Arch Chir* **1997**, *114*, 1346–1349. [PubMed]
13. Aygün, D.; Akçakaya, N.; Çokuğraş, H.; Camcıoğlu, Y. Evaluation of the Causes of Perianal Abscess in Childhood. *J. Pediatr. Infect.* **2019**, *13*, 21–27. [CrossRef]
14. Çelik, T.; Kaan, E.; Caglar, O. Perianal abscess in children: Are resistant microorganisms a real problem? *J. Pediatr. Inf.* **2022**, *16*, e106–e110.
15. Basaranoglu, S.T.; Ozsurekci, Y.; Cengiz, A.B.; Oncel, E.K.; Aykac, K.; Kara, A.; Ceyhan, M. Absceso perianal en niños: Perspectiva desde el campo de la infectología pediátrica [Perianal abscess in children: A pediatric infectious disease perspective]. *An. Pediatr.* **2019**, *90*, 370–375. [CrossRef] [PubMed]
16. Alabbad, J.; Raheem, F.A.; Alkhalifa, F.; Hassan, Y.; Al-Banoun, A.; Alfouzan, W. Retrospective Clinical and Microbiologic Analysis of Patients with Anorectal Abscess. *Surg. Infect.* **2019**, *20*, 31–34. [CrossRef] [PubMed]
17. Toyonaga, T.; Matsushima, M.; Tanaka, Y.; Shimojima, Y.; Matsumura, N.; Kannyama, H.; Nozawa, M.; Hatakeyama, T.; Suzuki, K.; Yanagita, K. Microbiological analysis and endoanal ultrasonography for diagnosis of anal fistula in acute anorectal sepsis. *Int. J. Color. Dis.* **2007**, *22*, 209–213. [CrossRef] [PubMed]
18. Yang, J.; Li, L.; Su, W.; Zhang, S.; Xu, H.; Wang, M.; Shen, W. Microbiomic signatures of anal fistula and putative sources of microbes. *Front. Cell. Infect. Microbiol.* **2024**, *14*, 1332490. [CrossRef] [PubMed]
19. Xu, R.W.; Tan, K.; Chong, C. Bacteriological study in perianal abscess is not useful and not cost-effective. *ANZ J. Surg.* **2016**, *86*, 782–784. [CrossRef] [PubMed]
20. Piazza, D.J.; Radhakrishnan, J. Perianal abscess and fistula-in-ano in children. *Dis. Colon Rectum* **1990**, *33*, 1014–1016. [CrossRef]
21. Macdonald, A.; Wilson-Storey, D.; Munro, F. Treatment of perianal abscess and fistula-in-ano in children. *Br. J. Surg.* **2003**, *90*, 220–221. [CrossRef] [PubMed]
22. Johnston, R.B.; Barton, L.L. Perianal Abscesses. *Pediatrics* **2008**, *121*, 447. [CrossRef]
23. Abercrombie, J.F.; George, B.D. Perianal abscess in children. *Ann. R. Coll. Surg. Engl.* **1992**, *74*, 385–386.
24. Zhu, Y.; Xu, F. The pathogens and curative effects analysis of perianal abscess of infants under 3 months. *Turk. J. Pediatr.* **2019**, *61*, 40–43. [CrossRef] [PubMed]
25. Tanny, S.P.T.; Wijekoon, N.; Nataraja, R.M.; Lynch, A.; Pacilli, M. Surgical management of perianal abscess in neonates and infants. *ANZ J. Surg.* **2020**, *90*, 1034–1036. [CrossRef]
26. Al-Salem, A.H.; Laing, W.; Talwalker, V. Fistula-in-ano in infancy and childhood. *J. Pediatr. Surg.* **1994**, *29*, 436–438. [CrossRef]
27. Ommer, A.; Herold, A.; Berg, E.; Fürst, A.; Sailer, M.; Schiedeck, T. Cryptoglandular anal fistulas. *Dtsch. Arztebl. Int.* **2011**, *108*, 707–713. [CrossRef]
28. Abcarian, H. Herand routine antibiotics for anorectal abscess: The answer is still no. *Dis. Colon Rectum* **2011**, *54*, 917–918. [CrossRef] [PubMed]
29. Steele, S.R.; Kumar, R.; Feingold, D.L.; Rafferty, J.L.; Buie, W.D.; Standards Practice Task Force of the American Society of Colon and Rectal Surgeons. Practice parameters for the management of perianal abscess and fistula-in-ano. *Dis. Colon Rectum* **2011**, *54*, 1465–1474. [CrossRef]
30. Ommer, A.; Herold, A.; Berg, E.; Fürst, A.; Sailer, M.; Schiedeck, T. German S3 guideline: Anal abscess. *Int. J. Color. Dis.* **2012**, *27*, 831–837. [CrossRef]
31. Amato, A.; Bottini, C.; De Nardi, P.; Giamundo, P.; Lauretta, A.; Luc, A.R.; Tegon, G.; Nicholls, R.J.; Italian Society of Colorectal Surgery. Evaluation and management of perianal abscess and anal fistula: A consensus statement developed by the Italian Society of Colorectal Surgery (SICCR). *Tech. Coloproctology* **2015**, *19*, 595–606. [CrossRef] [PubMed]

32. Ommer, A.; Herold, A.; Berg, E.; Farke, S.; Fürst, A.; Hetzer, F.; Köhler, A.; Post, S.; Ruppert, R.; Sailer, M.; et al. S3-Leitlinie: Analabszess. *Coloproctology* **2016**, *38*, 378–398. [CrossRef]
33. Shaughnessy, M.P.; Park, C.J.; Zhang, L.; Cowles, R.A. The Limited Utility of Routine Culture in Pediatric Pilonidal, Gluteal, and Perianal Abscesses. *J. Surg. Res.* **2019**, *239*, 208–215. [CrossRef] [PubMed]
34. Brar, M.S.; Remzi, F.; Warusavitarne, J.; Datta, I. Evidence Based Reviews in Surgery Group Does antibiotic therapy prevent fistula in-ano after incision and drainage of simple perianal abscess? *Can. J. Surg.* **2020**, *63*, E362–E364. [CrossRef] [PubMed]
35. Mocanu, V.; Dang, J.T.; Ladak, F.; Tian, C.; Wang, H.; Birch, D.W.; Karmali, S. Antibiotic use in prevention of anal fistulas following incision and drainage of anorectal abscesses: A systematic review and meta-analysis. *Am. J. Surg.* **2019**, *217*, 910–917. [CrossRef]
36. Sözener, U.; Gedik, E.; Aslar, A.K.; Ergun, H.; Elhan, A.H.; Memikoğlu, O.; Erkek, A.B.; Kuzu, M.A. Does adjuvant antibiotic treatment after drainage of anorectal abscess prevent development of anal fistulas? A randomized, placebo-controlled, double-blind, multicenter study. *Dis. Colon Rectum* **2011**, *54*, 923–929. [CrossRef]
37. Mandolfino, F.; Fornaro, R.; Stabilini, C.; Casaccia, M.; Testa, T.; Frascio, M. SECCA procedure for anal incontinence and antibiotic treatment: A case report of anal abscess. *BMC Surg.* **2018**, *18*, 53. [CrossRef]

antibiotics

MDPI

Review

Unseen Enemy: Mechanisms of Multidrug Antimicrobial Resistance in Gram-Negative ESKAPE Pathogens

Giedrė Valdonė Sakalauskienė *, Lina Malcienė, Edgaras Stankevičius and Aurelija Radzevičienė

Institute of Physiology and Pharmacology, Faculty of Medicine, Medical Academy, Lithuanian University of Health Sciences, 44307 Kaunas, Lithuania; lina.malciene@lsmu.lt (L.M.); edgaras.stankevicius@lsmu.lt (E.S.); aurelija.radzeviciene@lsmu.lt (A.R.)

* Correspondence: giedrevaldone.sakalauskiene@lsmu.lt

Abstract: Multidrug antimicrobial resistance (AMR) represents a formidable challenge in the therapy of infectious diseases, driven by the particularly concerning ESKAPE pathogens: the Gram-positive *Enterococcus faecium* and *Staphylococcus aureus*, as well as the Gram-negative *Klebsiella pneumoniae*, *Acinetobacter baumannii*, *Pseudomonas aeruginosa*, and *Enterobacter* spp. Designated as a "priority" in 2017, these bacteria continue to pose a significant threat in 2024, particularly during the worldwide SARS-CoV-2 pandemic, where coinfections with ESKAPE members contributed to worsened patient outcomes. The declining effectiveness of current treatments against these pathogens has led to an increased disease burden and an increase in mortality rates globally. This review explores the sophisticated mechanisms driving AMR in Gram-negative ESKAPE bacteria, focusing on *Acinetobacter baumannii*, *Klebsiella pneumoniae*, *Pseudomonas aeruginosa*, and *Enterobacter* spp. Key bacterial mechanisms contributing to resistance include limitations in drug uptake, production of antibiotic-degrading enzymes, alterations in drug target sites, and enhanced drug efflux systems. Comprehending these pathways is vital for formulating innovative therapeutic strategies and tackling the ongoing threat posed by these resistant pathogens.

Keywords: antimicrobial resistance; multidrug-resistant Gram-negative ESKAPE pathogens; mechanisms of antimicrobial resistance

1. Introduction

In 1945, during his Nobel Prize acceptance lecture, Sir Alexander Fleming (1881–1955) issued a strong warning to the medical community: "It is not difficult to make microbes resistant to penicillin in the laboratory by exposing them to concentrations not sufficient to kill them, and the same thing has occasionally happened in the body. The time may come when penicillin can be bought by anyone in the shops. Then, there is the danger that the ignorant man may easily underdose himself and by exposing his microbes to non-lethal quantities of the drug make them resistant" [1]. Unfortunately, the warning has not been properly taken into consideration. Although multidrug antimicrobial resistance (AMR) is a natural and inevitable process, widespread misuse and overuse of antimicrobial agents have provoked a rapid development of this phenomenon [2–5]. New classes of antibiotics and synthetic antibacterial agents were developed between 1940 and 1980, with particular intensity during the 1950s and the 1960s. Antibacterial drug development has faced a sharp decline in recent decades, parallelled by an alarming increase in drug-resistant pathogens (Figure 1) [3–6].

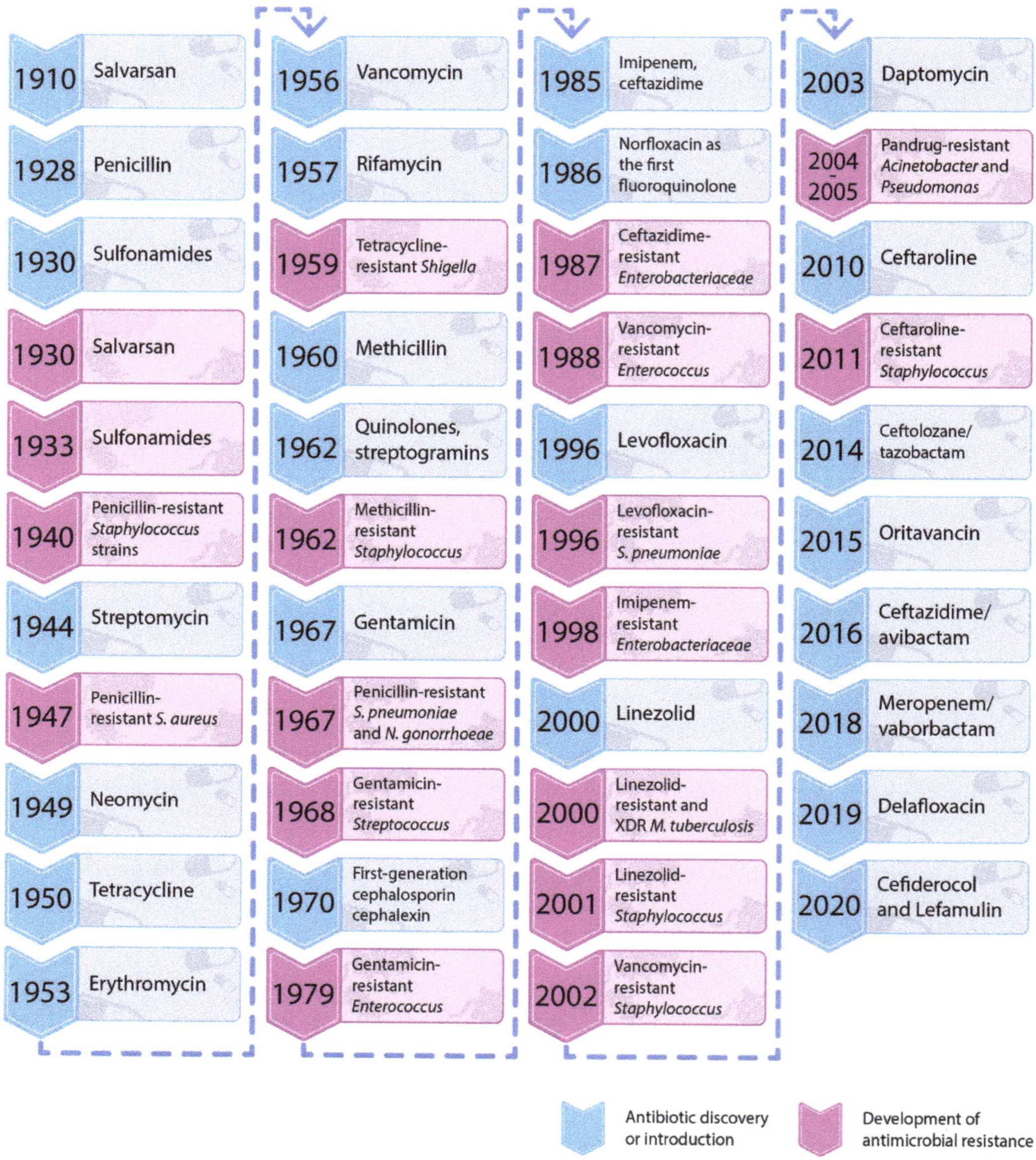

Figure 1. Progression of antibiotic discovery, clinical introduction, and antimicrobial resistance development (adapted and reused from the source [4] under the condition of the Creative Commons Attribution-Non-Commercial-Share Alike 3.0 License, also supplemented on the basis of data from the sources [3,6]).

Thus, since the early 21st century, decades after the Golden Age of antimicrobials, bacterial infections have re-emerged as a significant challenge to public health [7,8]. The World Health Organization (WHO) has recently issued a warning that the world has been "running out of antibiotics" and AMR has been reaching a new peak [9,10]. Based on these facts, and due to growing concerns about its potential impact on global health—which may not always receive the attention it warrants—AMR is widely referred to as the "silent pandemic". Throughout the COVID-19 crisis, reliance on broad-spectrum antibiotics without adequate diagnostic support became widespread, exacerbating the acceleration of antimicrobial resistance. Factors such as antibiotic overuse in agriculture, fewer novel drugs and vaccines in development, environmental toxicity, and limited funding for AMR research have compounded the issue. These practices drive structural and functional adaptations in bacteria, including enhanced efflux pumps, enzymatic modification, and biofilm formation, making infections harder to treat (Figure 2) [6,11–14].

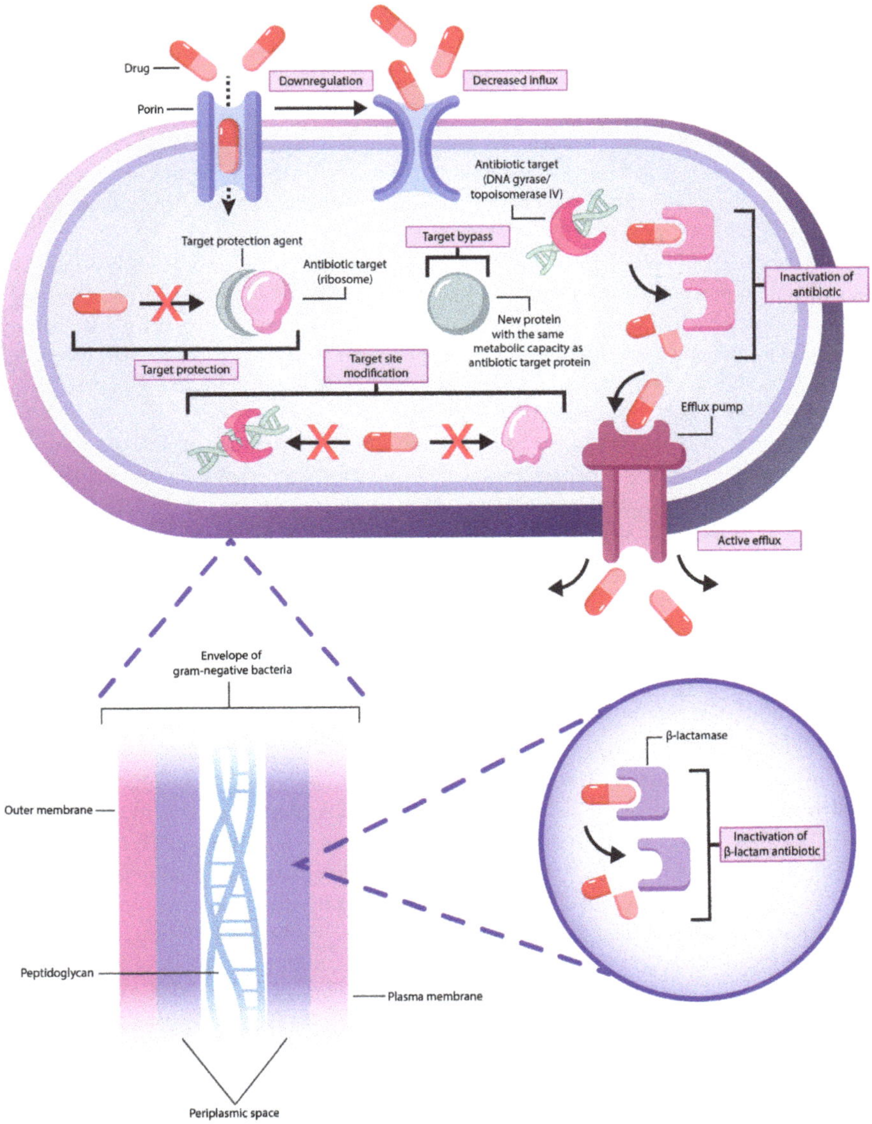

Figure 2. Bacterial mechanisms of AMR: structural and functional adaptation (reused from the source [14] under the condition of the Creative Commons CC BY license).

This insidious spread of resistance increasingly impacts vulnerable populations, such as cancer patients, and poses a global health crisis that advances largely unchecked [6,11–13]. This article examines the mechanisms of antimicrobial resistance (AMR) specifically in

Gram-negative bacteria such as *A. baumannii*, *K. pneumoniae*, *P. aeruginosa*, and *Enterobacter* spp., which are part of the ESKAPE group (*E. faecium*, *S. aureus*, *K. pneumoniae*, *A. baumannii*, *P. aeruginosa*, and *Enterobacter* spp.) pathogens. These pathogens warrant significant attention due to their exceptional ability to acquire and develop resistance to multiple antibiotic classes. They are leading causes of nosocomial infections, including bloodstream infections, particularly in hospital settings, where their prevalence continues to challenge global health. Their complex resistance mechanisms, including intrinsic resistance (from protective outer membranes and efflux pumps) and acquired resistance (via chromosomal mutations and horizontal gene transfer), make infections caused by these bacteria increasingly challenging to treat [15–17]. Other AMR mechanisms, such as target site mutation and modification, are shared with Gram-positive bacteria [14,18]. However, Gram-negative ESKAPE bacteria often produce hydrolytic enzymes like β-lactamases that degrade critical antibiotics, further driving multidrug-resistant (MDR) strains (Figure 2) [15–17]. Thus, Gram-negative ESKAPE bacteria exhibit a broader range of AMR mechanisms compared to Gram-positive pathogens from this group [18].

The COVID-19 pandemic added new challenges to managing AMR, especially in intensive care units (ICUs). Many ICU patients required intubation, increasing their risk of developing ventilator-associated pneumonia (VAP), which is often caused by drug-resistant bacteria and is difficult to treat. This risk is particularly high with Gram-negative pathogens such as *P. aeruginosa*, *Acinetobacter* spp., and *K. pneumoniae*, as well as with the Gram-positive bacterium *S. aureus* [19–21].

Thus, the persistence of these MDR isolates in healthcare settings, both prior to and throughout the COVID-19 crisis, emphasizes the need for improved stewardship, ongoing surveillance, and innovative treatment strategies to reduce their impact [16,17,22].

Considering the alarming global situation regarding AMR, this article analyzes the contribution of Gram-negative ESKAPE bacteria to AMR, focusing on the resistance mechanisms employed by these pathogens, including drug uptake limitation, drug target modification, enzymatic drug inactivation, and drug efflux, as well as the emergence of high-risk clones and novel treatment options.

2. The Worldwide Burden of AMR and the Critical Role of ESKAPE Bacteria

The global burden of a "silent pandemic" is enormous. In the United Kingdom, the number of individuals diagnosed with AMR-related infections rose from 61,946 in 2018 to 65,162 in 2019. In the European Union (EU) and European Economic Area (EEA), bacterial diseases were responsible for over 670,000 incidents annually, resulting in around 33,000 deaths directly attributed to these infections. In the Americas, which encompasses 35 countries, AMR was linked to approximately 569,000 deaths (95% UI: 406,000–771,000), with 141,000 deaths (99,900–196,000) directly caused by bacterial AMR [6,12,22]. In 2019, AMR was responsible for 1.27 million deaths and contributed to 4.95 million deaths globally. By 2050, it is anticipated that AMR could result in 10 million deaths per year worldwide [6,12,23,24]. Beyond its impact on mortality and health, AMR presents a significant economic strain. The World Bank has recently estimated that AMR could add US$1 trillion in healthcare costs by 2050 and lead to global GDP losses between US$1 trillion and US$3.4 trillion annually by 2030 [23,24]. AMR refers to the threatening phenomenon where pathogens such as bacteria, viruses, fungi, and parasites no longer respond to treatments they were once sensitive to, complicating the management of infections caused by these organisms. Bacteria are considered resistant if they stop responding to antibacterial agents and can grow, proliferate, and multiply under the exposure of inhibitory or bactericidal concentrations [2,5,9,23–25].

The most problematic microorganisms are multidrug-resistant (MDR), extensively drug-resistant (XDR), and pandrug-resistant (PDR) bacteria. MDR bacteria are resistant to multiple agents; XDR to nearly all drugs except colistin and tigecycline; and PDR to all antibacterial agents, including colistin and tigecycline. Classical resistant pathogens belong to the ESKAPE group, which includes *E. faecium*, methicillin-resistant *S. aureus*, *K. pneumoniae*, *A. baumannii*, *P. aeruginosa*, and *Enterobacter* spp. Currently, the most concerning among them are carbapenem-resistant Enterobacterales (CRE), carbapenem-resistant *K. pneumoniae* (CRKP), methicillin-resistant *S. aureus* (MRSA), ESBL-producing Enterobacterales, vancomycin-resistant *Enterococcus* (VRE), multidrug-resistant *P. aeruginosa*, and multidrug-resistant *Acinetobacter* [3,4,25–27].

Thus, in 2017, ESKAPE bacteria were designated as high-priority pathogens for the research and development of new antibacterial treatments aimed at countering them [2,27]. Next to the ESKAPE group, the WHO has identified other critically important bacteria for their resistance and clinical relevance. Significant changes in bacteria resistance have been noticed between 2017 and 2024 due to the dynamic nature of AMR (Figure 3) [28–30].

ESKAPE bacteria have become a worldwide menace during the SARS-CoV-2 pandemic. Soon after its declaration, half of the individuals with severe SARS-CoV-2 infection were found to have bacterial coinfections, mainly caused by pathogens from the ESKAPE group [18]. A systematic review analyzing data from 23 well-conducted studies published between 2019 and 2022 revealed high levels of AMR in *A. baumannii* and *K. pneumoniae*. The studies highlighting high levels of antimicrobial resistance were conducted across several countries, with the following distribution: four studies in India, four in Iran, two in Saudi Arabia, two in China, two in Turkey, two in Italy, and one each in Switzerland, USA, Serbia, Greece, Egypt, Pakistan, and Indonesia. Those aforementioned countries already had minor or moderate resistance rates, but during 2018–2020 reported an increased resistance to reserved antibiotics such as polymyxins, especially in India [31]. *A. baumannii* exhibited high resistance to levofloxacin (median [M] 97.05%; interquartile range [IQR] 91.92–100%), gentamicin (95.7%; 74.2–97.1%), cefepime (94.4%; 93–100%), and piperacillin/tazobactam (93.7%; 66.9–100%). For *K. pneumoniae*, the median of resistance to ampicillin, cefazolin, ceftazidime, and trimethoprim/sulfamethoxazole was 100% (IQR 90.5–100%), 93% (IQR 78–95.5%), 93.5% (IQR 83.7–97.9%), and 73.5% (IQR 32–74%), respectively. Although the overall AMR levels in *P. aeruginosa* were lower, resistance to ceftriaxone was observed at 75% (IQR 43.75–87.5%) (Figure 4).

Additionally, the SARS-CoV-2 pandemic contributed to a surge in the global use of personal protective devices and hygiene products, including soaps, handwashing solutions, and alcohol-based sanitizers, exacerbating the ongoing issue of AMR [26].

Despite the changes made in the WHO Bacterial Priority Pathogens List 2024, ESKAPE bacteria remain as a global alert, especially among critically ill and immunocompromised individuals [26,32]. These pathogens have reduced therapeutic choices for severe infections, increasing the severity of disease and mortality rates due to the acquisition of antimicrobial resistance genes [27]. For example, in 2017, a patient in the U.S. died from a *K. pneumoniae* strain resistant to every available antibiotic agent [33].

It has been established that ESKAPE pathogens have developed resistance to a wide range of antibiotics, including oxazolidinones, lipopeptides, macrolides, fluoroquinolones, tetracyclines, tigecycline, β-lactams, β-lactam/β-lactamase inhibitor combinations, carbapenems, glycopeptides, and polymyxins. However, resistance to lipoglycopeptides is less common and has only recently been observed [26,27]. Thus, to develop effective and targeted strategies to combat AMR of the ESKAPE group, the molecular mechanisms underlying this phenomenon must be fully clarified [34,35].

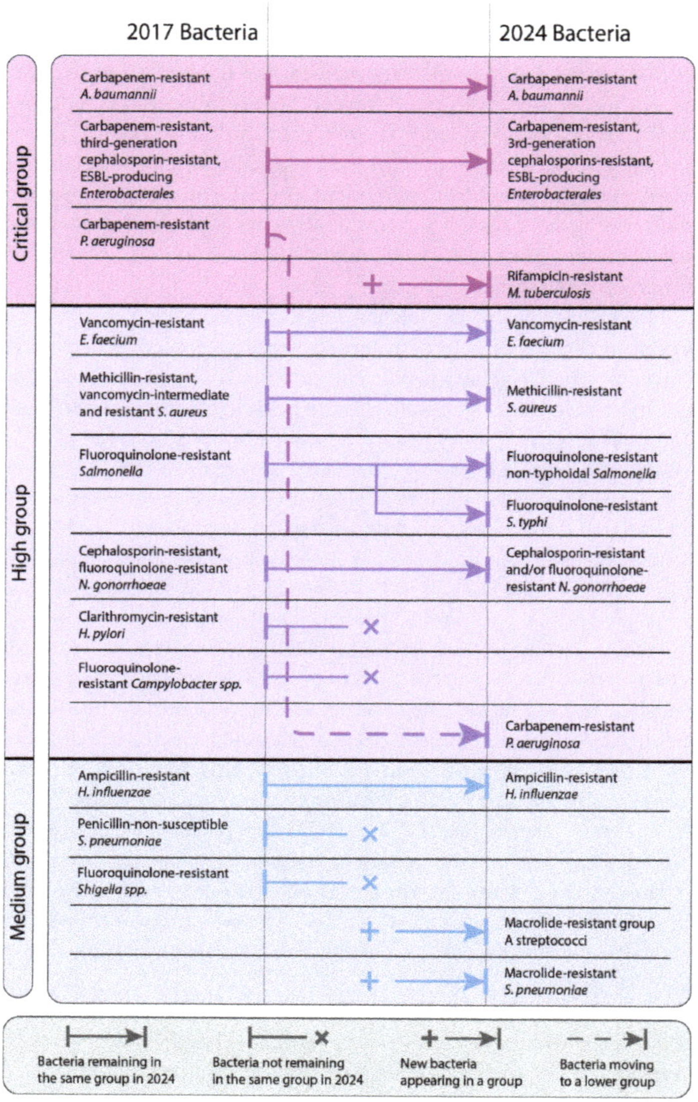

Figure 3. Changes in the WHO bacterial priority pathogens list during 2017–2024 (made on the basis of data from [3,4,6]).

Extremely specific cellular components secreted across the plasma membrane have recently generated interest due to being the structures widely involved in the transmission of biological information, killing of competitive bacteria, and exerting drug resistance. These components are outer membrane vesicles (OMV) [36,37].

Antibiotic	*A. baumannii* Median % (IQR)	*K. pneumonia* Median % (IQR)	*P. aeruginosa* Median % (IQR)
Amoxicillin/ clavulanate	-	81 (79.3–83.75)	-
Amikacin	84.6 (56.3–92.95)	69.85 (58.7–80.12)	25 (12–28)
Ampicillin	-	100 (90.5–100)	-
Aztreonam	-	84.7 (67.27–88.87)	-
Cefazolin	-	93 (78–95.5)	-
Cefuroxime	-	88.9 (79.6–91.42)	-
Cefepime	94.4 (93–100)	81.15 (71.7–87.25)	14.3 (12.5–47.8)
Ceftazidime	91.2 (50–100)	93.5 (83.7–97.9)	40 (23–41.7)
Cefoperazone/ sulbactam	-	76.2 (73.8–77.9)	-
Ceftriaxone	76.2 (54.75–95.55)	84 (77.55–93.4)	75 (43.75–87.5)
Colistin	2.5 (0–19.62)	21.1 (12.42–69.82)	4 (0–12.25)
Gentamicin	95.7 (74.2–97.1)	57.1 (33.45–86.6)	25 (19.75–58.75)
Levofloxacin	97.05 (91.92–100)	80.8 (78.55–90.85)	43.5 (28.6–80)
Meropenem	92.1 (64.02–95.65)	71.25 (55.37–77.37)	38 (18.37–42.17)
Imipenem	92.1 (80.65–95.72)	65.7 (19.25–72.87)	42.9 (19.75–52.9)
Ertapenem	-	71.4 (55.55–75.05)	-
Ciprofloxacin	91.2 (65–100)	87.8 (55.1–92.95)	50 (32.3–62.5)
Trimethroprim/ sulfamethoxazole	50 (46.8–84.2)	73.5 (32–74)	-
Tigecycline	9.5 (8.8–33.3)	31.4 (1.7–44)	-
Piperacillin/ tazobactam	93.7 (66.9–100)	77.7 (57.1–79.27)	11.25 (9.25–13.85)
Nitrofurantoin	-	51.8 (38.5–60.6)	-

Figure 4. Antibiotic resistance profiles of Gram-negative ESKAPE bacteria during the COVID-19 pandemic (adapted and reused from the source [19] under the condition of the Creative Commons CC BY license). Note: The picture reflects the data obtained from the aforementioned countries.

3. AMR Mechanisms in Gram-Negative ESKAPE Pathogens

3.1. *A. baumannii*

3.1.1. Drug Uptake Limitation

β-barrel-shaped monomeric or trimeric porins are mainly presented in the outer membrane (OM) of Gram-negative bacteria. These porins are called outer membrane proteins (Omps). Their function is to facilitate the diffusion of small molecules, such as nutrients and various antibiotics, into and out of the periplasmic space by creating tiny channels that allow for the passive transport of hydrophilic compounds and modulate cellular permeability [38,39].

Monomeric OmpA is the most expressed porin in the OM of *A. baumannii*. The OmpA porin is made up of an outer membrane β-barrel or N-terminal domain, as well as a periplasmic or C-terminal domain. While the N-terminal domain functions as transfer and shifts negatively charged β-lactams, such as β-lactamase inhibitor sulbactam and carbapenems like imipenem, into bacterial cells, the C-terminal domain interacts with peptidoglycan to support cell wall integrity and stability [40]. In wild strains, the C-terminal domain plays a key role in determining the sensitivity of *A. baumannii* to antibiotics such as nalidixic acid, chloramphenicol, the monobactam aztreonam, and the carbapenems imipenem and meropenem [38].

However, in mutant *A. baumannii* strains, due to the mutation in the *ompA* gene, the minimal inhibitory concentrations (MICs) of antibiotics, including nalidixic acid, chloramphenicol, trimethoprim, aztreonam, imipenem, meropenem as well as colistin are reduced. These findings indicate that additional mechanisms may contribute to a reduced susceptibility to these antibiotics, likely involving the expression of efflux pumps. Although the precise mechanism remains unclear, OmpA may be involved in the removal of compounds from the periplasmic space across the OM, potentially working in conjunction with the inner membrane (IM) efflux systems, such as the major facilitator family (MFS) pumps or resistance-nodulation-division (RND) systems that lack an Omp component [41].

It has also been also found that in mutants, permeability of cephalothin/cephaloridine is 2–3-fold reduced compared to the wild strains of *A. baumannii* due to alterations in OmpA [39]. Bacteria can also regulate the production of OMV via the interaction of the periplasmic domain of OmpA with peptidoglycan and preserve the stability of the membrane [39]. It has been established that OmpA plays a role in biofilm formation and contributes to the adhesion of abiotic surfaces as well as to cells of various lineages, such as epithelial cells. OmpA acts as a virulence factor, significantly contributing to the pathogenicity of *A. baumannii* and stimulating the innate immune response [38,40–42]. Other porins detected in *A. baumannii* are CarO, OprD, OprC, OprW, OprF, Omp 22–23 kDa, Omp 33–36 kDa, Omp 37 kDa, Omp 43 kDa, Omp 44 kDa, Omp 47 kDa, AbuO, TolB, DcaP, Oma87/BamA, NmRmpM, CadF, and CarOd. Many of them are the leading cause of resistance to carbapenems, especially CarO protein encoded by the *CarO* gene [38,43].

Alterations in the *CarO* gene may lead to significant structural modifications in the CarO porin, resulting in reduced OM permeability and the emergence of drug resistance. A complete loss of the CarO porin leads to bacterial insensitivity to imipenem and meropenem [32,44]. Thus, the synthesis of genetically altered porins that diminish OM permeability or decrease the expression of these proteins is primarily linked to β-lactam-related AMR, while the mechanism of reduced permeability is less significant for other antibiotics, such as fluoroquinolones [2,44].

3.1.2. Drug Target Modification

Lipopolysaccharides (LPSs) are the core components of the OM in the cells of *A. baumannii*. This component is targeted by polypeptide antibiotics such as colistin, a critical treatment option for severe infections induced by MDR bacteria, including *A. baumannii* [38,45]. The main two mechanisms of AMR involve alterations to the LPS complex and structural changes in lipid A, as well as the complete absence of LPS due to mutations in the genes responsible for lipid A synthesis. The modification of lipid A results from alterations in the *pmrA* and/or *pmrB* genes, with mutations being more prevalent in *pmrB*. It leads to the continuous activation of *pmrA* and enhances the expression of the *pmrCAB* operon. This results in the production and attachment of phosphoethanolamine (PEtN) to the 4′-phosphate or 1′-phosphate group of the lipid A chain [38,46].

LPSs are synthesized within the cytoplasm through the acetyltransferase Lpx pathway and then transported to the OM via the Lpt pathway, a system that moves LPSs across the periplasm for integration into the OM. Mutations, including nucleotide substitutions, deletions, or inactivation by the insertion sequence *ISAba11* in the *lpxA*, *lpxC*, and *lpxD* genes—encoding acetyltransferases critical for lipid A biosynthesis—result in the complete loss of LPSs. Meanwhile, mutations in the *lptD* gene result in the loss of the ability to transport LPSs from the cytosol to the OM [44,46,47]. Thus, the addition of PEtN to lipid A hinders polymyxins from penetrating the OM, while the absence of LPSs significantly diminishes their binding affinity to the OM [48].

Other two resistance mechanisms related to LPS synthesis include decreased expression of cofactors essential for LPS production and decreased levels of proteins responsible for exporting and stabilizing primary OM components. Biotin, which is crucial for fatty acid production, consequently facilitates LPS synthesis. The *lpsB* gene encodes an enzyme called glycosyltransferase, which is crucial for biotin synthesis in *A. baumannii*. Deletions in *lpsB* have been associated with resistance to colistin [44,46]. A linkage between resistance and *A. baumannii* isolates possessing the *vacJ*, *zndP*, and *pldA* genes has been identified. The *VacJ* gene is associated with the Vps–VacJ ABC transporter system responsible for maintaining LPSs in the outer leaflet and phospholipids in the inner leaflet of the OM [44].

Phospholipase D, known as PldA, is responsible for removing phospholipids from the outer leaflet of the OM, whereas a zinc-dependent peptidase ZndP acts prior to PldA and is essential for OM maintenance. Therefore, mutations in these genes contribute to colistin resistance [44,49]. Moreover, colistin resistance in *A. baumannii* is associated with a marked decrease in the production of osmoprotective amino acids, including proline, glycine, and aspartate, which are crucial for maintaining osmotic balance. Changes in the OM of colistin-resistant *A. baumannii* are typically associated with chromosomally encoded resistance [47].

Resistance to fluoroquinolones in *A. baumannii* primarily arises from spontaneous mutations in the *gyrA* and *gyrB* genes, which encode DNA gyrase, as well as in the *parC* gene, responsible for encoding topoisomerase IV. These mutations alter the structure of the DNA gyrase and topoisomerase IV complexes, preventing fluoroquinolones from binding, disrupting DNA replication, and thereby rendering the bacteria resistant to the drug. Mutations in the *gyrA* gene alone cause moderate resistance, while mutations in both the *gyrA* and *parC* genes lead to high resistance [38,44,50]. Another mechanism of resistance is the protection of DNR gyrase and topoisomerase IV by the proteins from the families of QrnA, QrnS, and QrnS [44].

An essential contributor to high-level resistance of *A. baumannii* to aminoglycosides is the *armA* gene, which encodes 16S rRNA methylase. This enzyme restricts the binding of aminoglycosides to bacterial ribosomes by methylating the ribosomal 30S subunit [38,44].

Ribosomal protection proteins, including TetM, TetW, TetO, and TetS, enable *A. baumannii* resistance to antibiotics by inducing a non-covalent alteration of ribosomes. Minocycline resistance is associated with the ribosomal protection protein encoded by the *tetM* gene. This protein facilitates the displacement of tetracycline from its binding site on the ribosome via a guanosine triphosphate (GTP)-dependent process, allowing protein synthesis to proceed despite the presence of tetracycline [44,50]. Resistance to tigecycline is mediated via the *rpsJ*, *trm*, and *rrf* genes, which subsequently cause an alteration in the drug target [51].

Infections caused by *A. baumannii* show resistance to rifampicin due to mutations in the *rpoB* gene, which encodes the β-subunit of rifamycin-sensitive RNA polymerase. These mutations disrupt RNA elongation shortly after the incorporation of the initial nucleotides. Spontaneous resistance to rifampicin is mainly caused by single-point mutations leading to amino acid substitutions, while it is more rarely linked to specific insertions or deletions [44].

Resistance of *A. baumannii* to macrolides is primarily attributed to the main 23S rRNA (adenine(2058)-N(6))-methyltransferases, encoded by the *erm(B)*, *erm(C)*, and *erm(F)* genes), as well as to the ABC-F-type ribosomal protection protein Msr(E), encoded by the *msr(E)* gene. Cross-resistance to macrolides, lincosamides, and streptogramin B can be explained by shared binding sites on the 23S rRNA [44,50].

Multidrug efflux pumps are the primary mechanisms of insensitivity to oxazolidinones. However, modifications to the target site, such as nucleotide alterations in the V domain of the 23S rRNA, and/or the presence of the transmissible Cfr(B) rRNA 23S methyltransferase, can also contribute to insensitivity. The Cfr-like rRNA 23S methyltransferase is additionally linked to resistance against lincosamides, streptogramins, phenicols, and pleuromutilins. Alterations in the P site of the 50S ribosomal subunit led to a resistance to linezolid, while the *poxtA* gene, encoding the ABC-F ATP-binding cassette ribosomal protection protein, plays a significant role in the development of resistance. The ABC-F ATP-binding cassette ribosomal protection protein is also implicated in A. baumannii resistance to bacterial protein synthesis inhibitors, such as tetracyclines, macrolides, lincosamides, streptogramins, phenicols, and pleuromutilins [44,50].

Trimethoprim insensitivity in *A. baumannii* is primarily mediated by enzymes called trimethoprim-resistant dihydrofolate reductases, including DfrA1, DfrA5, DfrA7, DfrA10, DfrA12, DfrA14, DfrA16, DfrA17, DfrA19, DfrA20, DfrA27, and DfrB1. These reductases are encoded by the corresponding genes *dfrA1*, *dfrA5*, *dfrA7*, *dfrA10*, *dfrA12*, *dfrA14*, *dfrA16*, *dfrA17*, *dfrA19*, *dfrA20*, *dfrA27*, and *dfrB1*. *A. baumannii*'s inherent insensitivity to sulfonamides is linked to sulfonamide-resistant dihydropteroate synthases encoded by the *sul1* and *sul2* genes [44].

Murein transpeptidases, known as penicillin-binding proteins (PBPs), are essential for the synthesis of the bacterial cell wall, as they catalyze the formation of cross-links between peptidoglycan strands. Peptidoglycan is a vital component that maintains the structural stability of the cell wall, protects the cell from osmotic stress, and ensures cell shape. β-lactams inhibit PBPs by mimicking the natural substrate, disrupting the cross-linking process. This interference destabilizes the peptidoglycan layer, leading to cell wall weakness and eventual bacterial lysis during growth [52]. PBPs are classified into two broad categories based on their molecular size: low-sized PBPs (class C PBPs or cPBPs, which include carboxypeptidases and endopeptidases) and high-sized PBPs [53,54]. The high-sized PBPs are further subdivided into class A and class B PBPs. Thus, resistance mediated by PBPs is a major factor contributing to β-lactam resistance in both Gram-negative and Gram-positive microorganisms. This mechanism involves alterations in the PBPs, which are the primary targets of β-lactam antibiotics [53]. *A. baumannii* contains seven distinct PBPs. These include several class A PBPs (PBP1a and PBP1b), several class B PBPs (PBP2 and PBP3), and three class C PBPs (PBP5/6, PBP6b, and PBP7/8). Additionally, this pathogen has a monofunctional transglycosylase [53]. However, most variations in PBPs were identified as silent mutations, which do not contribute to resistance against β-lactams [55].

Although alterations of PBPs are not an essential mechanism of resistance against β-lactams in *A. baumannii*, they should not be ignored [43,49]. Thus, mutations of the gene encoding PBP2 are related to imipenem resistance, while alterations of the gene encoding PBP3 are related to meropenem, sulbactam, and cefiderocol resistance [44,50]. Some PBPs, including PBP3-H370Y, PBP3-Q488K, and PBP3-Y528H, can be affected by the novel combination of sulbactam and durlobactam [53,56]. This intravenous β-lactam/β-lactamase inhibitor combination was approved by the US FDA in May 2023 for treating adults with hospital-acquired and ventilator-associated bacterial pneumonia caused by susceptible strains of the *Acinetobacter baumannii–calcoaceticus* complex, including carbapenem-resistant *A. baumannii* (CRAB) [57].

3.1.3. Enzymatic Drug Inactivation

The production of enzymes that inactivate antibiotics by adding specific chemical moieties or by degrading the drug molecule itself, thus preventing its interaction with the target, is one of the most potent mechanisms bacteria use to evade the effects of antimicrobial agents [58].

The intrinsic insensitivity of *A. baumannii* is closely linked to the inactivation of β-lactams. It has been established that this bacterium presents β-lactamases of Ambler classes A, B, C, and D [50]. The serine β-lactamases of molecular class A are TEM (Temoneira), SCO-1 (a novel plasmid-mediated class A β-lactamase from *E. coli*, exhibiting characteristics similar to carbenicillinase), CARB (carbenicillin-hydrolyzing β-lactamase), SHV (sulfhydryl reagent variable), GES (Guiana extended-spectrum β-lactamase), CTX-M (cefotaxime-hydrolyzing β-lactamase), PER (*Pseudomonas* extended resistant), and KPC (*K. pneumoniae* carbapenemase) [38,59–61]. Intrinsically, Ambler class A β-lactamases are narrow-spectrum enzymes responsible for the primary inactivation of antibiotics from the penicillin group and first- and second-generation cephalosporins, such as TEM-1, TEM-2, SCO-1, CARB-4, and SHV [44,59,62,63].

Usually, narrow-spectrum class A β-lactamases can be impeded by the classical β-lactamase inhibitors. However, these enzymes, through point mutations, can develop the capacity to hydrolyze extended-spectrum cephalosporins such as ceftazidime, ceftriaxone, cefotaxime, and aztreonam, the latter being currently the only available monobactam. Thus, due to their ability to destroy potently wide-spectrum β-lactams, these enzymes are named ESBL. Other ESBLs are GES-11, CTX-M, and KPC, including KPC-2, KPC-3, and KPC-5. ESBL encoding genes are widely distributed among Gram-negative bacteria by plasmids or mobile genetic elements (MGEs) [64,65].

Among the main genes encoding class A lactamases are bla_{TEM}-92, bla_{SHV}, bla_{GES}-11, bla_{GES}-14, bla_{PER}-1, bla_{PER}-7, and bla_{VEB}-1 [2,44,50,62].

Enzymes known as metallo-β-lactamases (MBLs) possess a Zn^{2+} ion at their active sites and are encoded by mobile genetic elements like plasmids and integrons. These enzymes catalyze the hydrolysis of nearly all β-lactams, including carbapenems, while sparing monobactams, thereby contributing to MDR [44]. MBLs belong to class B and are named as IMP (imipenemase), VIM (Verona integron-encoded MBL), NDM (New Delhi metallo-β-lactamase), and SIM (Seoul imipenemase) [35]. The main genes that encode MBLs are bla_{IMP}, bla_{VIM}, bla_{SIM}, $blaS_{PM}$, bla_{GIM}, bla_{DIM}, and bla_{NDM} [64,65].

Chromosomally encoded cephalosporinases, known as *Acinetobacter*-derived cephalosporinases (ADCs), are classified as class C β-lactamases and are present in all *A. baumannii* strains. The gene encoding ADC is bla_{ADC} (formerly known as bla_{AmpC}). The insertion of IS*Aba1* or IS*Aba125* sequences upstream of the genes that encode ADCs promotes their overexpression by introducing stronger promoters. ADC enzymes present extended-spectrum resistance to β-lactams, and their covalent modification leads to imipenem resistance [44,66].

Recently, enzymes known as class D β-lactamases, which are capable of inactivating carbapenems and cephalosporins, have been detected in *A. baumannii*. These enzymes are named oxacillinase (OXA)-type carbapenemases. The most predominant oxicillinases in *A. baumannii* include OXA-23-like, OXA-24/40, OXA-48, OXA-51, OXA-58, OXA-143-like, and OXA-235-like. Numerous bla_{OXA} genes have been identified such as bla_{OXA}-51, *bla*$_{OXA}$*-23*, *bla*$_{OXA}$*-24*, *bla*$_{OXA}$*-58*, *bla*$_{OXA}$*-143*, and *bla*$_{OXA}$*-235*. These genes are located on both chromosomes and plasmids. For instance, acquired carbapenemases include OXA-23, OXA-24, OXA-48, and OXA-58, while OXA-51 is an intrinsic carbapenemase in *A. baumannii*. OXA enzymes, characterized by low catalytic efficiency, along with porin deletion and various other antibiotic resistance mechanisms, can contribute to significantly high insensitivity to carbapenems [2,18,32,38].

Aminoglycoside-modifying enzymes (AMEs) are key factors responsible for bacterial insensitivity to aminoglycosides. Depending on the location of aminoglycoside alteration through *N*-acetylation, *O*-nucleotidylation, and *O*-phosphorylation, AMEs are classified as acetyl-, adenyl-, and phosphor-transferases [2,44].

The acetylation of -NH_2 groups is catalyzed by aminoglycoside N-acetyltransferases (AACs), which use acetyl coenzyme A as a donor substrate to transfer the acetyl group to the acceptor molecule. Aminoglycoside O-nucleotidyltransferases (ANTs) inhibit the activity of aminoglycosides by catalyzing the transfer of an adenosine monophosphate (AMP) molecule from adenosine triphosphate (ATP) to the -OH group on the antibiotic, with ATP serving as a donor substrate. Changes in the distribution of antibiotic charge and blockage of its interaction with the ribosome occur due to the addition of a phosphate group to the drug by aminoglycoside O-phosphotransferases (APHs). Thus, the drug becomes ineffective due to the action of enzymes that destroy the chemical structure of aminoglycosides, reducing their affinity for the target or impeding ribosomal binding [43,67]. The main families of genes responsible for resistance to aminoglycosides are *aadB*, *apa6*, *aadA*, and *aacc1*. Genes can be transferred by MGEs such as integrons, gene cassettes, transposons, and conjugated elements [38].

The enzymes known as fluoroquinolone-acetylating aminoglycoside 6′-N-acetyltransferase AAC (6′)-Ib-cr and fluoroquinolone-acetylating aminoglycoside 6′-N-acetyltransferase AAC, encoded by *aac(6′)-Ib-cr* and *aac(6′)-Ib-cr5I*, respectively, cause the inactivation of fluoroquinolones [38].

Rifampin is enzymatically modified and inactivated by rifampin ADP-ribosyltransferases, primarily including enzymes such as Arr, Arr-2, Arr-3, and Arr-4. In contrast, the inactivation of amphenicols involves chloramphenicol O-acetyltransferases, including type A-1, type A-2 CatII, type B-2 CatB11, type B-3 CatB3, and type B-3 CatB8, which are encoded by the genes *catA1*, *catA2*, *catB11*, *catB3*, and *catB8*, respectively. Another enzyme involved in the inactivation of amphenicols is the bifunctional type B-3 chloramphenicol O-acetyltransferase CatB8/aminoglycoside N-acetyltransferase AAC (6′)-Ib, or catB8/aac(6′)-Ib [44].

Two macrolide 2′-phosphotransferases, Mph(A) and Mph(E), encoded by the mph(A) and mph(E) genes, inactivate macrolides and, when specific regulatory proteins are present, result in resistance to erythromycin, clarithromycin, azithromycin, and oleandomycin [44].

Resistance to fosfomycin is attributed to glutathione transferases from the FosLL, FosA3, and FosA3/FosA4 families, encoded by the *fos*, *fosA3*, and *fosA* genes, respectively. Additionally, fosfomycin resistance is mediated by a bacillithiol transferase from the FosB1/FosB3 family, encoded by the *fosBOne* gene, which contributes to intrinsic *A. baumannii* resistance to fosfomycin. FosA facilitates the attachment of glutathione to the C1 position of fosfomycin, leading to its inactivation, while FosB catalyzes the cleavage of the drug's epoxide ring using either bacillithiol or L-cysteine [44].

It has been recently found that tetracyclines, including tigecycline, eravacycline, and omadacycline, can be also be inactivated by various Tet(X) monooxygenases, encoded by plasmid-mediated *tet(X3)*, *tet(X4)*, and *tet(X5)* genes [44,50,51].

3.1.4. Drug Efflux

The active efflux of antimicrobial medicines from bacterial compartments, along with diffusion across the membrane, serves as an important molecular and biochemical mechanism of both intrinsic and acquired MDR. It also interacts with various mechanisms, including the destruction or alteration of drugs and the modification or preservation of antibiotic targets, significantly elevating resistance levels and profiles [68].

Efflux pumps located in the IM or OM of bacteria are usually a complex of bacterial machineries that can expel toxic compounds out of the bacterial cell [43]. A critical characteristic of these complexes is their ubiquity across nearly all bacterial species, contributing to antibiotic resistance through their common underlying resistance mechanisms [43]. These efflux pumps have been classified into five major superfamilies: the ATP-binding cassette (ABC) superfamily, the RND family, the MFS, the multidrug and toxic compound extrusion (MATE) family, and the small multidrug resistance (SMR) family [43,50]. Recently, a new superfamily referred to as the proteobacterial antimicrobial compound efflux (PACE) was identified [43].

Efflux pumps are categorized based on their energy source into two groups: primary efflux pumps, which utilize ATP (e.g., the ABC superfamily), and secondary efflux pumps, which rely on the proton motive force (PMF) [43,50].

Most efflux pumps in Gram-negative pathogens are part of the RND superfamily and typically possess a tripartite structure consisting of an outward conduit or outer membrane protein in the OM and subsequently a membrane fusion protein (MFP) and an active pump or inner membrane protein in the IM [43,69]. The RND family is a class of secondary efflux pumps that are primarily arranged in an operon structure. These pumps play a significant role in AMR by enhancing insensitivity to multiple antibacterial drugs, including tetracyclines, fluoroquinolones, aminoglycosides, and the penicillin group [43].

In *A. baumannii*, insensitivity to β-lactams, particularly carbapenems and cephalosporins, is associated with the overexpression of the AdeABC efflux pump, which belongs to the RND superfamily. It consists of three components: AdeB, which expels antibiotics from the cell; AdeA, which functions as a membrane fusion protein; and AdeC, which serves as an outer membrane protein. AdeB can transport a wide range of substrates, from hydrophilic to hydrophobic, with either positive charges or neutral properties. The expression of this efflux pump is controlled by the AdeRS two-component regulatory system (TCRS) [38,44].

TCRSs enable bacteria to respond to environmental changes [70]. Point mutations within the *adeRS* operon can upregulate pump expression, thereby increasing antibiotic resistance. Recently detected RND AdeIJK and MATE AbeM efflux pumps may also play a role in the emergence of resistance to imipenem and cephalosporin [38,44].

The MATE AbeM efflux pump also removes fluoroquinolones, chloramphenicol, erythromycin, tetracyclines, and aminoglycosides [71]. *A. baumannii* is able to expel aminoglycosides through various drug efflux mechanisms, including the MDR efflux MFS transporter AmvA; the MDR efflux RND transporter AdeABC, which comprises the OM channel subunit AdeC, the periplasmic adaptor subunit AdeD, and the permease subunit AdeE; and the MDR efflux SMR superfamily transporter EmrE. Additionally, the expression of these transporters is regulated by the DNA-binding response regulator transcription factor AdeR and the two-component sensor histidine kinase AdeS. These components are encoded by the *amvA*, *adeC*, *adeD*, *adeE*, *adeR*, *adeS*, and *emrE* genes [44].

It has been established that another SMR AbeS transporter can extrude β-lactams, chloramphenicol, ciprofloxacin, and erythromycin. However, it has a limited impact on reducing the susceptibility of *A. baumannii* to these compounds [71].

Efflux pumps from the RND superfamily, encoded by the *adeA*, *adeB*, and *adeC* genes, along with the tetracycline MFS transporter, where TetA and TetB (encoded by *tetA* and *tetB*) play key roles, contribute to tetracycline resistance. The *tetA* gene may be contained in a transposon while *tetB* is carried by a plasmid. RND efflux pumps AdeABC, Ade-FGH:RND, and AdeIJK function synergistically with TetA. For example, TetA expels tigecycline into the periplasm, allowing the RND pumps to subsequently transport the drug across the OM. Other MFS pumps are TetC, TetD, TetG, and TetH [44].

The predominant RND efflux pump responsible for the effective elimination of tetracyclines is AdeABC. The AdeABC pump is controlled by a TCRS called AdeRS. Point mutations in the *adeRS* operon can upregulate the pump's expression, resulting in increased antibiotic resistance [44].

The integration of a sequence known as IS*Aba1* into the *adeS* gene leads to the overexpression of AdeABC. Although the RND efflux pump AdeIJK contributes less significantly to tetracycline resistance in *A. baumannii*, it can cooperate with other overexpressed pumps, such as AdeABC and AcrAB-TolC, to enhance tigecycline resistance [44,50]. Additionally, AdeIJK and AcrAB-TolC have been identified as potential mechanisms contributing to rifamycin resistance in *A. baumannii* strains [44].

Efflux pumps such as AdeABC, AdeDE, AdeFGH, and AdeIJK contribute to resistance against aminoglycosides, tetracyclines, fluoroquinolones, macrolides, lincosamides, and chloramphenicol. However, among them, AdeABC is the most predominant drug expulsion system in this bacterium [2,27,32].

LmrS, an MFS pump, is able to extrude trimethoprim as well as aminoglycosides, macrolides, oxazolidinones, and amphenicols. Resistance of *A. baumannii* to trimethoprim is associated with the RND efflux pump MexAB-OprM, which consists of MexA, a membrane fusion protein; MexB, a carrier in the IM; and OprM, a protein embedded in the OM. This transporter participates in the resistance of many antibiotics, including sulfonamides [66].

Another efflux pump that eliminates trimethoprim from the bacterial cell is the aforementioned AdeIJK. The RND plasmid-borne OqxAB transporter has been linked to resistance against trimethoprim, quinolones, tetracyclines, glycylcyclines, and nitrofurans, while the MFS AbaF efflux pump, encoded by the *abaF* gene, expels fosfomycin [44,50]. EmrAB-like efflux pumps in *A. baumannii* reduce its sensitivity to polymyxins due to an increased expression of *emr-B* like genes [47]. The upregulation of genes *adeI*, *adeC*, *emrB*, *mexB*, and *macAB*, which encode the protein components of efflux pumps, has been observed in colistin-resistant strains of *A. baumannii*. A recent study has linked an amino acid substitution (N104M) in the gene *ttg2C*, which encodes the toluene tolerance efflux pump Ttg2C, with high-level colistin resistance [66].

Thus, in this bacterium, the RND superfamily efflux pumps are related to insensitivity to aminoglycosides, β-lactams, macrolides, lincosamides, polymyxins, chloramphenicol, tetracyclines, fluoroquinolones, rifamycins, and folic acid inhibitors, while the MFS superfamily, to tetracyclines, macrolides, oxazolidinones, chloramphenicol, aminoglycosides, and fosfomycin; the MATE superfamily, to β-lactams, fluoroquinolones, chloramphenicol, erythromycin, tetracyclines, and aminoglycosides; and finally the SMR superfamily, to fluoroquinolones, aminoglycosides, chloramphenicol, erythromycin, and tetracyclines [44,71].

3.2. *P. aeruginosa*

The intrinsic and acquired resistance pathways enable *P. aeruginosa* to overcome the effects of antimicrobials [2,43]. However, the primary mechanisms of resistance include the overexpression of efflux pumps, reduction in OM integrity, and the acquisition or mutation of resistance genes that encode proteins involved in controlling the passive diffusion of antibiotics through the OM [2].

3.2.1. Drug Uptake Limitation

It has been established that *P. aeruginosa* lacks major porins, including OmpF and OmpC, leading to an OM resistance to antibiotic penetration that is 100-fold lower compared to *E. coli*. Moreover, a vast family of OprD (OccD1) porins, consisting of 19 representatives, has also been associated with a decreased permeability in this bacterium [72,73].

These porins are unique channel proteins that possess specific binding sites for a particular set of molecules, i.e., participate in the absorption of various nourishing substances, but mainly amino acids. Reduced permeability caused by the downregulation of OprD proteins is the key pathway of intrinsic resistance to carbapenems, including imipenem and meropenem [72,73]. Thus, carbapenem resistance has been associated with alterations that result in the downregulation of *oprD* manifestation, as well as with amino acid substitution mutations in OprD found in hypermutator isolates [73].

It has also been also established that epigenetic regulation of *opdQ* expression causes an elevation in the MIC of meropenem. This gene encodes the OpdQ porin, which is a member of the OprD family [73]. In addition, a large quantity of OprD proteins which lead to the carbapenem resistance have been found in biofilms and OMV [74].

Another protein found in the OM of *P. aeruginosa* is OprH, the smallest porin among all porins in this pathogen, belonging to the class of gated porins. OprH facilitates the absorption of bivalent cations, engages with LPSs via electrostatic bonds, and subsequently strengthens the OM. Due to it, OprH contributes to the antibiotic resistance, especially to polymyxins and aminoglycosides, which penetrate the bacterial cell membrane through interaction with LPSs. Overexpression of OprH due to Mg^{2+} starvation has been linked to increased insensitivity to polymyxin B and gentamicin, which is attributed to enhanced activity of the PhoP/PhoQ system [72,73,75].

The PhoP-PhoQ pathway is a TCRS that plays a role in polymyxin resistance by controlling the *arnBCADTEF-pmrE* operon, which encodes LPS alteration enzymes [76]. Non-specific porin OprF is responsible for preserving the OM structure. Alterations in OprF showed that its N-terminal is essential for protein synthesis and membrane integration, whereas the C-terminal is crucial for stable interaction with peptidoglycan and anchoring on the OM. It has been established that *P. aeruginosa* virulence and its protection against macrophages are related to OprF [73]. Overexpression of OprF has been found in the formation of *P. aeruginosa* biofilm and OMV, which consequently prevent antibiotic entry [43,74]. However, some investigations have suggested the total absence of OprF alone does not seem to be a significant pathway of antibiotic insensitivity in clinical strains [74]. Finally, other proteins such as the particular porins OprB, OprE, OprO, and OprP as well as the gated porin OprC are involved in metabolic and ion homeostasis, whereas proteins like OprM, OprN, and OprJ are classified as efflux porins [68].

3.2.2. Drug Target Modification

P. aeruginosa develops resistance to aminoglycosides with the help of RNA methylases encoded by the *rmtA* and *rmtB* genes, which are located on transposons and plasmids, as these enzymes methylate the 16S ribosomal RNA (rRNA). The 16S rRNA methylases methylate, a specific nucleotide in the A-site of 16S rRNA, thereby protecting the ribosome from the action of these drugs. Resistance to fluoroquinolones occurs through changes in topoisomerase IV and DNA gyrase due to target site alteration. For example, insensitivity to ciprofloxacin most commonly arises from alterations in the subunits GyrA and GyrB of DNA gyrase or the subunits ParC and ParD of topoisomerase. Thus, resistance to DNA synthesis inhibitors is linked to alterations in the motifs of the *gyrA* and *gyrB* genes within the quinolone-resistant determining region (QRDR motif), which forms part of the active site of the DNA gyrase enzyme [43,74].

Alterations in the amino acid sequences encoding the specific QRDR motif lead to modifications in the A and B subunits of the enzyme, resulting in a reduced binding affinity for fluoroquinolone. The alterations in topoisomerase IV have been identified as point mutations in the amino acid sequences encoded by the *parC* and *parE* genes, which form the ParC and ParE enzymatic subunits, leading to resistance to these antibacterial drugs. *P. aeruginosa* is considered highly resistant when it carries mutations in both the *gyrA* and *parC* genes or simultaneously harbors mutations in the *nalB*, *nfxB*, and *nfxC* genes, leading to their overexpression and, consequently, the overactivation of the MexAB-OprM, MexCD-OprJ, and MexEF-OprN efflux pumps. However, low insensitivity to fluoroquinolones, such as ciprofloxacin, has been observed in this pathogen when mutations in *gyrB*, *parC*, or *parD*, or the presence of efflux pumps, are the only factors involved [43,74].

This pathogen's insensitivity to β-lactams also relies on alterations in PBPs, primarily in the enzyme PBP3, or in the overexpression of genes encoding these proteins. As a result, β-lactams are unable to bind to their site of action [43]. Alterations in PBP3 lead to insensitivity to ceftazidime, ceftazidime-avibactam, cefepime, piperacillin-tazobactam, ceftolozane-tazobactam, and meropenem. Structural and functional studies have shown that PBP5 in *P. aeruginosa* exhibits β-lactamase activities, thus contributing to β-lactam insensitivity. Moreover, mutations in the nonessential *dacB* gene, encoding PBP4, are linked to the overexpression of AmpC. This leads to an increased resistance to β-lactam antibiotics, specifically lowering susceptibility to antibacterials such as piperacillin, cefotaxime, ceftazidime, cefepime, and aztreonam [77].

Polymyxin-induced resistance in *P. aeruginosa* results from changes in the bacterial LPSs. The electronegativity of LPSs is reduced due to modifications in the lipid A region through the insertion of 4-amino-4-deoxy-L-arabinose (L-Ara4N) and PEtN, which are upregulated by the respective operons *arnBCADTEF* and *pmrCAB*. The mentioned operons are regulated via TCRSs, such as PmrA/PmrB, PhoP/PhoQ, ParR/ParS, CprR/CprS, and ColR/ColS. However, in *P. aeruginosa* the primary pathway for polymyxin insensitivity is the function of the arnBCADTEF operon, responsible for the insertion of L-Ara4N into lipid A [43]. This occurs when L-Ara4N is added to the phosphate groups within lipid A and the main carbohydrate components of the LPSs [73]. As it has been aforementioned, the overproduction of the OMP OprH contributes to polymyxin resistance, as OprH interacts with divalent cation sites on LPSs, hindering the binding of polymyxin [43].

There are eight distinct PBPs in *P. aeruginosa*: class A PBPs, including PBP1a and PBP1b; class B PBPs, consisting of PBP2, PBP3, and PBP3x; and class C PBPs, such as PBP4, PBP5, and PBP7 [53,78]. Among these, PBP3, encoded by the *ftsl* gene is crucial for the growth of this pathogen [53,78–80]. Mutations in the *ftsI* gene lead to alterations in the PBP3 sequence, which are closely linked to insensitivity to β-lactams, especially cephalosporins and carbapenems, due to a reduced binding affinity for these drugs [79,81,82]. However, PBP1 and PBP3 in *P. aeruginosa* are still sensitive to the combination of ceftolozane and tazobactam [53,83,84].

Mutations in the enzyme PBP4 have been found to trigger the overexpression of AmpC β-lactamase, thereby contributing to resistance, primarily to penicillins and cephalosporins [53,78].

3.2.3. Enzymatic Drug Inactivation

As with *A. baumannii*, *P. aeruginosa* also generate various β-lactamases, which are classified according to Ambler classes A-D. The most common β-lactamases in *P. aeruginosa* are penicillinases from the molecular class A serine β-lactamases, including *Pseudomonas*-specific enzymes (PSE), CARB, and TEM, with PSE being the predominant one [85]. Chromosomally encoded Ambler class C β-lactamase AmpC plays a crucial role in the natural resistance of *P. aeruginosa* [74,85].

Although AmpC β-lactamase is naturally synthesized at low levels in *P. aeruginosa*, its overproduction can occur either due to induction of the *ampC* gene or through derepression, leading to constitutive high-level expression of the enzyme. Thus, chromosomal bla_{AmpC} gene activity is promoted by aminopenicillins and cephalosporins, leading to the immediate inactivation of β-lactams via hydrolysis by AmpC. AmpC activation is triggered by two distinct processes, impairment of the peptidoglycan recycling pathway and the depletion of PBP4. Normally, during peptidoglycan synthesis, the release of anhydromuropeptides (anhMPs) occurs [43,85].

Next, anhMPs are transported across the IM into the cytoplasm and are subsequently taken up for re-use in the peptidoglycan biosynthesis process. When anhMP processing is disrupted, these molecules accumulate in the cytoplasm, interact with AmpR, a key regulator of AmpC, and act as signaling molecules, triggering the activation of β-lactamase production in this bacterium [43,85].

Another mechanism associated with AmpC β-lactamase activity involves the inactivation of the *dacB* gene, which encodes PBP4. This inactivation results in the overexpression of the *ampC* gene and the activation of the CreBC (BlrAB) TCRSs. PBP4 plays an important role in producing an anhMP that regulates AmpR. When PBP4 is lost, the anhMP is excluded from the peptidoglycan recycling process, modifying AmpR activity and leading to the activation of *ampC* expression [74].

Consequently, antimicrobial insensitivity to β-lactam antibiotics, including drugs from the penicillin group, cephalosporins (such as cefepime and ceftazidime), tazobactam, avibactam, and aztreonam, emerges. AmpC β-lactamase has minimal effect on carbapenem susceptibility, unless other resistance mechanisms, such as efflux pump overproduction, decreased OprD, and/or the production of class A/class B carbapenemases, are involved [43,85].

Transferable class C β-lactamases, disseminated through horizontal gene transfer, are infrequent in microorganisms such as *P. aeruginosa*, which already possess chromosomally encoded *ampC*. Nonetheless, these transferable β-lactamases can still target penicillins, cephalosporins, and monobactams. It is believed that these enzymes originally derived from chromosomally encoded versions, which were later transferred to MGEs [79]. Mutations in ampC can reduce the binding affinity of β-lactamase inhibitors like avibactam, tazobactam, and relebactam to AmpC [79].

Class B enzymes, specifically MBLs produced by *P. aeruginosa*, catalyze the hydrolysis of β-lactam rings through pathways that require a zinc ion. High insensitivity of *P. aeruginosa* to carbapenems has been observed in the presence of MBLs, as these enzymes can degrade carbapenems and are critical in managing diseases caused by this pathogen [79]. The most prevalent MBLs are carbapenemases VIM and IMP that are horizontally acquired. It has been identified that these two enzymes have multiple variants, while other three enzymes—SPM (Sao Paulo MBL), NDM, and GIM—present only one variant each, i.e., SPM-1, NDM-1, and GIM-1, respectively [43,86]. The genes encoding VIM and IMP are often found on integrons, which enables these genes to integrate into either the chromosome or plasmids, facilitating their spread throughout bacterial populations [74]. β-lactamase inhibitors impede the activity of MBLs [77].

As mentioned earlier, ESBLs are a type of β-lactamases that inactivate β-lactams, especially oxyimino-β-lactams and monobactams, and are susceptible to inhibition by β-lactamase inhibitors such as clavulanic acid. They are encoded by plasmids and can easily be transmitted from one organism to another. The most prevalent ESBLs in *P. aeruginosa* are class A β-lactamases—TEM, CTX-M, and SHV—encoded by genes bla_{TEM},

bla_{CTX-M}, and bla_{SHV}, respectively [87]. Other Ambler class A ESBLs found in *P. aeruginosa* are VEB (Vietnam extended β-lactamase), PER, GES, PSE, and BEL (Belgium extended β-lactamase)-type enzymes [43,74].

Ambler class A ESBLs degrade oxyimino-aminothiazolyl cephalosporins (such as cefotaxime, cefuroxime, cefepime, ceftriaxone, and ceftazidime), along with other cephalosporins like cefpirome, as well as penicillins and aztreonam. However, they do not affect cephamycins (e.g., cefoxitin) or carbapenems [43,74]. TEM, SHV, CTX-M, BEL, PER, and VEB can be inhibited by clavulanic acid, sulbactam, tazobactam, avibactam, vaborbactam, and relebactam. Contrarily, the members of the PER family—PER-1 and PER-2—are the predominant variants and exhibit reduced susceptibility to inhibition by clavulanic acid, tazobactam, and avibactam [77].

A newly identified chromosomally encoded Ambler class A β-lactamase, PIB-1 (*Pseudomonas* imipenem β-lactamase), which degrades imipenem, has been discovered. However, PIB-1 is not able to inactivate other β-lactams, including cephalosporins [88,89]. Another novel β-lactamase, the German *Pseudomonas* carbapenemase (GPC-1), has recently been detected. Strains of *P. aeruginosa* strains that produce GPC-1 may demonstrate weak carbapenemase activity, indicating that they do not effectively hydrolyze carbapenems to a significant degree. Consequently, these strains may go undetected by standard diagnostic tests [90].

Moreover, *P. aeruginosa* can produce Ambler D class β-lactamases such as OXA-type β-lactamases. The majority of genes encoding these OXA-type β-lactamases are integrated into the bacterial genome via horizontal gene transfer, except for the native OXA-type enzyme OXA-50, which is capable of destroying carbapenems [43,79]. The OXA-type β-lactamases commonly identified in this pathogen include OXA-51, OXA-2, and OXA-10. These enzymes cause insensitivity to ureidopenicillins and carboxypenicillins. It has also been found that OXA-10-like β-lactamases can degrade ceftazidime, a third-generation cephalosporin [43,79].

The OXA-2 and OXA-10 β-lactamases, which arise from point mutations, are commonly found on integrons and plasmids. These MGEs promote the transfer of genes encoding these enzymes across various Gram-negative bacteria [43]. Both of these β-lactamases exhibit increased hydrolysis of cephalosporins such as cefepime and ceftazidime, as well as aztreonam, resulting in insensitivity to most β-lactamase inhibitors [43]. Contrary to *A. baumannii*, not all OXA enzymes produced by *P. aeruginosa* are able to inactivate carbapenems. Enzymes like OXA-23, OXA-40, and OXA-58, which can degrade carbapenems, have been identified [79,91].

Carbapenemases of the Ambler class A KPC group have also been identified in *P. aeruginosa*. The genes encoding KPC enzymes are acquired through horizontal gene transfer from *Enterobacteriaceae* spp. [43]. The production of AMEs such as APHs, AACs, and ANTs are among the resistance mechanisms to aminoglycosides in *P. aeruginosa* [92]. However, the most prevalent enzymes found in this pathogen are from the ANT and AAC classes [43].

Thus, *P. aeruginosa* APHs transfer a phosphoryl group to the 3′-hydroxyl of aminoglycosides such as kanamycin, neomycin, and streptomycin, effectively inactivating these antibiotics. Meanwhile, *P. aeruginosa* AACs transfer an acetyl group to the amino group at positions 3′ and 6′ of aminoglycosides, leading to the inactivation of gentamicin, tobramycin, netilmicin, kanamycin, and amikacin. Resistance to gentamicin, amikacin, and tobramycin in *P. aeruginosa* is mediated by ANTs, which transfer an adenylyl group to either the amino or hydroxyl group of these aminoglycosides [72]. The genes responsible for encoding AMEs are prevalent in *P. aeruginosa* isolates, with the primary genes being *aac(6′)-I*, *aac(6′)-II*, *ant(2″)*-I, and a*ph(3′)-VI* [93].

3.2.4. Drug Efflux

The principal efflux system in *P. aeruginosa* facilitating the removal of antibacterial agents is the RND superfamily of pumps, characterized by a tripartite structure. This pathogen contains at least 12 RND efflux pumps, with four, including MexAB-OprM, MexCD-OprJ, MexEF-OprN, and MexXY-OprM, serving pivotal roles in mediating antibacterial insensitivity via efflux mechanisms [43,74].

Efflux systems and the low-permeability OM collaborate to enhance antimicrobial resistance in bacteria; therefore, antibacterial drugs that successfully cross the OM are often rapidly expelled into the environment by efflux pumps. These pumps not only aid in removing antimicrobials but also play a role in bacterial stress responses. Stress signals, including host factors, detergents, and endogenous bacterial inducers, can promote the selection of efflux pump-expressing, antimicrobial-resistant mutants, even without direct exposure to antimicrobials [74].

Thus, in *P. aeruginosa*, the efflux pumps of the RND superfamily are composed of a periplasmic membrane fusion protein (e.g., MexC, MexA, MexX, and MexE), an RND transporter unit (such as MexD, MexB, MexY, and MexF), and an OM factor (including OprJ, OprM, and OprN). The most abundant contributor to antimicrobial resistance is the MexAB-OprM efflux pump [43]. This system is constitutively operational in natural strains of *P. aeruginosa* and typically does not result in high resistance to antimicrobial agents. However, its upregulation plays a significant role in enhancing *P. aeruginosa* resistance to various antibiotics, including macrolides, fluoroquinolones, a wide range of β-lactams (e.g., penicillins and carbapenems, with the exception of imipenem and biapenem), tetracyclines, chloramphenicol, lincomycin, novobiocin, as well as the antiseptic triclosan and the surfactant sodium dodecyl sulfate [43,74].

The proteins of the MexAB-OprM efflux pump are encoded by the *mexAB-oprM* operon, whose expression is controlled by the transcriptional regulator protein MexR. Thus, MexR acts as a local transcriptional repressor, suppressing the expression of both the *mexAB-oprM* operon and its own gene. Mutations in the *mexR* gene, along with premature truncation of its encoded peptide, are commonly observed in resistant strains of this bacterium. These mutations lead to an elevated expression level of the MexAB-OprM efflux pump [75]. Other repressor genes that regulate the *mexAB-oprM* operon are *nalC* and *nalD*. Single-nucleotide polymorphisms in these genes can also contribute to antimicrobial resistance by regulating the activity of this pump [43,94]. Increased expression of the MexAB-OprM efflux pump has been observed in *P. aeruginosa* strains resistant to carbapenems, as they also produce carbapenemases. The overexpression of the gene encoding the MexAB-OprM pump, coupled with AmpC activity, has been demonstrated to enhance antibiotic resistance synergistically; however, this drug expulsion system is not able to extrude imipenem, imipenem/relebactam, ceftolozane/tazobactam, or colistin [43].

The MexXY-OprM system contributes to the natural insensitivity of the pathogen to aminoglycosides, tetracycline, erythromycin, and cefepime [43]. The RND family pumps and the operon *mexXY*, which encodes the subunits of the MexXY-OprM efflux system, lack the gene for the OM channel. However, it has been discovered that MexXY can form a fully functional RND efflux pump when coupled to OprM, which typically forms part of the MexAB-OprM efflux system [74,75].

Interestingly, a distinct gene named *oprA*, positioned downstream of the *mexY* gene, has been identified as part of an operon with *mexXY* in certain strains of *P. aeruginosa*, encoding the OM channel OprA. OprA exhibits 47% sequence similarity to OprM; therefore, specific antibiotics, like carbenicillin and sulbenicillin, have been recognized as targets of the MexXY-OprA efflux system [69]. The MexXY-OprM efflux pump is controlled by the MexZ repressor protein, and mutations within or near the mexZ gene cause a loss of repression, leading to an overproduction of the efflux pump and increased expression [95].

Additionally, MexXY has been found to correlate with aminoglycoside-associated resistance as it works synergistically with AMEs. Overexpression of *mexXY* has been particularly noted in strains that produce these enzymes. Moreover, it has been established that MexXY plays a role in resistance to various antipseudomonal drugs, similar to the MexAB-OprM efflux system [43].

Other efflux systems, though not prominent, detected in *P. aeruginosa* strains include the MexCD-OprJ and MexEF-OprN pumps. However, their expression correlates with antimicrobial resistance [43]. The MexCD-OprJ efflux pump has been identified as a key factor in the resistance to fluoroquinolones, tetracyclines, chloramphenicol, as well as various detergents, dyes, and organic solvents. In natural-type strains of *P. aeruginosa*, this efflux system remains inactive under normal conditions [95]. However, exposure to agents like benzalkonium chloride, chlorhexidine gluconate, tetraphenylphosphonium chloride, ethidium bromide, rhodamine 6G, and acriflavine can induce its activity. The activity of the *mexCD-oprJ* operon is controlled by the local repressors NfxB and EsrC. Mutations in the *nfxB* gene are commonly observed in clinical strains of *P. aeruginosa*, driving overproduction of the MexCD-OprJ efflux pump and leading to high levels of insensitivity to multiple antibacterial drugs. The regulation of the *nfxB* promoter depends on the transcription factor VqsM. The absence of VqsM suppresses *nfxB* expression and enhances *P. aeruginosa* insensitivity to a wide range of antibacterials [69].

The EsrC protein functions as a second local repressor by binding to the promoter region of the *mexCD-oprJ* operon and modulating its expression in coordination with NfxB. Additionally, NfxB negatively regulates the transcription of the *esrC* gene [69]. As a result, the NfxB protein acts as a negative regulator of the MexCD-OprJ efflux pump. Mutations within the nfxB gene lead to different levels of MexCD-OprJ expression, which contributes to varying degrees of resistance to antibacterial agents in clinical isolates [95].

The MexEF-OprN system expels various antibacterial drugs, including trimethoprim, chloramphenicol, tetracycline, and imipenem, from the bacterial cell. In contrast to the local repressors regulating the MexAB-OprM and MexCD-OprJ efflux systems, the expression of the MexEF-OprN efflux pump is controlled by a LysR-like activator MexT. Encoded by the *mexT* gene, located upstream of the *mexEF-oprN* operon, MexT promotes the expression of *mexEF-oprN* by binding to the *nod* box in the promoter region. Similar to the MexCD-OprJ system, the MexEF-OprN pump remains inactive under normal conditions. However, mutations in MexT can activate this regulator, leading to the overproduction of the MexEF-OprN efflux pump and repression of the OprD porin. This results in the formation of nfxC-type MDR *P. aeruginosa* mutants [75,96].

A mutated *mexS* gene has been identified in these strains, which regulates the *mexEF-oprN* operon. The MexS protein functions as a repressor of the *mexT* gene, thus limiting the expression of the *mexEF-oprN* operon. Disruption of *mexS* results in an accumulation of electrophilic metabolites, which activate MexT, leading to the increased expression of *mexEF-oprN* [75,96].

Other proteins that activate the expression of the *mexEF-oprN* operon include BrlR and the transcriptional regulator CmrA. BrlR activates the expression by binding directly to the *mexEF-oprN* promoter, while CmrA enhances the expression of *mexT* and *mexEF-oprN* via its regulation of MexS and PA2048, a protein regulated by CmrA [75,96].

3.3. *K. pneumoniae*

K. pneumoniae naturally exhibits insensitivity to penicillin antibiotics, contributing to its role as a major global health issue. The emergence of MDR strains, which produce beta-lactamases such as ESBLs and carbapenemases, further complicates treatment options. Additionally, *K. pneumoniae* rapidly acquires genetic elements, leading to the emergence of two distinct pathotypes: classical *K. pneumoniae* (cKp) and hypervirulent *K. pneumoniae* (hvKp). The hypervirulent strain, more commonly associated with community-acquired infections, raises particular concern due to its increased virulence. The global spread of both the MDR *K. pneumoniae* and hvKp strains remains a pressing public health challenge [43,97].

3.3.1. Drug Uptake Limitation

OM porins, including OmpK35, OmpK36, OmpK37, OmpK38, OmpK26, and PhoE, are present in *K. pneumoniae*. Among these, OmpK35 (also known as OmpF) and OmpK36 (a member of the OmpC group) are classified as the major porins responsible for the uptake of antibiotics, such as β-lactams and fluoroquinolones, into the cell [43,98].

Although OmpK35 is not considered the primary pathway for *K. pneumoniae* resistance, studies have shown that tigecycline-resistant *K. pneumoniae* isolates exhibit reduced OmpK35 expression. This reduction is linked to the *K. pneumoniae* resistance to third-generation cephalosporins and carbapenems [43,99]. In certain instances, OmpK35 may be substituted by OmpK36, which features a smaller channel diameter. Strains lacking OmpK35 and expressing only OmpK36 typically display intermediate carbapenem susceptibility. In contrast, strains with no expression of either OmpK35 or OmpK36 are resistant to both carbapenems and cephalosporins, due to the absence of genes encoding these porins [27].

Mutations in OmpK36, combined with carbapenemase production, have been observed in multiple multi-drug resistant (MDR) *K. pneumoniae* strains. A key mutation involves the insertion of a glycine-aspartate di-amino sequence in extracellular loop three of OmpK36, which reduces the pore size and limits carbapenem entry. This mechanism, mediated by OmpK36, has been frequently detected in the highly virulent ST258/512 strain, which has spread across South and North America, as well as Europe [43]. Additionally, the loss of OmpK36, along with the production of ESBLs and/or AmpC, contributes to insensitivity to carbapenems in this bacterium [100].

The regulation of porin expression can occur through mutations in either the coding regions or promoter sites, as well as through insertional deletions, which may lead to altered or absent porins. For instance, OmpK36 deficiency has been linked to the occurrence of premature stop codons in the *ompK36* gene [99,100]. Furthermore, the reduced ability of the membrane to allow antibiotic passage in *K. pneumoniae* contributes to resistance against multiple drug classes, including β-lactams, fluoroquinolones, chloramphenicol, colistin, aminoglycosides, nalidixic acid, and tetracyclines [99].

3.3.2. Drug Target Modification

Resistance mechanisms involve modifications to LPSs, specifically by altering the phosphate moieties of lipid A with sugars or ethanolamine. Colistin resistance in *K. pneumoniae* is primarily attributed to genetic alterations in transcriptional regulatory systems that control LPS modifications. Key regulators include TCRs PhoPQ and PmrAB, which influence the expression of genes responsible for modifications such as PEtN addition (e.g., *pmrC*) and 4-amino-4-deoxy-l-arabinose (Ara4N) biosynthesis and transfer (e.g., *pbgP*, *pmrE*, and *pmrHFIJKLM*). Proteins such as PmrD and MgrB play pivotal roles in linking the PmrAB and PhoPQ pathways, with MgrB acting as a negative regulator of PhoPQ signaling. Additionally, the CrrAB system modulates PmrAB and PhoPQ expression. Alterations in genes such as *pmrAB*, *phoPQ*, *mgrB*, and *crrB* have been identified as contributors to colistin resistance. An alternative LPS modification pathway includes interactions between TupA-like glycosyltransferases and the CrrAB system [100–102].

Alterations in the *lpxM* and *ramA* genes disrupt lipid A maturation and neutralization, while changes in the *pbgP* and *pmrE* genes modify the addition of aminoarabinose. Modifications in the *pmrC* gene lead to an altered incorporation of PEtN, and variations in *pagP* impact the addition of palmitate. Additionally, mutations in regulatory genes such as *phoPQ*, *pmrA*, *pmrD*, and *mgrB* influence LPS modification, while alterations in the *mcr-1* gene affect phosphoethanolamine transferases involved in binding PEtN [97,102].

Fluoroquinolone resistance in *K. pneumoniae* is associated with point mutations in the chromosomal genes *gyrA*, *gyrB*, *parC*, and *parE*, which encode the subunits of DNA gyrase and topoisomerase IV. These mutations generally arise spontaneously in the *gyrA* and *parC* genes and lead to alterations in the 5′ quinolone-binding regions of the resulting enzymes. Additional mutations in subunit B contribute to higher insensitivity levels by affecting both enzymes, resulting in the highest fluoroquinolone resistance [43].

Another class of quinolone insensitivity genes includes plasmid-mediated quinolone resistance (PMQR) determinants, which play a critical role in fluoroquinolone resistance. These genes encode proteins such as QnrS, QnrA, and QnrB, which bind to DNA gyrase, protecting it from the effects of fluoroquinolones and thereby contributing to resistance [43,97].

RNA methylases, including ArmA, RmfH, RmfA, and NmpA, responsible for 16S rRNA methylation, have been detected in *K. pneumoniae*. Alterations to the aminoglycoside binding site contribute significantly to antimicrobial resistance in this pathogen. Genes encoding 16S rRNA methylases, such as rmtB, rmtC, and nmpA, are often disseminated between organisms through plasmids, which frequently carry additional resistance factors, such as ESBLs, MBLs, and carbapenemases, promoting the emergence of MDR strains. For example, the uncommon and threatening coexistence of *rmtB*, *armA*, *bla*$_{KPC\text{-}2}$, and *iuc* virulence operon-encoding plasmids in *K. pneumoniae* has been recently established. However, the chromosomal localization of these genes has not been determined yet [43,97,99].

Resistance to tetracyclines, particularly tigecycline, can arise due to point mutations in the *rpsJ* gene, responsible for ribosomal protein S10, a crucial part of the 30S ribosomal subunit. Alterations in this ribosomal protein may underlie this phenomenon. Similarly, modifications in ribosomal protein S3 can also contribute to insensitivity. Notably, mutations in the *rpsJ* gene can lead to tigecycline resistance independently of efflux pump involvement. Another pathway contributing to tigecycline insensitivity involves genes encoding 16S rRNA methylases [103].

There are few studies that thoroughly investigate the impact of alterations in PBPs on the resistance to beta-lactam antibiotics in *K. pneumoniae* [104]. However, recent evidence has linked PBPs to the growth and resistance of *K. pneumoniae* to β-lactam antibiotics. Specific enzymes known as zinc finger nucleases have been shown to significantly reduce β-lactam resistance in this pathogen. Furthermore, the knockdown of PBPs has been found to counteract the inhibitory impacts of zinc finger nucleases on bacterial growth during β-lactam therapy. These findings show that *K. pneumoniae* insensitivity to these antibacterial drugs is mediated through the ESBL signaling pathway, suggesting ESBL as a potential focus for combating resistance to β-lactam antibiotics [105].

Nonetheless, it has been determined that PBPs 1, 2, and 3 can still be targeted by certain β-lactam antibiotics such as cefepime, a combination of mecillinam and amoxicillin, piperacillin, aztreonam, and ceftazidime [53,104].

3.3.3. Enzymatic Drug Inactivation

K. pneumoniae produces several β-lactamases, with the most prominent being intrinsic narrow-profile Ambler class A β-lactamases, such as TEM-like enzymes and SHV-1, encoded by the *TEM* and *SHV* genes, respectively. These two enzymes mainly confer resistance to penicillins and cephalosporins. Moreover, due to evolution, these enzymes have been developed to ESBLs and are able to destroy aztreonam and oxyimino-β-lactams, e.g., third- and fourth-generation cephalosporins. The genes encoding ESBLs are located on MGEs and are transmitted among different Gram-negative bacteria via horizontal gene transfer [43,92,99].

Other class A ESBLs involved in the antimicrobial resistance of *K. pneumoniae* are enzymes from the CTX-M, GES, PER, and VEB families, with VEB standing for Vietnam extended-spectrum β-lactamase [43]. However, the most critical mechanism of β-lactam resistance in *K. pneumoniae* is carbapenem resistance, primarily mediated by serine carbapenemases from class A KPC, MBLs from class B (e.g., NDM, VIM, and IMP), and OXA-type carbapenemases from class D [43,92]. Genes responsible for carbapenemase production are predominantly plasmid-borne and disseminate through horizontal gene transfer [99]. It has been found that β-lactamases KPC-2 and KPC-3 are involved in outbreaks of *K. pneumoniae*-related diseases worldwide [106].

Carbapenem resistance related to KPCs in vivo is associated with alterations in the bla_{KPC} gene. Examples include mutations in the bla_{KPC} gene, acquisition of $bla_{KPC\text{-}2}$-harboring plasmids, mutations and reversions in bla_{KPC} genes, $bla_{KPC\text{-}2}$ duplication, increased gene expression of mutated bla_{KPC}, and various alterations in the $bla_{KPC\text{-}3}$ gene primarily leading to the bacterium's insensitivity to the ceftazidime-avibactam combination [100,107].

NDM-1, belonging to the MBLs, contributes to the insensitivity to nearly all β-lactam antibiotics except aztreonam, as well as rendering the new semi-synthetic aminoglycoside plazomicin ineffective. Moreover, bacteria acquiring this enzyme are linked to MDR strains, with *K. pneumoniae* serving as a primary source for the spread of NDM-1 [43].

Enzymes such as the class D β-lactamases OXA-48, OXA-51, OXA-181, and OXA-237, along with the most prominent ESBLs, including the classical oxacillinases OXA-11 and OXA-15, are involved in the resistance of this pathogen to antibacterial agents. It has been recently established that strains producing OXA-48, encoded by blaOXA-48 genes, are often recognized as a source of nosocomial infections in Middle Eastern and European in-patient settings [43]. Additionally, extensive genetic variation in $bla_{OXA\text{-}48}$-like-bearing plasmids has been detected in *K. pneumoniae* [108]. However, these β-lactamases are highly effective at breaking down carbapenems, while they exhibit low hydrolytic activity against clavulanic acid [43]. Genes such as bla_{GES}, bla_{SFO}, bla_{PER}, bla_{TLA}, bla_{VEB}, and $bla_{KLUC\text{-}5}$, which encode other class A ESBLs, are less commonly found [97].

An enzyme, AmpC from the class C β-lactamases, evolved in *K. pneumoniae* through the translocation of the *AmpC* gene from the chromosome to a plasmid. This enzyme hydrolyzes extended-spectrum β-lactams and is resistant to the suppressive effects of β-lactamase inhibitors, including clavulanic acid. However, it does not have the capability to degrade carbapenems. Concluding, due to the activity of various β-lactamases, *K. pneumoniae* exhibits resistance to carbapenems, penicillins, cephalosporins, and monobactams [99].

One of the mechanisms used by *K. pneumoniae* to confer resistance to aminoglycosides is through the action of AMEs, primarily AAC(6′)-Ib, and enzymes from the AAC(3)-II subclass, such as AAC(3)-IIa, AAC(3)-IIb, and AAC(3)-IIc. Among these, AAC(3)-IIa is the most commonly found in this bacterium. The genes that encode these enzymes are present on plasmids and transposons, facilitating their easy transfer between strains. It has been established that AAC(6′)-Ib is frequently found in strains possessing ESBL genes on the same plasmid, while the enzyme AAC(3)-IIa is typically detected in strains with the blaCTX-M gene. The enzyme AAC(6′)-Ib inactivates aminoglycosides like tobramycin and amikacin, but not gentamicin, whereas AAC(3)-II enzymes lead to insensitivity to gentamicin, netilmicin, and tobramycin [43,95,99].

Furthermore, the PMQR *aac(6′)-Ib-cr* gene is the only gene encoding an AAC that contributes to decreased quinolone efficacy through the *N*-acetylation of their piperazinyl amine group in *K. pneumoniae* [97,109]. This enzyme is capable of inactivating limited-spectrum quinolones like norfloxacin and ciprofloxacin, which have an unsubstituted piperazine group, the substrate for this enzyme. Originally found on plasmids in *K. pneumoniae*, this gene has also been recently discovered on chromosomes. Thus, insensitivity to aminoglycoside antibiotics is frequently accompanied by insensitivity to β-lactam agents and DNA gyrase inhibitors. The gene *aac(6′)-Ib-cr* contributes to resistance against antibacterial agents such as aminoglycosides like kanamycin, tobramycin, and amikacin, as well as the aforementioned quinolones [97].

3.3.4. Efflux Pumps

The most predominant efflux systems among efflux pumps in this pathogen are AcrAB-TolC and OqxAB, both from the RND family. The AcrAB-TolC system is composed of the membrane fusion protein AcrA, the efflux transporter AcrB, and the OM channel protein TolC. Various factors, including the AcrA, RamA, and RamR proteins, participate in the regulation of AcrAB-TolC expression. Alterations in the genes encoding these regulatory factors can contribute to increased expression of this pump, leading to an insensitivity to β-lactams, macrolides, fluoroquinolones, and tetracyclines [100,110]. In addition to these efflux pumps, tigecycline insensitivity in this bacterium also involves members of the SMR and MFS superfamilies, such as KpnEF and the MFS member Tet(A) [97,100].

Additionally, antimicrobial resistance to aminoglycosides in *K. pneumoniae* is complex, with changes in cell permeability linked to modifications in the KpnEF and AcrAB-TolC efflux pumps, as well as the potential absence of the KpnO porin. These alterations contribute to varying degrees of insensitivity to gentamicin, tobramycin, streptomycin, and spectinomycin. Additionally, these efflux systems play a role in resistance to polymyxin [97].

The genes encoding the OqxAB drug efflux system, named *oqxAB*, are typically plasmid-encoded and primarily regulated by the RarA protein, with additional regulation by RamA. Overexpression of *rarA* is associated with the MDR phenotypes of *K. pneumoniae*. When the *oqxAB* genes are transferred to a plasmid, it can lead to a tenfold increase in efflux pump manifestation, resulting in a significant reduction in the efficacy of various antimicrobial agents, including fluoroquinolones, chloramphenicol, quinoxaline compounds, and trimethoprim, thereby contributing to a fourfold or greater increase in drug resistance development [99,111].

In conclusion, efflux pumps in *K. pneumoniae* are responsible for resistance to tetracyclines, quinolones, fluoroquinolones, nitrofurantoin, chloramphenicol, polymyxins, macrolides, aminoglycosides, cephalosporins, chlorhexidine, rifampin, and some biocides [99].

3.4. *Enterobacter* spp.

MDR *E. aerogenes* and *E. cloacae* have emerged as significant threats over the past 35 years, particularly in causing nosocomial opportunistic infections in immunocompromised patients [2,27]. The bacterium *E. aerogenes* has recently been redefined as *K. aerogenes* due to its genetic similarity to the genus *Klebsiella*. The spread of *Enterobacter* spp. is facilitated by complementary regulatory pathways that effectively control membrane integrity, thereby providing bacterial defense and promoting the activity of detoxifying enzymes responsible for antibiotic degradation or neutralization [2,112].

Moreover, these bacterial species can obtain numerous MGEs that significantly contribute to resistance against antimicrobial agents [2,112]. Thus, *Enterobacter* spp. exhibit intrinsic insensitivity to aminopenicillins, such as ampicillin, and extended-spectrum cephalosporins. Additionally, through the possession of MGEs, they demonstrate insensitivity to many antibiotics, including third-generation cephalosporins and carbapenems, complicating the choice of an appropriate therapy [113].

3.4.1. Drug Uptake Limitation

The main OM proteins that significantly contribute to AMR in *Enterobacter* spp. have been identified as the predominant porins OmpE35, OmpE36, and OmpE37 in *E. cloacae*, and Omp35, Omp36, and Omp37 in *K. aerogenes*. Additionally, both strains share the porins LamB and PhoE, which also play roles in resistance mechanisms [114].

During antibacterial treatment, it has been observed that certain *Enterobacter* spp. sequentially modify the expression of various porins. For example, the Omp35 porin may be substituted with Omp36, which features a smaller channel size. Consequently, strains lacking Omp35, but producing Omp36 generally exhibit intermediate susceptibility to carbapenems, whereas strains lacking both porins demonstrate carbapenem resistance. Additionally, the overproduction of the LamB porin, along with strains exhibiting a lack of porin expression or decreased porin functionality, can further decrease β-lactam susceptibility. Mutations that cause structural alterations in the lumen eyelet region of *K. aerogenes* Omp36 have also been reported, leading to reduced β-lactam permeability [27].

Interestingly, a small protein called OmpX, found in the OM, regulates the synthesis of Omp35. Overproduction of OmpX results in a decreased expression of Omp35. Consequently, the reduced production, loss, or mutations of porins leads to a decreased OM permeability in *Enterobacter* spp. [112].

3.4.2. Drug Target Modification

Alterations in the OM can modify LPSs, further reducing the susceptibility of *Enterobacter* spp. to polymyxins [114]. Many various genes and operons are essential in modifying LPSs in *Enterobacter* spp., significantly influencing their susceptibility to antimicrobial agents. These genes encode enzymes directly involved in LPS modification. For instance, the *pmrA/pmrB* genes modify lipid A through the *arnBCADTEF* operon, alongside the *pmrC* and *pmrE* genes [115].

Additionally, genes like *phoP/phoQ* contribute by activating the *pmrHFIJKLM* operon and inducing the PmrAB TCRS through the *pmrD* gene. The *arnBCADTEF* operon is pivotal in lipid A modification, adding pEtN and L-4AraN [115]. In polymyxin-resistant bacteria, alterations in the *phoQ*, *pmrB*, and *pmrA* genes have been detected. Plasmid-mediated genes, such as *mcr-1*, *mcr-4*, and *mcr-5*, which encode pEtN transferases, also contribute to LPS modifications and polymyxin resistance in *Enterobacter* spp. These genetic mechanisms are essential for LPS structure alteration and confer resistance to polymyxins [114,115].

However, resistance mechanisms associated with drug target modification are mainly related to quinolones. In *Enterobacter* spp., mutations in the QRDRs of targeted enzymes, such as DNA gyrase and topoisomerase IV, lead to high-level insensitivity, primarily through point mutations in chromosomal genes like *gyrA*, *gyrB*, *parC*, and *parE*. PMQR is mediated by QnrA, QnrB, and QnrS, which belong to the Qnr family. These proteins shield bacterial enzymes, DNA gyrase, and topoisomerase IV, from the lethal effects of quinolones. These mechanisms have also been identified alongside genes encoding ESBLs and AmpC on the same plasmid, or through other ways. In *E. cloacae*, resistance to aminoglycosides is facilitated by genes such as *armA* and *rmtB* that encode 16S rRNA methyltransferases [27,114,116].

3.4.3. Enzymatic Drug Inactivation

The synthesis of β-lactamases is the primary process underlying β-lactam insensitivity. Both *K. aerogenes* and *E. cloacae*, in particular, possess a significant ability to regulate these resistance pathways. Typically, *Enterobacter* spp. produces low levels of cephalosporinase corresponding to chromosomal AmpC β-lactamase, which contributes to resistance to first-generation cephalosporins. However, resistance based on chromosomal acquisition can lead to a marked increase in the synthesis of AmpC cephalosporinase, especially upon exposure to subinhibitory concentrations of carbapenems [112,114].

The overproduction of AmpC β-lactamase, resulting from the loss of AmpR function or the acquisition of plasmid-encoded ampC, contributes to resistance against third-generation cephalosporins. *K. aerogenes* can also integrate large plasmids carrying the ampC gene, such as $bla_{CMY\text{-}10}$, which typically originates from the chromosome. Without the selective influence of antibiotics, this form of genetic material transfer can facilitate the widespread dissemination of resistance mechanisms. This AmpC-related resistance is often linked to the expression of ESBLs. *K. aerogenes* primarily produces TEM-type β-lactamases and those of the CTX-M family, particularly TEM-24 and CTX-M-2 [112,114].

In *E. cloacae*, a variety of ESBLs, including TEM, SHV, and CTX-M types, have been identified. Carbapenemases, such as NDM and VIM types, have been detected in both *E. aerogenes* and *E. cloacae*. Additionally, certain KPC-type or class D β-lactamases, such as OXA-48, which demonstrate carbapenemase activity, have been reported. Recently, MBLs, including IMP-type enzymes, NDM-, GIM-, and VIM-type carbapenemases, as well as serine carbapenemases like KPC and FRI (French imipenemase), have been identified. Among these, the OXA-48-type serine carbapenemase is the most prevalent [112,114,117]. Furthermore, the primary AMEs—AAC(3)-IIa, AAC(6′)-Ib, and ANT(2″)-I—encoded by the *aac(3)-IIa*, *aac(6′)-Ib*, and *ant(2″)-Ia* genes, respectively, have also been detected [112,114].

Clinical strains often exhibit multiple enzymes. Enzymatic insensitivity to fluoroquinolones has also been described due to the presence of a two-point mutation allele of *aac(6′)-Ib*, known as the *aac(6′)-Ib-cr* gene, which enables acetylation of ciprofloxacin and norfloxacin. This resistance mechanism spreads rapidly through its association with the *blaOXA-1* gene on various MGEs [112,114].

3.4.4. Efflux Pumps

The primary efflux mechanisms responsible for the extrusion of antimicrobial drugs, including fluoroquinolones, tetracycline, and chloramphenicol, involve the overproduction of the AcrAB-TolC pump, along with efflux proteins such as EefC and AcrZ [27]. Homologues of OqxAB have also been detected in some *Enterobacter* spp. [27]. The main genes encoding drug efflux pumps for fluoroquinolones are *qepA*, *acrAB-tolC*, *oqxAB*, *sugE*, and *emmdR*, while for aminoglycosides it is *acrAB-tolC* [114]. MarA, SoxS, and RamA are transcriptional regulators that trigger AMR by activating efflux pumps [112].

The summarizing data of AMR mechanisms is provided in Table 1.

Table 1. The mechanisms of AMR and their major contributors in Gram-negative ESKAPE bacteria.

Mechanisms of AMR	*P. aeruginosa*	*K. pneumoniae*	*A. baumannii*	*Enterobacter* spp.
Drug uptake limitation	OprD, OprH, OprF	OmpK35, OmpK36, OmpK37, OmpK38, OmpK26, PhoE	OmpA, CarO	OmpE35, OmpE36, and OmpE37 in *E. cloacae*; Omp35, Omp36, Omp37 in *K. aerogenes*; LamB and PhoE in both strains
Drug target modification	Methylation of 16S ribosomal RNA (rRNA) by RNA methylases encoded by the *rmtA* and *rmtB* genes; Alterations in the motifs of the *gyrA* and *gyrB* genes within the quinolone-resistant determining region (QRDR motif) in DNA gyrase; Alterations in topoisomerase IV via point mutations of *parC* and *parE* genes; Modification of PBPs, including PBP3, PBP4, and PBP5; Alteration of LPs via of lipid A remodeling; PBP3 is linked to resistance to cephalosporins and carbapenems, while PBP4—to resistance against penicillins and cephalosporins	Alterations of lipid A via mutations in genes such as *mcr-1*, *pbgP*, *pmrE*, *pmrHFIJKLM*, *phoPQ*, *pmrAB*, *mgrB*, *crrAB*, *lpxM*, *ramA*, and *pagP*; Mutations in the chromosomal genes *gyrA*, *gyrB*, *parC*, and *parE*; Synthesis of proteins QnrS, QnrA, and QnrB that shield DNA gyrase and topoisomerase IV; Production of RNA methylases such as ArmA, RmfH, RmfA, and NmpA; PBPs	Remodeling of lipid A via alterations in the *pmrA* and/or *pmrB* genes, with mutations being more prevalent in *pmrB*; Deprivation of LPS molecules due to alterations in the *lpxA*, *lpxC*, and *lpxD* genes; Mutations of *gyrA* and *gyrB* genes, which are responsible for DNA gyrase and *parC* gene, which is responsible for topoisomerase IV; Production of DNA gyrase and topoisomerase Iv protective proteins QrnA, QrnS, and QrnS; Production of ribosomal protective proteins, such as TetM, TetW, TetO, and TetS; Production of 23S rRNA (adenine(2058)-N(6))-methyltransferases, encoded by the *erm(B)*, *erm(C)*, and *erm(F)* genes, as well as the ABC-F-type ribosomal protection protein Msr(E), encoded by the *msr(E)* gene; Synthesis of trimethoprim-resistant dihydrofolate reductases, including DfrA1, DfrA5, DfrA7, DfrA10, DfrA12, DfrA14, DfrA16, DfrA17, DfrA19, DfrA20, DfrA27, DfrB1; PBP2 is linked to imipenem resistance, whereas PBP3 is associated with insensitivity to meropenem, sulbactam, and cefiderocol.	Alteration of lipid A via mutations in *mcr-1*, *pbgP*, *pmrE*, *pmrHFIJKLM*, *phoPQ*, *pmrAB*, *mgrB*, *crrAB*, *lpxM*, *ramA*, and *pagP* genes; Modification of the enzymes DNA gyrase and topoisomerase IV due to alterations in genes such as *gyrA*, *gyrB*, *parC*, and *parE*; Synthesis of proteins QnrS, QnrA, and QnrB, which protect DNA gyrase and topoisomerase IV; Alteration of LPS due to action of *mcr-1*, *mcr-4*, and *mcr-5* genes encoding pEtN transferases; Production of 16S rRNA methyltransferases, encoded by *armA* and *rmtB* genes
Enzymatic drug inactivation	AmpC, TEM, PSE, PER, VEB, GES, BEL, PIB-1, GPC-1, CARB, OXA-50, OXA-2, OXA-10, OXA-23, OXA-40, KPC, NDM-1, SPM-1,GIM-1, VIM, IMP; APHs, AACs, and ANTs	TEM, SHV-1, CTX-M, GES, PER, VEB, KPC-3, KPC-2, NDM-1, OXA-48, OXA-51, OXA-181, OXA-237, OXA-11, OXA-15; AACs	OXA-23, OXA-24/40, OXA-40, OXA-51, OXA-58, OXA-143, KPC-2, KPC-3, TEM-1,-2, CTX-M, SHV, PER, IMP, NDM, SIM, VIM; AACs, ANTs; ADP-ribosyltransferases, including, Arr-2, Arr-3, and Arr-4; Macrolide 2′-phosphotransferases, Mph(A) and Mph(E); Glutathione transferases from the FosLL, FosA3, and FosA3/FosA4 families; Tet(X) monooxygenases.	AmpC, TEM, SHV, CTX-M-2, TEM-24, IMP-type, NDM-type, VIM-type, GIM-type, OXA-48-type, KPC, FRI AACs, ANTs
Drug efflux	Mainly from RND family: MexCD-OprJ, MexAB-OprM, MexXY, MexEF-OprN	Mainly from RND family: AcrAB-TolC, OqxAB, KexEF, KexC, KexD, EefABC, and KdE; SMR superfamily member KpnEF; MFS superfamily member Tet(A)	AdeABC, MATE, MFS, RND, SMR, ABC	RND family members AcrAB-TolC and OqxAB

4. High Risk Clones of Gram-Negative ESKAPE Bacteria

The term "clone" describes isolates from different sources, locations, or times that exhibit extensive phenotypic and genotypic similarities, indicating a common origin. This definition is often applied to multidrug-resistant bacteria found in various locations [118,119].

The definition of "high-risk clone" refers to pathogen clones that play a significant role in the rapid spread of antibiotic resistance. These clones are alarming because they complicate treatment and efficiently spread MGEs carrying AMR genes [119–121].

In ESKAPE pathogens, high-risk clones play a substantial role in the worldwide dissemination of resistance mechanisms [18]. For instance, some carbapenem-resistant strains, such as *A. baumannii*, *P. aeruginosa*, and certain *Enterobacterales*, are linked to increased patient mortality and pose significant therapeutic challenges [122].

The global spread of MDR bacteria is linked to specific lineages: *A. baumannii* is classified into "international high-risk clones" (ICs), *K. pneumoniae* into clonal groups (CGs) defined by core-genome multilocus sequence typing (cgMLST) and sequence types (STs), while *Enterobacter cloacae* complex and *P. aeruginosa* are categorized primarily by STs (Table 2) [123–126].

Table 2. The most prevalent high-risk clones of ESKAPE Gram-negative bacteria.

Pathogen	Clonal Classification	Notes
A. baumannii	Spread worldwide: CC1P/CC109O, CC2P/CC92O, CC3P/CC187O, CC1P/CC109O (spread worlwide); Common in Europe and North America: CC2P/CC92O, CC3 P/CC187O.	Have caused outbreaks and persisted in ICUs globally for extended periods; Recently, ST136O and ST208O, which are part of CC2P/CC92O, as well as ST758O and ST1054O, which are part of CC636, have been reported; Sequence types ST58O and ST1054O represent a high-risk clone associated with multi-drug resistance; ST20O and ST136O are part of the most predominant and globally disseminated complex, linked to isolates with the capacity to acquire resistance.
K. pneumoniae	CG258 (ST258, ST11, ST512), CG14 (ST14, ST51). Recently identified non-clonal group 258 ST307 and ST147	*K. pneumoniae* strain ST258 is the dominant clone, mainly associated with the production of KPC-2 and KPC-3 carbapenemases; Additionally, ST11, ST340, and ST512 are single-locus variants of ST258, all of which carry arbapenemase genes; ST11, in particular, is genetically similar to ST258 and linked to the production of KPC, NDM, VIM, IMP, and OXA-48 carbapenemases; Strains of *K. pneumoniae* ST14 and ST15 are frequently producers of ESBLs or carbapenemases, which contribute to antibiotic insensitivity and hospital outbreaks globally; The ST307 and ST147 clones are believed to have originated in Europe during the early to mid-1990s and rapidly gained prominence as global pathogens. These clones have since spread across every continent, excluding Antarctica, and have been responsible for numerous global nosocomial outbreaks, including those linked to endoscopy procedures and long-term care facilities
E. cloacae complex (CREC)	ST66, ST78, ST108, ST114, ST171, ST105, ST108, ST93, ST90	ST171 and ST78 clones have shown epidemic potential, partly due to their occasional acquisition of plasmid-borne carbapenemases; Clones ST66, ST171, and ST78 are known to produce ESBLs, including CTX-M-15 and CTX-M-9, with 40% of these isolates also carrying the *mcr-9* gene; The presence of blaOXA-48 and blaCTX-M in MDR-resistant, high-risk *E. cloacae* clones is particularly concerning, as *mcr-9* may transfer to other bacteria or mutate, potentially resulting in colistin resistance; Effective surveillance is essential to identify and control these pathogens, which pose serious treatment challenges and risk the spread of resistance; A newly identified ST66 has exhibited the capacity to gain various antibiotic insensitivity genes and employ its secretion systems, highlighting its risk of developing into a high-risk clone.
P. aeruginosa	ST235, ST111, ST277, ST244, ST308, ST395	The primary global high-risk MDR/XDR clones are ST175, ST111, and ST235, with ST175 being the predominant, representing 68% of XDR isolates; ST235 possesses the DprA determinant, which facilitates homologous recombination, potentially enhancing its capacity to obtain and retain external resistance traits. In *P. aeruginosa*, resistance-associated mutations have been identified in ST111 and ST235 clones, including GyrA T83I, ParC S87L, and changes in the oprD protein; Compared to the other two high-risk clones, ST235 (serotype O11) is notably more prevalent and has been identified on all five continents; Similarly, ST111 (serotype O12) is widespread, though not reported in Oceania; In contrast, ST175 (serotype O4) is predominantly found in Europe.

Prepared on the basis of data from the sources [18,125,127–136].

Genomic analyses provide essential insights into the genetic basis of AMR in high-risk clones of *A. baumannii*, *K. pneumoniae*, *P. aeruginosa*, and *Enterobacter* spp. These analyses help uncover the complex clonal structures, resistance determinants, and virulence factors of these pathogens. Surveillance of these high-risk clones is critical for the prompt identification of new insensitivity patterns and for tracking the dissemination of resistance genes. Timely identification of emerging clones enables targeted control strategies, helping to maintain the effectiveness of existing antibiotics and enhance responses to AMR at the public health level [121,137,138].

5. Upcoming Novel Treatment Strategies to Combat AMR of Gram-Negative ESKAPE Bacteria

The treatment of infections generally relies on antibacterial agents, used alone or in polytherapy. However, the efficacy of these agents against ESKAPE pathogens continues to decline annually, raising concerns about the future due to the restricted use of effective therapeutic alternatives. A review of the Clinical and Laboratory Standards Institute (CLSI) guidelines reveals that, since 2010, numerous antibacterial agents effective against ESKAPE pathogens have been removed, with only a few new antibiotics or combinations introduced. This underscores the critical necessity to investigate alternative treatment strategies to manage diseases primarily triggered by ESKAPE pathogens [139]. To address this challenge, novel treatment approaches are being intensively studied, including combination antibiotic therapy, antimicrobial peptide (AMP) therapy, bacteriophage therapy, nanoparticles [140,141], including OMVs [14,36,142], and photodynamic therapy (Figure 5) [140,141,143]. These approaches seek to counteract resistance mechanisms and enhance the effectiveness of treatment options [140].

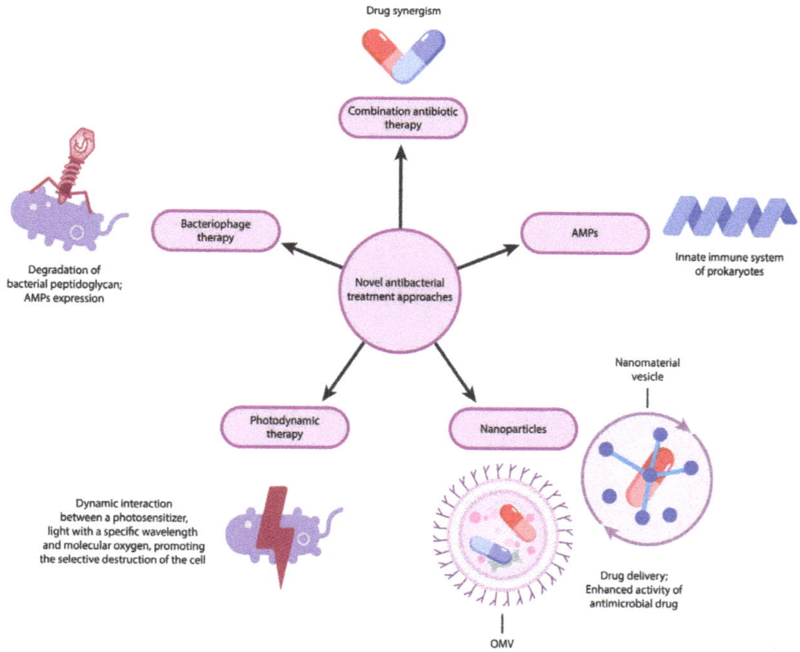

Figure 5. Innovative antimicrobial strategies and their mechanisms of action (adapted and reused from the source [141] under the condition of Creative Common CC BY license, also supplemented on the basis of data from sources [14,36,142]).

An example of antibacterial combination therapy includes the use of colistin and tigecycline in conjunction with other antibacterial drugs to treat infectious diseases caused by *K. pneumoniae* and *A. baumannii* infections [139,140]. Another promising combination involves vaborbactam, a beta-lactamase inhibitor, which helps restore meropenem's efficacy by inhibiting *K. pneumoniae* carbapenemases [140,144]. Beta-lactamases can be suppressed by metal chelators such as ethylenediaminetetraacetic acid (EDTA), deferasirox, and deferoxamine. These substances are essential for the enzymes' function [140,145–148]. Studies have demonstrated that combining certain antibacterial drugs, including imipenem, tobramycin, and vancomycin, with these chelating substances results in a reduction in bacterial burden in *S. aureus*, *P. aeruginosa*, and *E. coli* in mouse models [139]. Quorum quenchers are molecules that inhibit quorum sensing, thus preventing biofilm formation. Among these, 1-[(2,4-Dichlorophenethyl)amino]-3-Phenoxypropan-2-ol has emerged as one of the most promising microbicidal agents. It has been demonstrated to target both dormant cells of P. aeruginosa and non-dormant cells of all ESKAPE pathogens. Furthermore, it has shown the ability to eradicate both planktonic and biofilm-forming antibiotic-insensitive strains [140,149,150]. In combination with various antibiotics, this compound has proven effective against all ESKAPE pathogens, establishing it as a valuable adjuvant in antimicrobial therapy [140,151].

Bacteriophages are natural viruses found in various environmental sources that specifically target bacteria. While they are considered a novel therapeutic option, bacteriophage therapy has been explored since the 1920s. Phages are attractive for several reasons, including their ability to disrupt biofilms, their minimal impact on the commensal microbiota, and their efficacy in eliminating MDR bacteria. These characteristics make them a potential substitute for conventional antibacterial drugs in the fight against resistant infections [152,153]. In a study on bacteriophage therapy, a cocktail of bacteriophages (ECP311, KPP235, and ELP140) delivered through the last left proleg of *Galleria mellonella* resulted in a 100% survival rate of larvae infected with *K. pneumoniae* (KP235). However, multiple doses of bacteriophages were required for effective therapy in vivo [140,154,155]. Similarly, bacteriophage TSK1 was shown to reduce *K. pneumoniae* biofilm by 99–100% in vitro through the release of a depolymerase enzyme that targets the capsular polysaccharides within the biofilm [140,156]. Further studies on *A. baumannii* showed that local application of the virus vBGEC_Ab-M-G7 to wounds led to complete destruction of the bacteria in a rat wound infection model [140,157]. Additionally, the bacteriophage Bφ-C62 exhibited a multiplicity of infection-dependent efficacy against *A. baumannii* in both in vitro and in vivo settings [140,158]. Intranasal administration of the bacteriophage vB_AbaM-IMEAB2, along with intraperitoneal delivery of bacteriophage φkm18p, shielded mice from *A. baumannii* infection, reducing inflammation and decreasing levels of pro-inflammatory cytokines, such as TNF-α and IL-6 [140,159].

AMPs are natural bioactive molecules produced by insects, plants, and humans as part of their immune defense mechanisms against pathogens [152,160]. Given their strong ability to combat bacteria, AMPs are being investigated as potential treatments for bacterial infections [152,161]. These peptides are positively charged and bind to the negatively charged components of bacterial membranes, causing membrane disruption and resulting in bacterial cell death [17,152]. In addition to their membrane activity, certain AMPs can penetrate bacterial cells and interfere with vital intracellular functions, such as DNA and RNA synthesis, further contributing to their antimicrobial effects [17,152].

Evidence suggests that many AMPs exhibit significant efficacy against ESKAPE pathogens. For instance, the AMP S-thanatin and its analogs have shown enhanced antimicrobial effects against *K. pneumoniae* in both in vitro and in vivo models. S-thanatin displayed MIC values 2- to 8-fold lower than those of its analogs, demonstrating superior potency [140,162]. Similarly, Feleucin-K3 has exhibited strong activity against *P. aeruginosa*, including multidrug-resistant clinical strains [140,163].

Other AMP lactoferrin and its derived peptides have demonstrated antibacterial activity against *E. coli*, *S. aureus*, *Acinetobacter* spp., and *P. aeruginosa* in murine models [140,164]. Thus, these findings highlight the capability of AMPs as innovative therapeutic agents to combat resistant bacterial diseases [140,165].

Currently, nanomaterials, such as nanoparticles and nano-drug delivery systems, have shown significant promise in managing infections, particularly those caused by MDR pathogens. Nanoparticles enhance treatment effectiveness through two main mechanisms: enhancing the action of existing antibiotics and introducing novel bactericidal activities independent of conventional drugs [141,166,167].

By functioning as drug carriers, nanomaterials can bypass cellular barriers to deliver antibiotics directly to the bacterial cytoplasm. This targeted delivery allows for controlled drug release, reducing the required dosage, minimizing side effects, and improving the pharmacokinetics, therapeutic efficacy, and cost-effectiveness of treatments. These advancements hold significant potential in combating antibiotic resistance globally [141,166,167].

It has been established that zinc oxide-based nanoparticles exhibit notable antibacterial efficacy against a range of gram-negative pathogens, such as *E. coli* and *P. aeruginosa*, and gram-positive pathogens like *B. subtilis* and *S. aureus* [141,168]. Furthermore, one study has shown that chemically synthesized zinc oxide nanoparticles are effective in suppressing the growth of carbapenem-resistant *A. baumannii* strains [141,169].

Recently, significant attention has been directed towards OMVs, which are projected to be essential in drug delivery systems. These vesicles are being explored for their potential in transporting antibiotic molecules and for use in vaccine development. Their versatility and biocompatibility position them as strong candidates for enhancing the efficacy of various treatments, especially in overcoming challenges posed by antibiotic resistance and improving vaccine delivery. The growing interest in OMVs suggests a promising future in both therapeutic and preventive medicine [14,36,142].

Photodynamic therapy relies on the interaction between a photosensitizer, light of a specific wavelength, and molecular oxygen, which together enable the targeted destruction of microbial cells. The photosensitizer absorbs light, becoming activated and transferring energy to molecular oxygen, thereby generating reactive oxygen species (ROS). These ROS damage cellular components such as lipids, proteins, and nucleic acids, leading to microbial cell death [140,143].

The efficacy of antimicrobial photodynamic therapy (aPDT) is generally influenced by the type and quantity of photosensitizer, as well as the bacterial classification, particularly whether the bacteria are Gram-positive or Gram-negative [140,170–172]. For example, one study evaluated the effectiveness of three substances, acting as photosensitizers, including Azure A, Toluidine Blue O, and New Methylene Blue, on the Gram-positive *E. faecalis* and gram-negative *K. pneumoniae*. It was found that photoinactivation was significantly higher in *E. faecalis* than in *K. pneumoniae*, with an 8 log reduction in *E. faecalis* bacterial count after 80 s of irradiation, compared to a 3 log reduction in *K. pneumoniae* after 300 s. Additionally, a higher amount of photosensitizer adhered to *E. faecalis* compared to *K. pneumoniae*. This suggests that the potency of aPDT therapy depends on the quantity of photosensitizer that attaches to the bacterial cell membrane [140,173]

Another study demonstrated that Gram-positive bacteria, like MRSA and *E. faecalis*, exhibited greater sensitivity to aPDT than Gram-negative bacteria, such as *P. aeruginosa* and ESBL-producing *K. pneumoniae* [140,174]. The sensitivity varied with the duration and intensity of irradiation, with gram-positive bacteria showing greater reductions in colony counts, particularly *MRSA*, where colonies were reduced by 2 log CFU per ml after just 30 s of irradiation. However, the ESBL-producing *K. pneumoniae* strain showed no significant reduction under any of the irradiation conditions. The difference in efficacy between Gram-positive and Gram-negative pathogens can be attributed to the structural variations in their cell walls. The cell wall in Gram-positive bacteria is simpler and more abundant in peptidoglycan and teichoic acids, which makes these pathogens more vulnerable to aPDT. Gram-negative bacteria, on the other hand, have an additional outer membrane (OM) composed of LPSs, lipoproteins, and negatively charged proteins, which form a barrier that limits the penetration of photosensitizers and impedes the photo-oxidation process. These findings highlight the impact of bacterial cell wall structure on the efficacy of antimicrobial photodynamic therapy, with Gram-negative bacteria proving more resistant due to their additional protective layers [140,174–177].

With the increasing need for new antibacterial drugs, there is a growing emphasis on alternative technologies such as AMPs and aPDT in antibacterial research [140]. Notable progress has been made with bacteriophages, synthetic nanomaterials, and natural nanoparticles [140–142]. These strategies show great promise in combating multidrug-resistant bacteria and offer innovative solutions for both healthcare applications and industrial use [140,141].

6. Concluding Remarks

The mechanisms of AMR in Gram-negative ESKAPE bacteria, such as *K. pneumoniae, A. baumannii, P. aeruginosa*, and *Enterobacter* spp., underscore the sophisticated strategies these pathogens employ to evade antimicrobial treatments. AMR in these bacteria is driven by various mechanisms, including the generation of anti-biotic-degrading enzymes, alterations in drug binding sites, and the increased efflux of antibacterial drugs. These adaptations significantly diminish the effectiveness of current treatments, making diseases caused by Gram-negative ESKAPE bacteria increasingly challenging to manage.

High-risk clones emphasize the critical importance of global cooperation to combat the spread of resistance mechanisms. Continued monitoring and study of these clones are essential to mitigating their impact on healthcare and maintaining the effectiveness of existing antibacterial drugs.

To combat these formidable pathogens, it is crucial to explore and develop novel therapeutic strategies. Recent advances include the use of bacteriophages, which specifically target and kill bacteria, and nanoparticles, which can be engineered to deliver antibiotics more effectively or disrupt bacterial defenses. These innovative approaches, along with the development of new antimicrobial agents, offer promising solutions for overcoming resistance mechanisms.

Equally important in addressing AMR is antimicrobial stewardship. Stewardship programs promote the responsible administration and use of antibacterial drugs, ensuring their prescription only when necessary and at appropriate doses. By optimizing antibiotic use, stewardship efforts help to slow the development of resistance and preserve the effectiveness of current treatments, reducing the overall burden of AMR.

Addressing AMR in Gram-negative ESKAPE bacteria requires a multifaceted and innovative approach. By continuing to explore and integrate these new strategies, significant progress can be made in alleviating the global burden of drug-resistant infections and improving treatment outcomes.

Author Contributions: Conceptualization, G.V.S. and A.R.; writing—original draft preparation, G.V.S.; writing—review and editing, A.R., E.S. and L.M.; visualization, G.V.S.; supervision, A.R. All authors have read and agreed to the published version of the manuscript.

Funding: This research received no external funding.

Institutional Review Board Statement: Not applicable.

Informed Consent Statement: Not applicable.

Data Availability Statement: No new data were created or analyzed in this study.

Conflicts of Interest: The authors declare no conflicts of interest.

Abbreviations

AAC	aminoglycoside N-acetyltransferase
ABC	ATP-binding cassette
ADC	*Acinetobacter*-derived cephalosporinase
AME	aminoglycoside-modifying enzyme
AMP	adenosine monophosphate
AMPs	antimicrobial peptides
AMR	antimicrobial resistance
anhMP	anhydromuropeptide
ANTs	aminoglycoside O-nucleotidyltransferases
aPDT	antimicrobial pototdynamic therapy
APHs	aminoglycoside O-phosphotransferases
Ara4N	transfer of 4-amino-4-deoxy-1-arabinose
ATP	adenosine triphosphate
BEL	Belgium extended β-lactamase
CARB	carbenicillin-hydrolyzing β-lactamase
CFU	colony-forming unit
cgMLST	core-genome multilocus sequence typing
cKp	classical *K. pneumoniae*
CLSI	Clinical and Laboratory Standards Institute
CRE	carbapenem-resistant enterobacterales
CREC	*E. cloacea* complex
CRKP	carbapenem-resistant *K. pneumoniae*
CTX-M	cefotaxime-hydrolyzing β-lactamase
DNA	deoxyribonucleic acid
EEA	European Economic Area
EDTA	ethylenediaminetetraacetic acid
ESBL	extended-spectrum β-lactamases
ESKAPE	*E. faecium*, *S. aureus*, *K. pneumoniae*, *A. baumannii*, *P. aeruginosa*, and *Enterobacter* spp.
EU	European Union
GES	Guiana extended-spectrum β-lactamase
GPC-1	German *Pseudomonas* carbapenemase
GTP	guanosine triphosphate
hvKp	hypervirulent *K. pneumoniae*
ICs	international high-resistant clones
ICU	intensive care unit
IGs	clonal groups
IM	inner membrane
IMP	imipenemase
IQR	interquartile range
KPC	*K. pneumoniae* carbapenemase
L-Ara4N	4-amino-4-deoxy-L-arabinose
LPS	lipopolysaccharide
M	median
MATE	multidrug and toxic compound extrusion
MBL	metallo-β-lactamase
MDR	multidrug resistance

MFP	membrane fusion or an adaptor protein
MFS	major facilitator superfamily
MGE	mobile genetic element
MIC	minimal inhibitory concentration
MRSA	methicillin-resistant *S. aureus*
NDM	New Delhi metallo-β-lactamase
OM	outer membrane
Omp	outer membrane protein
OMV	outer membrane vesicle
OXA	oxacillinase
PACE	proteobacterial antimicrobial compound efflux
PBP	penicillin-binding protein
PDR	pandrug resistant
PER	*Pseudomonas* extended resistant
PEtN	phosphoethanolamine
PIB-1	*Pseudomonas* imipenem β-lactamase
PMF	proton motive force
PMQR	plasmid-mediated quinolone resistance
PSE	*Pseudomonas*-specific enzyme
QRDR	quinolone-resistant-determinating region
RND	resistance-nodulation-division
ROS	reactive oxygen species
SARS-CoV-2	a strain of corona virus that causes COVID-19
SCO-1	a novel plasmid-mediated class A β-lactamase with carbenicillinase characteristics from *E. coli*
SHV	sulfhydryl reagent variable
SIM	Seoul imipenemase
SMR	small multidrug resistance
SPM	Sao Paulo MBL
STs	sequence types
TCRS	two-component regulatory system
TEM	Temoneira serine β-lactamase of molecular class A
UI	uncertainty interval
US	United States
VAP	ventilator-associated pneumonia
VEB	Vietnam extended β-lactamase
VIM	Verona integron-encoded MBL
VRE	vancomycin-resistant *Enterococcus*
WHO	World Health Organization
XDR	extensively drug-resistant bacteria

Note: The names of genes and bacterial strains are written in italics

References

1. Fleming, A. *Nobel Lecture. Penicillin*; Nobel Prize Outreach AB: Stockholm, Sweden, 1945. Available online: https://www.nobelprize.org/ (accessed on 6 January 2025).
2. Mancuso, G.; Midiri, A.; Gerace, E.; Biondo, C. Bacterial Antibiotic Resistance: The Most Critical Pathogens. *Pathogens* **2021**, *10*, 1310. [CrossRef] [PubMed]
3. Salam, M.A.; Al-Amin, M.Y.; Salam, M.T.; Pawar, J.S.; Akhter, N.; Rabaan, A.A.; Alqumber, M.A.A. Antimicrobial Resistance: A Growing Serious Threat for Global Public Health. *Healthcare* **2023**, *11*, 1946. [CrossRef]
4. Cheesman, M.J.; Ilanko, A.; Blonk, B.; Cock, I.E. Developing New Antimicrobial Therapies: Are Synergistic Combinations of Plant Extracts/Compounds with Conventional Antibiotics the Solution? *Pharmacogn. Rev.* **2017**, *11*, 57–72. [CrossRef] [PubMed]
5. Ali, A.; Imran, M.; Sial, S.; Khan, A. Effective antibiotic dosing in the presence of resistant strains. *PLoS ONE* **2022**, *17*, e0275762. [CrossRef]
6. Tang, K.W.K.; Millar, B.C.; Moore, J.E. Antimicrobial Resistance (AMR). *Br. J. Biomed. Sci.* **2023**, *80*, 11387. [CrossRef] [PubMed]

7. Prestinaci, F.; Pezzotti, P.; Pantosti, A. Antimicrobial resistance: A global multifaceted phenomenon. *Pathog. Glob. Health* **2015**, *109*, 309–318. [CrossRef] [PubMed]
8. Ventola, C.L. The antibiotic resistance crisis: Part 1: Causes and threats. *Pharm. Ther.* **2015**, *40*, 277–283.
9. Chinemerem Nwobodo, D.; Ugwu, M.C.; Oliseloke Anie, C.; Al-Ouqaili, M.T.S.; Chinedu Ikem, J.; Victor Chigozie, U.; Saki, M. Antibiotic resistance: The challenges and some emerging strategies for tackling a global menace. *J. Clin. Lab. Anal.* **2022**, *36*, e24655. [CrossRef] [PubMed]
10. Zhu, Y.; Huang, W.E.; Yang, Q. Clinical Perspective of Antimicrobial Resistance in Bacteria. *Infect. Drug Resist.* **2022**, *15*, 735–746. [CrossRef]
11. Founou, R.C.; Blocker, A.J.; Noubom, M.; Tsayem, C.; Choukem, S.P.; Dongen, M.V.; Founou, L.L. The COVID-19 pandemic: A threat to antimicrobial resistance containment. *Future Sci. OA* **2021**, *7*, Fso736. [CrossRef] [PubMed]
12. Rayan, R.A. Flare of the silent pandemic in the era of the COVID-19 pandemic: Obstacles and opportunities. *World J. Clin. Cases* **2023**, *11*, 1267–1274. [CrossRef] [PubMed]
13. Akram, F.; Imtiaz, M.; Haq, I.U. Emergent crisis of antibiotic resistance: A silent pandemic threat to 21(st) century. *Microb. Pathog.* **2023**, *174*, 105923. [CrossRef] [PubMed]
14. Sakalauskienė, G.V.; Radzevičienė, A. Antimicrobial Resistance: What Lies Beneath This Complex Phenomenon? *Diagnostics* **2024**, *14*, 2319. [CrossRef] [PubMed]
15. Bereanu, A.S.; Bereanu, R.; Mohor, C.; Vintilă, B.I.; Codru, I.R.; Olteanu, C.; Sava, M. Prevalence of Infections and Antimicrobial Resistance of ESKAPE Group Bacteria Isolated from Patients Admitted to the Intensive Care Unit of a County Emergency Hospital in Romania. *Antibiotics* **2024**, *13*, 400. [CrossRef]
16. Barbu, I.C.; Gheorghe-Barbu, I.; Grigore, G.A.; Vrancianu, C.O.; Chifiriuc, M.C. Antimicrobial Resistance in Romania: Updates on Gram-Negative ESCAPE Pathogens in the Clinical, Veterinary, and Aquatic Sectors. *Int. J. Mol. Sci.* **2023**, *24*, 7892. [CrossRef] [PubMed]
17. Alcántar-Curiel, M.D.; Huerta-Cedeño, M.; Jarillo-Quijada, M.D.; Gayosso-Vázquez, C.; Fernández-Vázquez, J.L.; Hernández-Medel, M.L.; Zavala-Pineda, M.; Morales-Gil, M.; Hernández-Guzmán, V.A.; Bolaños-Hernández, M.I.; et al. Gram-negative ESKAPE bacteria bloodstream infections in patients during the COVID-19 pandemic. *PeerJ* **2023**, *11*, e15007. [CrossRef] [PubMed]
18. Loyola-Cruz, M.; Gonzalez-Avila, L.U.; Martínez-Trejo, A.; Saldaña-Padilla, A.; Hernández-Cortez, C.; Bello-López, J.M.; Castro-Escarpulli, G. ESKAPE and Beyond: The Burden of Coinfections in the COVID-19 Pandemic. *Pathogens* **2023**, *12*, 743. [CrossRef]
19. Sulayyim, H.J.A.; Ismail, R.; Hamid, A.A.; Ghafar, N.A. Antibiotic Resistance during COVID-19: A Systematic Review. *Int. J. Environ. Res. Public Health* **2022**, *19*, 11931. [CrossRef]
20. Dhesi, Z.; Enne, V.; Brealey, D.; Livermore, D.; High, J.; Russell, C.; Colles, A.; Kandil, H.; Mack, D.; Martin, D.; et al. Organisms causing secondary pneumonias in COVID-19 patients at 5 UK ICUs as detected with the FilmArray test. *medRxiv* **2020**. [CrossRef]
21. Hassan, M.E.; Al-Khawaja, S.A.; Saeed, N.K.; Al-Khawaja, S.A.; Al-Awainati, M.; Radhi, S.S.Y.; Alsaffar, M.H.; Al-Beltagi, M. Causative bacteria of ventilator-associated pneumonia in intensive care unit in Bahrain: Prevalence and antibiotics susceptibility pattern. *World J. Crit. Care Med.* **2023**, *12*, 165–175. [CrossRef] [PubMed]
22. European Centre for Disease Prevention and Control; World Health Organization Regional Office for Europe. *Antimicrobial Resistance Surveillance in Europe 2022—2020 Data*; World Health Organization Regional Office for Europe: Copenhagen, Denmark, 2022.
23. Jonas, O.B.; Irwin, A.; Berthe, F.C.J.; Le Gall, F.G.; Marquez, P.V. *Drug-Resistant Infections: A Threat to Our Economic Future*; Final Report (English); HNP/Agriculture Global Antimicrobial Resistance Initiative; World Bank Group: Washington, DC, USA, 2017; Volume 2.
24. Pabst, S. Antimicrobial Resistance 2023. Available online: https://www.who.int/news-room/fact-sheets/detail/antimicrobial-resistance (accessed on 6 June 2024).
25. Chiș, A.A.; Rus, L.L.; Morgovan, C.; Arseniu, A.M.; Frum, A.; Vonica-Țincu, A.L.; Gligor, F.G.; Mureșan, M.L.; Dobrea, C.M. Microbial Resistance to Antibiotics and Effective Antibiotherapy. *Biomedicines* **2022**, *10*, 1121. [CrossRef] [PubMed]
26. Catalano, A.; Iacopetta, D.; Ceramella, J.; Pellegrino, M.; Giuzio, F.; Marra, M.; Rosano, C.; Saturnino, C.; Sinicropi, M.S.; Aquaro, S. Antibiotic-Resistant ESKAPE Pathogens and COVID-19: The Pandemic beyond the Pandemic. *Viruses* **2023**, *15*, 1843. [CrossRef]
27. De Oliveira, D.M.P.; Forde, B.M.; Kidd, T.J.; Harris, P.N.A.; Schembri, M.A.; Beatson, S.A.; Paterson, D.L.; Walker, M.J. Antimicrobial Resistance in ESKAPE Pathogens. *Clin. Microbiol. Rev.* **2020**, *33*, e00181-19. [CrossRef] [PubMed]
28. Serna, C.; Gonzalez-Zorn, B. Antimicrobial resistance and One Health. *Rev. Esp. Quimioter.* **2022**, *35* (Suppl. S3), 37–40. [CrossRef] [PubMed]
29. World Health Organization. *Prioritization of Pathogens to Guide Discovery, Research and Development of New Antibiotics for Drug-Resistant Bacterial Infections, Including Tuberculosis*; WHO/EMP/IAU/2017.12; World Health Organization (WHO): Geneva, Switzerland, 2017; p. 87.
30. WHO. *WHO Bacterial Priority Pathogens List, 2024: Bacterial Pathogens of Public Health Importance to Guide Research, Development and Strategies to Prevent and Control Antimicrobial Resistance*; World Health Organization: Geneva, Switzerland, 2024; p. 72.

31. OneHealthTrust. ResistanceMap: Antibiotic Resistance. Available online: https://resistancemap.onehealthtrust.org/CountryPage.php?country=India (accessed on 31 May 2024).
32. Santajit, S.; Indrawattana, N. Mechanisms of Antimicrobial Resistance in ESKAPE Pathogens. *Biomed. Res. Int.* **2016**, *2016*, 2475067. [CrossRef]
33. Krapp, F.; Ozer, E.A.; Qi, C.; Hauser, A.R. Case Report of an Extensively Drug-Resistant *Klebsiella pneumoniae* Infection With Genomic Characterization of the Strain and Review of Similar Cases in the United States. *Open Forum Infect. Dis.* **2018**, *5*, ofy074. [CrossRef] [PubMed]
34. Singh, G.; Rana, A.; Smriti. Decoding antimicrobial resistance: Unraveling molecular mechanisms and targeted strategies. *Arch. Microbiol.* **2024**, *206*, 280. [CrossRef] [PubMed]
35. Jadimurthy, R.; Mayegowda, S.B.; Nayak, S.C.; Mohan, C.D.; Rangappa, K.S. Escaping mechanisms of ESKAPE pathogens from antibiotics and their targeting by natural compounds. *Biotechnol. Rep.* **2022**, *34*, e00728. [CrossRef]
36. Sartorio, M.G.; Pardue, E.J.; Feldman, M.F.; Haurat, M.F. Bacterial Outer Membrane Vesicles: From Discovery to Applications. *Annu. Rev. Microbiol.* **2021**, *75*, 609–630. [CrossRef] [PubMed]
37. Yao, L.; Wei, B.; Wang, Y.; Xu, B.; Yang, M.; Chen, X.; Chen, F. A critical role of outer membrane vesicles in antibiotic resistance in carbapenem-resistant *Klebsiella pneumoniae*. *Ann. Clin. Microbiol. Antimicrob.* **2023**, *22*, 95. [CrossRef] [PubMed]
38. Wu, H.J.; Xiao, Z.G.; Lv, X.J.; Huang, H.T.; Liao, C.; Hui, C.Y.; Xu, Y.; Li, H.F. Drug-resistant *Acinetobacter baumannii*: From molecular mechanisms to potential therapeutics (Review). *Exp. Ther. Med.* **2023**, *25*, 209. [CrossRef]
39. Zhou, G.; Wang, Q.; Wang, Y.; Wen, X.; Peng, H.; Peng, R.; Shi, Q.; Xie, X.; Li, L. Outer Membrane Porins Contribute to Antimicrobial Resistance in Gram-Negative Bacteria. *Microorganisms* **2023**, *11*, 1690. [CrossRef]
40. Skerniškytė, J.; Karazijaitė, E.; Lučiūnaitė, A.; Sužiedėlienė, E. OmpA Protein-Deficient *Acinetobacter baumannii* Outer Membrane Vesicles Trigger Reduced Inflammatory Response. *Pathogens* **2021**, *10*, 407. [CrossRef] [PubMed]
41. Smani, Y.; Fàbrega, A.; Roca, I.; Sánchez-Encinales, V.; Vila, J.; Pachón, J. Role of OmpA in the multidrug resistance phenotype of *Acinetobacter baumannii*. *Antimicrob. Agents Chemother.* **2014**, *58*, 1806–1808. [CrossRef]
42. Nie, D.; Hu, Y.; Chen, Z.; Li, M.; Hou, Z.; Luo, X.; Mao, X.; Xue, X. Outer membrane protein A (OmpA) as a potential therapeutic target for *Acinetobacter baumannii* infection. *J. Biomed. Sci.* **2020**, *27*, 26. [CrossRef] [PubMed]
43. Gauba, A.; Rahman, K.M. Evaluation of Antibiotic Resistance Mechanisms in Gram-Negative Bacteria. *Antibiotics* **2023**, *12*, 1590. [CrossRef] [PubMed]
44. Kyriakidis, I.; Vasileiou, E.; Pana, Z.D.; Tragiannidis, A. *Acinetobacter baumannii* Antibiotic Resistance Mechanisms. *Pathogens* **2021**, *10*, 373. [CrossRef]
45. Kamoshida, G.; Akaji, T.; Takemoto, N.; Suzuki, Y.; Sato, Y.; Kai, D.; Hibino, T.; Yamaguchi, D.; Kikuchi-Ueda, T.; Nishida, S.; et al. Lipopolysaccharide-Deficient *Acinetobacter baumannii* Due to Colistin Resistance Is Killed by Neutrophil-Produced Lysozyme. *Front. Microbiol.* **2020**, *11*, 573. [CrossRef] [PubMed]
46. Da Silva, G.J.; Domingues, S. Interplay between Colistin Resistance, Virulence and Fitness in *Acinetobacter baumannii*. *Antibiotics* **2017**, *6*, 28. [CrossRef]
47. Lima, W.G.; Alves, M.C.; Cruz, W.S.; Paiva, M.C. Chromosomally encoded and plasmid-mediated polymyxins resistance in *Acinetobacter baumannii*: A huge public health threat. *Eur. J. Clin. Microbiol. Infect. Dis.* **2018**, *37*, 1009–1019. [CrossRef] [PubMed]
48. Jiang, X.; Yang, K.; Han, M.L.; Yuan, B.; Li, J.; Gong, B.; Velkov, T.; Schreiber, F.; Wang, L.; Li, J. Outer Membranes of Polymyxin-Resistant *Acinetobacter baumannii* with Phosphoethanolamine-Modified Lipid A and Lipopolysaccharide Loss Display Different Atomic-Scale Interactions with Polymyxins. *ACS Infect. Dis.* **2020**, *6*, 2698–2708. [CrossRef] [PubMed]
49. Boulant, T.; Boudehen, Y.M.; Filloux, A.; Plesiat, P.; Naas, T.; Dortet, L. Higher Prevalence of PldA, a *Pseudomonas aeruginosa* Trans-Kingdom H2-Type VI Secretion System Effector, in Clinical Isolates Responsible for Acute Infections and in Multidrug Resistant Strains. *Front. Microbiol.* **2018**, *9*, 2578. [CrossRef] [PubMed]
50. Martínez-Trejo, A.; Ruiz-Ruiz, J.M.; Gonzalez-Avila, L.U.; Saldaña-Padilla, A.; Hernández-Cortez, C.; Loyola-Cruz, M.A.; Bello-López, J.M.; Castro-Escarpulli, G. Evasion of Antimicrobial Activity in *Acinetobacter baumannii* by Target Site Modifications: An Effective Resistance Mechanism. *Int. J. Mol. Sci.* **2022**, *23*, 6582. [CrossRef] [PubMed]
51. Shi, J.; Cheng, J.; Liu, S.; Zhu, Y.; Zhu, M. *Acinetobacter baumannii*: An evolving and cunning opponent. *Front. Microbiol.* **2024**, *15*, 1332108. [CrossRef] [PubMed]
52. Nauta, K.M.; Ho, T.D.; Ellermeier, C.D. The Penicillin-Binding Protein PbpP Is a Sensor of β-Lactams and Is Required for Activation of the Extracytoplasmic Function σ Factor σ(P) in *Bacillus thuringiensis*. *mBio* **2021**, *12*, e00179-21. [CrossRef] [PubMed]
53. Sethuvel, D.P.M.; Bakthavatchalam, Y.D.; Karthik, M.; Irulappan, M.; Shrivastava, R.; Periasamy, H.; Veeraraghavan, B. β-Lactam Resistance in ESKAPE Pathogens Mediated Through Modifications in Penicillin-Binding Proteins: An Overview. *Infect. Dis. Ther.* **2023**, *12*, 829–841. [CrossRef] [PubMed]
54. Sauvage, E.; Kerff, F.; Terrak, M.; Ayala, J.A.; Charlier, P. The penicillin-binding proteins: Structure and role in peptidoglycan biosynthesis. *FEMS Microbiol. Rev.* **2008**, *32*, 234–258. [CrossRef] [PubMed]

55. Cayô, R.; Rodríguez, M.C.; Espinal, P.; Fernández-Cuenca, F.; Ocampo-Sosa, A.A.; Pascual, A.; Ayala, J.A.; Vila, J.; Martínez-Martínez, L. Analysis of genes encoding penicillin-binding proteins in clinical isolates of *Acinetobacter baumannii*. *Antimicrob. Agents Chemother.* **2011**, *55*, 5907–5913. [CrossRef] [PubMed]
56. Findlay, J.; Poirel, L.; Bouvier, M.; Nordmann, P. In vitro activity of sulbactam-durlobactam against carbapenem-resistant *Acinetobacter baumannii* and mechanisms of resistance. *J. Glob. Antimicrob. Resist.* **2022**, *30*, 445–450. [CrossRef]
57. McLeod, S.M.; O'Donnell, J.P.; Narayanan, N.; Mills, J.P.; Kaye, K.S. Sulbactam-durlobactam: A β-lactam/β-lactamase inhibitor combination targeting *Acinetobacter baumannii*. *Future Microbiol.* **2024**, *19*, 563–576. [CrossRef]
58. Munita, J.M.; Arias, C.A. Mechanisms of Antibiotic Resistance. *Microbiol. Spectr.* **2016**, *4*, 481–511. [CrossRef]
59. Castanheira, M.; Simner, P.J.; Bradford, P.A. Extended-spectrum β-lactamases: An update on their characteristics, epidemiology and detection. *JAC Antimicrob. Resist.* **2021**, *3*, dlab092. [CrossRef] [PubMed]
60. Papagiannitsis, C.C.; Loli, A.; Tzouvelekis, L.S.; Tzelepi, E.; Arlet, G.; Miriagou, V. SCO-1, a novel plasmid-mediated class A beta-lactamase with carbenicillinase characteristics from *Escherichia coli*. *Antimicrob. Agents Chemother.* **2007**, *51*, 2185–2188. [CrossRef]
61. Tooke, C.L.; Hinchliffe, P.; Bragginton, E.C.; Colenso, C.K.; Hirvonen, V.H.A.; Takebayashi, Y.; Spencer, J. β-Lactamases and β-Lactamase Inhibitors in the 21st Century. *J. Mol. Biol.* **2019**, *431*, 3472–3500. [CrossRef] [PubMed]
62. Kaderabkova, N.; Bharathwaj, M.; Furniss, R.C.D.; Gonzalez, D.; Palmer, T.; Mavridou, D.A.I. The biogenesis of β-lactamase enzymes. *Microbiology* **2022**, *168*, 001217. [CrossRef]
63. Poirel, L.; Corvec, S.; Rapoport, M.; Mugnier, P.; Petroni, A.; Pasteran, F.; Faccone, D.; Galas, M.; Drugeon, H.; Cattoir, V.; et al. Identification of the novel narrow-spectrum beta-lactamase SCO-1 in *Acinetobacter* spp. from Argentina. *Antimicrob. Agents Chemother.* **2007**, *51*, 2179–2184. [CrossRef]
64. Amin, M.; Navidifar, T.; Saleh Shooshtari, F.; Goodarzi, H. Association of the genes encoding Metallo-β-Lactamase with the presence of integrons among multidrug-resistant clinical isolates of *Acinetobacter baumannii*. *Infect. Drug Resist.* **2019**, *12*, 1171–1180. [CrossRef] [PubMed]
65. Nikibakhsh, M.; Firoozeh, F.; Badmasti, F.; Kabir, K.; Zibaei, M. Molecular study of metallo-β-lactamases and integrons in *Acinetobacter baumannii* isolates from burn patients. *BMC Infect. Dis.* **2021**, *21*, 782. [CrossRef]
66. Novović, K.; Jovčić, B. Colistin Resistance in *Acinetobacter baumannii*: Molecular Mechanisms and Epidemiology. *Antibiotics* **2023**, *12*, 516. [CrossRef] [PubMed]
67. Zhang, G.; Zhang, L.; Sha, Y.; Chen, Q.; Lin, N.; Zhao, J.; Zhang, Y.; Ji, Y.; Jiang, W.; Zhang, X.; et al. Identification and characterization of a novel 6′-N-aminoglycoside acetyltransferase AAC(6′)-Va from a clinical isolate of *Aeromonas hydrophila*. *Front. Microbiol.* **2023**, *14*, 1229593. [CrossRef] [PubMed]
68. Li, X.Z.; Plésiat, P.; Nikaido, H. The challenge of efflux-mediated antibiotic resistance in Gram-negative bacteria. *Clin. Microbiol. Rev.* **2015**, *28*, 337–418. [CrossRef] [PubMed]
69. Gurvic, D.; Zachariae, U. Multidrug efflux in Gram-negative bacteria: Structural modifications in active compounds leading to efflux pump avoidance. *Npj Antimicrob. Resist.* **2024**, *2*, 6. [CrossRef]
70. Hirakawa, H.; Kurushima, J.; Hashimoto, Y.; Tomita, H. Progress Overview of Bacterial Two-Component Regulatory Systems as Potential Targets for Antimicrobial Chemotherapy. *Antibiotics* **2020**, *9*, 635. [CrossRef]
71. Zack, K.M.; Sorenson, T.; Joshi, S.G. Types and Mechanisms of Efflux Pump Systems and the Potential of Efflux Pump Inhibitors in the Restoration of Antimicrobial Susceptibility, with a Special Reference to *Acinetobacter baumannii*. *Pathogens* **2024**, *13*, 197. [CrossRef] [PubMed]
72. Pang, Z.; Raudonis, R.; Glick, B.R.; Lin, T.J.; Cheng, Z. Antibiotic resistance in *Pseudomonas aeruginosa*: Mechanisms and alternative therapeutic strategies. *Biotechnol. Adv.* **2019**, *37*, 177–192. [CrossRef] [PubMed]
73. Yin, C.; Alam, M.Z.; Fallon, J.T.; Huang, W. Advances in Development of Novel Therapeutic Strategies against Multi-Drug Resistant *Pseudomonas aeruginosa*. *Antibiotics* **2024**, *13*, 119. [CrossRef] [PubMed]
74. Langendonk, R.F.; Neill, D.R.; Fothergill, J.L. The Building Blocks of Antimicrobial Resistance in *Pseudomonas aeruginosa*: Implications for Current Resistance-Breaking Therapies. *Front. Cell Infect. Microbiol.* **2021**, *11*, 665759. [CrossRef]
75. Wu, W.; Huang, J.; Xu, Z. Antibiotic influx and efflux in *Pseudomonas aeruginosa*: Regulation and therapeutic implications. *Microb. Biotechnol.* **2024**, *17*, e14487. [CrossRef]
76. Yang, B.; Liu, C.; Pan, X.; Fu, W.; Fan, Z.; Jin, Y.; Bai, F.; Cheng, Z.; Wu, W. Identification of Novel PhoP-PhoQ Regulated Genes That Contribute to Polymyxin B Tolerance in *Pseudomonas aeruginosa*. *Microorganisms* **2021**, *9*, 344. [CrossRef] [PubMed]
77. Giovagnorio, F.; De Vito, A.; Madeddu, G.; Parisi, S.G.; Geremia, N. Resistance in *Pseudomonas aeruginosa*: A Narrative Review of Antibiogram Interpretation and Emerging Treatments. *Antibiotics* **2023**, *12*, 1621. [CrossRef] [PubMed]
78. Ropy, A.; Cabot, G.; Sánchez-Diener, I.; Aguilera, C.; Moya, B.; Ayala, J.A.; Oliver, A. Role of *Pseudomonas aeruginosa* low-molecular-mass penicillin-binding proteins in AmpC expression, β-lactam resistance, and peptidoglycan structure. *Antimicrob. Agents Chemother.* **2015**, *59*, 3925–3934. [CrossRef]

79. Glen, K.A.; Lamont, I.L. β-lactam Resistance in *Pseudomonas aeruginosa*: Current Status, Future Prospects. *Pathogens* **2021**, *10*, 1638. [CrossRef] [PubMed]
80. Chen, W.; Zhang, Y.M.; Davies, C. Penicillin-Binding Protein 3 Is Essential for Growth of *Pseudomonas aeruginosa*. *Antimicrob. Agents Chemother.* **2017**, *61*, e01651-16. [CrossRef]
81. Wardell, S.J.T.; Rehman, A.; Martin, L.W.; Winstanley, C.; Patrick, W.M.; Lamont, I.L. A large-scale whole-genome comparison shows that experimental evolution in response to antibiotics predicts changes in naturally evolved clinical *Pseudomonas aeruginosa*. *Antimicrob. Agents Chemother.* **2019**, *63*, e01619-19. [CrossRef] [PubMed]
82. Clark, S.T.; Sinha, U.; Zhang, Y.; Wang, P.W.; Donaldson, S.L.; Coburn, B.; Waters, V.J.; Yau, Y.C.W.; Tullis, D.E.; Guttman, D.S.; et al. Penicillin-binding protein 3 is a common adaptive target among *Pseudomonas aeruginosa* isolates from adult cystic fibrosis patients treated with β-lactams. *Int. J. Antimicrob. Agents* **2019**, *53*, 620–628. [CrossRef] [PubMed]
83. Zhanel, G.G.; Chung, P.; Adam, H.; Zelenitsky, S.; Denisuik, A.; Schweizer, F.; Lagacé-Wiens, P.R.; Rubinstein, E.; Gin, A.S.; Walkty, A.; et al. Ceftolozane/tazobactam: A novel cephalosporin/β-lactamase inhibitor combination with activity against multidrug-resistant gram-negative bacilli. *Drugs* **2014**, *74*, 31–51. [CrossRef]
84. Giancola, S.E.; Mahoney, M.V.; Bias, T.E.; Hirsch, E.B. Critical evaluation of ceftolozane-tazobactam for complicated urinary tract and intra-abdominal infections. *Ther. Clin. Risk Manag.* **2016**, *12*, 787–797. [CrossRef]
85. Kunz Coyne, A.J.; El Ghali, A.; Holger, D.; Rebold, N.; Rybak, M.J. Therapeutic Strategies for Emerging Multidrug-Resistant *Pseudomonas aeruginosa*. *Infect. Dis. Ther.* **2022**, *11*, 661–682. [CrossRef] [PubMed]
86. Oelschlaeger, P.; Kaadan, H.; Dhungana, R. Strategies to Name Metallo-β-Lactamases and Number Their Amino Acid Residues. *Antibiotics* **2023**, *12*, 1746. [CrossRef]
87. Hosu, M.C.; Vasaikar, S.D.; Okuthe, G.E.; Apalata, T. Detection of extended spectrum beta-lactamase genes in *Pseudomonas aeruginosa* isolated from patients in rural Eastern Cape Province, South Africa. *Sci. Rep.* **2021**, *11*, 7110. [CrossRef]
88. Fajardo, A.; Hernando-Amado, S.; Oliver, A.; Ball, G.; Filloux, A.; Martinez, J.L. Characterization of a novel Zn^{2+}-dependent intrinsic imipenemase from *Pseudomonas aeruginosa*. *J. Antimicrob. Chemother.* **2014**, *69*, 2972–2978. [CrossRef]
89. Medrano, F.J.; Hernando-Amado, S.; Martínez, J.L.; Romero, A. A new type of Class C β-lactamases defined by PIB-1. A metal-dependent carbapenem-hydrolyzing β-lactamase, from *Pseudomonas aeruginosa*: Structural and functional analysis. *Int. J. Biol. Macromol.* **2024**, *277*, 134298. [CrossRef]
90. Schauer, J.; Gatermann, S.G.; Hoffmann, D.; Hupfeld, L.; Pfennigwerth, N. GPC-1, a novel class A carbapenemase detected in a clinical *Pseudomonas aeruginosa* isolate. *J. Antimicrob. Chemother.* **2020**, *75*, 911–916. [CrossRef]
91. Evans, B.A.; Amyes, S.G. OXA β-lactamases. *Clin. Microbiol. Rev.* **2014**, *27*, 241–263. [CrossRef] [PubMed]
92. Kakoullis, L.; Papachristodoulou, E.; Chra, P.; Panos, G. Mechanisms of Antibiotic Resistance in Important Gram-Positive and Gram-Negative Pathogens and Novel Antibiotic Solutions. *Antibiotics* **2021**, *10*, 415. [CrossRef]
93. Panahi, T.; Asadpour, L.; Ranji, N. Distribution of aminoglycoside resistance genes in clinical isolates of *Pseudomonas aeruginosa* in north of Iran. *Gene Rep.* **2020**, *21*, 100929. [CrossRef]
94. Aguilar-Rodea, P.; Zúñiga, G.; Cerritos, R.; Rodríguez-Espino, B.A.; Gomez-Ramirez, U.; Nolasco-Romero, C.G.; López-Marceliano, B.; Rodea, G.E.; Mendoza-Elizalde, S.; Reyes-López, A.; et al. Nucleotide substitutions in the mexR, nalC and nalD regulator genes of the MexAB-OprM efflux pump are maintained in *Pseudomonas aeruginosa* genetic lineages. *PLoS ONE* **2022**, *17*, e0266742. [CrossRef] [PubMed]
95. Latifi, B.; Tajbakhsh, S.; Ahadi, L.; Yousefi, F. Coexistence of aminoglycoside resistance genes in CTX-M-producing isolates of *Klebsiella pneumoniae* in Bushehr province, Iran. *Iran. J. Microbiol.* **2021**, *13*, 161–170. [CrossRef] [PubMed]
96. Juarez, P.; Jeannot, K.; Plésiat, P.; Llanes, C. Toxic Electrophiles Induce Expression of the Multidrug Efflux Pump MexEF-OprN in *Pseudomonas aeruginosa* through a Novel Transcriptional Regulator, CmrA. *Antimicrob. Agents Chemother.* **2017**, *61*, e00585-17. [CrossRef]
97. Wang, G.; Zhao, G.; Chao, X.; Xie, L.; Wang, H. The Characteristic of Virulence, Biofilm and Antibiotic Resistance of *Klebsiella pneumoniae*. *Int. J. Environ. Res. Public Health* **2020**, *17*, 6278. [CrossRef] [PubMed]
98. Baran, A.; Kwiatkowska, A.; Potocki, L. Antibiotics and Bacterial Resistance—A Short Story of an Endless Arms Race. *Int. J. Mol. Sci.* **2023**, *24*, 5777. [CrossRef] [PubMed]
99. Li, Y.; Kumar, S.; Zhang, L. Mechanisms of Antibiotic Resistance and Developments in Therapeutic Strategies to Combat Klebsiella pneumoniae Infection. *Infect. Drug Resist.* **2024**, *17*, 1107–1119. [CrossRef]
100. Li, S.; Feng, X.; Li, M.; Shen, Z. In vivo adaptive antimicrobial resistance in *Klebsiella pneumoniae* during antibiotic therapy. *Front. Microbiol.* **2023**, *14*, 1159912. [CrossRef] [PubMed]
101. Wright, M.S.; Suzuki, Y.; Jones, M.B.; Marshall, S.H.; Rudin, S.D.; van Duin, D.; Kaye; Jacobs, M.R.; Bonomo, R.A.; Adams, M.D. Genomic and transcriptomic analyses of colistin-resistant clinical isolates of *Klebsiella pneumoniae* reveal multiple pathways of resistance. *Antimicrob. Agents Chemother.* **2015**, *59*, 536–543. [CrossRef] [PubMed]
102. Navon-Venezia, S.; Kondratyeva, K.; Carattoli, A. *Klebsiella pneumoniae*: A major worldwide source and shuttle for antibiotic resistance. *FEMS Microbiol. Rev.* **2017**, *41*, 252–275. [CrossRef]

103. Huy, T.X.N. Overcoming Klebsiella pneumoniae antibiotic resistance: New insights into mechanisms and drug discovery. *Beni-Suef Univ. J. Basic Appl. Sci.* **2024**, *13*, 13. [CrossRef]
104. Sutaria, D.S.; Moya, B.; Green, K.B.; Kim, T.H.; Tao, X.; Jiao, Y.; Louie, A.; Drusano, G.L.; Bulitta, J.B. First Penicillin-Binding Protein Occupancy Patterns of β-Lactams and β-Lactamase Inhibitors in *Klebsiella pneumoniae*. *Antimicrob. Agents Chemother.* **2018**, *62*, e00282-18. [CrossRef] [PubMed]
105. Jiang, W.; Yang, W.; Zhao, X.; Wang, N.; Ren, H. *Klebsiella pneumoniae* presents antimicrobial drug resistance for β-lactam through the ESBL/PBP signaling pathway. *Exp. Ther. Med.* **2020**, *19*, 2449–2456. [CrossRef]
106. Hobson, C.A.; Pierrat, G.; Tenaillon, O.; Bonacorsi, S.; Bercot, B.; Jaouen, E.; Jacquier, H.; Birgy, A. *Klebsiella pneumoniae* Carbapenemase Variants Resistant to Ceftazidime-Avibactam: An Evolutionary Overview. *Antimicrob. Agents Chemother.* **2022**, *66*, e0044722. [CrossRef] [PubMed]
107. Haidar, G.; Clancy, C.J.; Shields, R.K.; Hao, B.; Cheng, S.; Nguyen, M.H. Mutations in bla(KPC-3) That Confer Ceftazidime-Avibactam Resistance Encode Novel KPC-3 Variants That Function as Extended-Spectrum β-Lactamases. *Antimicrob. Agents Chemother.* **2017**, *61*, e02534-16. [CrossRef] [PubMed]
108. Li, W.; Guo, H.; Gao, Y.; Yang, X.; Li, R.; Li, S.; Sun, C.; Du, W.; Chen, S.; Xu, P.; et al. Comparative genomic analysis of plasmids harboring bla (OXA-48)-like genes in *Klebsiella pneumoniae*. *Front. Cell. Infect. Microbiol.* **2022**, *12*, 1082813. [CrossRef] [PubMed]
109. Gharbi, M.; Abbas, M.A.S.; Hamrouni, S.; Maaroufi, A. First Report of aac(6')-Ib and aac(6')-Ib-cr Variant Genes Associated with Mutations in gyrA Encoded Fluoroquinolone Resistance in Avian *Campylobacter coli* Strains Collected in Tunisia. *Int. J. Mol. Sci.* **2023**, *24*, 16116. [CrossRef] [PubMed]
110. Kim, J.Y.; Suh, J.W.; Kang, J.S.; Kim, S.B.; Yoon, Y.K.; Sohn, J.W. Gram-Negative Bacteria's Outer Membrane Vesicles. *Infect. Chemother.* **2023**, *55*, 1–9. [CrossRef]
111. Ni, R.T.; Onishi, M.; Mizusawa, M.; Kitagawa, R.; Kishino, T.; Matsubara, F.; Tsuchiya, T.; Kuroda, T.; Ogawa, W. The role of RND-type efflux pumps in multidrug-resistant mutants of *Klebsiella pneumoniae*. *Sci. Rep.* **2020**, *10*, 10876. [CrossRef]
112. Davin-Regli, A.; Pagès, J.M. *Enterobacter aerogenes* and *Enterobacter cloacae*; versatile bacterial pathogens confronting antibiotic treatment. *Front. Microbiol.* **2015**, *6*, 392. [CrossRef] [PubMed]
113. Intra, J.; Carcione, D.; Sala, R.M.; Siracusa, C.; Brambilla, P.; Leoni, V. Antimicrobial Resistance Patterns of Enterobacter cloacae and *Klebsiella aerogenes* Strains Isolated from Clinical Specimens: A Twenty-Year Surveillance Study. *Antibiotics* **2023**, *12*, 775. [CrossRef] [PubMed]
114. Davin-Regli, A.; Lavigne, J.P.; Pagès, J.M. *Enterobacter* spp.: Update on Taxonomy, Clinical Aspects, and Emerging Antimicrobial Resistance. *Clin. Microbiol. Rev.* **2019**, *32*, e00002-19. [CrossRef]
115. Aghapour, Z.; Gholizadeh, P.; Ganbarov, K.; Bialvaei, A.Z.; Mahmood, S.S.; Tanomand, A.; Yousefi, M.; Asgharzadeh, M.; Yousefi, B.; Kafil, H.S. Molecular mechanisms related to colistin resistance in Enterobacteriaceae. *Infect. Drug Resist.* **2019**, *12*, 965–975. [CrossRef]
116. Cano, M.E.; Rodríguez-Martínez, J.M.; Agüero, J.; Pascual, A.; Calvo, J.; García-Lobo, J.M.; Velasco, C.; Francia, M.V.; Martínez-Martínez, L. Detection of plasmid-mediated quinolone resistance genes in clinical isolates of *Enterobacter* spp. *in Spain*. *J. Clin. Microbiol.* **2009**, *47*, 2033–2039. [CrossRef]
117. Dortet, L.; Poirel, L.; Abbas, S.; Oueslati, S.; Nordmann, P. Genetic and Biochemical Characterization of FRI-1, a Carbapenem-Hydrolyzing Class A β-Lactamase from *Enterobacter cloacae*. *Antimicrob. Agents Chemother.* **2015**, *59*, 7420–7425. [CrossRef] [PubMed]
118. Orskov, F.; Orskov, I. From the national institutes of health. Summary of a workshop on the clone concept in the epidemiology, taxonomy, and evolution of the enterobacteriaceae and other bacteria. *J. Infect. Dis.* **1983**, *148*, 346–357. [CrossRef]
119. Woodford, N.; Turton, J.F.; Livermore, D.M. Multiresistant Gram-negative bacteria: The role of high-risk clones in the dissemination of antibiotic resistance. *FEMS Microbiol. Rev.* **2011**, *35*, 736–755. [CrossRef] [PubMed]
120. Baquero, F.; Coque, T.M. Multilevel population genetics in antibiotic resistance. *FEMS Microbiol. Rev.* **2011**, *35*, 705–706. [CrossRef]
121. de Lagarde, M.; Vanier, G.; Arsenault, J.; Fairbrother, J.M.M. High Risk Clone: A Proposal of Criteria Adapted to the One Health Context with Application to Enterotoxigenic *Escherichia coli* in the Pig Population. *Antibiotics* **2021**, *10*, 244. [CrossRef]
122. Silveira, M.C.; Rocha-de-Souza, C.M.; de Oliveira Santos, I.C.; Pontes, L.D.S.; Oliveira, T.; Tavares-Teixeira, C.B.; Cossatis, N.A.; Pereira, N.F.; da Conceição-Neto, O.C.; da Costa, B.S.; et al. Genetic Basis of Antimicrobial Resistant Gram-Negative Bacteria Isolated From Bloodstream in Brazil. *Front. Med.* **2021**, *8*, 635206. [CrossRef]
123. Shelenkov, A.; Akimkin, V.; Mikhaylova, Y. International Clones of High Risk of *Acinetobacter Baumannii*—Definitions, History, Properties and Perspectives. *Microorganisms* **2023**, *11*, 2115. [CrossRef]
124. Wyres, K.L.; Lam, M.M.C.; Holt, K.E. Population genomics of *Klebsiella pneumoniae*. *Nat. Rev. Microbiol.* **2020**, *18*, 344–359. [CrossRef] [PubMed]
125. Knecht, C.A.; García Allende, N.; Álvarez, V.E.; Prack Mc Cormick, B.; Massó, M.G.; Campos, J.; Fox, B.; Alonso, F.M.; Donis, N.; Canigia, L.F.; et al. New sequence type of an *Enterobacter cloacae* complex strain with the potential to become a high-risk clone. *J. Glob. Antimicrob. Resist.* **2022**, *31*, 162–164. [CrossRef]

126. Irum, S.; Naz, K.; Ullah, N.; Mustafa, Z.; Ali, A.; Arslan, M.; Khalid, K.; Andleeb, S. Antimicrobial Resistance and Genomic Characterization of Six New Sequence Types in Multidrug-Resistant *Pseudomonas aeruginosa* Clinical Isolates from Pakistan. *Antibiotics* **2021**, *10*, 1386. [CrossRef]
127. Cantón, R.; Gijón, D.; Ruiz-Garbajosa, P. Antimicrobial resistance in ICUs: An update in the light of the COVID-19 pandemic. *Curr. Opin. Crit. Care* **2020**, *26*, 433–441. [CrossRef]
128. Alcántar-Curiel, M.D.; Rosales-Reyes, R.; Jarillo-Quijada, M.D.; Gayosso-Vázquez, C.; Fernández-Vázquez, J.L.; Toledano-Tableros, J.E.; Giono-Cerezo, S.; Garza-Villafuerte, P.; López-Huerta, A.; Vences-Vences, D.; et al. Carbapenem-Resistant *Acinetobacter baumannii* in Three Tertiary Care Hospitals in Mexico: Virulence Profiles, Innate Immune Response and Clonal Dissemination. *Front. Microbiol.* **2019**, *10*, 2116. [CrossRef]
129. Peirano, G.; Chen, L.; Kreiswirth, B.N.; Pitout, J.D.D. Emerging Antimicrobial-Resistant High-Risk *Klebsiella pneumoniae* Clones ST307 and ST147. *Antimicrob. Agents Chemother.* **2020**, *64*, e01148-20. [CrossRef] [PubMed]
130. Caliskan-Aydogan, O.; Alocilja, E.C. A Review of Carbapenem Resistance in Enterobacterales and Its Detection Techniques. *Microorganisms* **2023**, *11*, 1491. [CrossRef]
131. Rodrigues, C.; Lanza, V.F.; Peixe, L.; Coque, T.M.; Novais, Â. Phylogenomics of Globally Spread Clonal Groups 14 and 15 of *Klebsiella pneumoniae*. *Microbiol. Spectr.* **2023**, *11*, e0339522. [CrossRef] [PubMed]
132. Annavajhala, M.K.; Gomez-Simmonds, A.; Uhlemann, A.C. Multidrug-Resistant *Enterobacter cloacae* Complex Emerging as a Global, Diversifying Threat. *Front. Microbiol.* **2019**, *10*, 44. [CrossRef]
133. Ordonez, A.A.; Wintaco, L.M.; Mota, F.; Restrepo, A.F.; Ruiz-Bedoya, C.A.; Reyes, C.F.; Uribe, L.G.; Abhishek, S.; D'Alessio, F.R.; Holt, D.P.; et al. Imaging *Enterobacterales* infections in patients using pathogen-specific positron emission tomography. *Sci. Transl. Med.* **2021**, *13*, eabe9805. [CrossRef]
134. Lumbreras-Iglesias, P.; de Toro, M.; Vázquez, X.; García-Carús, E.; Rodicio, M.R.; Fernández, J. High-risk international clones ST66, ST171 and ST78 of *Enterobacter cloacae* complex causing blood stream infections in Spain and carrying bla(OXA-48) with or without mcr-9. *J. Infect. Public Health* **2023**, *16*, 272–279. [CrossRef] [PubMed]
135. Horcajada, J.P.; Montero, M.; Oliver, A.; Sorlí, L.; Luque, S.; Gómez-Zorrilla, S.; Benito, N.; Grau, S. Epidemiology and Treatment of Multidrug-Resistant and Extensively Drug-Resistant *Pseudomonas aeruginosa* Infections. *Clin. Microbiol. Rev.* **2019**, *32*, e00031-19. [CrossRef]
136. Mendes, G.; Ramalho, J.F.; Bruschy-Fonseca, A.; Lito, L.; Duarte, A.; Melo-Cristino, J.; Caneiras, C. Whole-Genome Sequencing Enables Molecular Characterization of Non-Clonal Group 258 High-Risk Clones (ST13, ST17, ST147 and ST307) among Carbapenem-Resistant *Klebsiella pneumoniae* from a Tertiary University Hospital Centre in Portugal. *Microorganisms* **2022**, *10*, 416. [CrossRef] [PubMed]
137. Feng, Y.; Yang, Y.; Hu, Y.; Xiao, Y.; Xie, Y.; Wei, L.; Wen, H.; Zhang, L.; McNally, A.; Zong, Z. Population genomics uncovers global distribution, antimicrobial resistance, and virulence genes of the opportunistic pathogen *Klebsiella aerogenes*. *Cell Rep.* **2024**, *43*, 114602. [CrossRef]
138. Mosquera-Rendón, J.; Moreno-Herrera, C.X.; Robledo, J.; Hurtado-Páez, U. Genome-Wide Association Studies (GWAS) Approaches for the Detection of Genetic Variants Associated with Antibiotic Resistance: A Systematic Review. *Microorganisms* **2023**, *11*, 2866. [CrossRef] [PubMed]
139. Mulani, M.S.; Kamble, E.E.; Kumkar, S.N.; Tawre, M.S.; Pardesi, K.R. Emerging Strategies to Combat ESKAPE Pathogens in the Era of Antimicrobial Resistance: A Review. *Front. Microbiol.* **2019**, *10*, 539. [CrossRef]
140. Aloke, C.; Achilonu, I. Coping with the ESKAPE pathogens: Evolving strategies, challenges and future prospects. *Microb. Pathog.* **2023**, *175*, 105963. [CrossRef] [PubMed]
141. Berger, I.; Loewy, Z.G. Antimicrobial Resistance and Novel Alternative Approaches to Conventional Antibiotics. *Bacteria* **2024**, *3*, 171–182. [CrossRef]
142. Collins, S.M.; Brown, A.C. Bacterial Outer Membrane Vesicles as Antibiotic Delivery Vehicles. *Front. Immunol.* **2021**, *12*, 733064. [CrossRef] [PubMed]
143. Correia, J.H.; Rodrigues, J.A.; Pimenta, S.; Dong, T.; Yang, Z. Photodynamic Therapy Review: Principles, Photosensitizers, Applications, and Future Directions. *Pharmaceutics* **2021**, *13*, 1332. [CrossRef] [PubMed]
144. Jorgensen, S.C.J.; Rybak, M.J. Meropenem and Vaborbactam: Stepping up the Battle against Carbapenem-resistant Enterobacteriaceae. *Pharmacotherapy* **2018**, *38*, 444–461. [CrossRef] [PubMed]
145. Aoki, N.; Ishii, Y.; Tateda, K.; Saga, T.; Kimura, S.; Kikuchi, Y.; Kobayashi, T.; Tanabe, Y.; Tsukada, H.; Gejyo, F.; et al. Efficacy of calcium-EDTA as an inhibitor for metallo-β-lactamase in a mouse model of *Pseudomonas aeruginosa* pneumonia. *Antimicrob. Agents Chemother.* **2010**, *54*, 4582–4588. [CrossRef] [PubMed]
146. Santos, A.L.; Sodre, C.L.; Valle, R.S.; Silva, B.A.; Abi-Chacra, E.A.; Silva, L.V.; Souza-Goncalves, A.L.; Sangenito, L.S.; Goncalves, D.S.; Souza, L.O.; et al. Antimicrobial action of chelating agents: Repercussions on the microorganism development, virulence and pathogenesis. *Curr. Med. Chem.* **2012**, *19*, 2715–2737. [CrossRef] [PubMed]

147. Yoshizumi, A.; Ishii, Y.; Livermore, D.M.; Woodford, N.; Kimura, S.; Saga, T.; Harada, S.; Yamaguchi, K.; Tateda, K. Efficacies of calcium-EDTA in combination with imipenem in a murine model of sepsis caused by *Escherichia coli* with NDM-1 β-lactamase. *J. Infect. Chemother. Off. J. Jpn. Soc. Chemother.* **2013**, *19*, 992–995. [CrossRef]
148. Finnegan, S.; Percival, S.L. EDTA: An Antimicrobial and Antibiofilm Agent for Use in Wound Care. *Adv. Wound Care* **2015**, *4*, 415–421. [CrossRef]
149. Balamurugan, P.; Hema, M.; Kaur, G.; Sridharan, V.; Prabu, P.C.; Sumana, M.N.; Princy, S.A. Development of a biofilm inhibitor molecule against multidrug resistant *Staphylococcus aureus* associated with gestational urinary tract infections. *Front. Microbiol.* **2015**, *6*, 832. [CrossRef] [PubMed]
150. Chatterjee, M.; Anju, C.P.; Biswas, L.; Anil Kumar, V.; Gopi Mohan, C.; Biswas, R. Antibiotic resistance in *Pseudomonas aeruginosa* and alternative therapeutic options. *Int. J. Med. Microbiol. IJMM* **2016**, *306*, 48–58. [CrossRef]
151. Defraine, V.; Liebens, V.; Loos, E.; Swings, T.; Weytjens, B.; Fierro, C.; Marchal, K.; Sharkey, L.; O'Neill, A.J.; Corbau, R.; et al. 1-((2,4-Dichlorophenethyl)Amino)-3-Phenoxypropan-2-ol Kills *Pseudomonas aeruginosa* through Extensive Membrane Damage. *Front. Microbiol.* **2018**, *9*, 129. [CrossRef]
152. Kelly, J.B.; Nolan, A.C.; Zeden, M.S. How can we escape the ESKAPEs: Antimicrobial resistance mechanisms and what lies ahead? *PLoS Pathog.* **2024**, *20*, e1012270. [CrossRef] [PubMed]
153. Chang, R.Y.K.; Nang, S.C.; Chan, H.K.; Li, J. Novel antimicrobial agents for combating antibiotic-resistant bacteria. *Adv. Drug Deliv. Rev.* **2022**, *187*, 114378. [CrossRef] [PubMed]
154. Manohar, P.; Nachimuthu, R.; Lopes, B.S. The therapeutic potential of bacteriophages targeting gram-negative bacteria using *Galleria mellonella* infection model. *BMC Microbiol.* **2018**, *18*, 97. [CrossRef] [PubMed]
155. Manohar, P.; Tamhankar, A.J.; Lundborg, C.S.; Nachimuthu, R. Therapeutic Characterization and Efficacy of Bacteriophage Cocktails Infecting *Escherichia coli*, *Klebsiella pneumoniae*, and Enterobacter Species. *Front. Microbiol.* **2019**, *10*, 574. [CrossRef] [PubMed]
156. Tabassum, R.; Shafique, M.; Khawaja, K.A.; Alvi, I.A.; Rehman, Y.; Sheik, C.S.; Abbas, Z.; Rehman, S.U. Complete genome analysis of a *Siphoviridae phage* TSK1 showing biofilm removal potential against *Klebsiella pneumoniae*. *Sci. Rep.* **2018**, *8*, 17904. [CrossRef]
157. Kusradze, I.; Karumidze, N.; Rigvava, S.; Dvalidze, T.; Katsitadze, M.; Amiranashvili, I.; Goderdzishvili, M. Characterization and Testing the Efficiency of *Acinetobacter baumannii* Phage vB-GEC_Ab-M-G7 as an Antibacterial Agent. *Front. Microbiol.* **2016**, *7*, 1590. [CrossRef]
158. Jeon, J.; Ryu, C.M.; Lee, J.Y.; Park, J.H.; Yong, D.; Lee, K. In Vivo Application of Bacteriophage as a Potential Therapeutic Agent To Control OXA-66-Like Carbapenemase-Producing *Acinetobacter baumannii* Strains Belonging to Sequence Type 357. *Appl. Environ. Microbiol.* **2016**, *82*, 4200–4208. [CrossRef]
159. Wang, Y.; Mi, Z.; Niu, W.; An, X.; Yuan, X.; Liu, H.; Li, P.; Liu, Y.; Feng, Y.; Huang, Y.; et al. Intranasal treatment with bacteriophage rescues mice from *Acinetobacter baumannii*-mediated pneumonia. *Future Microbiol.* **2016**, *11*, 631–641. [CrossRef]
160. Browne, K.; Chakraborty, S.; Chen, R.; Willcox, M.D.; Black, D.S.; Walsh, W.R.; Kumar, N. A New Era of Antibiotics: The Clinical Potential of Antimicrobial Peptides. *Int. J. Mol. Sci.* **2020**, *21*, 7047. [CrossRef]
161. Chen, N.; Jiang, C. Antimicrobial peptides: Structure, mechanism, and modification. *Eur. J. Med. Chem.* **2023**, *255*, 115377. [CrossRef] [PubMed]
162. Wu, G.; Wu, P.; Xue, X.; Yan, X.; Liu, S.; Zhang, C.; Shen, Z.; Xi, T. Application of S-thanatin, an antimicrobial peptide derived from thanatin, in mouse model of *Klebsiella pneumoniae* infection. *Peptides* **2013**, *45*, 73–77. [CrossRef] [PubMed]
163. Xie, J.; Li, Y.; Li, J.; Yan, Z.; Wang, D.; Guo, X.; Zhang, J.; Zhang, B.; Mou, L.; Yang, W.; et al. Potent effects of amino acid scanned antimicrobial peptide Feleucin-K3 analogs against both multidrug-resistant strains and biofilms of *Pseudomonas aeruginosa*. *Amino Acids* **2018**, *50*, 1471–1483. [CrossRef]
164. Sánchez-Gómez, S.; Ferrer-Espada, R.; Stewart, P.S.; Pitts, B.; Lohner, K.; Martínez de Tejada, G. Antimicrobial activity of synthetic cationic peptides and lipopeptides derived from human lactoferricin against *Pseudomonas aeruginosa* planktonic cultures and biofilms. *BMC Microbiol.* **2015**, *15*, 137. [CrossRef]
165. Liu, S.; Long, Q.; Xu, Y.; Wang, J.; Xu, Z.; Wang, L.; Zhou, M.; Wu, Y.; Chen, T.; Shaw, C. Assessment of antimicrobial and wound healing effects of Brevinin-2Ta against the bacterium *Klebsiella pneumoniae* in dermally-wounded rats. *Oncotarget* **2017**, *8*, 111369–111385. [CrossRef]
166. Gupta, R.; Sharma, S. Role of alternatives to antibiotics in mitigating the antimicrobial resistance crisis. *Indian J. Med. Res.* **2022**, *156*, 464–477. [CrossRef]
167. Gao, W.; Zhang, L. Nanomaterials arising amid antibiotic resistance. *Nat. Reviews. Microbiol.* **2021**, *19*, 5–6. [CrossRef]
168. León-Buitimea, A.; Garza-Cárdenas, C.R.; Garza-Cervantes, J.A.; Lerma-Escalera, J.A.; Morones-Ramírez, J.R. The Demand for New Antibiotics: Antimicrobial Peptides, Nanoparticles, and Combinatorial Therapies as Future Strategies in Antibacterial Agent Design. *Front. Microbiol.* **2020**, *11*, 1669. [CrossRef] [PubMed]
169. Tiwari, V.; Mishra, N.; Gadani, K.; Solanki, P.S.; Shah, N.A.; Tiwari, M. Mechanism of Anti-bacterial Activity of Zinc Oxide Nanoparticle Against Carbapenem-Resistant *Acinetobacter baumannii*. *Front. Microbiol.* **2018**, *9*, 1218. [CrossRef]

170. Tavares, A.; Carvalho, C.M.; Faustino, M.A.; Neves, M.G.; Tomé, J.P.; Tomé, A.C.; Cavaleiro, J.A.; Cunha, A.; Gomes, N.C.; Alves, E.; et al. Antimicrobial photodynamic therapy: Study of bacterial recovery viability and potential development of resistance after treatment. *Mar. Drugs* **2010**, *8*, 91–105. [CrossRef]
171. Hamblin, M.R.; Hasan, T. Photodynamic therapy: A new antimicrobial approach to infectious disease? *Photochem. Photobiol. Sci. Off. J. Eur. Photochem. Assoc. Eur. Soc. Photobiol.* **2004**, *3*, 436–450. [CrossRef] [PubMed]
172. de Freitas, L.M.; Lorenzón, E.N.; Santos-Filho, N.A.; Zago, L.H.P.; Uliana, M.P.; de Oliveira, K.T.; Cilli, E.M.; Fontana, C.R. Antimicrobial Photodynamic therapy enhanced by the peptide aurein 1.2. *Sci. Rep.* **2018**, *8*, 4212. [CrossRef] [PubMed]
173. Demidova, T.N.; Hamblin, M.R. Effect of cell-photosensitizer binding and cell density on microbial photoinactivation. *Antimicrob. Agents Chemother.* **2005**, *49*, 2329–2335. [CrossRef] [PubMed]
174. Malá, Z.; Žárská, L.; Malina, L.; Langová, K.; Večeřová, R.; Kolář, M.; Henke, P.; Mosinger, J.; Kolářová, H. Photodynamic effect of TPP encapsulated in polystyrene nanoparticles toward multi-resistant pathogenic bacterial strains: AFM evaluation. *Sci. Rep.* **2021**, *11*, 6786. [CrossRef] [PubMed]
175. Carrera, E.T.; Dias, H.B.; Corbi, S.C.T.; Marcantonio, R.A.C.; Bernardi, A.C.A.; Bagnato, V.S.; Hamblin, M.R.; Rastelli, A.N.S. The application of antimicrobial photodynamic therapy (aPDT) in dentistry: A critical review. *Laser Phys.* **2016**, *26*, 123001. [CrossRef] [PubMed]
176. Wong, T.-W.; Wu, E.-C.; Ko, W.-C.; Lee, C.-C.; Hor, L.-I.; Huang, I.H. Photodynamic inactivation of methicillin-resistant *Staphylococcus aureus* by indocyanine green and near infrared light. *Dermatol. Sin.* **2018**, *36*, 8–15. [CrossRef]
177. Paramanantham, P.; Siddhardha, B.; Lal Sb, S.; Sharan, A.; Alyousef, A.A.; Al Dosary, M.S.; Arshad, M.; Syed, A. Antimicrobial photodynamic therapy on *Staphylococcus aureus* and *Escherichia coli* using malachite green encapsulated mesoporous silica nanoparticles: An in vitro study. *PeerJ* **2019**, *7*, e7454. [CrossRef]

pharmaceutics

MDPI

Article

Chitosan Nanoparticles Loaded with *Capparis cartilaginea* Decne Extract: Insights into Characterization and Antigenotoxicity In Vivo

Asmaa S. Salman [1,2], Shaza N. Alkhatib [1], Fatimah M. Ahmed [1] and Ragaa A. Hamouda [1,3,*]

1 Biology Department, College of Science and Arts at Khulis, University of Jeddah, Jeddah 21959, Saudi Arabia; drasmaasalman@gmail.com (A.S.S.); snalkhatib@uj.edu.sa (S.N.A.); mothanna.fatima@gmail.com (F.M.A.)
2 Genetic and Cytology Department, Biotechnology Research Institute, National Research Center, Cairo 12622, Egypt
3 Microbial Biotechnology Department, Genetic Engineering and Biotechnology Research Institute (GEBRI), University of Sadat City, Sadat City 32897, Egypt
* Correspondence: ragaahom@yahoo.com

Abstract: Plant-based foods may enhance the prevention of cancer. The present investigation aimed to assess the antigenotoxic effects of chitosan nanoparticles (CNPs) when loaded with the ethanol extract of *C. cartilaginea* (CNPs/Cc). Synthesis of CNPs and CNPs/Cc and their characterization were carried out using TEM, EDS, DSC, and Zeta potential. For in vivo experiments, animal groups were treated in the following groups: negative control, ethyl methanesulfonate (EMS) (240 mg/kg), CNPs (350 mg/kg), high and low doses of CNPs/Cc, CNPs plus EMS, high dose of CNPs/Cc plus EMS, and low dose of CNPs/Cc plus EMS. Bone marrow chromosomal aberrations and sperm shape abnormalities were examined. TEM results showed that CNPs and CNPs/Cc are spherical particles. CNPs' physical stability was observed to be lower than that of CNPs/Cc due to the presence of more positive charges on CNPs/Cc. EMS significantly enhanced chromosomal abnormalities and sperm shape abnormalities. CNPs showed powerful antigenotoxic properties. For the first time, it could be concluded that loading chitosan nanoparticles with *C. cartilaginea* extract significantly promotes its protective properties.

Keywords: chitosan; nanoparticles; *Capparis cartilaginea*; genotoxicity; sperm

1. Introduction

Cancer is the second leading cause of death worldwide [1,2]. Environmental and dietary chemicals can induce endogenous and exogenous DNA damage, which is the first step in the process of carcinogenesis [3]. Given the relationship between food, nutrition, and cancer, plant-based diets have been reported to decrease overall cancer risk [4].

Medicinal plants with preventive and therapeutic properties play a vital role in healthcare systems. *Capparaceae* comprises a medium-sized family of about 45 genera and 700–900 species, whose members exhibit considerable variation in their habitat [5]. *Capparis* is the largest genus of *Capparidaceae*, comprising about 250–400 species of shrubs, trees, and woody climbers [6], represented by four species in Saudi Arabia. *Capparis C. cartilaginea* Decne is used in traditional medicine for treatment of various illnesses. These plants have shown different biological effects, including antioxidant, antihyperglycemic, hypolipidemic, and analgesic effects [7,8]. There is a growing body of literature that recognizes the immunostimulant, antitumoral, antidiabetic, antisclerotic, antimicrobial, anti-inflammatory, immunomodulatory, and antiviral activities of the *Capparis* species [9].

Currently, the development of new therapeutic drugs with improved efficiency is gaining much interest in the field of plant-based medicines. Nanoscale systems for drug delivery reduce the drug particles into the sub-micron range, thus enhancing the solubility, permeability, absorption, and bioavailability of their active ingredients [10]. One such system uses nanoparticles formed from chitosan, which is the second most abundant natural polymer after cellulose. Chitosan is a nontoxic, biocompatible, biodegradable, and cationic polysaccharide, which can be easily crosslinked with tripolyphosphate (TPP polyanions) under mild conditions to form nanoparticles [11,12]. The use of chitosan nanoparticles, with a tunable size and the possibility of surface modification, is a very promising and versatile strategy to overcome the bioavailability and stability issues of diverse natural active ingredients [13]. Chitosan-based nanosystems are among the most important and advanced drug delivery systems due to their remarkable physicochemical and biological characteristics. Thus, it has been widely used in the medical and pharmaceutical fields, and its potential as a drug carrier has been considered [14].

Ethyl methanesulfonate (EMS) is a dangerous chemical substance: it is mutagenic, teratogenic, and carcinogenic, and neither occurs naturally nor has any commercial uses. However, EMS is a known trace impurity in medications made of mesylate salts. Thus, it is regarded as an impurity rather than a contamination. EMS can cause gene mutations and chromosomal abnormalities through nucleotide substitution, especially through G:C-to-A:T transitions brought on by guanine alkylation [15–17]. In biochemical and medical research, EMS is generally utilized as a model alkylating agent, particularly in studies of DNA repair mechanisms. Usually, the mutations are severe loss-of-function or null alleles. Deletions or other chromosomal rearrangements account for around 13% of the EMS lesions [18].

Multiple phenotypes, metabolic byproducts, and biotic/abiotic stress tolerance are displayed in the EMS-induced mutants [19]. Guruprasad [20] observed that *Brahmarasayana* significantly reduced Ethylmethanesulfonate induced chromosomal aberrations in mouse bone marrow cells. The genotoxicity of EMS in mammals like mouse and rats was accomplished with changes in sperm morphology and limb defects in the embryos of pregnant rats [21].

Therefore, the present study aims to synthesize and characterize chitosan nanoparticles loaded with *C. cartilaginea* leaf extract and to establish whether loading chitosan nanoparticles with *C. cartilaginea* extract improves antigenotoxic activity against ethyl methanesulfonate (EMS)-induced genotoxicity in mice.

2. Materials and Methods

2.1. Materials

The leaves of *C. cartilaginea* were gathered in December 2021 from Al-Hada, Saudi Arabia, and were identified according to Chaudhary [22]. The chemicals used in this study were of analytical grade and used without further purification. Chemicals were purchased from the Saudi Chemical Company (PanReac AppliChem, Ar Riyad, Saudi Arabia).

2.1.1. Preparation of *C. cartilaginea* Leaf Extract

A total of 1031.0 g of *C. cartilaginea* leaves were cleaned and air-dried in the shade, then ground using a suitable grinder. The powdered leaves were soaked in absolute ethanol for 72 h. The extract was filtered, then evaporated using a rotary evaporator (IKA RV 8 Basic V-C, Leicestershire, UK). The leaf extract was lyophilized using a freeze-drying lyophilizer until complete dryness was achieved and preserved at −20 °C. The residue yield was 8.2% of *C. cartilaginea*.

2.1.2. *C. cartilaginea* Characterization

The characterization of *C. cartilaginea* was performed via Fourier-transform infrared spectroscopy (FTIR) using a Thermo Fisher Nicolet IS10 (Waltham, MA, USA) spectrometer.

GC–MS Analysis

The method of GC–MS analysis was used on the plant according to Rautela et al. [23].

2.2. Synthesis and Optimization Process of CNPs/Cc

2.2.1. Synthesis of CNPs

Chitosan was obtained from Sigma-Aldrich. The ionic gelation of CNPs was conducted as described by Masarudin et al. [24]. Three grams of chitosan (medium molecular weight) was mixed with 10 mL of 1.0% acetic acid, and then 990 mL of distilled water was added to the mixture. The solution was stirred using a magnetic stirrer for 2 h to dissolve the chitosan. For the cross-linker, 1 g of TPP was dissolved in 1 L of distilled water at pH 5.0. This was added to the chitosan solution and stirred continuously with a magnetic stirrer until complete dissolution. The mixture was centrifuged at 5000× *g* for 20 min, washed, and then lyophilized until complete dryness (Figure 1).

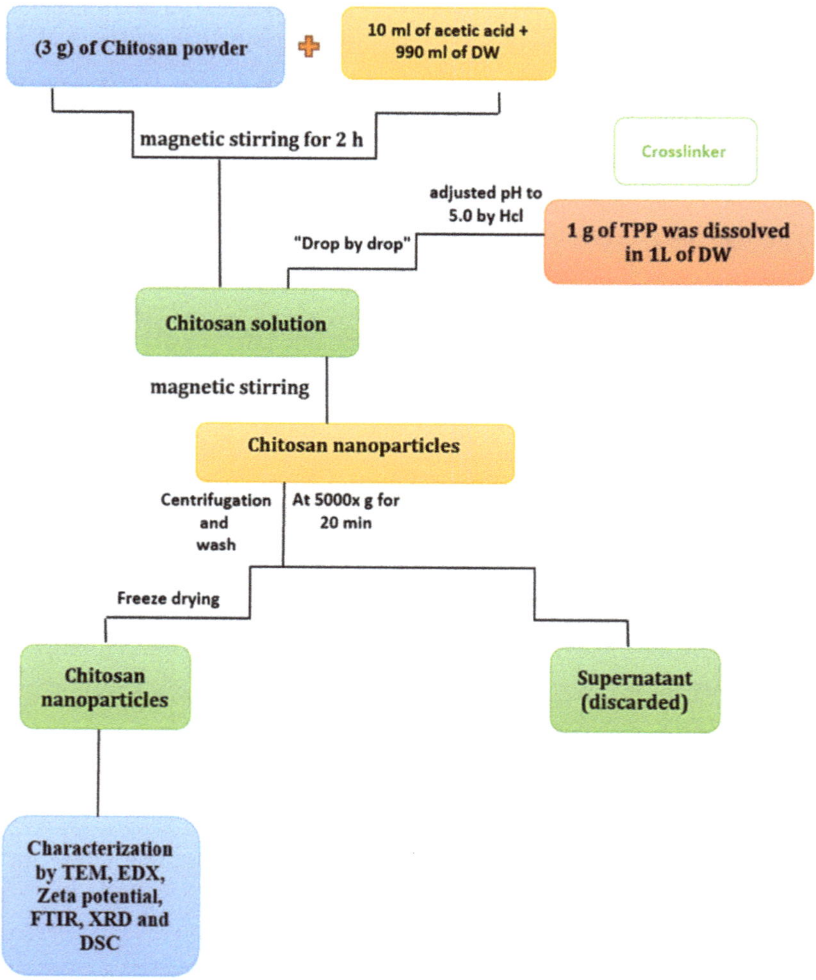

Figure 1. Schematic representation of the CNP synthesis process.

2.2.2. Synthesis of CNPs/Cc

CNPs and CNPs/Cc were synthesized according to Masarudin et al. [24]. Three grams of chitosan (medium molecular weight) was dissolved in 10 mL of 1.0% acetic acid, then 990 mL of distilled water was added. The chitosan solution was stirred using a magnetic stirrer for two hours to guarantee complete dissolution. The plant extract (3 g) was dissolved in 100 mL of dimethyl sulfoxide (DMSO, purity $\geq$ 99.5%), using a magnetic stirrer to ensure that the plant extract was completely dissolved. For the cross-linker, 1 g of TPP was dissolved in 900 mL of distilled water at pH 5.0. TPP solution and plant extract were mixed and added drop-by-drop to the dissolved chitosan, then stirred for two hours. The mixture was refined by centrifugation for 20 min at 5000 rpm, then washed three times with distilled water, followed by lyophilization until it was completely dry (Figure 2). This extract was used for pharmacological testing.

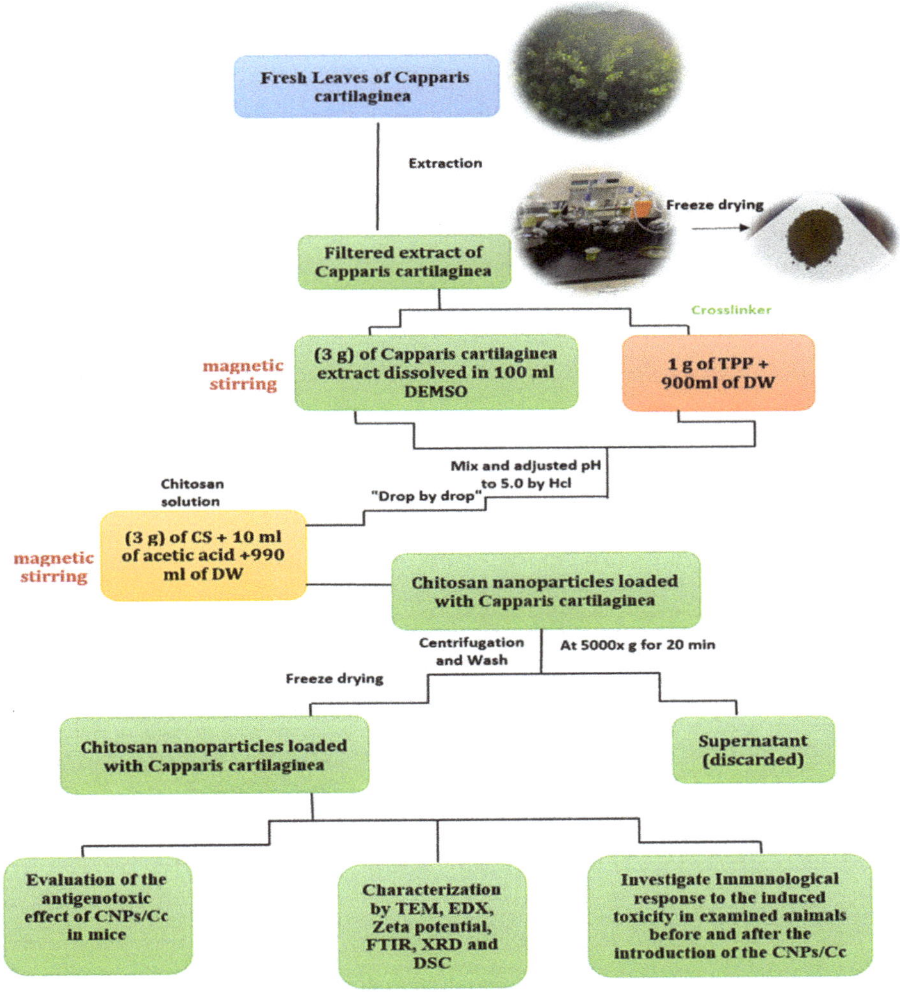

Figure 2. Schematic representation of the CNP/Cc synthesis process.

2.2.3. Characterization of CNPs and CNPs/Cc

Characterization of CNPs and CNPs/Cc was accomplished using an FTIR (Waltham, MA, USA) Spectrometer, Thermo Fisher Nicolet IS10, to produce the FTIR spectrum (4000 and 400 cm^{-1}). The size and morphology of the nanoparticles at the nanoscale were characterized by means of transmission electron microscopy (TEM) (JEOL JSM-6510/v, Tokyo, Japan). The chemical contents of nanoparticles were determined via energy-dispersive spectroscopy (EDS) (JEOL JSM-6510/v, Tokyo, Japan), and a Zeta potential analyzer was used to determine the surface charge and stability [25] (Malvern Zeta size

Nano-Zs90, Malvern, PA, USA). Thermographic analysis was performed via differential scanning calorimetry (DSC) testing of 5–10 mg using a DSC131 EVO (Lyon, France).

2.3. *In Vivo Antigenotoxicity*

2.3.1. Experimental Animals

Male SWR mice (9–12 weeks old, 25–27 g each) were obtained from the Animal House Colony at King Fahad Medical Research Centre. Animals were kept in plastic cages under normal conditions of a night–day cycle (12/12 h). Food and water were supplied ad libitum.

Animals were kept in artificially illuminated (12 h dark/light cycle) and thermally controlled (25 ± 1 °C) conditions. Humane care was applied to all animal groups in compliance with the guidelines of the Animal Care and Use Committee of the University of Jeddah (Approval # UJ-21-DR-41).

2.3.2. Experimental Design

Animals were divided into the groups listed in Table 1.

Table 1. Experimental design and animal groups.

Groups	Treatment and Doses	Treatment Day(s)
Control	Control (negative)	
EMS:	intraperitoneal single injection with 240 mg/kg b.w	
CNPs:	oral administration with 350 mg/kg b.w. of chitosan nanoparticle	
HD of CNPs/Cc:	oral administration with 700 mg/kg b.w. of CNPs loaded with *C. cartilaginea*	
LD of CNPs/Cc:	oral administration with 350 mg/kg b.w. of CNPs loaded with *C. cartilaginea*	1
CNPs + EMS	CNPs (350 mg/kg b.w.) + EMS (single injection with 240 mg/kg b.w)	
HD of CNPs/Cc +EMS	(single injection with 240 mg/kg b.w)	
LD of CNPs/Cc + EMS	(single injection with 240 mg/kg b.w)	
Control	Control (negative)	
EMS:	intraperitoneal single injection with 240 mg/kg b.w, 24 h before the experiment.	
CNPs:	oral administration with 350 mg/kg b.w. of chitosan nanoparticle	
HD of CNPs/Cc:	oral administration with 700 mg/kg b.w. of CNPs loaded with *C. cartilaginea*	7
LD of CNPs/Cc:	oral administration with 350 mg/kg b.w. of CNPs loaded with *C. cartilaginea*	
CNPs + EMS	CNPs (350 mg/kg b.w.) + EMS (single injection with 240 mg/kg b.w)	
HD of CNPs/Cc +EMS	(single injection with 240 mg/kg b.w)	
LD of CNPs/Cc + EMS	(single injection with 240 mg/kg b.w)	

2.3.3. Chromosome Abnormalities Assay

Chromosomal aberration was assessed as described by Moore et al. [26]. For this, animals were injected i.p. with colchicine, 2 h before sacrifice. The method of Yosida and Amano [27] was applied in bone marrow chromosome preparations. One hundred metaphases were examined per animal, and the metaphases with aberrations such as gaps, chromosome or chromatid breakage, and fragments were recognized.

2.3.4. Sperm Morphology Assay

Rasgele's sperm morphology testing was carried out [28]. Cauda epididymides were crushed in an isotonic sodium citrate solution. Sperms were smeared on slides, fixed, and stained with Eosin Y. In total, 1000 sperms were examined per animal. The head and tail abnormalities of sperm were detected.

2.4. *Statistical Analysis*

All data are provided as mean values with standard deviations (SDs) based on five replicates and probability values ($p \leq 0.05$). Tukey's method and one-way ANOVA (version 16) were used to evaluate the significance of data.

3. Results and Discussion

3.1. Fourier-Transform Infrared Spectroscopy (FTIR)

FTIR was performed to compare the functional groups of the *C. cartilaginea* extract, the CNPs, and the CNPs/Cc. In total, eight band regions were observed in *C. cartilaginea* (3231, 2924, 1598, 1508, 1396, 1267, 1038, and 614), in the CNPs (3335, 2126, 1636, 1558, 1540, 1457, 1096, and 623 cm^{-1}), and in the CNPs/Cc (3346, 2132, 1636, 1540, 1418, 1077, 1011, and 950). The results showed similarity between the chemical structures of CNPs and CNPs/Cc; the wavenumber 1636 was the peak in the CNPs/Cc, and the CNPs presented CHO stretching of the carbonyl group, typical saccharide absorption, and a peak of 1540 cm^{-1}, which relates to the amide groups that are present in chitosan (Table 2 and Figure 3).

Table 2. *C. cartilaginea*, CNPs, and CNPs/Cc analysis by FTIR.

Wavenumber cm^{-1}	Plant	CNP/Cc	CNPs	Functional Groups	Ref.
3346.88	Nd	D	Nd	Hydroxyl groups	[29]
3335.45	Nd	Nd	D	Stretching N-H asymmetric	[30]
3231.22	D	Nd	Nd	O-H bond stretching	[31]
2132.35	Nd	D	−6	Si–H stretching	[29]
1636.33	−38	D	D	CHO stretching of carbonyl group	[30]
1558.81	Nd	Nd	D	C–C stretch aromatic rings (phenolic)	[32]
1540.41	−22	D	D	Amide II	[30]
1418.53	−21	D	+39	Deformation C–H	[30]
1267.27	+	Nd	Nd	C-O stretching	[33]
1077.43	Nd	D	+19	C=C bond	[34]
1011.96	+17	D	Nd	C-F groups	[28]
950.72	Nd	D	Nd	Amines	[35]
614.09	D	Nd	+9	C–S stretch	[36]

D: detected, Nd: not detected, (−): shifted wavenumber by minus, (+): shifted wavenumber by addition.

3.2. GC–MS Analysis of *C. cartilaginea* Extract

The GC–MS analysis of *Capparis cartilaginea* showed the presence of 15 compounds (Figure 4). The prevailing compounds were 9,12,15-octadecatrienoic acid, methyl ester (14.99%); Hexadecanoic acid, methyl ester (11.79%); D-Pyroglutamic acid (9.03), hydroxy-bicyclo[3.3.1]n on-2-en-9-one (6.15%); 7-hydroxy-bicyclo[3.3.1]n on-2-en-9-one (4.25%); octadecanoic acid, methyl ester (3.24%); CIS-5,8,11,14,17-eicosapentaenoic acid (3.22%); 9-octadecenoic acid (z)- hexadecanoic ACID (3.21%); hexadecanoic acid, ethyl ester (3.09%); 2-aminoethanethiol hydrogen sulfate (ester) (2.98%); 9-octadecenoic ACID (Z)- 2-aminoethanethiol (2.91%); 9,12-Oetadecadienoyl chloride, (Z,Z)- (2.76%); 9-Octadecenoic acid (Z)-, methyl ester (2.09%); acethydrazide, n2-[1-(2,3-dihydro-6-methyl pyran-2-yl) ethylideno]- (2.05%); 3,5-Heptadienal and 2-ethylidene-6-methyl (2.00%). All these compounds are bioactive and possess many activities such as antibacterial, anticancer, antiviral, and antioxidant, as presented in Table 3.

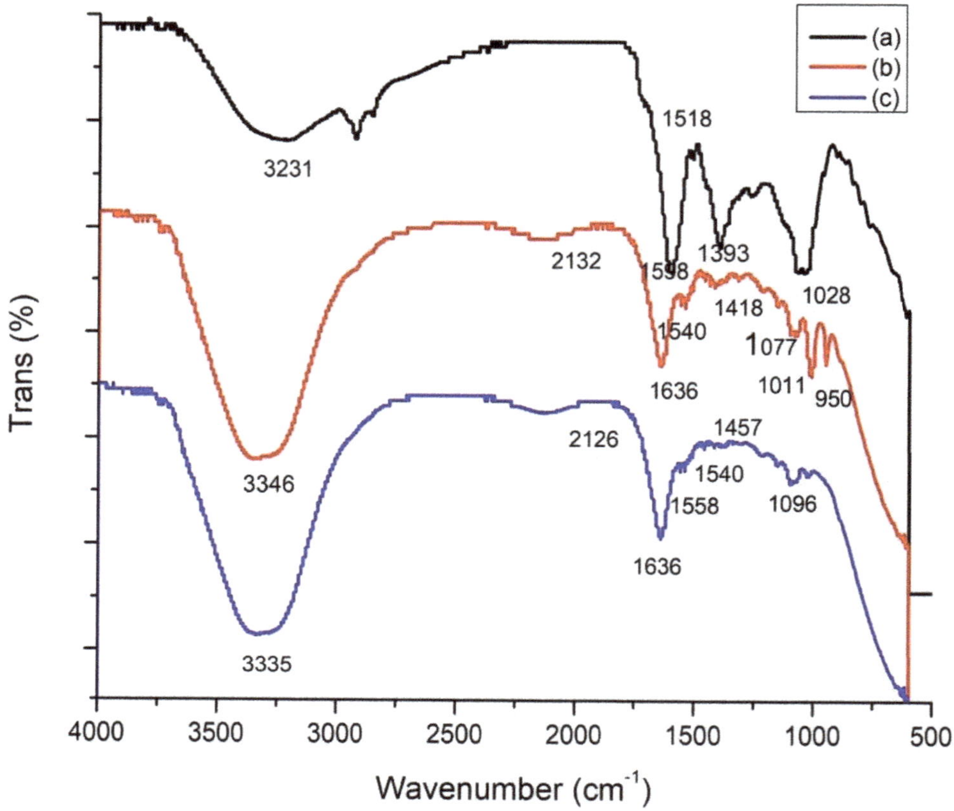

Figure 3. FTIR spectroscopy of *C. cartilaginea* (a), CNPs (b), and CNPs/Cc (c).

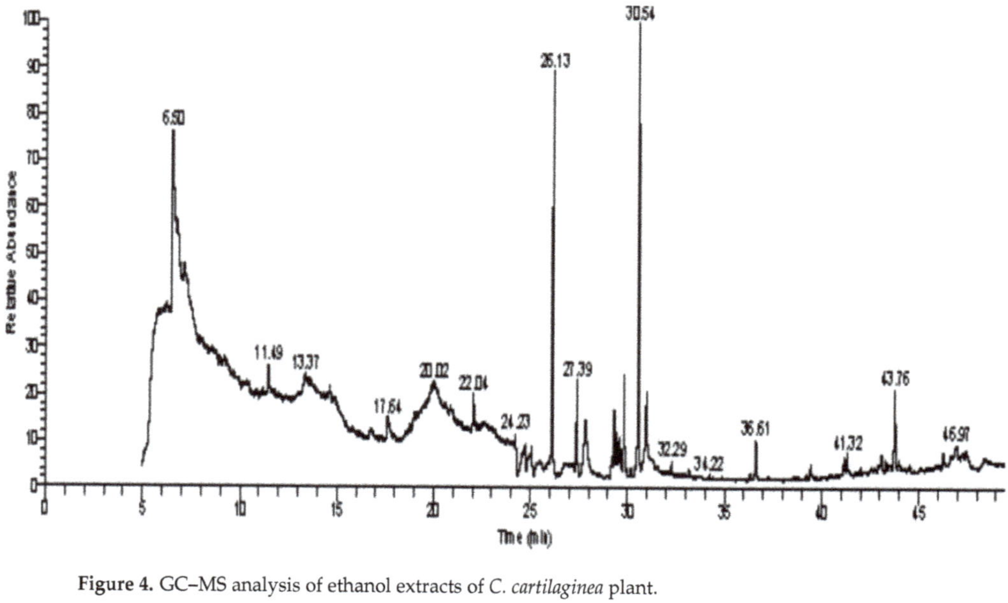

Figure 4. GC–MS analysis of ethanol extracts of *C. cartilaginea* plant.

Table 3. GC–MS analysis of *C. cartilaginea* extract.

Biological Activity	Molecular Weight	Molecular Formula	Chemical Structure	Area %	Compound Name	RT
Antimicrobial, Anti-inflammatory, Antioxidant [37]	152	$C_9H_{12}O_2$	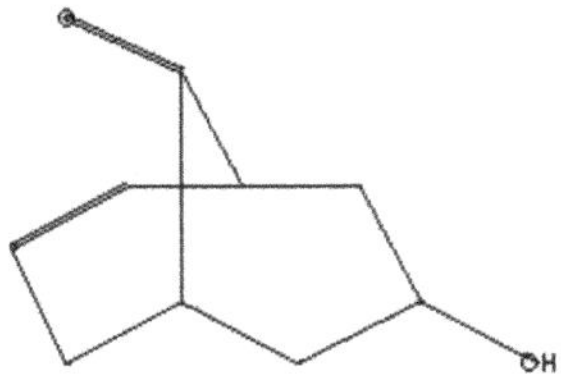	4.25	7-Hydroxy-bicyclo[3.3.1]non-2-en-9-one	5.53
Antimicrobial, Anti-inflammatory, Antioxidant [37]	152	$C_9H_{12}O_2$	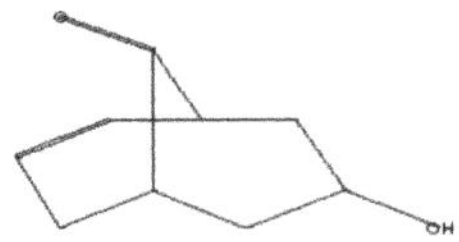	6.15	7-HYDROXY-BICYCLO[3.3.1]NON-2-EN-9-ONE	5.58
Anticancer agent [38]	129	$C_5H_7NO_3$	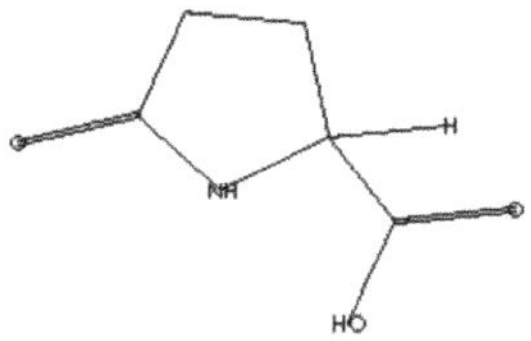	9.03	D-Pyroglutamic acid	6.50
Acetohydrazides and acetamides possessed anticancer agents [39]	196	$C_{10}H_{16}N_2O_2$	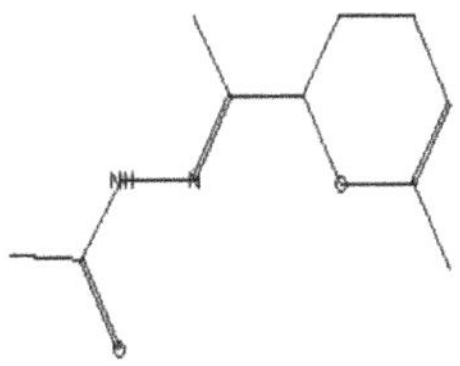	2.05	Acethydrazide, n2-[1-(2,3-dihydro-6-methyl pyran-2-yl)ethylideno]-	7.14
Analgesic, digestive, and wound healing [40] Agonist activity at human TRPA1 channel expressed in HEK293 cell [41]	150	$C_{10}H_{14}O$	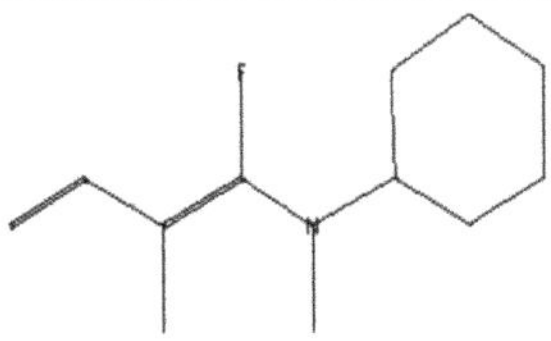	2.00	3,5-Heptadienal, 2-ethylidene-6-methyl-	13.36
The main compounds are found in Kei Apple fruits and have antioxidant and anticancer activities [42]	157	$C_2H_7NO_3S_2$	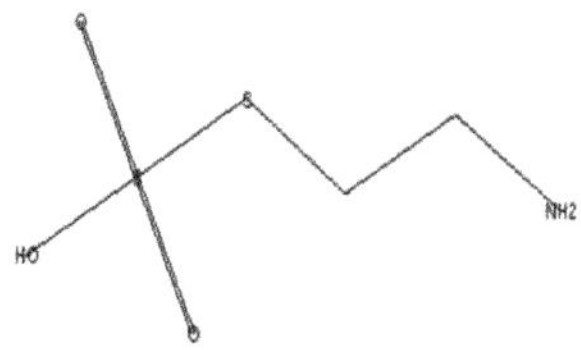	2.98	2-AMINOETHANETHIOL HYDROGEN SULFATE (ESTER)	24.73
Hepatoprotective Antiandrogenic, Antihistaminic, Anticoronary, Insectifuge, Anticancer [43]	298	$C_{18}H_{31}ClO$	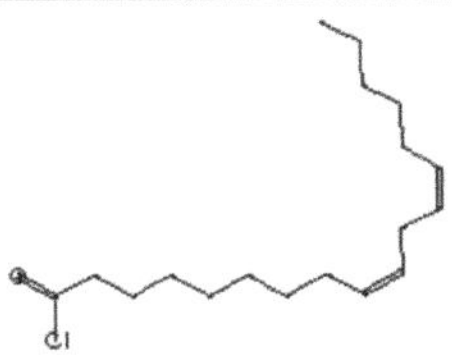	2.76	9,12-Octadecadienoyl chloride, (Z,Z)-	25.07

Table 3. *Cont.*

Biological Activity	Molecular Weight	Molecular Formula	Chemical Structure	Area %	Compound Name	RT
Anti-oxidant, decreases blood cholesterol, anti-inflammatory [44]	270	$C_{17}H_{34}O_2$	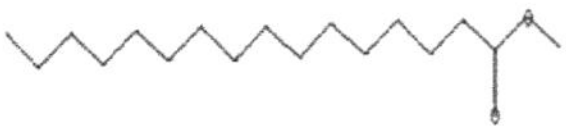	11.79	Hexadecanoic acid, methyl ester	26.13
Antimicrobial [45]	284	$C_{18}H_{36}O_2$	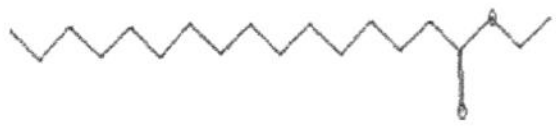	3.09	Hexadecanoic acid, ethyl ester	27.39
Antibacterial [46]	282	$C_{18}H_{34}O_2$		3.21	9-Octadecenoic acid (z)-, hexadecanoic acid	27.84
Antioxidant, anticancer [44,47]	296	$C_{19}H_{36}O_2$		2.09	9-Octadecenoic acid (Z)-, methyl ester	29.31
Antimicrobial [48]	298	$C_{19}H_{38}O_2$		3.24	Octadecanoic acid, methyl ester	29.82
Antibacterial, Anti-inflammatory, cancer preventive, hepatoprotective, nematicide, insectifuge [45]	292	$C_{19}H_{32}O_2$	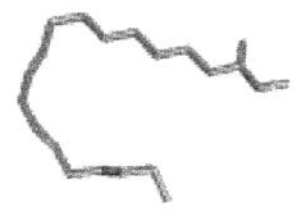	14.99	9,12,15-Octadecatrienoic acid, methyl ester	30.54
Antibacterial [46]	282	$C_{18}H_{34}O_2$	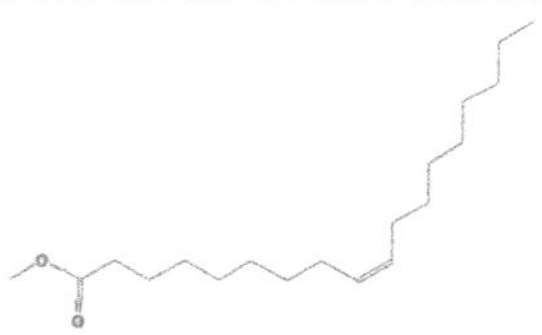	2.91	9-OCTADECENOIC ACID (Z)- 2-AMINOETHANETHIOL	30.96
Anti-atherosclerotic, anti-inflammatory, and anti-proliferative effects [49]	302	$C_{20}H_{30}O_2$	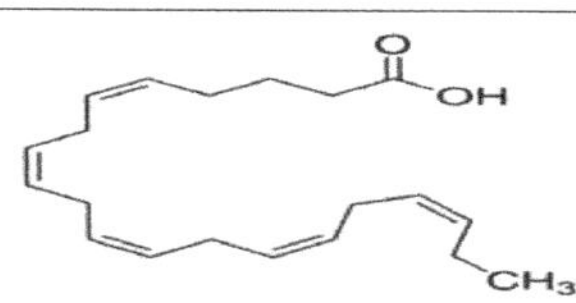	3.22	cis-5,8,11,14,17-Eicosapentaenoic acid	43.76
			21,736,059.79			

3.3. Zeta Potential Characterization

Zeta potential determination is a crucial method for characterizing the surface charge and comprehending the physical stability of nanosuspensions [50]. The nanoparticles examined in this investigation had surfaces that were positively charged (Figure 5a,b). The results showed that the particles generated in this investigation were comparable to those described in the literature [51–56]. The CNPs/Cc had a zeta potential of about +34.49 mV (Figure 5a), while the CNPs had a zeta potential of about +52.78 mV (Figure 5b). This showed the CNPs/Cc's incipient instability and the CNPs' moderate stability. The positive surface charges of chitosan were affected by the addition of bioactive components, altering the stability of the nanoparticles. If the value of the zeta potential ranges between 0 and ±5 mV, it is an indicator of rapid coagulation; ±10 to ±30 mV indicates incipient instability; ±30 to ±40 mV indicates moderate stability; ±40 to ±60 mV indicates good stability; and >±61 mV indicates excellent stability [57].

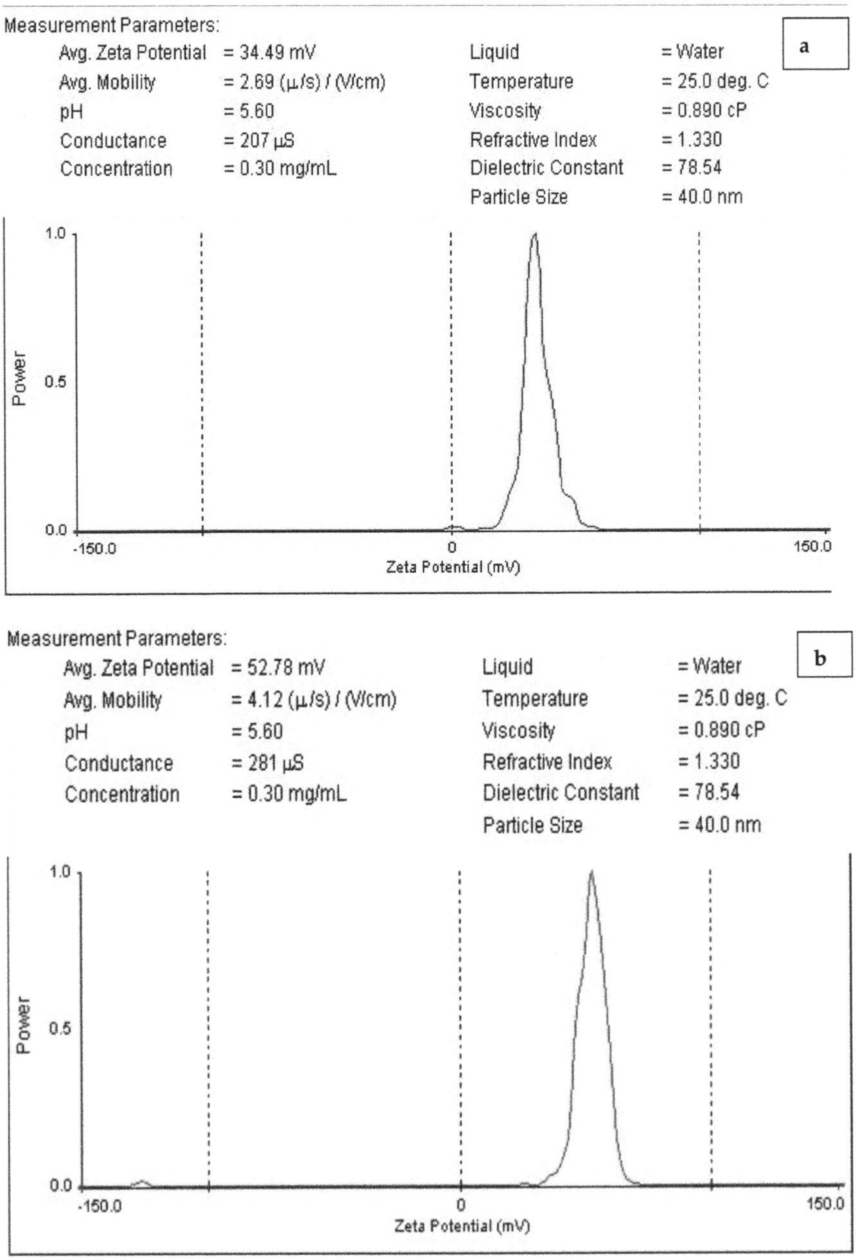

Figure 5. Zeta potential of CNPs/Cc (**a**) and CNPs (**b**).

3.4. TEM Analysis

A TEM micrograph showed the CNPs to have a spherical particle shape (Figure 6a) and a size range of 9–25 nm [54,56,58], the CNPs/Cc had a spherical shape and a size range of 18–30.1 nm (Figure 6b). The size difference between the CNPs and the CNPs/Cc may be due to the effect of the plant extract being loaded with nanoparticles. Using cross-linked chitosan is necessary to have better control over its shape and size [59]. The TEM image showed the aggregation of the chitosan nanoparticles [60]. The chitosan nanoparticles appeared as tiny and individual spheres, with a diameter ranging from 30 to 40 nm. Larger particles are caused by the aggregation of single tiny particles that manage to fuse, producing a larger entity [61,62].

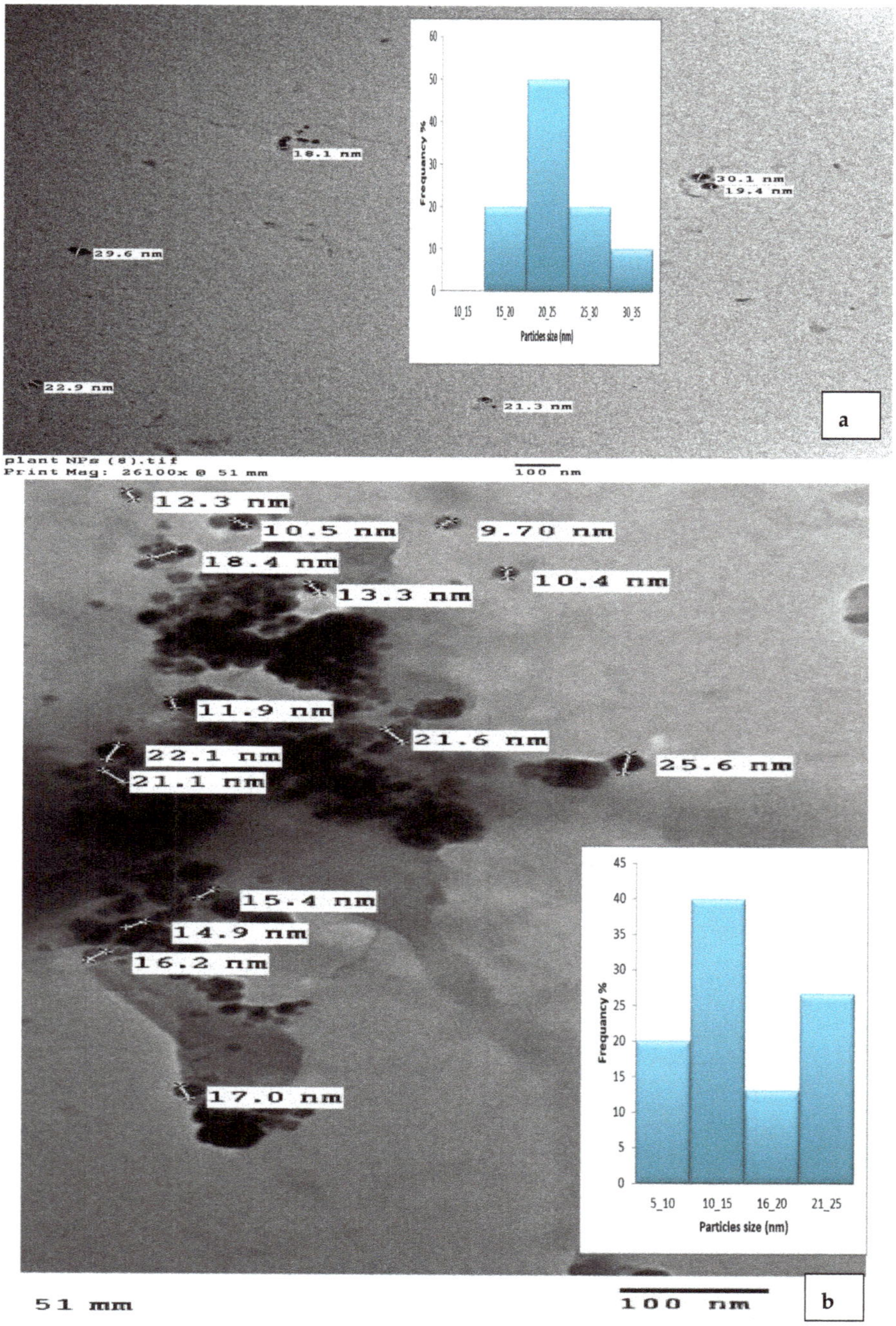

Figure 6. TEM image of CNPs (**a**) and CNPs/Cc (**b**).

3.5. Energy-Dispersive Spectroscopy (EDS)

Figure 7a,b depict the EDS analysis of the CNPs and the CNPs/Cc. The EDS of the CNPs revealed a high-intensity metallic peak of elements such as oxygen (O) with an atomic weight of 52.72%, 37.82% for carbon (C), 8.37% for phosphate (P), and low-intensity peaks of aluminum (Al). The EDS of the CNPs/Cc also confirmed the presence of carbon with an atomic weight of 30.40%, 49.72% for oxygen, 13.83% for nitrogen, 3.91% for phosphate, and low-intensity peaks of aluminum (Al), sodium (Na), and sulfur (S). It was possible to observe an increase in the intensities of the C and O signals relative to the other metals' signals, which was likely caused by the chitosan nanoparticles [28]. EDS was employed to confirm the presence of TPP as cross-linked junctions in the CS particles, by monitoring the content of phosphorus in CS-TPP [63]. New signals of Na and P were obtained in the CNPs/Cc samples, due to the cross-linking of TPP with CS [64].

3.6. Differential Scanning Calorimetry (DSC)

Figure 8 shows the DSC thermogram of the CNPs and the CNPs/Cc. The temperature ranged from 20 to 600 °C. A sharp endothermic peak was observed between 75.60 and 186.73 °C, showing the nanoparticles' potential for heat breakdown. The exothermic peak of the CNP at 196.82 and 219.39.73 °C followed by an exothermic peak at 219.91 and 256.44 °C could be due to water evaporation. The CNPs exhibited a sharp endothermic peak at 45.46 and 163.82, after a tiny endothermic peak at 169.23 and 207.41, while an exothermic peak was observed at 214.66 and 247.90 °C. The result is in agreement with Yousef et al. [65], whose DSC thermogram of chitosan showed an endothermic peak at 180 °C and an exothermic peak at 370 °C. Chitosan DSC thermogram showed two peaks. The loss of water molecules was connected to the first one (endothermic, below 100 °C); the second signal, which was exothermic and occured at a temperature of about 300 °C, was connected to the breakdown of the chitosan pyranose ring [66,67]. The results showed that exothermic peaks, also known as crystallization peaks, were present throughout the DSC analyses' heating scan [25].

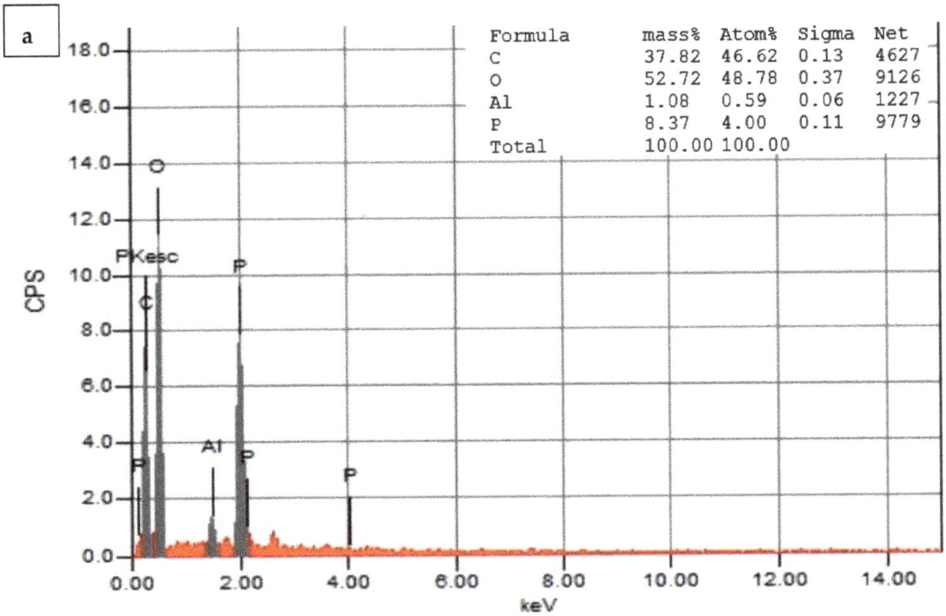

Figure 7. *Cont.*

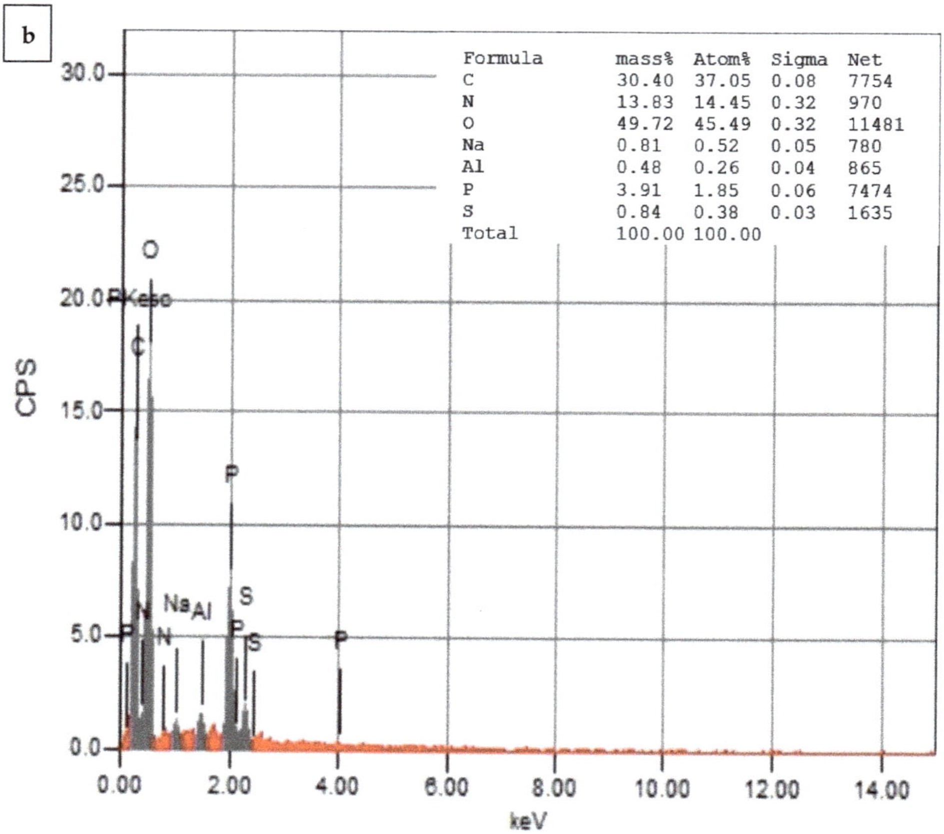

Figure 7. Energy-dispersive X-ray spectrophotometry analysis of CNPs (**a**) and CNPs/Cc (**b**).

3.7. Chromosomal Aberrations Observed in Bone Marrow Cells

Table 4 presents the number and percent of chromosomal aberrations in control and treatment animals. Treatment with EMS induced a significant increase in chromosomal aberrations in the bone marrow cells, reaching 24.4% ($p < 0.01$). The present results are in agreement with the findings of Norizadeh Tazehkand et al. [68], who documented that treatment of mice with 240 mg/kg EMS significantly increased chromosomal aberrations. There are large and diverse groups of chemicals that can induce DNA damage, and they are classified as carcinogens. Humans maybe exposed to these substances directly or indirectly in the environment or through diet. EMS has a potential negative effect, including cancer, aberrant birth outcomes, and heritable impacts, which are seen as natural consequences of its genotoxic activity [69]. EMS is an alkylating agent commonly used as a positive agent in genotoxicity studies [70] with the potential to cause cancer by altering DNA nucleotides, resulting in point mutations as the initial G:C base pair undergoes a mutation to A:T during replication base-pair insertions or deletions [71].

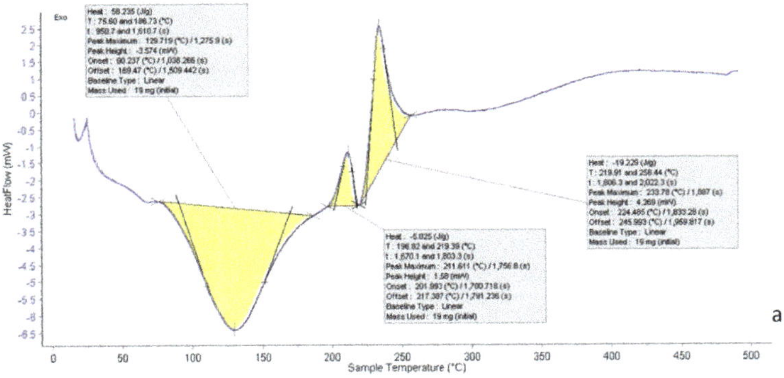

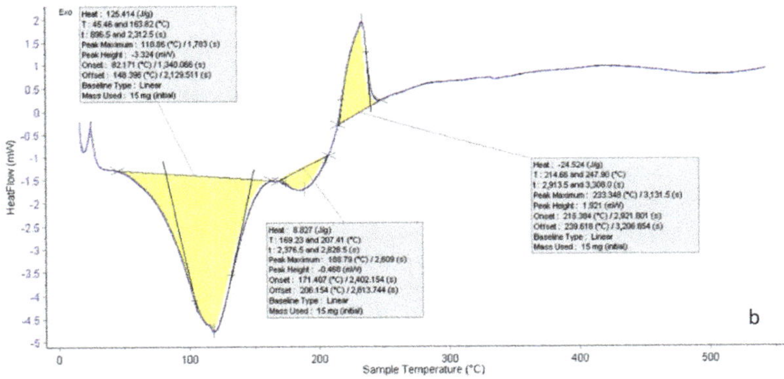

Figure 8. DSC of CNPs (**a**) and CNPs/Cc (**b**).

Table 4. Number and mean percentage of the different types of chromosomal aberrations in mouse bone marrow cells after treatment with different doses of CNPs/Cc and CNPs alone or in combination with EMS.

Treatment	No. of Metaphases with Aberrations						Chromosomal Aberrations		
	Gap	Frag. and/or Break	Gap + (Frag. or Break)	Deletion	Ring	Total No.	Excluding Gaps Mean ± S.E.	Including Gaps Mean ± S.E.	Inhibition %
One-day treatments									
Control	2	4	0	0	0	6	0.8 ± 0.4	1.2 ± 0.4	
EMS	20	62	28	10	2	122	20.4 ± 0.7 a	24.4 ± 0.4 a	
CNPs	10	9	5	0	0	24	3.6 ± 0.48	4.8 ± 0.4	
HD of CNPs/Cc	12	5	2	2	0	21	4.6 ± 0.2	4.2 ± 0.2	
LD of CNPs/Cc	13	3	6	1	0	23	4.2 ± 0.3	4.106 ± 0.4	
CNPs + EMS	14	50	33	11	0	108	18.8 ± 0.37	21.6 ± 0.7 a	11.4
HD of CNPs/Cc + EMS	11	61	15	9	1	97	17.2 ± 0.37	19.4 ± 0.6 a	20.49
LD of CNPs/Cc + EMS	22	51	33	0	0	106	16.8 ± 0.6 a	21.2 ± 0.6 a	13.1
Seven-day treatments									
Control	2	4	0	0	0	6	0.8 ± 0.4	1.2 ± 0.1	
EMS	20	62	28	10	2	122	20.4 ± 0.7 a	24.4 ± 0.4 a	
CNPs	12	3	5	2	0	23	2.2 ± 0.2	4.6 ± 0.6	
HD of CNPs/Cc	12	5	3	0	0	20	1.6 ± 0.4	4.0 ± 0.1	
LD of CNPs/Cc	12	10	0	0	0	22	2.0 ± 0.4	4.4 ± 0.4	
CNPs + EMS	20	39	11	12	0	82	12.4 ± 0.6 ab	16.4 ± 0.7 ab	32.7
HD of CNPs/Cc + EMS	12	27	4	7	0	50	7.6 ± 0.5 ab	10 ± 0.8 ab	59.01
LD of CNPs/Cc + EMS	18	24	32	4	0	78	12 ± 0.7 ab	15.6 ± 0.7 ab	36.06

a: Significant at 0.01 level (*t*-test) compared to control (non-treated). b: Significant at 0.01 level (*t*-test) compared to treatment (EMS).

No significant aberrations were observed in the animal groups treated with CNPs or CNPs/Cc at their low tested doses. Moreover, pretreatment with CNPs/Cc at two tested doses for seven days succeeded in reducing ($p < 0.01$) the EMS-induced abnormalities significantly in a dose-dependent manner. The percentage of reduction reached 36.06% and 59.01% after treatment with low and high doses of CNPs/Cc, respectively (Table 4). Nanocurcumin and nanochitosan have antioxidant and antigenotoxic properties against the toxicity of potassium dichromate [72]. *Salix subserrata* bark extract-loaded chitosan nanoparticles exhibited promising potent neuroprotective and antioxidative efficiencies against arsenic-induced oxidative threats [73].

3.8. Sperm Shape Abnormalities

As shown in Table 5, male mice exposed to EMS developed defective sperm in a statistically highly significant ($p < 0.01$) percentage that reached 14.44%. The most prevalent anomalies were triangular, amorphous, and lacking hook heads and coiled tails. Chromosomal aberrations, testicular DNA alterations, and point mutations may contribute to the induction of sperm shape abnormalities [28].

The percentage of aberrant sperm was decreased when mice were treated with cCNP and CNPs simultaneously, reaching 10.02, 9.7, and 7.26% ($p < 0.01$). Following the treatment of mice with LD cCNP, HD cCNP and CNPs, the percentage of reduction reached 30.6, and 32.8, 49.7%. A previous study proved that pre-treatment with ZnO-Alg/NCMs for seven days significantly reduced chromosomal aberrations in somatic cells and spermatocytes and reduced the percentage of morphological sperm abnormalities [74].

Several studies on the phytochemistry of the genus *Capparis* have revealed the presence of terpenoids, flavonoids, alkaloids, glucosinolates, isothiocyanates, sterols, and fatty acids in various plant sections [9]. The methanol extract of *Capparis cartilaginea* leaves contains seven types of antioxidants [75] that could inhibit various human oxidative stress pathologies and possess DNA-protective attributes [76]. According to Moharram et al. [77], the anti-inflammatory activity is probably due to the antioxidant active constituents of the plant such as flavonoids, alkaloids, phenolic, coumarins, and tannin compounds.

Table 5. Number and percentage of different types of sperm shape abnormalities in male mice after treatment with different doses of CNPs/Cc and CNPs alone or in combination with EMS.

Treatment and Doses (mg/kg b.wt.)	Sperm No.	No. of Sperm with Abnormalities in						Abnormal Sperm No.	Abnormal Sperm Mean % ± S.E.	Inhibition %
		Head			Tail					
		Amorphous	Tringle	Without Hook	Small	Big	Coiled			
Control	5000	29	36	28	0	0	6	99	1.98 ± 0.1	
EMS	5000	259	183	47	3	1	229	722	14.4 ± 1.78 a	
CNPs	5000	31	25	6	0	0	24	86	1.72 ± 0.24	
HD of CNPs/Cc	5000	25	32	5	0	0	21	83	1.66 ± 0.2	
LD of CNPs/Cc	5000	32	21	7	0	0	35	95	1.9 ± 0.9	
CNPs + EMS	5000	204	109	19	2	0	167	501	10.02 ± 0.98 ab	30.6
HD of CNPs/Cc + EMS	5000	158	93	23	7	2	80	363	7.26 ± 0.56 ab	49.7
LD of CNPs/Cc + EMS	5000	106	130	30	4	1	214	485	9.7 ± 0.22 ab	32.8

a: Significant at 0.01 level (*t*-test) compared to control (non-treated). b: Significant at 0.01 level (*t*-test) compared to treatment (EMS).

Many recent studies emphasized the utilization of nanoparticles as a phytochemical carrier molecule to improve their solubility, bio-accessibility, and cellular bioavailability [78,79]. Several studies proved the protective capabilities of chitosan nanoparticles. Abdel-Wahhab et al. [80] demonstrated that chitosan nanoparticles modulate DNA fragmentation and suppress mycotoxin-induced genotoxicity in rats. The antioxidant and antigenotoxic properties of chitosan nanoparticles against potassium-dichromate-induced toxicity in male albino mice were documented by Mohamed et al. [72].

4. Conclusions

CNPs were synthesized, loaded with *C. cartilaginea* extract, and investigated via Fourier-transform infrared spectroscopy. The Zeta potential analyses proved that CNPs/Cc are more positively charged than CNPs. In vivo experiments illustrated that oral treatment with CNPs and CNPs/Cc is safe. Moreover, chitosan nanoparticles showed powerful antigenotoxic properties. Loading the chitosan nanoparticles with *C. cartilaginea* extract greatly improved their anticancer properties, mainly due to the plant's active constituents and the small nanoparticle size that encouraged its cellular bioavailability.

Author Contributions: This work was carried out through the collaboration of all authors. Author A.S.S. wrote the protocol, carried out the experimental work of the genetic part, analyzed the study data, managed the literature searches, and wrote the final draft of the manuscript. Author S.N.A. carried out the experimental work, writing, and review. Author F.M.A. performed the statistical analysis, carried out the experimental work of plant materials, nanoparticle synthesis and characterization, and the in vivo part of the work. Author R.A.H. carried out the experimental work and analyses of nanoparticle synthesis and characterization and managed the literature searches. All authors have read and agreed to the published version of the manuscript.

Funding: This research received no external funding.

Institutional Review Board Statement: The animal study protocol was approved by the Animal Care and Use Committee of the University of Jeddah (Approval # UJ-21-DR-41).

Informed Consent Statement: Not applicable.

Data Availability Statement: The datasets used and/or analyzed during this study are available from the corresponding author on reasonable request.

Conflicts of Interest: The authors declare no conflict of interest.

References

1. Barnes, J.L.; Zubair, M.; John, K.; Poirier, M.C.; Martin, F.L. Carcinogens and DNA damage. *Biochem. Soc. Trans.* **2018**, *46*, 1213–1224. [PubMed]
2. Bray, F.; Laversanne, M.; Weiderpass, E.; Soerjomataram, I. The ever-increasing importance of cancer as a leading cause of premature death worldwide. *Cancer* **2021**, *127*, 3029–3030. [CrossRef] [PubMed]
3. Basu, A.K. DNA damage, mutagenesis, and cancer. *Int. J. Mol. Sci.* **2018**, *19*, 970. [CrossRef] [PubMed]
4. DeClercq, V.; Nearing, J.T.; Sweeney, E. Plant-based diets and cancer risk: What is the evidence? *Curr. Nutr. Rep.* **2022**, *11*, 354–369.
5. Alshehri, K.M. Anticancer Plants Naturally Growing in Al-Baha Region, Saudi Arabia. *Int. J. Pharm. Res. Allied Sci.* **2020**, *9*, 92–101.
6. Inocencio, C.; Rivera, D.; Obón, M.C.; Alcaraz, F.; Barreña, J.A. A systematic revision of capparis section Capparis (*Capparaceae*). *Ann. Mo. Bot. Gard.* **2006**, *93*, 122–149. [CrossRef]
7. Eddouks, M.; Lemhadri, A.; Michel, J.B. Hypolipidemic activity of aqueous extract of *Capparis spinosa* L. in normal and diabetic rats. *J. Ethnopharmacol.* **2005**, *98*, 345–350. [CrossRef]
8. Ghule, B.; Murugananthan, G.; Yeole, P. Analgesic and antipyretic effects of *Capparis zeylanica* leaves. *Fitoterapia* **2007**, *78*, 365–369. [CrossRef]
9. Tlili, N.; Elfalleh, W.; Saadaoui, E.; Khaldi, A.; Triki, S.; Nasri, N. The caper (*Capparis* L.): Ethnopharmacology, phytochemical and pharmacological properties. *Fitoterapia* **2011**, *82*, 93–101.
10. Peltonen, L.; Hirvonen, J. Drug nanocrystals–versatile option for formulation of poorly soluble materials. *Int. J. Pharm.* **2018**, *537*, 73–83.
11. Aghili-Khorasani, H.; Makhzan, A. Source of Plant Research Institute for Islamic and complementary medicine Press. *Iran Univ. Med. Sci.* **2008**, *729*, 30.
12. Soutter, W. *Chitosan Nanoparticles—Properties and Applications*, 2nd ed.; AZoNano: Sydney, Australia, 2013.
13. Mohammed, M.A.; Syeda, J.T.; Wasan, K.M.; Wasan, E.K. An overview of chitosan nanoparticles and its application in non-parenteral drug delivery. *Pharmaceutics* **2017**, *9*, 53. [CrossRef]
14. Sharifi-Rad, J.; Quispe, C.; Butnariu, M.; Rotariu, L.S.; Sytar, O.; Sestito, S.; Rapposelli, S.; Akram, M.; Iqbal, M.; Krishna, A.; et al. Chitosan nanoparticles as a promising tool in nanomedicine with particular emphasis on oncological treatment. *Cancer Cell Int.* **2021**, *21*, 318. [CrossRef]
15. Jacinto, F.V.; Esteller, M. MGMT hypermethylation: A prognostic foe, a predictive friend. *DNA Repair* **2007**, *6*, 1155–1160. [CrossRef] [PubMed]

16. Gerber, C.; Toelle, H.-G. What happened: The chemistry side of the incident with EMS contamination in Viracept tablets. *Toxicol. Lett.* **2009**, *190*, 248–253. [CrossRef] [PubMed]
17. Flibotte, S.; Edgley, M.L.; Chaudhry, I.; Taylor, J.; Neil, S.E.; Rogula, A.; Zapf, R.; Hirst, M.; Butterfield, Y.; Jones, S.J.; et al. Whole-genome profiling of mutagenesis in *Caenorhabditis elegans*. *Genetics* **2010**, *185*, 431–441. [CrossRef] [PubMed]
18. Anderson, P. Mutagenesis. *Methods Cell Biol.* **1995**, *48*, 31–58. [PubMed]
19. Saito, T.; Ariizumi, T.; Okabe, Y.; Asamizu, E.; Hiwasa-Tanase, K.; Fukuda, N.; Mizoguchi, T.; Yamazaki, Y.; Aoki, K.; Ezura, H. TOMATOMA: A novel tomato mutant database distributing Micro-Tom mutant collections. *Plant Cell Physiol.* **2011**, *52*, 283–296. [CrossRef]
20. Guruprasad, K.P.; Subramanian, A.; Singh, V.J.; Sharma, R.S.K.; Gopinath, P.M.; Sewram, V.; Satyamoorthy, K. Brahmarasayana protects against Ethyl methanesulfonate or Methyl methanesulfonate induced chromosomal aberrations in mouse bone marrow cells. *BMC Complement. Altern. Med.* **2012**, *12*, 113. [CrossRef]
21. Kurniadi, A.S.; Irawati, F.; Putra, S.E.D.; Hardjo, P.H. Induction of Protocorm-Like Bodies (PLBs) Phalaenopsis spp. Hybrids Mutation through Ultraviolet Irradiation (UV254) and Ethyl Methane Sulfonate (EMS). *Agriprima J. Appl. Agric. Sci.* **2023**, *7*, 1–15. [CrossRef]
22. Chaudhary, S.A. *Flora of the Kingdom of Saudi Arabia*, 2nd ed.; Ministry of Agriculture & Water: Riyadh, Saudi Arabia, 2001; pp. 455–457.
23. Rautela, I.; Dheer, P.; Thapliyal, P.; Joshi, T.; Sharma, N.; Sharma, M.D. GC-MS analysis of plant leaf extract of Datura stramonium in different solvent system. *Eur. J. Biomed. Pharm. Sci.* **2018**, *5*, 236–245.
24. Masarudin, M.J.; Cutts, S.M.; Evison, B.J.; Phillips, D.R.; Pigram, P.J. Factors determining the stability, size distribution, and cellular accumulation of small, monodisperse chitosan nanoparticles as candidate vectors for anticancer drug delivery: Application to the passive encapsulation of [14C]-doxorubicin. *Nanotechnol. Sci. Appl.* **2015**, *8*, 67–80. [CrossRef]
25. Iglesias-Montes, M.L.; Soccio, M.; Siracusa, V.; Gazzano, M.; Lotti, N.; Cyras, V.P.; Manfredi, L.B. Chitin nanocomposite based on plasticized poly (lactic acid)/poly (3-hydroxybutyrate) (PLA/PHB) blends as fully biodegradable packaging materials. *Polymers* **2022**, *14*, 3177. [CrossRef] [PubMed]
26. Moore, P.D.; Patlolla, A.K.; Tchounwou, P.B. Cytogenetic evaluation of malathion-induced toxicity in Sprague-Dawley rats. *Mutat. Res./Genet. Toxicol. Environ. Mutagen.* **2011**, *725*, 78–82. [CrossRef] [PubMed]
27. Yosida, T.H.; Amano, K. Autosomal polymorphism in laboratory bred and wild Norway rats, Rattus norvegicus, found in Misima. *Chromosoma* **1965**, *16*, 658–667. [CrossRef] [PubMed]
28. Rasgele, P.G. Abnormal sperm morphology in mouse germ cells after short-term exposures to acetamiprid, propineb, and their mixture. *Arh. Hig. Rada Toksikol.* **2014**, *65*, 47–55. [CrossRef]
29. Tokarek, K.; Hueso, J.L.; Kuśtrowski, P.; Stochel, G.; Kyzioł, A. Green synthesis of chitosan-stabilized copper nanoparticles. *Eur. J. Inorg. Chem.* **2013**, *2013*, 4940–4947. [CrossRef]
30. Wang, B.; Song, Y.; Zhang, X.; Chen, K.; Liu, M.; Hu, X.; He, L.; Huang, Q. Polymer derived SiBCN (O) ceramics with tunable element content. *Ceram. Int.* **2022**, *48*, 10280–10287. [CrossRef]
31. Talari, A.C.S.; Martinez, M.A.G.; Movasaghi, Z.; Rehman, S.; Rehman, I.U. Advances in Fourier transform infrared (FTIR) spectroscopy of biological tissues. *Appl. Spectrosc. Rev.* **2017**, *52*, 456–506. [CrossRef]
32. Umamaheswari, S.; Murali, M. FTIR spectroscopic study of fungal degradation of poly (ethylene terephthalate) and polystyrene foam. *Chem. Eng.* **2013**, *64*, 159.
33. Trivedi, M.; Branton, A.; Trivedi, D.; Shettigar, H.; Bairwa, K.; Jana, S. Fourier transform infrared and ultraviolet-visible spectroscopic characterization of biofield treated salicylic acid and sparfloxacin. *Nat. Prod. Chem. Res.* **2015**, *3*, 1000186. [CrossRef]
34. Shi, J.; Xing, D.; Lia, J. FTIR studies of the changes in wood chemistry from wood forming tissue under inclined treatment. *Energy Procedia* **2012**, *16*, 758–762. [CrossRef]
35. Bianchi, C.L.; Stucchi, M.; Pirola, C.; Cerrato, G.; Sacchi, B.; Vitali, S.; Di Michele, A.; Capucci, V. Micro-sized TiO_2 catalyst in powder form and as coating on porcelain grès tile for the photodegradation of phenol as model pollutant for water phase. *Adv. Mater. Sci.* **2017**, *2*, 1–6. [CrossRef]
36. Sahoo, S.; Chakraborti, C.K.; Mishra, S.C.; Nanda, U.; Naik, S. FTIR and XRD investigations of some fluoroquinolones. *Int. J. Pharm. Pharm. Sci.* **2011**, *3*, 165–170.
37. Volkov, D.S.; Rogova, O.B.; Proskurnin, M.A. Organic matter and mineral composition of silicate soils: FTIR comparison study by photoacoustic, diffuse reflectance, and attenuated total reflection modalities. *Agronomy* **2021**, *11*, 1879. [CrossRef]
38. Dutta, S.; Ray, S.; Nagarajan, K. Glutamic acid as anticancer agent: An overview. *Saudi Pharm. J.* **2013**, *21*, 337–343. [CrossRef] [PubMed]
39. Aslam, S.; Wang, F.; Ahmad, M.; Zahoor, A.F.; Mansha, A.; Rasul, A.; Fu, L. Benzothiazine based acetohydrazides and acetamides as anticancer agents. *Pak. J. Pharm. Sci.* **2019**, *32*, 2795–2800.
40. Maduabuchi, E.K.; Tobechukwu, O.P. Advanced phytochemistry and chemo-metric profiling of the bioactive medicinal components of n-hexane seed extract of *Xylopia aethiopica* using FTIR and GC-MS techniques. *GSC Biol. Pharm. Sci.* **2023**, *22*, 247–256. [CrossRef]
41. Jan, N.; Andrabi, K.I.; John, R. *Calendula officinalis*—An important medicinal plant with potential biological properties. *Proc. Indian Natl. Sci. Acad.* **2017**, *83*, 769–787.

42. Qanash, H.; Yahya, R.; Bakri, M.M.; Bazaid, A.S.; Qanash, S.; Shater, A.F.; TM, A. Anticancer, antioxidant, antiviral and antimicrobial activities of Kei Apple (*Dovyalis caffra*) fruit. *Sci. Rep.* **2022**, *12*, 5914. [CrossRef]
43. Vijayabaskar, G.; Elango, V. Determination of phytocompounds in *Withania somnifera* and Smilax China using GC-MS technique. *J. Pharmacogn. Phytochem.* **2018**, *7*, 554–557.
44. Sumathi, R. GC-MS Analysis of Methanol Extracts of Flowers of *Allamanda neriifolia* Hook. *Indian J. Curr. Res.* **2015**, *2*, 49–53.
45. Mujeeb, F.; Bajpai, P.; Pathak, N. Phytochemical evaluation, antimicrobial activity, and determination of bioactive components from leaves of *Aegle marmelos*. *BioMed Res. Int.* **2014**, *2014*, 497606. [CrossRef] [PubMed]
46. Awa, E.P.; Ibrahim, S.; Ameh, D.A. GC/MS analysis and antimicrobial activity of diethyl ether fraction of methanolic extract from the stem bark of Annona senegalensis Pers. *Int. J. Pharm. Sci. Res.* **2012**, *3*, 4213–4218.
47. Belakhdar, G.; Benjouad, A.; Abdennebi, E.H. Determination of some bioactive chemical constituents from *Thesium humile* Vahl. *J. Mater. Environ. Sci.* **2015**, *6*, 2778–2783.
48. Akpuaka, A.; Ekwenchi, M.M.; Dashak, D.A.; Dildar, A. Biological activities of characterized isolates of n-hexane extract of *Azadirachta indica* A. Juss (Neem) leaves. *Nat. Sci.* **2013**, *11*, 141–147.
49. Brooks, J.D.; Milne, G.L.; Yin, H.; Sanchez, S.C.; Porter, N.A.; Morrow, J.D. Formation of highly reactive cyclopentenone isoprostane compounds (A3/J3-isoprostanes) in vivo from eicosapentaenoic acid. *J. Biol. Chem.* **2008**, *283*, 12043–12055. [CrossRef]
50. Jiang, J.; Oberdörster, G.; Biswas, P. Characterization of size, surface charge, and agglomeration state of nanoparticle dispersions for toxicological studies. *J. Nanoparticle Res.* **2009**, *11*, 77–89. [CrossRef]
51. Ali, S.W.; Joshi, M.; Rajendran, S. Synthesis and characterization of chitosan nanoparticles with enhanced antimicrobial activity. *Int. J. Nanosci.* **2011**, *10*, 979–984. [CrossRef]
52. Santo Pereira, A.E.; Silva, P.M.; Oliveira, J.L.; Oliveira, H.C.; Fraceto, L.F. Chitosan nanoparticles as carrier systems for the plant growth hormone gibberellic acid. *Colloids Surf. B. Biointerfaces* **2017**, *150*, 141–152. [CrossRef]
53. Hejjaji, E.M.; Smith, A.M.; Morris, G.A. Evaluation of the mucoadhesive properties of chitosan nanoparticles prepared using different chitosan to tripolyphosphate (CS: TPP) ratios. *Int. J. Biol. Macromol.* **2018**, *120*, 1610–1617. [CrossRef]
54. Barrera-Necha, L.L.; Correa-Pacheco, Z.N.; Bautista-Baños, S.; Hernández-López, M.; Jiménez, J.E.M.; Mejía, A.F.M. Synthesis and characterization of chitosan nanoparticles loaded botanical extracts with antifungal activity on *Colletotrichum gloeosporioides* and *Alternaria* species. *Adv. Microbiol.* **2018**, *8*, 286. [CrossRef]
55. Guo, H.; Li, F.; Qiu, H.; Liu, J.; Qin, S.; Hou, Y.; Wang, C. Preparation and characterization of chitosan nanoparticles for chemotherapy of melanoma through enhancing tumor penetration. *Front. Pharmacol.* **2020**, *11*, 317. [CrossRef] [PubMed]
56. Elisa, F.R.; Taís, T.; de Barros, A.; Assis, O.B.G.; Junior, A.C.; Quiles, A.; Hernando, I.; Nicoletti, V.R. Chitosan and crosslinked chitosan nanoparticles: Synthesis, characterization and their role as *Pickering emulsifiers*. *Carbohydr. Polym.* **2020**, *250*, 116878.
57. Müller, R.H.; Jacobs, C.; Kayser, O. Nanosuspensions as particulate drug formulations in therapy: Rationale for development and what we can expect for the future. *Adv. Drug Deliv. Rev.* **2001**, *47*, 3–19. [CrossRef]
58. Kumar, A.; Dixit, C.K. Methods for Characterization of Nanoparticles. In *Advances in Nanomedicine for the Delivery of Therapeutic Nucleic Acids*; Elsevier: Amsterdam, The Netherlands, 2017; pp. 43–58.
59. Huang, J.; Li, Q.; Sun, D.; Lu, Y.; Su, Y.; Yang, X.; Wang, H.; Wang, Y.; Shao, W.; He, N.; et al. Biosynthesis of silver and gold nanoparticles by novel sundried *Cinnamomum camphora* leaf. *Nanotechnology* **2007**, *18*, 105104. [CrossRef]
60. Chandra Hembram, K.; Prabha, S.; Chandra, R.; Ahmed, B.; Nimesh, S. Advances in preparation and characterization of chitosan nanoparticles for therapeutics. *Artif. Cells Nanomed. Biotechnol.* **2016**, *44*, 305–314. [CrossRef]
61. Rampino, A.; Borgogna, M.; Blasi, P.; Bellich, B.; Cesàro, A. Chitosan nanoparticles: Preparation, size evolution and stability. *Int. J. Pharm.* **2013**, *455*, 219–228. [CrossRef]
62. Gokce, Y.; Cengiz, B.; Yildiz, N.; Calimli, A.; Aktas, Z. Ultrasonication of chitosan nanoparticle suspension: Influence on particle size. *Colloids Surf. A Physicochem. Eng. Asp.* **2014**, *462*, 75–81. [CrossRef]
63. Khoerunnisa, F.; Nurhayati, M.; Dara, F.; Rizki, R.; Nasir, M.; Aziz, H.A.; Opaprakasit, P. Physicochemical properties of TPP-crosslinked chitosan nanoparticles as potential antibacterial agents. *Fibers Polym.* **2021**, *22*, 2954–2964. [CrossRef]
64. Koukaras, E.N.; Papadimitriou, S.A.; Bikiaris, D.N.; Froudakis, G.E. Insight on the formation of chitosan nanoparticles through ionotropic gelation with tripolyphosphate. *Mol. Pharm.* **2012**, *9*, 2856–2862. [CrossRef]
65. Yousefpour, P.; Atyabi, F.; Vasheghani-Farahani, E.; Movahedi, A.M.; Dinarvand, R. Targeted delivery of doxorubicin-utilizing chitosan nanoparticles surface-functionalized with anti-Her2 trastuzumab. *Int. J. Nanomed.* **2011**, *6*, 1977–1990. [CrossRef] [PubMed]
66. Corazzari, I.; Nisticò, R.; Turci, F.; Faga, M.G.; Franzoso, F.; Tabasso, S.; Magnacca, G. Advanced physicochemical characterization of chitosan by means of TGA coupled on-line with FTIR and GCMS: Thermal degradation and water adsorption capacity. *Polym. Degrad. Stab.* **2015**, *112*, 1–9. [CrossRef]
67. Kittur, F.; Prashanth, K.H.; Sankar, K.U.; Tharanathan, R. Characterization of chitin, chitosan and their carboxymethyl derivatives by differential scanning calorimetry. *Carbohydr. Polym.* **2002**, *49*, 185–193. [CrossRef]
68. Norizadeh Tazehkand, M.; Topaktas, M.; Yilmaz, M.B.; Hajipour, O.; Valipour, E. Delineating the antigenotoxic and anticytotoxic potentials of 4-methylimidazole against ethyl methanesulfonate toxicity in bone marrow cell of Swiss albino mice. *Bratisl. Med. J.* **2016**, *9248*, 290–294. [CrossRef]
69. Müller, L.; Gocke, E.; Lavé, T.; Pfister, T. Ethyl methanesulfonate toxicity in Viracept—A comprehensive human risk assessment based on threshold data for genotoxicity. *Toxicol. Lett.* **2009**, *190*, 317–329. [CrossRef] [PubMed]

70. Zhu, X.; Huo, J.; Zeng, Z.; Liu, Y.; Li, R.; Chen, Y.; Zhang, L.; Chen, J. Determination of potential thresholds for N-ethyl-N-nitrosourea and ethyl methanesulfonate based on a multi-endpoint genotoxicity assessment platform in rats. *Environ. Sci. Pollut. Res.* **2022**, *29*, 85128–85142. [CrossRef]
71. Yan, W.; Deng, X.W.; Yang, C.; Tang, X. The genome-wide EMS mutagenesis bias correlates with sequence context and chromatin structure in rice. *Front. Plant Sci.* **2021**, *12*, 579675. [CrossRef]
72. Mohamed, N.R.; Badr, T.M.; Elnagar, M.R. Efficiency of curcumin and chitosan nanoparticles against toxicity of potassium dichromate in male mice. *Int. J. Pharm. Pharm. Sci.* **2021**, *13*, 22159. [CrossRef]
73. Hamdan, D.I.; Tawfeek, N.; El-Shiekh, R.A.; Khalil, H.M.; Mahmoud, M.Y.; Bakr, A.F.; Zaafar, D.; Farrag, N.; Wink, M.; El-Shazly, A.M. Salix subserrata bark extract-loaded chitosan nanoparticles attenuate neurotoxicity induced by sodium arsenate in rats in relation with HPLC–PDA-ESI–MS/MS profile. *AAPS PharmSciTech* **2022**, *24*, 15. [CrossRef]
74. Hamouda, R.A.; Salman, A.S.; Alharbi, A.A.; Alhasani, R.H.; Elshamy, M.M. Assessment of the Antigenotoxic Effects of Alginate and ZnO/Alginate–Nanocomposites Extracted from Brown Alga *Fucus vesiculosus* in Mice. *Polymers* **2021**, *13*, 3839. [CrossRef] [PubMed]
75. Sinitsky, M.Y.; Kutikhin, A.G.; Tsepokina, A.V.; Shishkova, D.K.; Asanov, M.A.; Yuzhalin, A.E.; Minina, V.I.; Ponasenko, A.V. Mitomycin C induced genotoxic stress in endothelial cells is associated with differential expression of proinflammatory cytokines. *Mutat. Res./Genet. Toxicol. Environ. Mutagen.* **2020**, *858*, 503252. [CrossRef] [PubMed]
76. Sharifi-Rad, M.; Anil Kumar, N.V.; Zucca, P.; Varoni, E.M.; Dini, L.; Panzarini, E.; Rajkovic, J.; Tsouh Fokou, P.V.; Azzini, E.; Peluso, I.; et al. Lifestyle, oxidative stress, and antioxidants: Back and forth in the pathophysiology of chronic diseases. *Front. Physiol.* **2020**, *11*, 694. [CrossRef]
77. Moharram, B.A.; Al-Mahbashi, H.M.; Saif-Ali, R.; AliAqlan, F. Phytochemical, anti-inflammatory, antioxidant, cytotoxic and antibacterial study of *Capparis cartilaginea* decne from Yemen. *Int. J. Pharm. Pharm. Sci.* **2018**, *10*, 38. [CrossRef]
78. He, J.R.; Zhu, J.J.; Yin, S.W.; Yang, X.Q. Bioaccessibility and intracellular antioxidant activity of phloretin embodied by gliadin/sodium carboxymethyl cellulose nanoparticles. *Food Hydrocoll.* **2022**, *122*, 107076. [CrossRef]
79. Hu, Y.; Lin, Q.; Zhao, H.; Li, X.; Sang, S.; McClements, D.J.; Long, J.; Jin, Z.; Wang, J.; Qiu, C. Bioaccessibility and bioavailability of phytochemicals: Influencing factors, improvements, and evaluations. *Food Hydrocoll.* **2022**, 108165. [CrossRef]
80. Abdel-Wahhab, M.A.; Aljawish, A.; El-Nekeety, A.A.; Abdel-Aziem, S.H.; Hassan, N.S. Chitosan nanoparticles plus quercetin suppress the oxidative stress, modulate DNA fragmentation and gene expression in the kidney of rats fed ochratoxin A-contaminated diet. *Food Chem. Toxicol.* **2017**, *99*, 209–221. [CrossRef] [PubMed]

pharmaceutics

MDPI

Review

Progress of Antimicrobial Mechanisms of Stilbenoids

Xiancai Li [1,*], Yongqing Li [2], Binghong Xiong [1] and Shengxiang Qiu [1,*]

[1] Key Laboratory of Plant Resources Conservation and Sustainable Utilization, South China Botanical Garden, Chinese Academy of Sciences, Guangzhou 510650, China; xiongbh@scbg.ac.cn

[2] Key Laboratory of South China Agricultural Plant Molecular Analysis and Genetic Improvement, South China Botanical Garden, Chinese Academy of Sciences, Guangzhou 510650, China; liyongqing@scbg.ac.cn

* Correspondence: lixiancai@scbg.ac.cn (X.L.); sxqiu@scbg.ac.cn (S.Q.)

Abstract: Antimicrobial drugs have made outstanding contributions to the treatment of pathogenic infections. However, the emergence of drug resistance continues to be a major threat to human health in recent years, and therefore, the search for novel antimicrobial drugs is particularly urgent. With a deeper understanding of microbial habits and drug resistance mechanisms, various creative strategies for the development of novel antibiotics have been proposed. Stilbenoids, characterized by a C_6–C_2–C_6 carbon skeleton, have recently been widely recognized for their flexible antimicrobial roles. Here, we comprehensively summarize the mode of action of stilbenoids from the viewpoint of their direct antimicrobial properties, antibiofilm and antivirulence activities and their role in reversing drug resistance. This review will provide an important reference for the future development and research into the mechanisms of stilbenoids as antimicrobial agents.

Keywords: stilbenoids; antimicrobial mechanisms; conventional targets; antibiofilm; antivirulence; reversing drug resistance

1. Introduction

Microorganisms are life forms with the widest distribution, the most abundant species and the largest biomass on earth. Some microorganisms, such as intestinal beneficial microorganisms, industrial microbial strains and agricultural microbial agents, are widely used in the human health industry and agricultural production and have greatly promoted the development of the economy and society. Others, however, cause pathogenic diseases. Bacterial infections were fatal until the 1940s, when penicillin was discovered. In the years that followed, the discovery of antibiotics released people from the hegemony of infection and improved our quality of life and life expectancy. However, with the abuse of antimicrobial drugs, many new challenges in the treatment of pathogenic infections have emerged recently. First of all, new pathogens continue to emerge, such as SARS-CoV and SARS-CoV-2, and there is still a lack of effective drugs against these pathogens. Second, the development of drug resistance makes it difficult to treat previously curable infections. According to statistics, an estimated 4.95 million people died from bacterial AMR in 2019, of which 1.27 million died directly [1]. Third, many existing antimicrobials have defects. For example, the antifungal drugs azole, polyene and echinomycin can cause a variety of side effects in the human body. Therefore, it is urgent to find more efficient and safer antimicrobials.

Stilbenoid, characterized by a C_6–C_2–C_6 carbon skeleton, possess a wide range of pharmacological activities. For example, resveratrol has demonstrated anti-inflammatory, antioxidant, hypolipidemic, hypoglycemic and anticancer activities [2–4]. It can also be used as a supplement to prevent cardiovascular system diseases, nervous system diseases and cancer [5]. In addition, the antimicrobial activity of stilbenoids is also of great concern to phytochemists and pharmacologists. Plant-derived stilbenoids are a class of important phytoalexins produced by plants to protect against pathogen infections and toxins. Mattio et al. summarized the sources, chemical structures and the antibacterial mechanism of natural stilbenoids [6]. The role of stilbenoid-abundant extracts in the mitigation of mycotoxins in food and feedstuffs has also attracted great interest [7]. However, there is no review providing a comprehensive overview of the antimicrobial (antibacterial and antifungal) mechanisms of stilbenoids from the perspective of their direct microbicidal activity, antibiofilm activity and/or regulation of the pathogenicity of microbes. Here, this review will emphasize recent achievements in these aspects of stilbenoids.

2. The Family of Stilbenoids

Resveratrol was the first stilbenoid to be discovered. It was originally isolated from the root of *Veratrum grandiflorum* in 1939 by Takaoka [8]. Subsequently, with the introduction of more accurate metabolite identification/isolation techniques and chemical synthesis, more and more complex stilbenoids were identified. These stilbenoids can be divided into three categories according to their chemical structure, including monomers and oligomers (Figure 1).

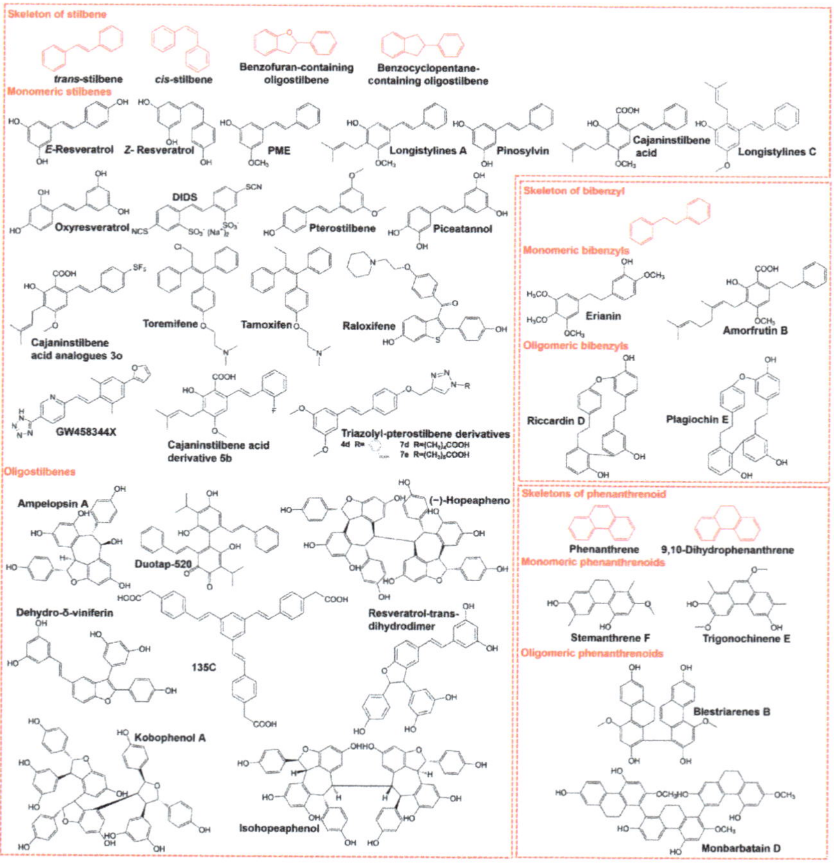

Figure 1. Skeleton characteristics and representatives of three kinds of stilbenoids. Stilbenoids can be divided into three categories according to their chemical structure, including stilbene, bibenzyl and phenanthrenoid.

Monomeric stilbenes, such as resveratrol, piceatannol, pterostilbene and pinosylvin, are characterized by two phenyl rings, joined by an ethylene bridge. Oligomeric stilbenes are a group of compounds polymerized from monomeric stilbene. Resveratrol oligomers, which consist of two to eight or more resveratrol subunits, are the largest group of the oligomeric stilbenes [9]. Hopeaphenol, a tetramer of resveratrol isolated from *Hopea odorata* and *Balanocarpus heimti*, was the first oligomeric stilbenoid to be discovered in 1951 [10]. Subsequently, more and more oligomeric stilbenes from plants were identified. Oligomers can be polymerized from either homogenous or heterogeneous monomeric stilbenes. For example, hopeaphenol and ε-viniferin are polymerized from resveratrol units, while Gnetuhianin Q is a mixed dimer that is polymerized from both isorhapotigenin and resveratrol. Gnetuhainin K is the mixed dimer of gnetol and isorhapotigenin [11]. Those oligomers can be divided into benzofuran oligomers and benzocyclopentane oligomers.

Bibenzyls (e.g., amorfrutin B and erianin, etc.) are comprised of two aromatic rings linked by an ethyl bridge. Monomeric bibenzyls can be polymerized into bisbibenzyls. However, no oligomers polymerized by more than three bibenzyl units have been found.

Phenanthrenoids include the phenanthrenes, 9,10-dihydrophenanthrenes and their oligomers. More than one hundred dimers have been identified [12]. However, only three triphenanthrenes have been reported [13–15].

3. Structural Diversity and Antimicrobial Activity of Stilbenoids

Simple monomeric stilbenes (including resveratrol, piceatannol and pinosylvin) or bibenzyls (including 5-hydroxy-lunularic acid and 3,3′,5-trihydroxybibenzyl), synthesized through the phenylpropanoid pathway, are the precursor of many other stilbenoids in plants. It is generally believed that phenanthrenes and 9,10-dihydrophenanthrenes are formed by the oxidative coupling of the aromatic rings of stilbene and bibenzyl precursors, respectively [16,17]. However, other views on the biosynthesis of phenanthrenoids have been discussed in another review [16]. Subsequently, the diversity of stilbenoids is greatly increased via enzymatic modifications of side groups on the precursors, such as methylation, prenylation, oligomerization, hydroxylation, glycosylation and isomerization. The chemical diversity of stilbenoids provides them with a wide range of antimicrobial activities (Table 1), especially drug-resistant microbes.

3.1. Methylation

Methylation is the most common modification in plants. Methylation alters the chemical properties of hydroxyl stilbenes and therefore affects their antimicrobial activity. For example, rapid absorption and metabolism result in the limited oral bioavailability of resveratrol, which restricts the application of this potentially valuable compound in clinical trials [18,19]. However, pterostilbene, a dimethoxy derivative of resveratrol, shows better bioavailability and stronger antifungal activity [20]. In addition, stilbenes such as pinosylvin, piceatannol and pinostilbene, which contain a varied number of methylated groups, also exhibit higher antimicrobial activity compared to resveratrol [21–24]. Li et al. shows that the chemical affinity balance of substituent groups of stilbenes is key for their antifungal activity [25].

3.2. Isomerization

Structurally, stilbenes can exist as *cis* (*Z*) and *trans* (*E*) isomers, based on the configuration of the ethylene bridge, such as *trans*-resveratrol and *cis*-resveratrol. Generally, the two isomers exhibit different chemical characteristics and thereby different biological activities. The *E*-type configuration is more common in nature. This configuration is not sterically hindered, being therefore more stable. For example, research showed that *trans*-resveratrol exhibited stronger antibacterial effects compared to *cis*-resveratrol [26].

3.3. Prenylation

Prenyl groups can be attached to the skeleton at different positions and in different configurations (e. g., chain or ring-closed prenylation). One isoprenyl (3,3-dimethylallyl) moiety bound to the aromatic ring of stilbenes is the most common form in nature. In addition, prenyl patterns such as 3-methyl-but-1-enyl and geranyl are also often introduced to various positions of the stilbene backbone, along with oxidation, methylation and cyclization moieties, increasing the structural diversity of stilbenoids. Prenylation of compounds improves the binding affinity to receptors and is beneficial to membrane partitioning, consequently enhancing bioactivity [27]. For example, prenylated stilbenoids generally exhibit stronger antibacterial activity with MICs (minimal inhibitory concentration) in the μg/mL range. Improved properties, such as hydrophobicity, charge or molecular geometry, may contribute to their enhanced activity [28]. However, the rule is not absolute. For example, isopentenylated longistylin C shows lower antifungal activity than pinosylvin monomethyl ether (PME) [25]. This is closely related to their mechanism of action.

3.4. Oligomerization

Oligomeric stilbenoids are formed by oxidative coupling between homogenous or heterogeneous monomeric stilbenoids. A series of oligomeric stilbenoids has been identified in recent years, of which some are superior in bioactivity, stability and selectivity compared to the parental monomers [29,30]. For example, tapinarof is a stilbene drug isolated from the gammaproteobacterial *Photorhabdus* genus [31]. Duotap-520, isolated from *Photorhabdus* gammaproteobacteria, is the dimer of tapinarof and exhibits enhanced activity against Gram-positive bacteria (methicillin-resistant *Staphylococcus aureus* and vancomycin-resistant *Enterococcus faecalis*), with no resistance development even after daily non-lethal exposure for a duration of three months [31]. Interestingly, this metabolic dimer of tapinarof has gained antibacterial activity but has lost its anti-inflammatory activity [31]. Another dimer is dehydro-δ-viniferin, which shows enhanced antibacterial activity compared to monomer resveratrol [32].

Table 1. Antimicrobial activities and mechanisms of stilbenoids.

Targets	Compounds	Details	Determination Methods	Refs
Cell membrane	Longistylin A	MRSA, MIC = 1.56 μg/mL; disturbing membrane potential and increasing permeability.	Micro-well dilution.	[33]
	Toremifene	*P. gingivalis* and *S. mutans*, MICs = 12.5–25 μM; disrupting the cell membrane.	Micro-well dilution.	[34]
	Dehydro-δ-viniferin	*L. monocytogenes*, MIC = 2 μg/mL.	Micro-well dilution.	[35]
	Pterostilbene	*F. nucleatum*, MIC = 20 μg/mL utilizing 2-hydroxypropyl-β-cyclodextrin as a solubilizer.	Micro-well dilution.	[36]
	Resveratrol	*S. aureus* ATCC 25923, MIC = 512 μg/mL; *P. aeruginosa* ATCC 27853, MIC > 512 μg/mL.	Micro-well dilution.	[32]
	Cajaninstilbene acid derivative 5b	*S. aureus* ATCC25923, MIC = 4 μg/mL; *S. epidermidis* ATCC12228, MIC = 1 μg/mL; *B. subtilis* ATCC6633, MIC = 0.5 μg/mL; interferring in PG synthesis pathway by targeting PgsA.	Micro-well dilution.	[37,38]
	PME	*A. flavus*, IC50 = 260 μg/mL; binding the phospholipids of cell membrane.	Agar drug plate growth assay.	[25]

Table 1. *Cont.*

Targets	Compounds	Details	Determination Methods	Refs
Cell wall	Duotap-520	MRSAs, MICs = 4 μM; VRE, MIC = 6 μM; binding to lipid II.	Micro-well dilution.	[39]
	GW458344X	Inhibiting MurC activity, IC_{50} = 368 μM; MurD, IC50 = 104 μM; MurE, IC_{50} = 49 μM; MurF, IC_{50} = 59 μM.	Enzyme activity test.	[40]
	135C	*S. aureus*, MICs = 0.12–0.5 μg/mL; targeting cell wall teichoic acids.	Micro-well dilution.	[41]
	Plagiochin E	Inhibiting the activity of chitin synthases.	Enzyme activity test.	[42]
	Tamoxifen	*S. pombe*, MIC = 32 μg/mL; inhibiting Ccr1 NADPH-cytochrome P450 reductase activitie.	Micro-well dilution.	[43]
DNA	Resveratrol-trans-dihydrodimer	*B. cereus*, MIC = 15.0 μM; *L. monocytogenes*, MIC = 125 μM; *S. aureus*, MIC = 62.0 μM; *E. coli*, MIC = 123 μM, upon addition of the efflux pump inhibitor; inhibiting DNA gyrase.	Micro-well dilution.	[44]
	Triazolyl-pterostilbene derivative 4d	MRSAs, MICs = 1.2–2.4 μg/mL, MBCs = 19.5–39 μg/mL; inhibiting the activity of DNA polymerase.	Micro-well dilution.	[45]
	Oxyresveratrol	*C. albicans* ATCC90028, MIC = 5.0 μg/mL; *C. parapsilosis* ATCC22019, MIC = 5 μg/mL; inflicting cleavage on DNA.	Micro-well dilution.	[46]
	Resveratrol	*S. typhimurium*, MIC = 5 μg/mL; inducing DNA disruption.	Micro-well dilution.	[47]
Mitochondria	Resveratrol	*C. albicans* ATCC 90028, MIC = 20 μM; inducing mitochondria-dependent apoptosis.	Micro-well dilution.	[48]
	Plagiochin E	*C. albicans* CA2, MIC = 16 μg/mL; inducing mitochondria-dependent apoptosis.	-	[49,50]
ATPase	Resveratrol and piceatannol	Inhibiting ATPase activity; piceatannol, IC_{50} = 14 μM; resveratrol, IC_{50} = 94 μM.	Enzyme activity test.	[51]
Cell-division	Resveratrol	*E. coli*, MIC = 456 μg/mL; preventing Z-ring formation.	-	[52]
PTS system	Cajaninstilbene acid	Sensitive *Enterococcus* strains and VRE strains, MICs = 0.5–2 μg/mL; inhibiting carbohydrate specific type II transporters of PTS system.	Micro-well dilution; proteomics; q-PCR.	[53]
Calmodulin–calcineurin pathway	Tamoxifen	*S. pombe*, MIC = 35 μg/mL; *Candida* spp. and *C. neoformans*, MICs = 8–64 μg/mL; directly binding to calmodulin.	Agar drug plate growth assay; micro-well dilution	[54–56]

Table 1. *Cont.*

Targets	Compounds	Details	Determination Methods	Refs
Virulence factors	Erianin	*S. aureus* ATCC25904, MIC = 512 μg/mL; inhibiting the activity of SrtA with a IC_{50} = 20.91 ± 2.31 μg/mL.	Micro-well dilution.	[57]
	Resveratrol	Reducing the secretion of α-hemolysin by downregulating saeRS.	Red blood hemolysis assay.	[58,59]
	DIDS	Inhibiting *V. vulnificus* toxicity to HeLa cells at 10–300 μM with no influence on host cell viability and bacterial growth; reducing the expression of *TolCV1*.	-	[60]
	Raloxifene	Enhancing the survival percentage of *C. elegans* infected with *P. aeruginosa* PA14 at 12.5–100 μg/mL; inhibiting pyocyanin production by binding and inhibiting PhzB2.	-	[61]
	(-)-Hopeaphenol	Reducing cell entry and subsequent intracellular growth of bacteria; inhibiting the expression of *YopE* with a IC_{50} = 8.8 μM; inhibiting the activity of YopH with a IC_{50} = 2.9 μM.	A luminescent reporter-gene assay (*YopE*); an enzyme-based YopH assay.	[62,63]
	Hopeaphenol, isohopeaphenol, kobophenol A and ampelopsin A	Reducing the pathogenicity of *P. syringae* pv. tomato DC3000 on leaves; reducing the expression of *hrpA*, *hrpL* and *hopP1* genes without influence on bacterial growth.		[64,65]
Biofilm	Resveratrol	Inhibiting swarming of *P. mirabilis* at 15 μg/mL, and completely inhibited swarming at 60 μg/mL.	-	[66]
		S. typhimurium SL1344, MIC > 512 μg/mL; inhibiting adhesion of *S. typhimurium* to HeLa cells; downregulating the expression of flagella genes.	-	[67]
		P. gingivalis, MICs= 78.12–156.25 μg/mL; reducing the expression fimbriae genes.	-	[68]
		V. cholerae, MIC = 60 μg/mL; antibiofilm at 10–30 μg/mL; inhibiting the activity of AphB.	Crystal violet assay; confocal laser scanning microscopy.	[69]
		S. mutans, MIC = 800 μg/mL; inhibiting biofilm formation at 50–400 μg/mL; reducing the biosynthesis of polysaccharide.	Growth curve assay; crystal violet assay; confocal laser scanning microscopy.	[70,71]
		Inhibiting pyocyanin production of *P. aeruginosa* PAO1 by directly binding LasR.	-	[72]
		Reducing MN production.	-	[73]
	Resveratrol and oxyresveratrol	Antibiofilm at 10 μg/mL and 100 μg/mL; reducing fimbriae production and the swarming motility of UPEC.	Crystal violet assay; confocal laser scanning microscopy.	[74]

Table 1. *Cont.*

Targets	Compounds	Details	Determination Methods	Refs
Biofilm	Trans-stilbene	Reducing biofilm of *S. aureus* ATCC6538 at 50–200 μg/mL; decreasing the expression of intercellular adhesion locus.	Crystal violet assay; confocal laser scanning microscopy.	[58]
	Piceatannol	*S. mutans*, MIC_{50} = 564 ± 38 μM; antibiofilm, IC_{50} = 52 ± 6 μM; inhibiting the activity of GtfB and GtfC.	Crystal violet assay; confocal laser scanning microscopy.	[75]
	Oxyresveratrol	*S. mutans*, MIC = 500 μg/mL; reducing biofilm formation at 62.5–250 μg/mL; suppressing the expression of *gtfB* and *gtfC*.	Micro-well dilution; confocal laser scanning microscopy.	[76]
	Amorfrutin B	Antibiofilm activity against *P. aeruginosa* PAO1 with a biofilm inhibition ratio of 50.3 ± 2.7 at 50 μM; binding the receptors of signal molecule.	Crystal violet assay.	[77]
	Cajaninstilbene acid analogue 3o	Inhibiting biofilm formation of *P. aeruginosa* with inhibition ratio of 49.50 ± 1.35% at 50 μM; suppressing the expression of *lasB* and *pqsA*.	Crystal violet assay.	[78]
	Riccardin D	Inhibiting biofilm formation at 16 μg/mL and 64 μg/mL using central venous catheter (CVC)-associated *C. albicans* biofilms in an infectious rabbit model; reducing the expression of hypha-specific genes.	XTT reduction assay; scanning electron microscopy; confocal laser scanning microscopy.	[79,80]
	Pterostilbene	Inhibiting the formation of *C. albicans* biofilms in vitro at 1–32 μg/mL; downregulating the expression of filamentation-related genes.	XTT reduction assays; confocal laser scanning microscopy; scanning electron microscopy.	[81]
Reversing antibiotic resistance	Resveratrol	Enhancing the efficacy of aminoglycosides against Gram-positive pathogens; inhibiting the activity of ATP synthase.	-	[82]
		Increasing susceptibility of *P. aeruginosa* PAO1 biofilm to aminoglycosides; inhibiting the expression of signaling molecule synthase genes *lasI* and *rhlI*.	-	[83,84]
	Pterostilbene	Restoring the effectiveness of meropenem against NDM-expressing strains; inhibiting NDM-1 hydrolysis activity at 4–32 μg/mL.	-	[85]
	Cajaninstilbene acid	Restoring the susceptibility of polymyxin B to *mcr-1* positive Gram-negative bacteria; inhibiting the enzymatic activity of MCR-1.	-	[86]
	Resveratrol, pterostilbene, and pinosylvin	Increasing sensitivity of *A. butzleri* strains to chloramphenicol, erythromycin and ciprofloxacin by acting as EPIs.	-	[87–89]
	Piceatannol	Increasing sensitivity of *S. aureus* to ciprofloxacin by decreasing PMF.	-	[90]

4. Antimicrobial Mechanisms of Stilbenoids

Stilbenoids exert antimicrobial activities via various modes (Figure 2). The antimicrobial mechanisms mainly include (I) acting on conventional targets, (II) acting on nonconventional targets and (III) reversing antibiotic resistance. Firstly, stilbenoids can inhibit conventional targets, such as the cell membrane, cell wall, DNA, the calmodulin–calcineurin pathway, mitochondria, cell division and the phosphoenolpyruvate (PEP)-dependent phosphotransferase system. Secondly, stilbenoids can function as antibiofilm or antivirulence agents. Stilbenoids inhibit biofilm formation and assist in the eradication of preformed biofilms. Moreover, stilbenoids can reverse drug resistance by inhibiting alternative targets, target-modifying enzymes or antibiotic-modifying enzymes.

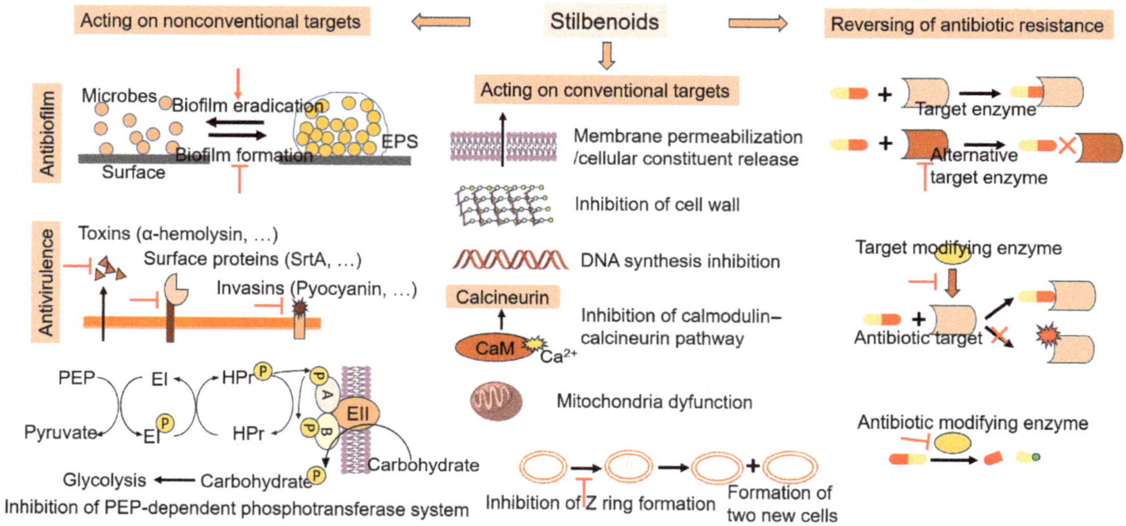

Figure 2. Overview of the antimicrobial mechanisms of stilbenoids. The red arrow represents activation. The red T represents inhibition. The red cross means the reaction is blocked.

4.1. Direct Antimicrobial Mechanisms of Stilbenoids

4.1.1. Targeting the Cell Membrane

The antimicrobial mechanism elucidates the membrane-disruption effects of stilbenoids, resulting in the leakage of cell contents and subsequent cell death. For example, the anti-MRSA (methicillin-resistant *Staphylococcus aureus*) activity of longistylin A, a stilbene isolated from the leaves of *Cajanus cajan* (L.) Millsp, is mediated by disturbing the bacterial membrane potential and increasing permeability [33]. Li et al. showed that PME exerted anti-*Aspergillus flavus* activity by binding the phospholipids, leading to decreased fluidity and increased permeability of the cell membrane [25]. In addition, toremifene (an FDA-approved anticancer agent synthesized in 1981), pterostilbene, dehydro-δ-viniferin (a stilbene dimer chemically semi-synthesized from resveratrol), resveratrol and so on, are all shown to exert antimicrobial activity by disrupting the cell membrane [32,34,36].

Membrane-associated protein PgsA catalyzes glycerolphosphate to replace cytidine monophosphate to produce phosphatidylglycerol phosphate (PG-P). Subsequently, PG-P is dephosphorylated by PgpP to yield phosphatidylglycerol (PG), a major component of cell membrane [91]. Research has shown that a chemically synthesized cajaninstilbene acid-derivative, 5b, interferes in the PG synthesis pathway by targeting PgsA [37].

4.1.2. Targeting the Cell Wall

The bacterial cell wall is a peptidoglycan polymer network that is composed of N-acetylglucosamine (GlcNAc) and N-acetylmuramic acid (MurNAc), with a pentapeptide attached. Firstly, the UDP-MurNAc-pentapeptide is coupled with bactoprenyl-phosphate (lipid I) on the cytosolic side of the cell membrane, followed by the coupling of the GlcNAc sugar by the enzyme MurG to produce lipid II. Next, lipid II, containing the complete peptidoglycan subunit, is translocated to the outside of the cell membrane. The subunits are inserted into the cell wall by penicillin-binding proteins (PBPs) to make the cell wall grow further. Finally, the lipid anchor is returned to the cell for the next round of synthesis [92]. Duotap-520, a natural stilbene dimer isolated from *Photorhabdus*, can bind to lipid II, preventing the role of lipid II [39]. Bacterial peptidoglycan biosynthesis is catalyzed by a series of Mur ligases in the intracellular steps. Research has shown that chemically synthesized aza-stilbene GW458344X is a competitive inhibitor of MurD, an enzyme catalyzing the reaction from UDP-MurNAc-Ala to UDP-MurNAc-dipeptide [40]. Wall teichoic acids (WTAs) are anionic glycopolymers anchored in the cell walls of Gram-positive bacteria and have critical functions in bacterial physiology. The first-committed step of WAT biosynthesis is catalyzed by TagA, a membrane-associated glycosyltransferase [93]. The two-component transporter TagGH is responsible for the exportation of WATs [94]. A mutation in genes (*tagH, tagA* and *tagG*) has conferred drug resistance to *S. aureus* against the chemically synthesized tris-stilbene 135C. Therefore, 135C showed promising activity against Gram-positive bacteria, probably via targeting cell wall teichoic acids [41].

The fungal cell wall, a complex matrix with no mammalian counterpart, plays a role in maintaining cell shape and integrity, and therefore presents an ideal drug target. The cell wall consists of complex components, mainly polysaccharides (glucans, mannans and chitins), proteins and lipids. Studies showed that the macrocyclic bis(bibenzyl) plagiochin E caused damage in cell wall of *C. albicans* by inhibiting chitin synthetase activities [42]. The NADPH-cytochrome P450 reductase Ccr1 plays a role in the cell wall assembly of budding yeast [95]. The deletion of Ccr1 could cause defects in yeast cell wall integrity. Tamoxifen, an estrogen receptor antagonist used for treating breast cancer, has been found to inhibit Ccr1 NADPH-cytochrome P450 reductase activities of fission yeast and *C. albicans* in a dose-dependent manner, thereby causing defects in cell wall integrity [43].

4.1.3. Targeting the DNA

Stilbenoids can inhibit the synthesis of DNA. For example, resveratrol-trans-dihydrodimer, obtained by the oligomerization of resveratrol catalyzed by soybean peroxidase in vitro, has been shown to inhibit DNA synthesis by reducing DNA gyrase activity via blocking the ATP-binding site of the enzyme [44]. Another study demonstrated that the chemically synthesized triazolyl-pterostilbene derivatives (**4d**, **7d** and **7e**) exhibited potent anti-MRSA activity by inhibiting the activity of DNA polymerase [45]. Stilbenoids can also damage DNA. For example, oxyresveratrol inflicted cleavage on DNA by directly binding to the DNA, which in turn led to mitochondria-mediated apoptosis in *C. albicans* [46]. Resveratrol induces DNA disruption via pro-oxidant activity (including increasing ROS and malondialdehyde accumulation and depleting glutathione) against *Salmonella typhimurium* [47].

4.1.4. Targeting the Mitochondria

Mitochondria not only power life via their varied metabolic functions, but also play a central role in apoptotic cell death. Mitochondrial membrane permeabilization upon stimulation usually causes cytochrome c release from the mitochondria and subsequent metacaspase activation, and eventually commits a cell to die [96]. As such, targeting mitochondrial membrane to manipulate cell death holds tremendous antimicrobial potential. Studies have demonstrated that resveratrol caused the loss of mitochondrial membrane potential ($\Delta\Psi m$), leading to metacaspase activation, cytochrome c release and eventually apoptosis of *C. albicans* [48]. Another study showed that plagiochin E could induce mitochondria-dependent apoptosis in yeast, mainly including the activation of F(0)F(1)-ATPase, inhibition of dehydrogenase and release of cytochrome c, leading to metacaspase activation [49,50].

4.1.5. Actions on other Conventional Targets

The cell membrane and wall, the DNA and mitochondria mentioned above are all conventional antimicrobial targets. In addition, stilbenoids can also act on other conventional targets. For example, resveratrol and piceatannol inhibited both the ATPase activity and ATP synthesis of *Escherichia coli* reversibly [51]. Resveratrol can also exert antibacterial activity by downregulating *FtsZ* expression and preventing Z-ring formation. FtsZ can polymerize to form a dynamic ring (Z-ring) to promote cell division in prokaryotes. The Z-ring plays a role in recruiting division-related proteins and directing septal peptidoglycan synthesis to initiate division [52,97].

The bacterial phosphoenolpyruvate (PEP)-dependent phosphotransferase system (PTS) consists of the coupled carbohydrate-specific transporters (called "EIIs") and two general components, EI and HPr, encoded by *ptsI* and *ptsH*, respectively [98]. The PTS catalyzes the coupled phosphorylation of carbohydrates and their transport into cells. Firstly, the phosphoryl groups derived from PEP are transferred from the EI to an EII with the mediation of HPr. Finally, the phosphoryl groups are transferred to a carbohydrate substrate by its cognate EII [53]. Given that this system plays an important role in catalytic transport and has extensive regulatory functions, it presents an important drug target for the development of antibacterial agents [99]. Cajaninstilbene acid has been shown to inhibit vancomycin-resistant *Enterococcus* by inhibiting the carbohydrate-specific type II transporters of the PTS system, as confirmed by proteomics and fluorescence quantitative PCR [53].

Tamoxifen is a known inhibitor of mammalian calmodulin. Recent studies have found that tamoxifen could exert antifungal activity through targeting the calmodulin–calcineurin pathway. Evidence suggests that tamoxifen could induce a consistent phenotype in *C. albicans*, resembling yeasts that have lost the calmodulin function (e. g., disrupting cell integrity, blocking new bud emergence, interfering with the polarization of the actin cytoskeleton and inhibiting germ tube formation). Moreover, the increased expression of calmodulin inhibited the antifungal activity of tamoxifen. The strains with mutations in calmodulin are hypersensitive to tamoxifen [54,55]. MYO2 is a calmodulin-binding protein involved in various cellular polarized growth processes [100]. Studies demonstrated that tamoxifen can interfere with the interaction between Myo2p and calmodulin in yeast [55]. Another study also showed that toremifene and tamoxifen exert the anti-cryptococcal activity by directly binding to calmodulin, leading to suppressed calmodulin-mediated calcineurin activation and nuclear localization of the transcription factor Crz1/SP-1 [56,101].

4.2. *Stilbenoids Targeting Virulence Factors*

With the continuous emergence of multidrug-resistant microbes, conventional antibiotics are becoming increasingly ineffective at treating microbial infections. Virulence determines the pathogenicity of pathogens. Pathogens achieve host colonization, tissue damage and immune evasion/modulation with the help of virulence factors. In recent years, there has been a growing interest in disrupting the microbial virulence mechanisms that disarm pathogenic microbes rather than killing them using anti-infective/anti-virulence drugs. Anti-virulence agents aim to inhibit the virulence factors (adhesins, invasins, enzymes and toxins) or other surface proteins that allow the pathogens to survive in adverse conditions, which, in turn, slows the infection rate [102]. It has been greatly encouraged to research and develop agents for clinical use that specifically target the production of virulence factors, without inhibiting planktonic cell growth, to control infection in the resistance era. Firstly, an anti-virulence treatment strategy places less selective pressure on microbes to evolve novel antibiotic resistance mechanisms [103]. Secondly, anti-virulence agents can, in theory, distinguish between the endogenous microbiome and infectious pathogens, which is something that conventional antibiotics do not have.

Surface proteins can assist the bacteria to adhere to the surface of host organ tissues and evade the host's immune defense with the assistance of transpeptidase [104,105]. For example, the deletion of sortase A (SrtA), an important transpeptidase in *S. aureus*, could downregulate the level of surface proteins and thereby relieve infection symptoms of *S. aureus* in mouse models, without any influence on the bacterial growth [106]. Erianin, a bibenzyl isolated from *Dendrobium chrysotoxum* [107], could inhibit the activity of SrtA at subminimum inhibitory concentrations by binding to SrtA, leading to the reduced adhesion of *S. aureus* to fibrinogen without influencing bacterial growth and thereby improving survival in mice infected with *S. aureus* [57].

The toxins have been implicated in the pathogenesis, such as hemolysins, leukocidins, superantigens and surface proteins. For example, α-hemolysin (Hla) is secreted by *S. aureus* and can lyse erythrocyte by directly binding to the cell membrane, forming perforations. Studies have shown that resveratrol and trans-stilbene can markedly inhibit the hemolysis of *S. aureus* at subinhibitory concentrations by repressing the expression of the *hla* gene, thereby attenuating *S. aureus* virulence in the nematode *Caenorhabditis elegans* [58]. The two-component system SaeRS plays a crucial role in the production of virulence factors [108]. Studies showed that resveratrol could reduce the secretion of α-hemolysin by downregulating *saeRS* [59]. Moreover, the toxin RtxA1 of *Vibrio vulnificus* can cause membrane permeabilization upon host contact. It has been shown that although 4,4′-diisothiocyanatostilbene-2,2′-disulfonic acid disodium salt hydrate (DIDS) could not exert antibacterial activity against *V. vulnificus*, it significantly inhibited the secretion of RtxA1 by reducing the expression of *TolCV1* [60]. The outer membrane portal TolC (designated TolCV1 and TolCV2 in *V. vulnificus*), coupled with tripartite efflux pumps, is crucial for the export of RtxA1 [109].

Pyocyanin, produced by *Pseudomonas aeruginosa*, can assist bacteria in escaping the host's immune system and successfully establishing infection by preventing T-cells from effectively responding against *P. aeruginosa* and by inhibiting the activity of monocytes and macrophages, thus reducing the production of cytokines [110]. Therefore, pyocyanin biosynthesis is an important target for novel antimicrobial drug discovery. Pyocyanin synthesis is catalyzed by a series of enzymes encoded by the homologous *phzA1B1C1D1E1F1G1* and *phzA2B2C2D2E2F2G2* operons. The estrogen receptor modulator, raloxifene, was found to inhibit pyocyanin production by binding and inhibiting PhzB2, and thereby attenuating *P. aeruginosa* virulence in a *Caenorhabditis elegans* model [61].

The type III secretion system (T3SS) is a complex "molecular syringe" consisting of a base embedded within the bacterial membrane, a needle filament that connects the bacteria to the host cell, a tip that functions as a platform for the assembly of the translocon, and a transposon that spans the host cell membrane [111]. The T3SS functions to inject effectors of bacteria into host cells, leading to damaged cell structure and compromised immune response and eventually assisting in bacterial infection. The T3SS is an attractive drug target for developing novel antibacterial agents, since it is crucial in virulence and evolutionarily conserved in Gram-negative pathogens. In addition, the endogenous gut microflora lacks the T3SS and therefore is not likely to be affected by anti-T3SS agents. (−)-Hopeaphenol, a natural tetramer of resveratrol, was found to reduce cell entry and subsequent intracellular growth of bacteria by inhibiting the translocation of effector protein YopE from *Y. pseudotuberculosis* and ExoS from *P. aeruginosa* into HeLa cells. (−)-Hopeaphenol could inhibit the expression and secretion of the translocator protein YopD of T3SS [62,63]. The type III secretion system (also called the Hrp secretion system) is encoded by *hrp* (which stands for hypersensitive reaction and pathogenicity) genes in plant pathogenic bacteria. HrpL can activate *hrp* and *avr* gene expression by recognizing consensus motifs in the promotors of these genes [112]. HrpA encodes components of the type III secretion system [113]. Research showed that resveratrol-derived oligomers (hopeaphenol, isohopeaphenol, kobophenol A and ampelopsin A), isolated from grapevine roots, could reduce the pathogenicity of *Pseudomonas syringae* pv. *tomato* DC3000 on tomato leaves by reducing the expression of *hrpA*, *hrpL* and *hopP1* genes without influencing bacterial growth [64,65].

4.3. Stilbenoids Targeting Biofilms

It is estimated that 40–80% of bacteria and archaea on earth reside in biofilms [114]. A biofilm is a macrocolony of microorganisms that are attached to a surface, and a key factor that cause chronic infections (e.g., cystic fibrosis, dental plaque, chronic wounds, urinary infection, prosthetic joint infection, chronic otitis media, middle ear cholesteatoma, chronic adenoiditis, chronic prostatitis and cardiac valve infection) [115,116]. According to the National Institutes of Health (NIH), more than 80% of human microbial infections are related to biofilms [117]. Biofilms offer microorganisms strong competitive advantages under various environmental challenges [118]. For example, it has been estimated that bacteria in biofilms are 10,000 times more resistant to antibiotics than plankton bacteria [119,120]. Several reasons could account for the antibiotic tolerance of biofilms. Firstly, it is difficult for antibiotics to invade into the deeper layers of biofilm due to the protection of the extracellular polymeric substance (EPS) matrix [121]. In addition, the biofilm increases the transfer rate of horizontal genes, which are responsible for antibiotic resistance [122]. Moreover, the microbes in biofilm have a relatively low rate of cell growth and reproduction. Reduced metabolic activity will lead to decreased antibiotic effectiveness [123]. Moreover, biofilms can also colonize medical devices (for example, catheters and implants). Thus, biofilm is an attractive target for the development of new antibiotics to treat aggressive pathogenic infections.

Biofilm formation can be described in five progressive stages: initial reversible attachment; irreversible attachment; the first layer formation; formation of mushroom- or tower-like structures; and dispersion and reattachment [115]. Diverse studies have shown that stilbenoids can target the major events in the formation processes to prevent biofilm formation and induce mature biofilm clearance.

4.3.1. Prevention of Initial Attachment

It is essential to move close enough to a physical surface for the formation of microbial biofilms. Studies have shown that flagella and fimbriae are crucial for bacterial attachment to surfaces. Stilbenoids can inhibit microbial attachment by reducing these factors. For example, flagella-mediated motility is essential for overcoming the electrostatic repulsion of microbial cells and surfaces, while fimbriae-mediated twitching motility assists bacterial cells to aggregate to form microcolonies [124–126]. Resveratrol inhibited flagellin production and the swarming of *Proteus mirabilis*, thereby inhibiting their ability to invade human urothelial cells [66]. Another study showed that resveratrol downregulated the expression of flagella genes, leading to reduced swimming motility and, therefore, compromised adhesion of *Salmonella typhimurium* to HeLa cells [67].

Fimbriae, encoded by the *fimA* gene, can be classified into six genotypes (type I, Ib, II, III, IV and V). Studies have found that resveratrol could suppress the expression of *fimA* and *xadA* (which encode an afimbrial adhesion), thereby reducing cell surface adhesion of the Gram-negative bacterium *Xylella fastidiosa*, which can cause Pierce's disease in grapevines [127]. Resveratrol dose-dependently prevented the biofilm formation of *P. gingivalis* by reducing the expression of genes which encode fimbriae (type II and IV) [68]. Resveratrol and oxyresveratrol reduced fimbriae production and the swarming motility of uropathogenic *Escherichia coli* (UPEC), leading to reduced biofilm formation and hemagglutinating ability, reducing the defense of UPEC against human whole blood [74]. In addition, type IV toxin-coregulated pilus (TCP) is an important attachment factor [128]. AphB controls the expression of TcpP. TcpP activates TCP by controlling the expression of ToxT, a direct activator of TCP [129]. Resveratrol can suppress biofilm formation of *V. cholerae* by inhibiting the activity of AphB [69].

The slime substance polysaccharide intercellular adhesin (PIA) is critical for adhesion on hydrophilic surfaces. It can be synthesized by products of the intercellular adhesion (ica) locus [130]. Studies showed that trans-stilbene could reduce the biofilm by decreasing the expression of the intercellular adhesion locus (*icaA* and *icaD*) in *S. aureus* [58].

4.3.2. Targeting Biofilm Maturation

After initial attachment and aggregation, the pathogens will further become embedded and secured in the matrix. This process is called biofilm maturation. Biofilm maturation includes cell–cell interaction, hyphal differentiation (fungi) and extracellular matrix production [131]. The development of the microcolony into a mature biofilm is regulated by various regulatory systems, such as the quorum-sensing (QS) system, the two-component regulatory systems and the type III secretion system [132]. Stilbenoids can target extracellular polymeric substance (EPS), signaling regulatory systems, as well as other surface properties of pathogens that are required for the development of the biofilm.

Microbes in biofilms are wrapped in a self-produced EPS matrix. The components of the EPS mainly include extracellular DNA, lipids, polysaccharides and proteins [133]. Among them, polysaccharides can provide a number of benefits to the cells in the biofilm, such as assisting in intercellular adhesion and protection and providing structural support [134]. Studies found that resveratrol could inhibit the biofilm formation of *Streptococcus mutans* by reducing the biosynthesis of polysaccharides [70,71]. Water-insoluble glucans form the structural scaffold of biofilms. Extracellular glucosyltransferases GtfB and GtfC are involved in the biosynthesis of water-insoluble glucans. Studies showed that piceatannol could inhibit the activity of GtfB and GtfC, leading to the compromised biofilm formation of *S. mutans* in a dose-dependent manner, both in vitro and in vivo [75]. Another study showed that oxyresveratrol reduced *S. mutans* biofilm formation by suppressing the expression of *gtfB* and *gtfC* [76,135].

Acetic acid, as a signaling molecule, can stimulate bacterial biofilm formation [136]. *Streptococcus mutans* is a major inducer of dental caries. An effective way to treat this disease is to replace *S. mutans* by *Lactobacillus casei* within the dental plaque. Studies have found that oxyresveratrol could promote *L. casei* to produce acetic acid, which inhibited the biofilm formation of *S. mutans* and therefore facilitated *L. casei* to compete with *S. mutans* [76,135].

Quorum sensing (QS) is a kind of communication between microbial cells. Bacteria can secrete signaling molecules into their surroundings. Signaling molecules can be imported into cells after the bacterial population density reaches a certain threshold, directly binding to the receptor LasR to regulate biofilm formation or the production of virulence factors [83,137]. *Pseudomonas aeruginosa* has four QS systems: *las*, *rhl*, *pqs* and *iqs*. *pqs* QS regulates the biosynthesis of pyocyanin [138]. Pyocyanin can facilitate extracellular electron transfer in *P. aeruginosa* biofilms, therefore supporting the growth of biofilms [139,140]. Research has found that amorfrutin B, a bibenzyl isolated from *Amorpha fruticose* [141], can reduce the pyocyanin production and biofilm formation of *P. aeruginosa* by competitively binding the receptors of the signal molecules ODdHL and PQS, thereby inhibiting the expression of downstream genes such as *lasB*, *rhlA* and *pqsA* [77]. Cajaninstilbene acid analogue 3o also inhibited the QS systems by suppressing the expression of *lasB* and *pqsA*, leading to the inhibition of biofilm formation [78]. Moreover, resveratrol could inhibit the biofilm formation and pyocyanin production of *P. aeruginosa* PAO1 by directly binding LasR [72].

Hyphal differentiation is required for robust biofilm formation. Research showed that although mutants of hyphal formation can develop into biofilm, these biofilms are loose, rather than stable, compared with wild-type biofilms [131]. The Ras/cAMP/Efg1 pathway plays a key role in fungal morphological transitions by regulating the expression of hyphae-specific genes, such as *HGC1* (functioning in the polarized growth of hyphae), *ALS3* (encoding a cell wall surface protein related to adhesion), *HWP1* (encoding a cell wall mannose protein, which is essential for the normal growth of the mycelium) and *ECE1* (encoding a membrane protein, which is related to the extension of hyphal). Efg1 is a transcription factor, regulating the expression of hyphae-specific genes [142]. Stilbenoids can target the morphological transition of *C. albicans*, thereby inhibiting biofilm formation. For example, the macrocyclic bisbibenzyl riccardin D has been reported to inhibit the biofilm formation of *C. albicans* both in vitro and in vivo by reducing the expression of hypha-specific genes (*ALS1*, *ALS3*, *ECE1*, *EFG1*, *HWP1* and *CDC35*) and therefore hypha formation [79,80]. Another study showed that pterostilbene could inhibit biofilm formation and destroy mature biofilms both in vitro and in vivo. Pterostilbene could downregulate the expression of filamentation-related genes in the Ras/cyclic AMP (cAMP) pathway, including *ECE1*, *ALS3*, *HWP1*, *HGC1* and *RAS1*. The addition of exogenous cAMP reverted the defect in pterostilbene-induced filamentous growth [81].

4.3.3. Disarming Pathogens within the Biofilm

It is clear that biofilms continuously release planktonic cells or small clusters of pathogens by dispersal, seeding new sites of infection and ensuring further replication of biofilms [143]. Stilbenoids can interfere with this process by targeting the enzymes involved in it, thereby eradicating the preformed biofilms. For example, *S. aureus* could escape neutrophil extracellular traps by cleaving their DNA-backbone through secreting micrococcal nuclease (MN). MN also inhibits biofilm development and adhesion by cleaving the extracellular DNA. Reports have shown that resveratrol could reduce MN production, which assists the host immune system to clear the biofilm of *S. aureus* in clinical situations [73].

4.4. *Reversing Antibiotic Resistance*

Antimicrobial resistance is a global health problem. Microbes use various mechanisms to resist the action of antimicrobials [144]. Some are intrinsic mechanisms, by which bacteria can use genes they already possess to avoid antibiotic exposure, and some are acquired mechanisms, by which bacteria evolve new genetic materials under antibiotic exposure in order to survive. Inhibiting resistance mechanisms is a potential approach to combat drug resistance.

Studies have shown that ATP synthase is an intrinsic resistance factor of bacteria. For example, the inactivation of genes encoding the subunits of ATP synthase could significantly increase the efficacy of gentamicin against *S. aureus* [145]. Another study showed that the deletion of the ATP synthase subunit *atpG* increased the efficacy of aminoglycosides against *E. coli* [146]. Resveratrol could enhance the efficacy of aminoglycosides against various Gram-positive pathogens. Mechanism research has revealed that resveratrol could inhibit the activity of ATP synthase [82].

Biofilm reduces antibiotic permeability due to the protection of the EPS matrix. For example, the minimum biofilm eradication concentrations of clinical *Acinetobacter baumannii* isolates were 44, 407 and 364 times higher than the minimum bactericidal concentrations for colistin, ciprofloxacin and imipenem, respectively [147]. Considering the roles of biofilm in antimicrobial resistance, inhibiting the formation of biofilm theoretically enhances the effect of antibiotics. *P. aeruginosa* infection is a serious disease in patients with cystic fibrosis and is difficult to treat due to the persistence of biofilms. A study has shown that the biofilms of *P. aeruginosa* PAO1 were more susceptible to aminoglycosides in the presence of resveratrol. Mechanism research has revealed that resveratrol could function as a QS inhibitor. QS deficiency leads to thin and less developed biofilms [83]. Resveratrol could inhibit the expression of the signaling molecule synthase genes *lasI* and *rhlI* in *P. aeruginosa* PAO1 biofilms [84].

Antimicrobial resistance can be conferred by drug-modifying enzymes. The enzymatic degradation of antibiotics involves the hydrolysis of their functional groups, thereby rendering them ineffective. For example, the New Delhi metallo-β-lactamase-1 (NDM-1) can hydrolyze almost all β-lactam antibiotics. Liu et al. discovered that pterostilbene could significantly inhibit the NDM-1 hydrolysis activity by binding to the catalytic pocket of NDM-1, thereby preventing the substrate from binding to this protein, effectively restoring the effectiveness of meropenem against NDM-expressing strains and protecting mice from pneumonia caused by *Klebsiella pneumoniae* [85].

Bacteria can prevent antibiotics from reaching to corresponding target by enzymatic modification of the target protein. For example, polymyxin disrupts membrane integrity by binding to the lipid A moiety of lipopolysaccharides (LPS), ultimately causing cell death. Polymyxin is the last line of defense against Gram-negative bacterial infections.

However, the plasmid-borne mobilized colistin resistance gene *mcr-1* (encoding phosphoethanolamine transferase) significantly decreases the efficiency of polymyxin against Gram-negative bacteria. *mcr-1* is widely spread throughout the world and significantly threatens the usefulness of polymyxin. It can append phosphoethanolamine to a phosphate on lipid A, reducing the electrostatic interaction between LPS and polymyxin, thereby increasing bacteria resistance to polymyxin [148]. Recent studies have demonstrated that Cajaninstilbene acid from *Cajanus cajan* (L.) Millsp can significantly restore the susceptibility of polymyxin B to *mcr-1* positive Gram-negative bacteria through inhibiting the enzymatic activity of MCR-1 via binding the active center [86].

Active efflux is mediated by the upregulation of efflux pumps that extrude antibiotics from bacteria to reduce drug concentrations in the cell to non-cytotoxic levels. Recently, efflux pump inhibitors (EPIs) that can reverse antibiotic resistance have received great attention. The AcrAB-TolC multidrug efflux pump, consisting of the inner membrane protein AcrB and the outer membrane protein TolC, linked by AcrA, can transport various toxic compounds out of bacteria, contributing to drug resistance [149,150]. Research has shown that resveratrol could significantly inhibit *tolC*-promoter activity and decrease *tolC* expression [87]. Moreover, stilbenes (resveratrol, pterostilbene and pinosylvin) enhance the susceptibility of *A. butzleri* strains to chloramphenicol, erythromycin and ciprofloxacin by acting as EPIs [88,89]. Proton motive force (PMF) is essential for efflux pumps to export drugs [151]. Piceatannol could decrease PMF of *S. aureus*, particularly the $\Delta\psi$ component, leading to increased sensitivity of bacteria to ciprofloxacin [90].

5. Conclusions

The heavy use of antibiotics to treat pathogenic infections in humans and animals, as well as in agriculture has promoted the development of drug resistance in microbial populations. Presently, we are entering an era of ever-increasing resistance, in which the treatment of pathogen infection through the use of some conventional antibiotics can no longer be taken for granted, and the need for new antibiotics has never been more urgent [152]. The development of effective therapies against resistant microorganisms and emerging pathogenic microorganisms is a public health priority. The chemical structure of stilbenoids is new in the field of antimicrobial drugs and has received great attention. Stilbenoids demonstrate a diverse range of functions, including direct microbicidal activity, a wide range of effects on biofilms, virulence factors and the reversal of drug resistance, highlighting the considerable possibility of developing stilbenoids as a potential antimicrobial agent. Drug repurposing has emerged as a novel paradigm to find antimicrobial agents. Several stilbenoids with antimicrobial activities have been approved for clinical use, such as toremifene, raloxifene and resveratrol, which are easier to obtain approval for in the treatment of pathogenic infections. However, the practical application of stilbenoids needs to consider challenges such as toxicity, solubility, stability or other factors in advance. Some of these problems could be addressed by chemical modification or by developing delivery devices such as nanocapsules, liposomes and nano-carrier systems [153]. We expect major breakthroughs and revolutionary technologies in the research on stilbenoids. This will serve as a new driving force for improving people's health.

Author Contributions: Writing—original draft preparation, X.L.; writing—review and editing, X.L., Y.L. and B.X.; financial supporting, X.L. and S.Q. All authors have read and agreed to the published version of the manuscript.

Funding: This work was supported by the Basic and Applied Basic Research Youth Fund Project of Guangdong Province, China (Grant No. 2022A1515110442), the Science and Technology Planning Project of Guangzhou (Grant No. 2023A04J0716), and the National Natural Science Foundation of China (NSFC) (No. 82073733).

Conflicts of Interest: The authors declare no conflicts of interest.

References

1. Murray, C.J.; Ikuta, K.S.; Sharara, F.; Swetschinski, L.; Aguilar, G.R.; Gray, A.; Han, C. Global burden of bacterial antimicrobial resistance in 2019: A systematic analysis. *Lancet* **2022**, *399*, 629–655. [CrossRef]
2. Meng, T.; Xiao, D.; Muhammed, A.; Deng, J.; Chen, L.; He, J. Anti-Inflammatory Action and Mechanisms of Resveratrol. *Molecules* **2021**, *26*, 229. [CrossRef] [PubMed]
3. Szkudelski, T.; Szkudelska, K. Resveratrol and diabetes: From animal to human studies. *Biochim. Biophys. Acta BBA Mol. Basis Dis.* **2015**, *1852*, 1145–1154. [CrossRef] [PubMed]
4. Szczepańska, P.; Rychlicka, M.; Groborz, S.; Kruszyńska, A.; Ledesma-Amaro, R.; Rapak, A.; Gliszczyńska, A.; Lazar, Z. Studies on the Anticancer and Antioxidant Activities of Resveratrol and Long-Chain Fatty Acid Esters. *Int. J. Mol. Sci.* **2023**, *24*, 7167. [CrossRef]
5. Bonnefont-Rousselot, D. Resveratrol and Cardiovascular Diseases. *Nutrients* **2016**, *8*, 250. [CrossRef]
6. Mattio, L.M.; Catinella, G.; Dallavalle, S.; Pinto, A. Stilbenoids: A Natural Arsenal against Bacterial Pathogens. *Antibiotics* **2020**, *9*, 336. [CrossRef] [PubMed]
7. Tran, T.M.; Atanasova, V.; Tardif, C.; Richard-Forget, F. Stilbenoids as Promising Natural Product-Based Solutions in a Race against Mycotoxigenic Fungi: A Comprehensive Review. *J. Agric. Food Chem.* **2023**, *71*, 5075–5092. [CrossRef]
8. Takaoka, M. Resveratrol, a new phenolic compound, from *Veratrum grandiflorum*. *Chem. Soc. Jpn.* **1939**, *60*, 1090–1100.
9. Xue, Y.Q.; Di, J.M.; Luo, Y.; Cheng, K.J.; Wei, X.; Shi, Z. Resveratrol oligomers for the prevention and treatment of cancers. *Oxid. Med. Cell. Longev.* **2014**, *2014*, 765832. [CrossRef]
10. Lin, M.; Yao, C.-S. Natural Oligostilbenes. In *Studies in Natural Products Chemistry*; Attaur, R., Ed.; Elsevier: Amsterdam, The Netherlands, 2006; Volume 33, pp. 601–644.
11. Wang, Y.H.; Huang, K.S.; Lin, M. Four new stilbene dimers from the lianas of *Gnetum hainanense*. *J. Asian Nat. Prod. Res.* **2001**, *3*, 169–176. [CrossRef]
12. De Natale, A.; Pollio, A.; De Marco, A.; Luongo, G.; Di Fabio, G.; Zarrelli, A. Phenanthrene Dimers: Promising Source of Biologically Active Molecules. *Curr. Top. Med. Chem.* **2022**, *22*, 939–956. [CrossRef]
13. Xu, D.L.; Pan, Y.C.; Li, L.; ShangGuan, Y.N.; Zhang, S.B.; Liu, G.Y.; Cheng, L.; Xiao, S.J. Chemical constituents of *Bletilla striata*. *J. Asian Nat. Prod. Res.* **2019**, *21*, 1184–1189. [CrossRef]
14. Xue, Z.; Li, S.; Wang, S.; Wang, Y.; Yang, Y.; Shi, J.; He, L. Mono-, Bi-, and triphenanthrenes from the tubers of *Cremastra appendiculata*. *J. Nat. Prod.* **2006**, *69*, 907–913. [CrossRef]
15. Yang, M.; Cai, L.; Tai, Z.; Zeng, X.; Ding, Z. Four new phenanthrenes from *Monomeria barbata* Lindl. *Fitoterapia* **2010**, *81*, 992–997. [CrossRef]
16. Tóth, B.; Hohmann, J.; Vasas, A. Phenanthrenes: A Promising Group of Plant Secondary Metabolites. *J. Nat. Prod.* **2018**, *81*, 661–678. [CrossRef]
17. Preisig-Müller, R.; Gnau, P.; Kindl, H. The inducible 9,10-dihydrophenanthrene pathway: Characterization and expression of bibenzyl synthase and S-adenosylhomocysteine hydrolase. *Arch. Biochem. Biophys.* **1995**, *317*, 201–207. [CrossRef]
18. Wenzel, E.; Somoza, V. Metabolism and bioavailability of trans-resveratrol. *Mol. Nutr. Food Res.* **2005**, *49*, 472–481. [CrossRef]
19. Gambini, J.; Inglés, M.; Olaso, G.; Lopez-Grueso, R.; Bonet-Costa, V.; Gimeno-Mallench, L.; Mas-Bargues, C.; Abdelaziz, K.M.; Gomez-Cabrera, M.C.; Vina, J.; et al. Properties of Resveratrol: In Vitro and In Vivo Studies about Metabolism, Bioavailability, and Biological Effects in Animal Models and Humans. *Oxid. Med. Cell. Longev.* **2015**, *2015*, 837042. [CrossRef]
20. Jeandet, P.; Douillet-Breuil, A.C.; Bessis, R.; Debord, S.; Sbaghi, M.; Adrian, M. Phytoalexins from the Vitaceae: Biosynthesis, phytoalexin gene expression in transgenic plants, antifungal activity, and metabolism. *J. Agric. Food Chem.* **2002**, *50*, 2731–2741. [CrossRef]
21. Lee, S.K.; Lee, H.j.; Min, H.y.; Park, E.j.; Lee, K.m.; Ahn, Y.h.; Cho, Y.j.; Pyee, J.H. Antibacterial and antifungal activity of pinosylvin, a constituent of pine. *Fitoterapia* **2005**, *76*, 258–260. [CrossRef]
22. Zakova, T.; Rondevaldova, J.; Bernardos, A.; Landa, P.; Kokoska, L. The relationship between structure and in vitro antistaphylococcal effect of plant-derived stilbenes. *Acta Microbiol. Immunol. Hung.* **2018**, *65*, 467–476. [CrossRef]
23. Pastorkova, E.; Zakova, T.; Landa, P.; Novakova, J.; Vadlejch, J.; Kokoska, L. Growth inhibitory effect of grape phenolics against wine spoilage yeasts and acetic acid bacteria. *Int. J. Food Microbiol.* **2012**, *161*, 209–213. [CrossRef]
24. Yang, S.C.; Tseng, C.H.; Wang, P.W.; Lu, P.L.; Weng, Y.H.; Yen, F.L.; Fang, J.Y. Pterostilbene, a Methoxylated Resveratrol Derivative, Efficiently Eradicates Planktonic, Biofilm, and Intracellular MRSA by Topical Application. *Front. Microbiol.* **2017**, *8*, 1103. [CrossRef]
25. Li, X.; Yao, L.; Xiong, B.; Wu, Y.; Chen, S.; Xu, Z.; Qiu, S.X. Inhibitory Mechanism of Pinosylvin Monomethyl Ether against *Aspergillus flavus*. *J. Agric. Food Chem.* **2022**, *70*, 15840–15847. [CrossRef]
26. Kukrić, Z.; Topalic-Trivunovic, N.L. Antibacterial activity of cis- and trans-resveratrol isolated from *Polygonum cuspidatum* rhizome. *Acta Period. Technol.* **2006**, *37*, 131–136. [CrossRef]
27. Brents, L.K.; Medina-Bolivar, F.; Seely, K.A.; Nair, V.; Bratton, S.M.; Ñopo-Olazabal, L.; Patel, R.Y.; Liu, H.; Doerksen, R.J.; Prather, P.L.; et al. Natural prenylated resveratrol analogs arachidin-1 and -3 demonstrate improved glucuronidation profiles and have affinity for cannabinoid receptors. *Xenobiotica* **2012**, *42*, 139–156. [CrossRef]

28. Araya-Cloutier, C.; den Besten, H.M.; Aisyah, S.; Gruppen, H.; Vincken, J.P. The position of prenylation of isoflavonoids and stilbenoids from legumes (Fabaceae) modulates the antimicrobial activity against Gram positive pathogens. *Food Chem.* **2017**, *226*, 193–201. [CrossRef]
29. Yuan, S.W.; Chen, S.H.; Guo, H.; Chen, L.T.; Shen, H.J.; Liu, L.; Gao, Z.Z. Elucidation of the Complete Biosynthetic Pathway of Phomoxanthone A and Identification of a Para-Para Selective Phenol Coupling Dimerase. *Org. Lett.* **2022**, *24*, 3069–3074. [CrossRef]
30. Gabaston, J.; Cantos-Villar, E.; Biais, B.; Waffo-Teguo, P.; Renouf, E.; Corio-Costet, M.F.; Richard, T.; Mérillon, J. Stilbenes from *Vitis vinifera* L. Waste: A Sustainable Tool for Controlling *Plasmopara viticola*. *J. Agric. Food Chem.* **2017**, *65*, 2711–2718. [CrossRef]
31. Park, H.B.; Goddard, T.N.; Oh, J.; Patel, J.; Wei, Z.; Perez, C.E.; Mercado, B.Q.; Wang, R.; Wyche, T.P.; Piizzi, G.; et al. Bacterial Autoimmune Drug Metabolism Transforms an Immunomodulator into Structurally and Functionally Divergent Antibiotics. *Angew. Chem. Int. Ed.* **2020**, *59*, 7871–7880. [CrossRef]
32. Mattio, L.M.; Dallavalle, S.; Musso, L.; Filardi, R.; Franzetti, L.; Pellegrino, L.; D'Incecco, P.; Mora, D.; Pinto, A.; Arioli, S. Antimicrobial activity of resveratrol-derived monomers and dimers against foodborne pathogens. *Sci. Rep.* **2019**, *9*, 19525. [CrossRef]
33. Wu, J.; Li, B.; Xiao, W.; Hu, J.; Xie, J.; Yuan, J.; Wang, L. Longistylin A, a natural stilbene isolated from the leaves of *Cajanus cajan*, exhibits significant anti-MRSA activity. *Int. J. Antimicrob. Agents* **2020**, *55*, 105821. [CrossRef]
34. Gerits, E.; Defraine, V.; Vandamme, K.; De Cremer, K.; De Brucker, K.; Thevissen, K.; Cammue, B.P.; Beullens, S.; Fauvart, M.; Verstraeten, N.; et al. Repurposing Toremifene for Treatment of Oral Bacterial Infections. *Antimicrob. Agents Chemother.* **2017**, *61*, e01846-16. [CrossRef]
35. Catinella, G.; Mattio, L.M.; Musso, L.; Arioli, S.; Mora, D.; Beretta, G.I.; Zaffaroni, N.; Pinto, A.; Dallavalle, S. Structural Requirements of Benzofuran Derivatives Dehydro-δ- and Dehydro-ε-viniferin for Antimicrobial Activity Against the Foodborne Pathogen *Listeria monocytogenes*. *Int. J. Mol. Sci.* **2020**, *21*, 2168. [CrossRef]
36. Lim, Y.R.I.; Preshaw, P.M.; Lim, L.P.; Ong, M.M.A.; Lin, H.S.; Tan, K.S. Pterostilbene complexed with cyclodextrin exerts antimicrobial and anti-inflammatory effects. *Sci. Rep.* **2020**, *10*, 9072. [CrossRef]
37. Lu, K.; Hou, W.; Xu, X.F.; Chen, Q.; Li, Z.; Lin, J.; Chen, W.M. Biological evaluation and chemoproteomics reveal potential antibacterial targets of a cajaninstilbene-acid analogue. *Eur. J. Med. Chem.* **2020**, *188*, 112026. [CrossRef]
38. Geng, Z.Z.; Zhang, J.J.; Lin, J.; Huang, M.Y.; An, L.K.; Zhang, H.B.; Sun, P.H.; Ye, W.C.; Chen, W.M. Novel cajaninstilbene acid derivatives as antibacterial agents. *Eur. J. Med. Chem.* **2015**, *100*, 235–245. [CrossRef]
39. Goddard, T.N.; Patel, J.; Park, H.B.; Crawford, J.M. Dimeric Stilbene Antibiotics Target the Bacterial Cell Wall in Drug-Resistant Gram-Positive Pathogens. *Biochemistry* **2020**, *59*, 1966–1971. [CrossRef]
40. Hrast, M.; Rožman, K.; Ogris, I.; Škedelj, V.; Patin, D.; Sova, M.; Barreteau, H.; Gobec, S.; Grdadolnik, S.G.; Zega, A. Evaluation of the published kinase inhibitor set to identify multiple inhibitors of bacterial ATP-dependent mur ligases. *J. Enzym. Inhib. Med. Chem.* **2019**, *34*, 1010–1017. [CrossRef]
41. Man, N.Y.T.; Knight, D.R.; Stewart, S.G.; McKinley, A.J.; Riley, T.V.; Hammer, K.A. Spectrum of antibacterial activity and mode of action of a novel tris-stilbene bacteriostatic compound. *Sci. Rep.* **2018**, *8*, 6912. [CrossRef]
42. Wu, X.Z.; Cheng, A.X.; Sun, L.M.; Lou, H.X. Effect of plagiochin E, an antifungal macrocyclic bis(bibenzyl), on cell wall chitin synthesis in *Candida albicans*. *Acta Pharmacol. Sin.* **2008**, *29*, 1478–1485. [CrossRef]
43. Liu, Q.; Guo, X.; Jiang, G.; Wu, G.; Miao, H.; Liu, K.; Chen, S.; Sakamoto, N.; Kuno, T.; Yao, F.; et al. NADPH-Cytochrome P450 Reductase Ccr1 Is a Target of Tamoxifen and Participates in Its Antifungal Activity via Regulating Cell Wall Integrity in Fission Yeast. *Antimicrob. Agents Chemother.* **2020**, *64*, e00079-20. [CrossRef]
44. Mora-Pale, M.; Bhan, N.; Masuko, S.; James, P.; Wood, J.; McCallum, S.; Linhardt, R.J.; Dordick, J.S.; Koffas, M.A. Antimicrobial mechanism of resveratrol-trans-dihydrodimer produced from peroxidase-catalyzed oxidation of resveratrol. *Biotechnol. Bioeng.* **2015**, *112*, 2417–2428. [CrossRef]
45. Tang, K.W.; Yang, S.C.; Tseng, C.H. Design, Synthesis, and Anti-Bacterial Evaluation of Triazolyl-Pterostilbene Derivatives. *Int. J. Mol. Sci.* **2019**, *20*, 4564. [CrossRef]
46. Kim, S.; Lee, D.G. Oxyresveratrol-induced DNA cleavage triggers apoptotic response in *Candida albicans*. *Microbiology* **2018**, *164*, 1112–1121. [CrossRef]
47. Lee, W.; Lee, D.G. Resveratrol induces membrane and DNA disruption via pro-oxidant activity against *Salmonella typhimurium*. *Biochem. Biophys. Res. Commun.* **2017**, *489*, 228–234. [CrossRef]
48. Lee, J.; Lee, D.G. Novel antifungal mechanism of resveratrol: Apoptosis inducer in *Candida albicans*. *Curr. Microbiol.* **2015**, *70*, 383–389. [CrossRef]
49. Wu, X.Z.; Cheng, A.X.; Sun, L.M.; Sun, S.J.; Lou, H.X. Plagiochin E, an antifungal bis(bibenzyl), exerts its antifungal activity through mitochondrial dysfunction-induced reactive oxygen species accumulation in *Candida albicans*. *Biochim. Biophys. Acta* **2009**, *1790*, 770–777. [CrossRef]
50. Wu, X.Z.; Chang, W.Q.; Cheng, A.X.; Sun, L.M.; Lou, H.X. Plagiochin E, an antifungal active macrocyclic bis(bibenzyl), induced apoptosis in *Candida albicans* through a metacaspase-dependent apoptotic pathway. *Biochim. Biophys. Acta BBA Gen. Subj.* **2010**, *1800*, 439–447. [CrossRef]
51. Dadi, P.K.; Ahmad, M.; Ahmad, Z. Inhibition of ATPase activity of *Escherichia coli* ATP synthase by polyphenols. *Int. J. Biol. Macromol.* **2009**, *45*, 72–79. [CrossRef]

52. Hwang, D.; Lim, Y.H. Resveratrol antibacterial activity against *Escherichia coli* is mediated by Z-ring formation inhibition via suppression of *FtsZ* expression. *Sci. Rep.* **2015**, *5*, 10029. [CrossRef] [PubMed]
53. Tan, S.; Hua, X.; Xue, Z.; Ma, J. Cajanin Stilbene Acid Inhibited Vancomycin-Resistant *Enterococcus* by Inhibiting Phosphotransferase System. *Front. Pharmacol.* **2020**, *11*, 473. [CrossRef] [PubMed]
54. Zhang, X.; Fang, Y.; Jaiseng, W.; Hu, L.; Lu, Y.; Ma, Y.; Furuyashiki, T. Characterization of Tamoxifen as an Antifungal Agent Using the Yeast *Schizosaccharomyces pombe* Model Organism. *Kobe J. Med. Sci.* **2015**, *61*, e54–e63. [PubMed]
55. Dolan, K.; Montgomery, S.; Buchheit, B.; Didone, L.; Wellington, M.; Krysan, D.J. Antifungal activity of tamoxifen: In vitro and in vivo activities and mechanistic characterization. *Antimicrob. Agents Chemother.* **2009**, *53*, 3337–3346. [CrossRef] [PubMed]
56. Butts, A.; Koselny, K.; Chabrier-Roselló, Y.; Semighini, C.P.; Brown, J.C.; Wang, X.; Annadurai, S.; DiDone, L.; Tabroff, J.; Childers-We, J.R.; et al. Estrogen receptor antagonists are anti-cryptococcal agents that directly bind EF hand proteins and synergize with fluconazole in vivo. *mBio* **2014**, *5*, e00765-13. [CrossRef] [PubMed]
57. Ouyang, P.; He, X.; Yuan, Z.W.; Yin, Z.Q.; Fu, H.; Lin, J.; He, C.; Liang, X.; Lv, C.; Shu, G.; et al. Erianin against *Staphylococcus aureus* Infection via Inhibiting Sortase A. *Toxins* **2018**, *10*, 385. [CrossRef] [PubMed]
58. Lee, K.; Lee, J.h.; Ryu, S.y.; Cho, M.h.; Lee, J. Stilbenes reduce *Staphylococcus aureus* hemolysis, biofilm formation, and virulence. *Foodborne Pathog. Dis.* **2014**, *11*, 710–717. [CrossRef] [PubMed]
59. Duan, J.; Li, M.; Hao, Z.; Shen, X.; Liu, L.; Jin, Y.; Wang, S.; Guo, Y.; Yang, L.; Wang, L.; et al. Subinhibitory concentrations of resveratrol reduce alpha-hemolysin production in *Staphylococcus aureus* isolates by downregulating saeRS. *Emerg. Microbes Infect.* **2018**, *7*, 136. [CrossRef] [PubMed]
60. Guo, R.H.; Gong, Y.; Kim, S.Y.; Rhee, J.H.; Kim, Y.R. DIDS inhibits *Vibrio vulnificus* cytotoxicity by interfering with TolC-mediated RtxA1 toxin secretion. *Eur. J. Pharmacol.* **2020**, *884*, 173407. [CrossRef] [PubMed]
61. Ho Sui, S.J.; Lo, R.; Fernandes, A.R.; Caulfield, M.D.; Lerman, J.A.; Xie, L.; Bourne, P.E.; Baillie, D.l.; Brinkman, F.S. Raloxifene attenuates *Pseudomonas aeruginosa* pyocyanin production and virulence. *Int. J. Antimicrob. Agents* **2012**, *40*, 246–251. [CrossRef] [PubMed]
62. Zetterström, C.E.; Hasselgren, J.; Salin, O.; Davis, R.A.; Quinn, R.J.; Sundin, C.; Elofsson, M. The resveratrol tetramer (-)-hopeaphenol inhibits type III secretion in the gram-negative pathogens *Yersinia pseudotuberculosis* and *Pseudomonas aeruginosa*. *PLoS ONE* **2013**, *8*, e81969. [CrossRef] [PubMed]
63. Davis, R.A.; Beattie, K.D.; Xu, M.; Yang, X.; Yin, S.; Holla, H.; Healy, P.C.; Sykes, M.; Shelper, T.; Avery, V.M.; et al. Solving the supply of resveratrol tetramers from Papua New Guinean rainforest anisoptera species that inhibit bacterial type III secretion systems. *J. Nat. Prod.* **2014**, *77*, 2633–2640. [CrossRef] [PubMed]
64. Kang, J.E.; Yoo, N.; Jeon, B.J.; Kim, B.S.; Chung, E.H. Resveratrol Oligomers, Plant-Produced Natural Products With Anti-virulence and Plant Immune-Priming Roles. *Front. Plant Sci.* **2022**, *13*, 885625. [CrossRef] [PubMed]
65. Kang, J.E.; Jeon, B.J.; Park, M.Y.; Yang, H.J.; Kwon, J.; Lee, D.H.; Kim, B.s. Inhibition of the type III secretion system of *Pseudomonas syringae* pv. tomato DC3000 by resveratrol oligomers identified in *Vitis vinifera* L. *Pest Manag. Sci.* **2020**, *76*, 2294–2303. [CrossRef] [PubMed]
66. Wang, W.B.; Lai, H.C.; Hsueh, P.R.; Chiou, R.Y.; Lin, S.B.; Liaw, S.J. Inhibition of swarming and virulence factor expression in *Proteus mirabilis* by resveratrol. *J. Med. Microbiol.* **2006**, *55*, 1313–1321. [CrossRef] [PubMed]
67. Lou, F.; Wang, K.; Hou, Y.; Shang, X.; Tang, F. Inhibitory effect of resveratrol on swimming motility and adhesion ability against *Salmonella enterica* serovar Typhimurium infection. *Microb. Pathog.* **2023**, *184*, 106323. [CrossRef] [PubMed]
68. Kugaji, M.S.; Kumbar, V.M.; Peram, M.R.; Patil, S.; Bhat, K.G.; Diwan, P.V. Effect of Resveratrol on biofilm formation and virulence factor gene expression of *Porphyromonas gingivalis* in periodontal disease. *APMIS* **2019**, *127*, 187–195. [CrossRef] [PubMed]
69. Augustine, N.; Goel, A.K.; Sivakumar, K.C.; Kumar, R.A.; Thomas, S. Resveratrol—A potential inhibitor of biofilm formation in *Vibrio cholerae*. *Phytomedicine* **2014**, *21*, 286–289. [CrossRef] [PubMed]
70. Li, J.; Wu, T.; Peng, W.; Zhu, Y. Effects of resveratrol on cariogenic virulence properties of *Streptococcus mutans*. *BMC Microbiol.* **2020**, *20*, 99. [CrossRef] [PubMed]
71. Guo, R.; Peng, W.; Yang, H.; Yao, C.; Yu, J.; Huang, C. Evaluation of resveratrol-doped adhesive with advanced dentin bond durability. *J. Dent.* **2021**, *114*, 103817. [CrossRef] [PubMed]
72. Vasavi, H.S.; Sudeep, H.V.; Lingaraju, H.B.; Shyam Prasad, K. Bioavailability-enhanced Resveramax™ modulates quorum sensing and inhibits biofilm formation in *Pseudomonas aeruginosa* PAO1. *Microb. Pathog.* **2017**, *104*, 64–71. [CrossRef] [PubMed]
73. Rosman, C.W.K.; van der Mei, H.C.; Sjollema, J. Influence of sub-inhibitory concentrations of antimicrobials on micrococcal nuclease and biofilm formation in *Staphylococcus aureus*. *Sci. Rep.* **2021**, *11*, 13241. [CrossRef] [PubMed]
74. Lee, J.H.; Kim, Y.G.; Raorane, C.J.; Ryu, S.Y.; Shim, J.J.; Lee, J. The anti-biofilm and anti-virulence activities of trans-resveratrol and oxyresveratrol against uropathogenic *Escherichia coli*. *Biofouling* **2019**, *35*, 758–767. [CrossRef] [PubMed]
75. Nijampatnam, B.; Zhang, H.; Cai, X.; Michalek, S.M.; Wu, H.; Velu, S.E. Inhibition of *Streptococcus mutans* Biofilms by the Natural Stilbene Piceatannol Through the Inhibition of Glucosyltransferases. *ACS Omega* **2018**, *3*, 8378–8385. [CrossRef] [PubMed]
76. Wu, J.; Fan, Y.; Wang, X.; Jiang, X.; Zou, J.; Huang, R. Effects of the natural compound, oxyresveratrol, on the growth of *Streptococcus mutans*, and on biofilm formation, acid production, and virulence gene expression. *Eur. J. Oral Sci.* **2020**, *128*, 18–26. [CrossRef] [PubMed]

77. Xu, X.J.; Zeng, T.; Huang, Z.X.; Xu, X.F.; Lin, J.; Chen, W.M. Synthesis and Biological Evaluation of Cajaninstilbene Acid and Amorfrutins A and B as Inhibitors of the *Pseudomonas aeruginosa* Quorum Sensing System. *J. Nat. Prod.* **2018**, *81*, 2621–2629. [CrossRef] [PubMed]
78. Huang, Z.X.; Yu, J.H.; Xu, X.J.; Xu, X.F.; Zeng, T.; Lin, J.; Chen, W.M. Cajaninstilbene acid analogues as novel quorum sensing and biofilm inhibitors of *Pseudomonas aeruginosa*. *Microb. Pathog.* **2020**, *148*, 104414. [CrossRef] [PubMed]
79. Li, Y.; Ma, Y.; Zhang, L.; Guo, F.; Ren, L.; Yang, R.; Li, Y.; Lou, H. In vivo inhibitory effect on the biofilm formation of *Candida albicans* by liverwort derived riccardin D. *PLoS ONE* **2012**, *7*, e35543. [CrossRef] [PubMed]
80. Cheng, A.; Sun, L.; Wu, X.; Lou, H. The inhibitory effect of a macrocyclic bisbibenzyl riccardin D on the biofilms of *Candida albicans*. *Biol. Pharm. Bull.* **2009**, *32*, 1417–1421. [CrossRef]
81. Li, D.D.; Zhao, L.X.; Mylonakis, E.; Hu, G.H.; Zou, Y.; Huang, T.K.; Yan, L.; Wang, Y.; Jiang, Y.Y. In vitro and in vivo activities of pterostilbene against *Candida albicans* biofilms. *Antimicrob. Agents Chemother.* **2014**, *58*, 2344–2355. [CrossRef]
82. Nøhr-Meldgaard, K.; Ovsepian, A.; Ingmer, H.; Vestergaard, M. Resveratrol enhances the efficacy of aminoglycosides against *Staphylococcus aureus*. *Int. J. Antimicrob. Agents* **2018**, *52*, 390–396. [CrossRef]
83. Nelson, L.K.; D'Amours, G.H.; Sproule-Willoughby, K.M.; Morck, D.W.; Ceri, H. *Pseudomonas aeruginosa* las and rhl quorum-sensing systems are important for infection and inflammation in a rat prostatitis model. *Microbiology* **2009**, *155*, 2612–2619. [CrossRef]
84. Zhou, J.W.; Chen, T.T.; Tan, X.J.; Sheng, J.Y.; Jia, A.Q. Can the quorum sensing inhibitor resveratrol function as an aminoglycoside antibiotic accelerant against *Pseudomonas aeruginosa*? *Int. J. Antimicrob. Agents* **2018**, *52*, 35–41. [CrossRef]
85. Liu, S.; Zhang, J.; Zhou, Y.; Hu, N.; Li, J.; Wang, Y.; Niu, X.; Deng, X.; Wang, J. Pterostilbene restores carbapenem susceptibility in New Delhi metallo-β-lactamase-producing isolates by inhibiting the activity of New Delhi metallo-β-lactamases. *Br. J. Pharmacol.* **2019**, *176*, 4548–4557. [CrossRef]
86. Jia, Y.; Liu, J.; Yang, Q.; Zhang, W.; Efferth, T.; Liu, S.; Hua, X. Cajanin stilbene acid: A direct inhibitor of colistin resistance protein MCR-1 that restores the efficacy of polymyxin B against resistant Gram-negative bacteria. *Phytomedicine* **2023**, *114*, 154803. [CrossRef]
87. Hwang, D.; Lim, Y.H. Resveratrol controls *Escherichia coli* growth by inhibiting the AcrAB-TolC efflux pump. *FEMS Microbiol. Lett.* **2019**, *366*, fnz030. [CrossRef]
88. Sousa, V.; Luís, Â.; Oleastro, M.; Domingues, F.; Ferreira, S. Polyphenols as resistance modulators in *Arcobacter butzleri*. *Folia Microbiol.* **2019**, *64*, 547–554. [CrossRef]
89. Ferreira, S.; Silva, F.; Queiroz, J.A.; Oleastro, M.; Domingues, F.C. Resveratrol against Arcobacter butzleri and *Arcobacter cryaerophilus*: Activity and effect on cellular functions. *Int. J. Food Microbiol.* **2014**, *180*, 62–68. [CrossRef]
90. Shi, M.; Bai, Y.; Qiu, Y.; Zhang, X.; Zeng, Z.; Chen, L.; Cheng, F.; Zhang, J. Mechanism of Synergy between Piceatannol and Ciprofloxacin against *Staphylococcus aureus*. *Int. J. Mol. Sci.* **2022**, *23*, 15341. [CrossRef]
91. Yang, B.; Yao, H.; Li, D.; Liu, Z. The phosphatidylglycerol phosphate synthase PgsA utilizes a trifurcated amphipathic cavity for catalysis at the membrane-cytosol interface. *Curr. Res. Struct. Biol.* **2021**, *3*, 312–323. [CrossRef]
92. Breukink, E.; de Kruijff, B. Lipid II as a target for antibiotics. *Nat. Rev. Drug Discov.* **2006**, *5*, 321–332. [CrossRef]
93. Kattke, M.D.; Gosschalk, J.E.; Martinez, O.E.; Kumar, G.; Gale, R.T.; Cascio, D.; Sawaya, M.R.; Philips, M.; Brown, E.D.; Clubb, R.T. Structure and mechanism of TagA, a novel membrane-associated glycosyltransferase that produces wall teichoic acids in pathogenic bacteria. *PLoS Pathog.* **2019**, *15*, e1007723. [CrossRef]
94. Yang, C.S.; Huang, W.C.; Ko, T.P.; Wang, Y.C.; Wang, A.H.; Chen, Y. Crystal structure of the N-terminal domain of TagH reveals a potential drug targeting site. *Biochem. Biophys. Res. Commun.* **2021**, *536*, 1–6. [CrossRef]
95. Tiedje, C.; Holland, D.G.; Just, U.; Höfken, T. Proteins involved in sterol synthesis interact with Ste20 and regulate cell polarity. *J. Cell Sci.* **2007**, *120*, 3613–3624. [CrossRef]
96. Bock, F.J.; Tait, S.W.G. Mitochondria as multifaceted regulators of cell death. *Nat. Rev. Mol. Cell Biol.* **2020**, *21*, 85–100. [CrossRef]
97. Barrows, J.M.; Goley, E.D. FtsZ dynamics in bacterial division: What, how, and why? *Curr. Opin. Cell Biol.* **2021**, *68*, 163–172. [CrossRef]
98. Min, H.; Seok, Y.J. Phosphotransferase system sugars immediately induce mutations of Cra in an *Escherichia coli* ptsH mutant. *Environ. Microbiol.* **2022**, *24*, 5425–5436. [CrossRef]
99. Huang, K.J.; Lin, S.H.; Lin, M.R.; Ku, H.; Szkaradek, N.; Marona, H.; Hsu, A.; Shiuan, D. Xanthone derivatives could be potential antibiotics: Virtual screening for the inhibitors of enzyme I of bacterial phosphoenolpyruvate-dependent phosphotransferase system. *J. Antibiot.* **2013**, *66*, 453–458. [CrossRef]
100. Brockerhoff, S.E.; Stevens, R.C.; Davis, T.N. The unconventional myosin, Myo2p, is a calmodulin target at sites of cell growth in *Saccharomyces cerevisiae*. *J. Cell Biol.* **1994**, *124*, 315–323. [CrossRef]
101. Butts, A.; Martin, J.A.; DiDone, L.; Bradley, E.K.; Mutz, M.; Krysan, D.J. Structure-activity relationships for the antifungal activity of selective estrogen receptor antagonists related to tamoxifen. *PLoS ONE* **2015**, *10*, e0125927. [CrossRef]
102. Kong, C.; Neoh, H.M.; Nathan, S. Targeting Staphylococcus aureus Toxins: A Potential form of Anti-Virulence Therapy. *Toxins* **2016**, *8*, 72. [CrossRef]
103. Allen, R.C.; Popat, R.; Diggle, S.P.; Brown, S.P. Targeting virulence: Can we make evolution-proof drugs? *Nat. Rev. Microbiol.* **2014**, *12*, 300–308. [CrossRef]
104. Lowy, F.D. *Staphylococcus aureus* infections. *N. Engl. J. Med.* **1998**, *339*, 520–532. [CrossRef]

105. Mazmanian, S.K.; Ton-That, H.; Su, K.; Schneewind, O. An iron-regulated sortase anchors a class of surface protein during *Staphylococcus aureus* pathogenesis. *Proc. Natl. Acad. Sci. USA* **2002**, *99*, 2293–2298. [CrossRef]
106. Mazmanian, S.K.; Liu, G.; Jensen, E.R.; Lenoy, E.; Schneewind, O. *Staphylococcus aureus* sortase mutants defective in the display of surface proteins and in the pathogenesis of animal infections. *Proc. Natl. Acad. Sci. USA* **2000**, *97*, 5510–5515. [CrossRef]
107. Ng, T.B.; Liu, F.; Wang, Z.T. Antioxidative activity of natural products from plants. *Life Sci.* **2000**, *66*, 709–723. [CrossRef]
108. Liu, Q.; Yeo, W.S.; Bae, T. The SaeRS Two-Component System of *Staphylococcus aureus*. *Genes* **2016**, *7*, 81. [CrossRef]
109. Lee, S.; Song, S.; Lee, M.; Hwang, S.; Kim, J.S.; Ha, N.C.; Lee, K. Interaction between the α-barrel tip of *Vibrio vulnificus* TolC homologs and AcrA implies the adapter bridging model. *J. Microbiol.* **2014**, *52*, 148–153. [CrossRef]
110. Willcox, M.D.P.; Zhu, H.; Conibear, T.C.R.; Hume, E.B.H.; Givskov, M.; Kjelleberg, S.; Rice, S.A. Role of quorum sensing by *Pseudomonas aeruginosa* in microbial keratitis and cystic fibrosis. *Microbiology* **2008**, *154*, 2184–2194. [CrossRef]
111. Shaulov, L.; Gershberg, J.; Deng, W.; Finlay, B.B.; Sal-Man, N. The Ruler Protein EscP of the Enteropathogenic *Escherichia coli* Type III Secretion System Is Involved in Calcium Sensing and Secretion Hierarchy Regulation by Interacting with the Gatekeeper Protein SepL. *mBio* **2017**, *8*, e01733-16. [CrossRef]
112. Xiao, Y.; Hutcheson, S.W. A single promoter sequence recognized by a newly identified alternate sigma factor directs expression of pathogenicity and host range determinants in *Pseudomonas syringae*. *J. Bacteriol.* **1994**, *176*, 3089–3091. [CrossRef] [PubMed]
113. Kim, J.F.; Wei, Z.M.; Beer, S.V. The hrpA and hrpC operons of *Erwinia amylovora* encode components of a type III pathway that secretes harpin. *J. Bacteriol.* **1997**, *179*, 1690–1697. [CrossRef] [PubMed]
114. Flemming, H.C.; Wuertz, S. Bacteria and archaea on Earth and their abundance in biofilms. *Nat. Rev. Microbiol.* **2019**, *17*, 247–260. [CrossRef] [PubMed]
115. Rabin, N.; Zheng, Y.; Opoku-Temeng, C.; Du, Y.; Bonsu, E.; Sintim, H.O. Biofilm formation mechanisms and targets for developing antibiofilm agents. *Future Med. Chem.* **2015**, *7*, 493–512. [CrossRef] [PubMed]
116. Ramadan, H.H. Chronic rhinosinusitis and bacterial biofilms. *Curr. Opin. Otolaryngol. Head Neck Surg.* **2006**, *14*, 183–186. [CrossRef] [PubMed]
117. Davies, D. Understanding biofilm resistance to antibacterial agents. *Nat. Rev. Drug Discov.* **2003**, *2*, 114–122. [CrossRef] [PubMed]
118. Hall-Stoodley, L.; Costerton, J.w.; Stoodley, P. Bacterial biofilms: From the natural environment to infectious diseases. *Nat. Rev. Microbiol.* **2004**, *2*, 95–108. [CrossRef] [PubMed]
119. Costerton, J.W.; Lewandowski, Z.; Caldwell, D.E.; Korber, D.R.; Lappin-Scott, H.M. Microbial biofilms. *Annu. Rev. Microbiol.* **1995**, *49*, 711–745. [CrossRef]
120. Nickel, J.C.; Ruseska, I.; Wright, J.B.; Costerton, J.W. Tobramycin resistance of *Pseudomonas aeruginosa* cells growing as a biofilm on urinary catheter material. *Antimicrob. Agents Chemother.* **1985**, *27*, 619–624. [CrossRef]
121. Venkatesan, N.; Perumal, G.; Doble, M. Bacterial resistance in biofilm-associated bacteria. *Future Microbiol.* **2015**, *10*, 1743–1750. [CrossRef]
122. Hausner, M.; Wuertz, S. High rates of conjugation in bacterial biofilms as determined by quantitative in situ analysis. *Appl. Environ. Microbiol.* **1999**, *65*, 3710–3713. [CrossRef] [PubMed]
123. Yang, J.H.; Wright, S.N.; Hamblin, M.; McCloskey, D.; Alcantar, M.A.; Schrübbers, L.; Lopatkin, A.J.; Satish, S.; Nili, A.; Palsson, B.O.; et al. A White-Box Machine Learning Approach for Revealing Antibiotic Mechanisms of Action. *Cell* **2019**, *177*, 1649–1661. [CrossRef] [PubMed]
124. O'Toole, G.A.; Kolter, R. Flagellar and twitching motility are necessary for *Pseudomonas aeruginosa* biofilm development. *Mol. Microbiol.* **1998**, *30*, 295–304. [CrossRef] [PubMed]
125. Pratt, L.A.; Kolter, R. Genetic analysis of *Escherichia coli* biofilm formation: Roles of flagella, motility, chemotaxis and type I pili. *Mol. Microbiol.* **1998**, *30*, 285–293. [CrossRef] [PubMed]
126. Walker, S.L.; Redman, J.a.; Elimelech, M. Role of Cell Surface Lipopolysaccharides in *Escherichia coli* K12 adhesion and transport. *Langmuir* **2004**, *20*, 7736–7746. [CrossRef]
127. Lee, S.A.; Wallis, C.M.; Rogers, E.E.; Burbank, L.p. Grapevine phenolic compounds influence cell surface adhesion of *Xylella fastidiosa* and bind to lipopolysaccharide. *PLoS ONE* **2020**, *15*, e0240101. [CrossRef] [PubMed]
128. Krebs, S.J.; Taylor, R.K. Protection and attachment of *Vibrio cholerae* mediated by the toxin-coregulated pilus in the infant mouse model. *J. Bacteriol.* **2011**, *193*, 5260–5270. [CrossRef] [PubMed]
129. Chen, J.; Byun, H.; She, Q.; Liu, Z.; Ruggeberg, K.G.; Pu, Q.; Jung, I.J.; Zhu, D.; Brockett, M.R.; Hsiao, A.; et al. S-Nitrosylation of the virulence regulator AphB promotes *Vibrio cholerae* pathogenesis. *PLoS Pathog.* **2022**, *18*, e1010581. [CrossRef] [PubMed]
130. Götz, F. Staphylococcus and biofilms. *Mol. Microbiol.* **2002**, *43*, 1367–1378. [CrossRef]
131. Blankenship, J.R.; Mitchell, A.P. How to build a biofilm: A fungal perspective. *Curr. Opin. Microbiol.* **2006**, *9*, 588–594. [CrossRef]
132. Rasamiravaka, T.; Vandeputte, O.M.; Pottier, L.; Huet, J.; Rabemanantsoa, C.; Kiendrebeogo, M.; Andriantsimahavandy, A.; Rasamindrakotroka, A.; Stévigny, C.; Duez, P.; et al. *Pseudomonas aeruginosa* Biofilm Formation and Persistence, along with the Production of Quorum Sensing-Dependent Virulence Factors, Are Disrupted by a Triterpenoid Coumarate Ester Isolated from *Dalbergia trichocarpa*, a Tropical Legume. *PLoS ONE* **2015**, *10*, e0132791. [CrossRef] [PubMed]
133. Flemming, H.C.; Wingender, J. The biofilm matrix. *Nat. Rev. Microbiol.* **2010**, *8*, 623–633. [CrossRef] [PubMed]
134. Limoli, D.H.; Jones, C.J.; Wozniak, D.J. Bacterial Extracellular Polysaccharides in Biofilm Formation and Function. *Microbiol. Spectr.* **2015**, *3*, 10. [CrossRef] [PubMed]

135. Wu, J.; Jiang, X.; Yang, Q.; Zhang, Y.; Wang, C.; Huang, R. Inhibition of *Streptococcus mutans* Biofilm Formation by the Joint Action of Oxyresveratrol and *Lactobacillus casei*. *Appl. Environ. Microbiol.* **2022**, *88*, e0243621. [CrossRef] [PubMed]
136. Chen, Y.; Gozzi, K.; Yan, F.; Chai, Y. Acetic Acid Acts as a Volatile Signal to Stimulate Bacterial Biofilm Formation. *mBio* **2015**, *6*, e00392-15. [CrossRef] [PubMed]
137. Ng, W.L.; Bassler, B.L. Bacterial quorum-sensing network architectures. *Annu. Rev. Genet.* **2009**, *43*, 197–222. [CrossRef] [PubMed]
138. Liu, X.; Jia, M.; Wang, J.; Cheng, H.; Cai, Z.; Yu, Z.; Liu, Y.; Ma, L.; Zhang, L.; Zhang, Y.; et al. Cell division factor ZapE regulates *Pseudomonas aeruginosa* biofilm formation by impacting the pqs quorum sensing system. *mLife* **2023**, *2*, 28–42. [CrossRef]
139. Saunders, S.H.; Tse, E.C.M.; Yates, M.D.; Otero, F.J.; Trammell, S.A.; Stemp, E.D.A.; Barton, J.K.; Tender, L.M.; Newman, D.K. Extracellular DNA Promotes Efficient Extracellular Electron Transfer by Pyocyanin in *Pseudomonas aeruginosa* Biofilms. *Cell* **2020**, *182*, 919–932. [CrossRef]
140. Costa, K.C.; Glasser, N.R.; Conway, S.J.; Newman, D.K. Pyocyanin degradation by a tautomerizing demethylase inhibits *Pseudomonas aeruginosa* biofilms. *Science* **2017**, *355*, 170–173. [CrossRef] [PubMed]
141. Mitscher, L.A.; Park, Y.H.; Alshamma, A.; Hudson, P.B.; Haas, T. Amorfrutin A and B, bibenzyl antimicrobial agents from *Amorpha fruticosa*. *Phytochemistry* **1981**, *20*, 781–785. [CrossRef]
142. Fan, Y.; He, H.; Dong, Y.; Pan, H. Hyphae-specific genes HGC1, ALS3, HWP1, and ECE1 and relevant signaling pathways in *Candida albicans*. *Mycopathologia* **2013**, *176*, 329–335. [CrossRef] [PubMed]
143. Uppuluri, P.; Chaturvedi, A.k.; Srinivasan, A.; Banerjee, M.; Ramasubramaniam, A.k.; Köhler, J.r.; Kadosh, D.; Lopez-Ribot, J.L. Dispersion as an important step in the *Candida albicans* biofilm developmental cycle. *PLoS Pathog.* **2010**, *6*, e1000828. [CrossRef] [PubMed]
144. Darby, E.M.; Trampari, E.; Siasat, P.; Gaya, M.S.; Alav, I.; Webber, M.A.; Blair, J.M.A. Molecular mechanisms of antibiotic resistance revisited. *Nat. Rev. Microbiol.* **2023**, *21*, 280–295. [CrossRef] [PubMed]
145. Vestergaard, M.; Leng, B.; Haaber, J.; Bojer, M.S.; Vegge, C.S.; Ingmer, H. Genome-Wide Identification of Antimicrobial Intrinsic Resistance Determinants in *Staphylococcus aureus*. *Front. Microbiol.* **2016**, *7*, 2018. [CrossRef] [PubMed]
146. Liu, A.; Tran, L.; Becket, E.; Lee, K.; Chinn, L.; Park, E.; Tran, K.; Miller, J.H. Antibiotic sensitivity profiles determined with an *Escherichia coli* gene knockout collection: Generating an antibiotic bar code. *Antimicrob. Agents Chemother.* **2010**, *54*, 1393–1403. [CrossRef] [PubMed]
147. Shenkutie, A.M.; Yao, M.Z.; Siu, G.K.; Wong, B.K.C.; Leung, P.H. Biofilm-Induced Antibiotic Resistance in Clinical *Acinetobacter baumannii* Isolates. *Antibiotics* **2020**, *9*, 817. [CrossRef] [PubMed]
148. Sun, J.; Zhang, H.; Liu, Y.H.; Feng, Y. Towards Understanding MCR-like Colistin Resistance. *Trends Microbiol.* **2018**, *26*, 794–808. [CrossRef] [PubMed]
149. Pérez, A.; Poza, M.; Fernández, A.; Fernández-Mdel, C.; Mallo, S.; Merino, M.; Rumbo-Feal, S.; Cabral, M.P.; Bou, G. Involvement of the AcrAB-TolC efflux pump in the resistance, fitness, and virulence of *Enterobacter cloacae*. *Antimicrob. Agents Chemother.* **2012**, *56*, 2084–2090. [CrossRef] [PubMed]
150. Du, D.; Wang, Z.; James, N.R.; Voss, J.E.; Klimont, E.; Ohene-Agyei, T.; Venter, H.; Chiu, W.; Luisi, B.F. Structure of the AcrAB-TolC multidrug efflux pump. *Nature* **2014**, *509*, 512–515. [CrossRef] [PubMed]
151. Lowrence, R.C.; Subramaniapillai, S.G.; Ulaganathan, V.; Nagarajan, S. Tackling drug resistance with efflux pump inhibitors: From bacteria to cancerous cells. *Crit. Rev. Microbiol.* **2019**, *45*, 334–353. [CrossRef] [PubMed]
152. Brown, E.D.; Wright, G.D. Antibacterial drug discovery in the resistance era. *Nature* **2016**, *529*, 336–343. [CrossRef] [PubMed]
153. Barani, M.; Bilal, M.; Sabir, F.; Rahdar, A.; Kyzas, G.Z. Nanotechnology in ovarian cancer: Diagnosis and treatment. *Life Sci.* **2021**, *266*, 118914. [CrossRef] [PubMed]

pharmaceutics

MDPI

Review

Bacterial Efflux Pump Inhibitors Reduce Antibiotic Resistance

Lan Zhang [1,2], Xiaoyuan Tian [1,2], Lei Sun [1,2], Kun Mi [1,2], Ru Wang [1,3], Fengying Gong [1,3] and Lingli Huang [1,2,3,4,*]

[1] College of Veterinary Medicine, Huazhong Agricultural University, Wuhan 430070, China; zl-0118@webmail.hzau.edu.cn (L.Z.); shitxy@webmail.hzau.edu.cn (X.T.); sunlei23@webmail.hazu.edu.cn (L.S.); mikun@webmail.hzau.edu.cn (K.M.); w1665790632@webmail.hzau.edu.cn (R.W.); gongfengying@webmail.hzau.edu.cn (F.G.)

[2] National Reference Laboratory of Veterinary Drug Residues, Huazhong Agricultural University, Wuhan 430070, China

[3] MAO Key Laboratory for Detection of Veterinary Drug Residues, Huazhong Agricultural University, Wuhan 430070, China

[4] MOA Laboratory for Risk Assessment of Quality and Safety of Livestock and Poultry Products, Huazhong Agricultural University, Wuhan 430070, China

* Correspondence: huanglingli@mail.hzau.edu.cn

Abstract: Bacterial resistance is a growing problem worldwide, and the number of deaths due to drug resistance is increasing every year. We must pay great attention to bacterial resistance. Otherwise, we may go back to the pre-antibiotic era and have no drugs on which to rely. Bacterial resistance is the result of several causes, with efflux mechanisms widely recognised as a significant factor in the development of resistance to a variety of chemotherapeutic and antimicrobial medications. Efflux pump inhibitors, small molecules capable of restoring the effectiveness of existing antibiotics, are considered potential solutions to antibiotic resistance and have been an active area of research in recent years. This article provides a review of the efflux mechanisms of common clinical pathogenic bacteria and their efflux pump inhibitors and describes the effects of efflux pump inhibitors on biofilm formation, bacterial virulence, the formation of bacterial persister cells, the transfer of drug resistance among bacteria, and mismatch repair. Numerous efforts have been made in the past 20 years to find novel efflux pump inhibitors which are known to increase the effectiveness of medicines against multidrug-resistant strains. Therefore, the application of efflux pump inhibitors has excellent potential to address and reduce bacterial resistance.

Keywords: efflux pump inhibitor; drug resistance; biofilm; virulence; bacterial persister cells

1. Introduction

Bacterial infections seriously threaten food safety and public health, causing substantial economic losses and increasing the risk of zoonosis [1]. These infections can also lead to genomic instability of host cells and eventually to cancer [2]. As a result, bacterial infections remain a major challenge for modern medicine. Due to a lack of cross-protection from vaccines, the primary treatment for bacterial infections is still antibiotics. These drugs are used extensively in the medical, agricultural, and industrial fields and have reduced the severity and mortality of various infectious diseases caused by bacteria. However, their irrational clinical use has decreased drug efficacy and caused serious drug resistance problems.

According to survey data, each year, there are about 700,000 deaths worldwide caused by bacterial resistance [3]. If not controlled, this number will reach about 10 million by 2050. Moreover, bacterial resistance is expected to reduce the global gross domestic product by 2–3.5%. Global losses and consumption in antibiotic-resistance–related industries will reach USD 105 billion [4,5]. The World Health Organization first published a global priority list of antibiotic-resistant microorganisms in the beginning of 2017. Among them, *Pseudomonas aeruginosa*, *Enterobacteriaceae*, *Acinetobacter baumannii*, *Enterococcus faecium*, *Staphylococcus aureus*, and *Campylobacter* spp. constitute the main causes of life-threatening nosocomial infections throughout the world and are recognised as key, high-priority pathogens [6]. The antimicrobial resistance of pathogenic microorganisms is becoming increasingly serious, and the control of drug-resistant bacteria has become an urgent public health problem.

Over the past two decades, a variety of therapeutic treatments have been used to slow the development of antibiotic resistance, including (i) new drug development, (ii) multiple antibiotic combinations, (iii) phage therapy, (iv) antimicrobial peptide therapy, and (v) antibiotic–adjuvant combination approaches. However, new drug development is costly, has numerous difficulties in preparation, and the products may present structural instability. If we cannot address the growing problem of bacterial resistance, we may return to the pre-antibiotic era with no drugs to rely on.

Inhibitors have been widely used in recent years to delay bacterial resistance due to their high safety, low cost, and high efficiency, and they have great developmental value and application potential. The selection of inhibitors requires understanding the specific mechanisms of bacterial resistance, especially efflux pumps, which are widely expressed in Gram-positive and Gram-negative bacteria and mycobacteria. The overexpression of efflux pumps increases bacterial resistance to antibiotics, allowing them to survive at the injection site. Therefore, an essential way to inhibit drug-resistant bacteria is to search for and discover efflux pump inhibitors.

2. Mechanisms of Bacterial Resistance

2.1. The Main Resistance Mechanisms of Common Clinical Pathogens

Antibiotic resistance is the result of a complicated process involving multiple mechanisms. Bacteria have many different complementary mechanisms through which they can resist the effects of antibiotics [7]. Our summary of the resistance mechanisms of common clinical pathogens found that there are four main mechanisms through which bacteria become resistant to antibiotics: (i) by altering cell permeability to prevent antibiotics from entering cells, (ii) by altering an antibiotic's molecular target such that it can no longer interact with the antibiotic, (iii) antibiotic inactivation through enzymatic modification, and (iv) the expression of an efflux pump, which keeps the antibiotic out of the cell's surroundings (Table 1). These elements can be inherent or acquired through several processes. Additionally, free movement to human carriers and resistance determinants on mobile genetic components like plasmids and transposons might result in the spread of resistance to multiple microbial genera.

Table 1. The mechanisms of drug resistance in common clinical pathogens.

Bacteria	Drug Resistance Mechanism	References
Staphylococcus aureus	Reduced affinity for penicillin-binding proteins and the simultaneous production of β-lactamases to destroy the antibacterial activity of the drug. An eEnhanced efflux effect; the most effective multidrug resistance system is NorA. A decrease in cell membrane permeability affects the energy metabolism of bacteria, reduces the absorption of drugs, and leads to drug resistance.	[8–10]
Klebsiella pneumoniae	The generation of 16S rRNA methylesterase and β-lactamase, the mutation of target genes, the generation of multidrug resistance efflux pumps, and the modification of enzymes and target protection proteins.	[11]
Pseudomonas aeruginosa	Metallo-β-lactamase-producing enzymes, target mutations, alterations in membrane pore proteins, and active exocytosis systems.	[12]
Escherichia coli	The production of β-lactamase, the inhibition of bacterial DNA topoisomerase, and a reduction in drug uptake and increase in efflux through outer membrane alteration.	[13]
Campylobacter jejuni	Alteration of antibiotic targets or their expression, exocytosis (e.g., CmeABC), and the modification or inactivation of antibiotics.	[14]

In response to exposure to various antimicrobial drugs, these pathogens have the capacity to quickly acquire or develop resistance mechanisms. Common to these pathogenic bacterial resistance mechanisms is a range of efflux pumps that efficiently exclude or reduce the intracellular concentration of large amounts of antibiotics, resulting in a significant increase in pathogenic resistance. In several bacteria, the upregulation of efflux pumps induced by antibiotic treatment underpins the initial intermediate resistance phenotype, which can result in higher-resistance mutants. Thus, efflux pumps are a target for creating innovative adjuvant medicines, and it is necessary to understand their purpose and contribution to drug resistance [7,15].

2.2. Bacterial Efflux System

On bacterial cells, membranes contain proteins called efflux pumps that control the flow of toxic materials from the internal cellular environment to the exterior cellular environment. Bacterial efflux pumps are the quickest acting and most efficient resistance mechanism for bacteria in response to stress, and they are present in almost all bacteria [16]. Efflux pumps improve the viability of microorganisms in challenging conditions and increase their resistance to antibiotics and antimicrobial agents [17].

With the development of molecular microbiology, many bacterial efflux pumps have been identified, including in methicillin-resistant *Staphylococcus aureus* (MRSA), *Klebsiella pneumoniae, Streptococcus pneumoniae, Pseudomonas aeruginosa, Listeria monocytogenes, Acinetobacter baumannii, Enterococcus* spp., *Escherichia coli, Campylobacter jejuni*, and *Salmonella* spp. [18]. Six bacterial drug efflux pump families have been identified to be involved in the efflux pathway, including the ATP-binding cassette (ABC) family, the resistance-nodulation-cytosis (RND) superfamily, the small multidrug resistance (SMR) family, the multidrug and toxin extrusion (MATE) family, the major facilitator superfamily (MFS), and the proteobacterial antimicrobial complex efflux (PACE) family. The RND family is found only in Gram-negative bacteria. Among them, the five other families of secondary active transporters are driven by electrochemical energy captured by transmembrane ion gradients; the ATP-binding cassette (ABC) family uses ATP as a source of energy for transport (Figure 1) [19]. An efflux pump can excrete a variety of substrates, often antibiotics, thus leading to efflux-pump-expressing bacterial antibiotic resistance phenotypes. Efflux pumps can also excrete detergents, dyes, toxins, and waste metabolites [20,21]. Table 2 presents examples of the main relevant efflux pumps for common clinical bacteria.

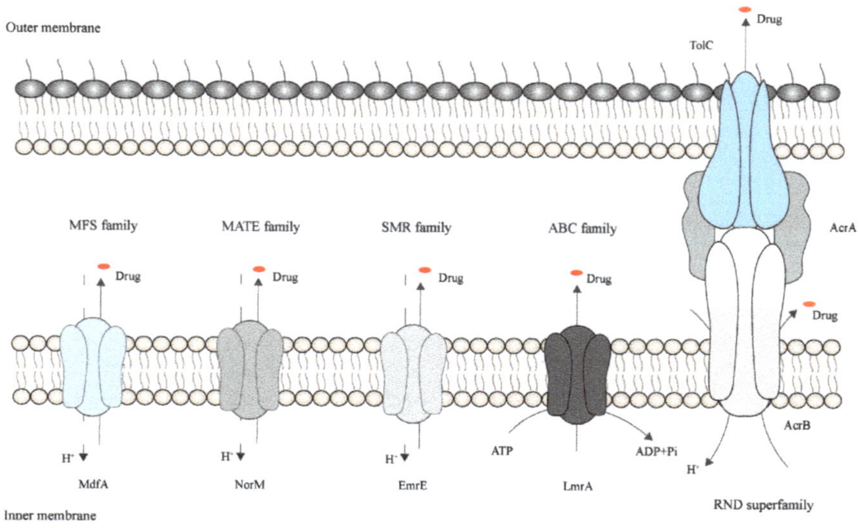

Figure 1. Representative structure diagram of five types of efflux pumps.

Table 2. Examples of relevant efflux pumps for common clinical pathogens.

Bacteria	Efflux Pump	Family	Efflux Substrate	References
Staphylococcus aureus	Nor A	MFS	β-lactams, fluoroquinolones, tetracyclines, fungicides, dyes, quaternary ammonium compounds, and preservatives	[22–24]
Klebsiella pneumoniae	AcrAB-TolC	RND	β-lactams, macrolides, fluoroquinolones, tetracycline, chloramphenicol, cationic dyes, and detergents	[25]
	OqxAB	RND	Quinolones, chloramphenicol, olaquindox, and tigecycline	
Pseudomonas aeruginosa	MexAB-OprM	RND	Quinolones, tetracyclines, macrolides, chloramphenicol, neomycin, and lincomycin, but not the β-lactam imipenem	[6,26]
	MexCD-OprJ		Quinolones, tetracyclines, macrolides, neomycin, lincomycin, chloramphenicol, flomoxef, and meropenem	
Escherichia coli	AcrAB-TolC	RND	Fluoroquinolones, tetracyclines, macrolides, rifampin, linezolid, oxazolidinone, novobiocin, clindamycin, and chloramphenicol	[27]
Campylobacter jejuni	CmeABC	RND	Macrolides, fluoroquinolones, tetracyclines, rifampin, toxic chemicals, and bile acids	[28–30]

3. The Role of Efflux Pump Inhibitors in Reversing Drug Resistance

3.1. The Status of Drug Resistance Reversal by Efflux Pump Inhibitors

Efflux pump inhibitors block efflux pumps through one or more processes, which can inactivate drug transport. To increase the action of antibiotics against bacteria expressing efflux pumps and restore the effectiveness of antibiotics that are no longer active against a pathogen, efflux pump inhibitors can be used as adjuvants with antibiotics [21]. Efflux pump inhibitors promote the intracellular accumulation of multiple antibiotics at the same time, making bacteria more susceptible to antimicrobial drugs. The effect of an inhibitor usually has two consequences: first, it will re-sensitise resistant microorganisms to currently used antibiotics. Second, it will lessen the opportunity to select resistant mutants [31]. Bacterial efflux pump inhibitors are available from a wide variety of sources, both natural and synthetic compounds (Table 3). Based on the global priority list of antibiotic-resistant microorganisms published by the World Health Organization, the discovery of efflux pump inhibitors for high-priority pathogens is urgent and significant.

Table 3. Types of efflux pump inhibitors and their sources.

Source	Inhibitor Type	Representative Drugs	Suppressed Efflux Pumps	References
Plants	Alkaloids	Reserpine	NorA	[32]
		Berberine	MexXY-OprM and NorA	[33]
		Piperine	NorA, MdeA, and Rv1258c	[34–36]
	Flavonoids	Astragalusin, kaempferol, silymarin	NorA	[32,37,38]
		Chalcone	NorA and MepA	[39]
		Lignocaine	MsrA	[40]
	Phenolic metabolites	5′-MHC	NorA	[41]
	Fermentation products	EA-371α and EA-371δ	MexAB-OprM	[15]
	Gallic acid	Epigallocatechin gallate	TetK, MexAB-OprM, and CmeABC	[42,43]

Table 3. *Cont.*

Source	Inhibitor Type	Representative Drugs	Suppressed Efflux Pumps	References
Chemical synthesis	Pyranopyridine	MBX2319	AcrAB-TolC	[44]
	Peptides	Phe-Arg-β-naphthylamide	MexAB-OprM, MexEF-OprN, and MexCD-OprJ	[45]
	Phenothiazines	Chlorpromazine	AcrB	[46]
	Aryl piperazine derivatives	NMP	AcrAB and AcrEF	[47]
	Pyridopyrimidine derivatives	D13-9001	AcrB and MexB	[48]

3.1.1. *Staphylococcus aureus* Efflux Pump Inhibitors

Staphylococcus aureus is a common zoonotic pathogen that causes tissue-limited suppurative infections, and poorly controlled infections may result in adverse outcomes such as extension of infection and bacteraemia. *S. aureus* is the most common Gram-positive bacterium in clinical infections. Antibiotic treatment is the first choice for *S. aureus* infections, but it is difficult to treat *S. aureus* infections because of its high resistance and because the disease recurs easily. Determining how to treat infections caused by drug-resistant *S. aureus* is a hotspot in current research. Efflux-mediated resistance has recently attracted more attention in *S. aureus* than other resistance mechanisms, and the inhibition of the activity of drug efflux pumps reduces the bacterial efflux of drugs and increases bacterial susceptibility to drugs. Thus, efflux pump inhibitors are important substances in combatting drug-resistant *S. aureus*.

Capsaicin (Figure 2), a novel NorA efflux pump inhibitor, not only increases the inherent susceptibility of *S. aureus* to ciprofloxacin but also remarkably reduces the emergence of ciprofloxacin-resistant mutant strains. The combination of ciprofloxacin and capsaicin has a bactericidal effect and enhances the post-antibiotic effect (PAE) of ciprofloxacin in a concentration-dependent manner [49]. Due to its poor water solubility, significant first-pass effect, and low oral bioavailability, capsaicin is currently used as a topical agent for a wide range of diseases, including sciatica, rheumatism, and lumbago [50]. The Food and Drug Administration (FDA) and the European Medicines Agency (EMA) approved a prescription drug containing capsaicin as the active ingredient in 2009 for the treatment of peripheral neuropathic pain in adults, either alone or in combination with other pain medications. Considering the potential of capsaicin to inhibit the NorA efflux pump, researchers prepared a range of capsaicin-based 1,3,4 oxadiazole conjugates via the single-site modification of the hydroxyl group of the vanilloid group of capsaicin and evaluated their potentiation of ciprofloxacin activity. Of all the compounds, 17i (Figure 2) showed robust activity, and the minimum effective concentration (MEC) of this inhibitor against SA1199B, a NorA-overexpressing strain, was 12.5 μg/mL, which is superior to that of its parent coupler, capsaicin (MEC 50 μg/mL). Compound 17i enhanced the activity of ciprofloxacin fourfold at an MEC of 12.5 μg/mL and reduced the mutation prevention concentration (MPC) of ciprofloxacin. In addition, 17i reduced EtBr efflux, confirming the inhibitory effect on NorA [51].

Piperine (Figure 2) is an important natural compound with a variety of health benefits. Despite its low absorption and bioavailability due to hydrophobicity, its ability to participate in herb–drug or herb–herb interactions also contributes to the pharmacokinetic and dynamic properties of other drugs/nutraceuticals [52]. The activity of ciprofloxacin against *S. aureus* is increased by piperine and its analogues, which also act as NorA inhibitors by lowering the minimum inhibitory concentration (MIC) of ciprofloxacin in *S. aureus* 1199B. Piperine also acts as an inhibitor of the *S. aureus* MdeA efflux pump, reducing the mupirocin MIC by 2–4-fold in *S. aureus* ATCC 29213 and MRSA isolates. The relative probability of selecting resistant mutants for the combination of mupirocin and piperine is significantly lower compared with mupirocin alone. Piperine is on the FDA's list of "Generally Regarded as Safe" (GRAS) compounds. Therefore, there should be no toxicity issues with topical preparations of piperine and mupirocin, and piperine may prove to be useful for nasal debridement and skin infections [35].

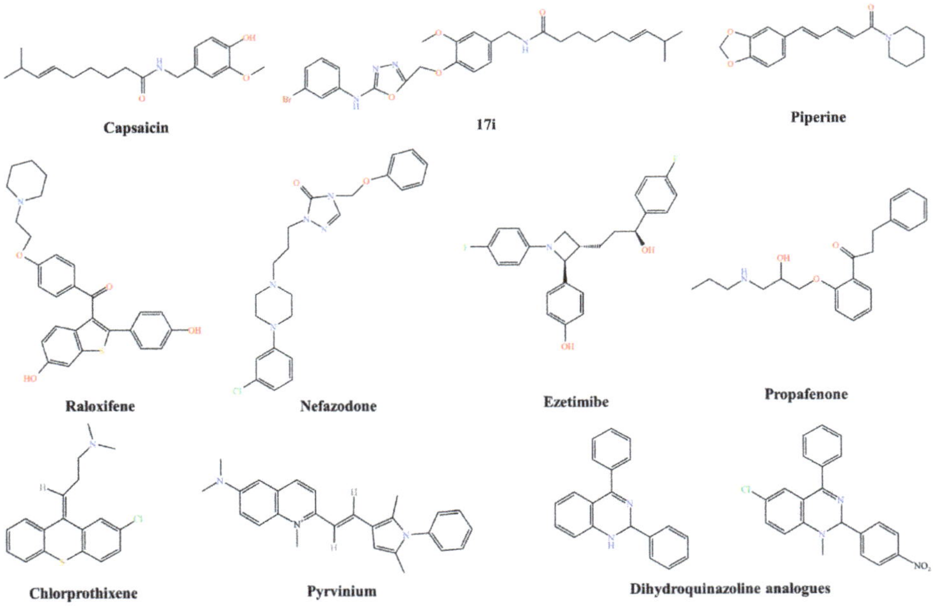

Figure 2. Inhibitor structural formula of *Staphylococcus aureus*.

Mahey N et al. screened a library of FDA-approved small-molecule drugs and determined six potential compounds using molecular docking which are chlorprothixene, ezetimibe, raloxifene, nefazodone, propafenone, and pyridinium (Figure 2). These compounds acted synergistically with ciprofloxacin and reduced its MIC to less than or equal to the clinical resistance breakpoint in overexpressed norA as well as in several clinical *S. aureus* strains. Furthermore, raloxifene, ezetimibe, and pyridinium showed synergistic effects with norfloxacin. In norA-overexpressing strains, the MIC was reduced to the clinical resistance breakpoint. Many of the early EPIs discovered had some toxicity concerns due to their off-target effects. In the current study, none of the working concentrations of EPIs interfered with membrane function or depleted the overall bacterial ATP pool, and the synergistic effect embodied by the drugs suggests that the EPIs of the drugs have a genuine similarity rather than being the results of any off-target effects [22].

Dihydroquinazoline, a complete class of heterocyclic compounds, has synergistic effects with norfloxacin and ethidium bromide. Of the seventeen dihydroquinazolines synthesised and screened for possible efflux pump inhibitor activity against *S. aureus* 1199B, eight proved to be effective inhibitors, significantly reducing the MICs of norfloxacin and ethidium bromide against *S. aureus* 1199B by up to 16-fold. Dihydroquinazoline analogues (Figure 2) also significantly reduce norA gene expression in the presence of subinhibitory concentrations of norfloxacin [53].

3.1.2. *Klebsiella pneumoniae* Efflux Pump Inhibitors

Klebsiella pneumoniae is a Gram-negative opportunistic pathogen that can cause serious hospital-acquired infections such as sepsis, pneumonia, urinary tract infections, soft tissue infections, and surgical site infections [54]. The increasing mortality rate due to drug-resistant *K. pneumoniae* infections has caused global alarm. Drug efflux pumps are a critical mechanism of drug resistance in *K. pneumoniae*. They continuously pump out drugs that have infiltrated the bacteria, resulting in low enterobacterial drug concentrations that are insufficient to exert antibacterial effects. To address the severe problem of efflux-pump-mediated multidrug resistance, many efflux pump inhibitors have been identified and tested.

One of the best characterized efflux pump inhibitors for Gram-negative bacteria is β-naphthylamide (PAβN) (Figure 3). PAβN is a competitive EPI and is considered a broad-spectrum pump substrate. This compound significantly reduces the MICs of chloramphenicol, tetracycline, and quinolones against multidrug-resistant *K. pneumoniae*. Chloramphenicol accumulation assays revealed low levels of chloramphenicol accumulation in cells in the absence of an inhibitor. In contrast, the use of the inhibitor resulted in a 3–4-fold increase in intracellular drug concentration, and the addition of PAβN significantly inhibited the activity of the efflux mechanism [55].

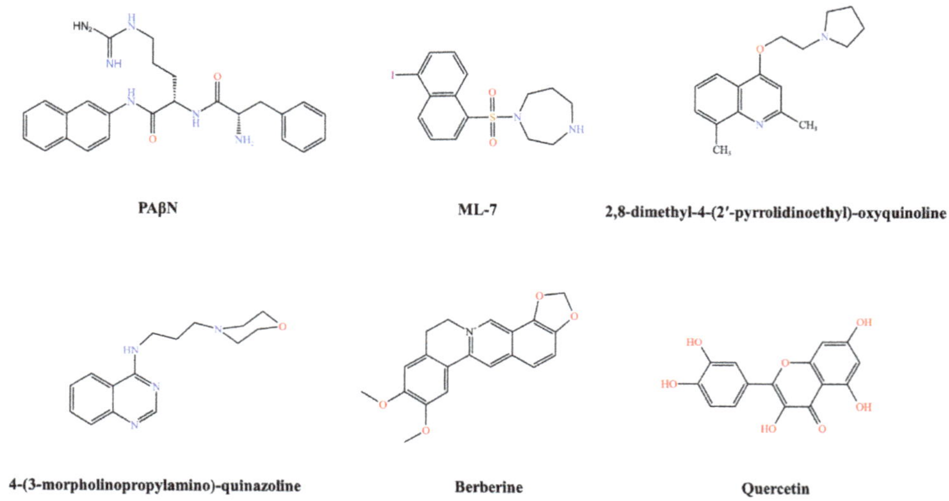

Figure 3. *Klebsiella pneumoniae* inhibitor structural formulae.

Tigecycline is regarded as the final line of defence in treating severe infections caused by multidrug-resistant (MDR) Gram-negative bacteria, especially extensively drug-resistant (XDR) Enterobacteriaceae bacteria. Sun et al. conducted a preliminary screening of 2711 compounds in combination with tigecycline using clinical isolates of *K. pneumoniae*. The results indicated that ML-7 (Figure 3) and tigecycline had synergistic effects on 11 strains of clinically isolated *K. pneumoniae* and could reduce the MIC of tigecycline by 4–128-fold. These *K. pneumoniae* belonged to different sequence types, suggesting that the ML-7-induced reduction in the MIC of tigecycline may be a common pattern for isolates of different sequence types of *K. pneumoniae*. ML-7 could disrupt the function of the multidrug efflux pump and promote the accumulation of tigecycline in cells. The potential applicability of ML-7, an inhibitor of myosin light chain kinase (MLCK), in bacterial infections has not been reported to date, and studies on ML-7 are still in the in vitro phase, but its potential as a novel tigecycline adjuvant to reverse drug resistance should not be overlooked [56].

Quinoline and quinazoline derivatives have been reported to be useful chemical sensitisers in efflux-pump-overexpressing strains. Alkylaminoquinolines, alkoxyquinolines, and chloroquinolines are considered broad-spectrum efflux pump inhibitors that increase susceptibility to chloramphenicol, norfloxacin, and tetracycline-resistant *K. pneumoniae* [31]. 2,8-dimethyl-4-(2′-pyrrolidinoethyl)-oxyquinoline (Figure 3) reduced the MIC of chloramphenicol by 16-fold, and 4-(3-morpholinopropylamino)-quinazoline (Figure 3) reduced the MICs of chloramphenicol, nalidixic acid, and tigecycline by 32-, 32-, and 8-fold, respectively [57,58]. Quinazoline derivatives are also effective at increasing the susceptibility of *K. pneumoniae* to chloramphenicol and nalidixic acid [59].

Berberine (Figure 3) has antibacterial, antiviral, anti-yeast, antiparasitic, antifungal, and anti-*Candida albicans* activity and is considered an antibacterial agent in Chinese and Western medicine. Berberine has low oral absorption into the blood and large intestinal excretion [60]. Previous studies have shown that berberine can be extensively metabolised and that metabolites also contribute to its therapeutic effect [61]. In one study, combining berberine with ciprofloxacin reduced ciprofloxacin concentrations by one-half to three-fourths in *K. pneumoniae*. There was a synergistic effect in 18.18% of the isolates. In addition, combining ciprofloxacin with berberine resulted in bacterial stagnation in a time–kill assay. These findings suggest that berberine enhances bacterial susceptibility to ciprofloxacin [62]. Berberine is currently approved for marketing in China and is mostly used in tablets. Berberine hydrochloride tablets belong to the group of digestive and metabolic drugs used as antidiarrhoeal, intestinal anti-inflammatory, and anti-infective medications.

Quercetin is a flavonoid found in fruits and vegetables that is poorly absorbed, extensively metabolised, and rapidly eliminated. Its unique biological profile forms the basis for potential benefits to overall health and disease resistance, including anticancer, anti-inflammatory, antibacterial, antiviral, and antioxidant properties [63]. Many virtual screening and molecular docking-based investigations, as well as particular fluorescence and mass spectrometry-based techniques, have shown quercetin to be an active efflux pump inhibitor over the past 20 years. Pal et al. evaluated carbapenemase and the efflux pump inhibition potential of quercetin (Figure 3) using an enzyme inhibition assay and an efflux pump inhibition assay, respectively. Quercetin at 1/2 MIC (64 μg/mL) showed an approximately 1.7-fold inhibition of AcrB efflux activity in *K. pneumoniae* overexpressing carbapenemase, and its MIC was reduced when combined with meropenem [64]. Quercetin is currently in the clinical research phase, and clinical trials have been conducted to evaluate quercetin in prostate cancer (NCT01912820), COPD (NCT03989271), and COVID-19 infection (NCT04468139) [65].

3.1.3. *Pseudomonas aeruginosa* Efflux Pump Inhibitors

Pseudomonas aeruginosa is a common conditionally pathogenic Gram-negative bacterium mainly found in intensive care units which can cause diseases such as intraoperative port infections and chronic respiratory infections [66]. The misuse of antibiotics in recent years has made *P. aeruginosa* widely resistant to cephalosporins, polymyxins, and sulphonamides, increasing the difficulty of clinical anti-infective treatment [67]. Efflux pumps are the material basis of their multi- and pan-drug resistance, mediating natural and acquired resistance. The discovery of efflux pump inhibitors is the key to solving the *P. aeruginosa* drug resistance problem.

Epigallocatechin-3-gallate (EGCG) (Figure 4) is a polyphenol compound extracted from green tea leaves that has antimicrobial activity and several health benefits. Subinhibitory concentrations of EGCG significantly enhanced antibiotic susceptibility, especially to chloramphenicol and tetracycline (≥4-fold). The enhanced antimicrobial activity and EGCG-mediated antibiotic susceptibility of EGCG in externally pumped (MexAB-OPrM) mutant strains compared to parental strains suggest a potent externally pumped inhibitory effect [42]. Like other phenolic compounds, EGCG has low bioavailability. Researchers have proposed that the relevant effects of EGCG in the body may not be caused by itself but rather by its metabolites, as it is metabolised by intestinal flora before it is absorbed into the bloodstream [68]. EGCG has promising applications with important protective effects against metabolic diseases, cerebral ischaemic injury, neurodegenerative diseases, tumours, and cardiovascular diseases. In addition, EGCG was the first botanical drug approved by the FDA to play a vital role in the treatment of genital warts as an anti-infective prescription [69].

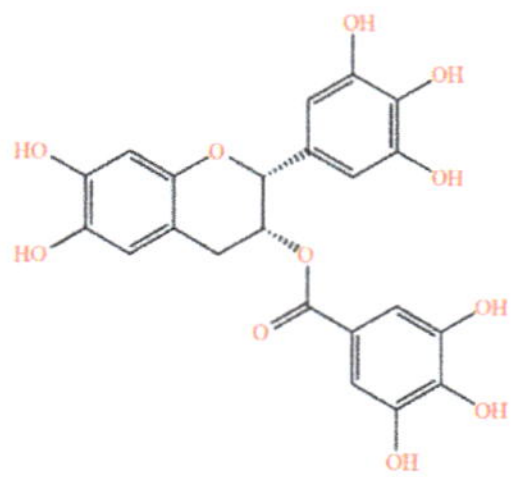

Epigallocatechin-3-gallate

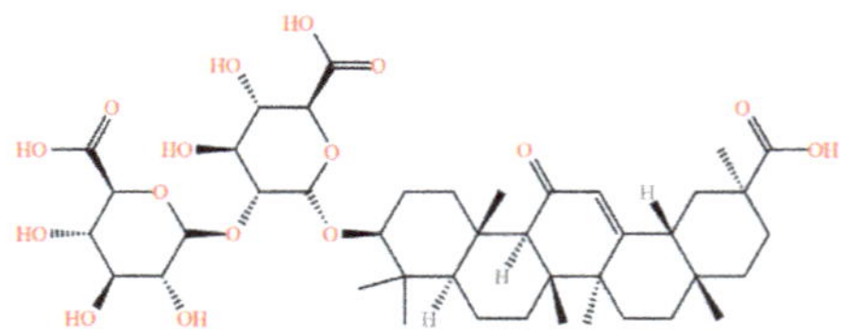

Glycyrrhizin

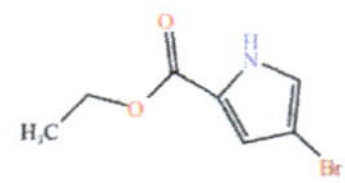

RP1

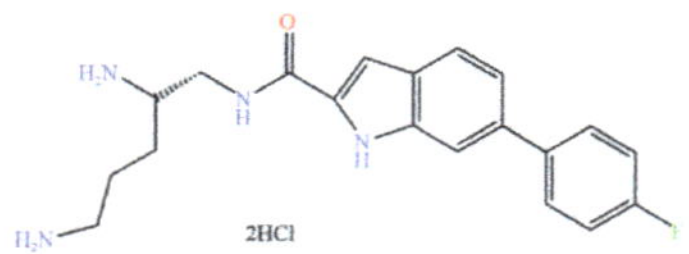

TXA01182

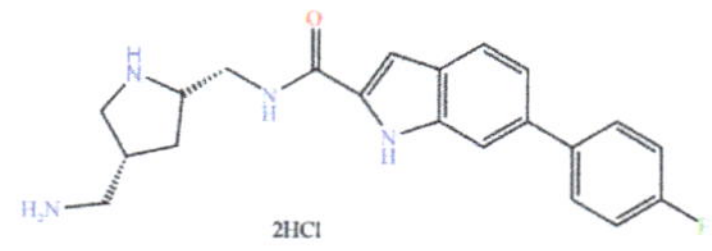

TXA09155

Figure 4. *Pseudomonas aeruginosa* inhibitor structural formulae.

Glycyrrhizin (GLY) (Figure 4) is a glycyrrhizic triterpenoid extracted from liquorice root (Glycyrrhiza glabra) which has a variety of properties and is used in the clinical management of chronic hepatitis. GLY is poorly absorbed in the human gastrointestinal tract [70], but GLY has been shown to increase the bioavailability of antibiotics, making them more effective at lower concentrations. The combination of ciprofloxacin and GLY with ciprofloxacin reduced its MIC. GLY altered the permeability of bacterial cell membranes and reduced bacterial survival. GLY significantly reduced the mRNA expression levels of the RND efflux pumps tested, and the fluorescence accumulation of EtBr was enhanced by the addition of the inhibitor, which suggests reduced efflux activity. In mouse treatment trials, GLY and ciprofloxacin treatment significantly reduced clinical scores, plate counts, and Myeloperoxidase (MPO) compared to ciprofloxacin [71].

Ethyl 4-bromopyrrole-2-carboxylate (RP1) (Figure 4) was isolated from the soil bacterium Streptomyces IMTB 2501. RP1 acted synergistically with an antibiotic by reducing its MIC in strains overexpressing the prototypical RND transporter protein (MexAB-OprM). The antibiotic-RP1 combination prolonged the after-effects of the antibiotic and reduced the mutation-preventive concentration of the antibiotic [72].

Yuan et al. evaluated the potential of heterocyclic formamide as an efflux pump inhibitor for *P. aeruginosa*. Among several types of thickened heterocyclic formamides, the aryl indole formamide compound TXA01182 (Figure 4) at 6.25 μg/mL showed an eight-fold enhancement of levofloxacin with synergistic effects on several multidrug-resistant *P. aeruginosa* clinical isolates, reducing the frequency of antimicrobial agent resistance and enhancing the killing kinetics of levofloxacin. The authors also reported synergistic activity with other antibacterial classes (fluoroquinolones, sulphonamides, and tetracyclines) against *P. aeruginosa*, resulting in a 4–32-fold reduction in the MIC [73]. Later, Zhang et al. improved the efflux pump inhibitor activity and overall pharmacological properties of TXA01182 and discovered TXA09155 (Figure 4). This novel efflux pump inhibitor shows robust potentiation when combined with multiple antibiotics. In both wild-type and multidrug-resistant clinical isolates of *P. aeruginosa*, it increases the effectiveness of numerous antimicrobial drugs with efflux capacity while contributing little to membrane disruption and the prevention of resistance flare-ups. TXA09155 at 6.25 μg/mL showed an 8–256-fold enhancement of levofloxacin, doxycycline, moxifloxacin, chloramphenicol, cefpirome, cotrimoxazole, and minocycline which reduced the prevalence of levofloxacin resistance and improved the killing kinetics of moxifloxacin [74].

3.1.4. *Escherichia coli* Efflux Pump Inhibitors

Foodborne pathogens are an essential cause of food poisoning and other related diseases and pose an important threat to food safety. *Escherichia coli*, as an extremely widely distributed pathogen, is one of the main factors causing foodborne diseases, posing a great threat to human health, and controlling the contamination of food with *E. coli* is particularly important. Of bacteria, *E. coli* has the most types of efflux pumps, and drug-resistant strains often carry antimicrobial drug-resistance genes against aminoglycosides, quinolones, chloramphenicol, and sulphonamides, among others. A bacterium can have multiple resistance mechanisms and show characteristics of multidrug resistance. We can solve the problem of *E. coli* drug resistance by discovering novel efflux pump inhibitors or structural modifications to known antibiotics.

MFS efflux system inhibitors work by reducing the level of resistance of *E. coli* to different fluoroquinolones. Verapamil (Figure 5) reduced the MICs of norfloxacin, levofloxacin, and gatifloxacin by 2–6-fold, and chlorpromazine (Figure 5) had the same effect on norfloxacin, levofloxacin, and gatifloxacin [75]. Currently, Verapamil is approved by the FDA for marketing as a prescription drug in the form of tablets, capsules, injections, and solutions for the treatment of cardiovascular diseases. Similarly, chlorpromazine was approved by the FDA for marketing as a prescription drug in tablet form for the treatment of neurological disorders. MBX2319 (Figure 5) is a new pyrano-pyridine that acts as a potent efflux pump inhibitor of the RND efflux system in *Enterobacteriaceae* bacteria. MBX2319 reduced the MICs of ciprofloxacin, levofloxacin, and piperacillin against *E. coli* strains by two, four, and eight times, respectively, but shows no activity against strains lacking AcrAB-TolC [44]. Zeng et al. created synthetic indole derivatives based on the TolC structure. They reported that 3-amino-6-carboxyindole and loxarsenyl-6-aminoindole (Figure 5) had synergistic antibacterial activity with ciprofloxacin, tetracycline, chloramphenicol, and erythromycin against *E. coli* strains overexpressing the AcrAB pump, resulting in a 2–64-fold reduction in the MIC [76].

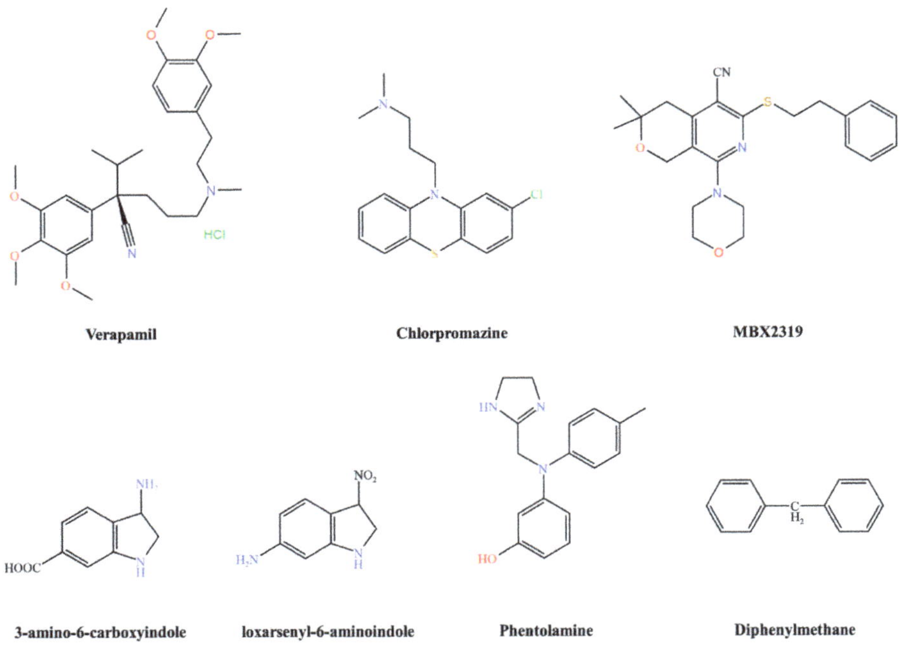

Figure 5. *Escherichia coli* inhibitor structural formulae.

Cui et al. combined phentolamine (Figure 5) with the macrolide antibiotics azithromycin, clarithromycin, and erythromycin and found a synergistic effect on *E. coli* test strains, as indicated by a fractional concentration inhibition index (FICI) of 0.375–0.5. The kinetic time–kill assay revealed that the bactericidal activity of all three macrolides against the strains increased in the presence of phentolamine. The combination of antibiotics and phentolamine reduced bacterial density to above $2\log_{10}$ CFU/mL, again indicating synergy. Phentolamine is a reversible, non-selective alpha-adrenergic antagonist that has been approved for marketing by the FDA. The compound has been successfully used in the treatment of cocaine-induced cardiovascular complications to counteract serious peripheral vasoconstriction and pain caused by the infusion of vasopressors. Numerous investigations and clinical data confirm its safety profile, and its reversal of resistance to macrolide antibiotics highlights its potential as an efflux pump inhibitor [77].

Seaweed is rich in polysaccharides, pigments, fatty acids, polyphenols, and peptides which have antioxidant, antibacterial, and anti-inflammatory effects. The seaweed compound diphenylmethane (DPM) (Figure 5) could regulate the activity of macrolides and fluoroquinolones against drug-resistant *E. coli* and reduce the half-maximal inhibitory concentration (IC_{50}) values of clarithromycin, ciprofloxacin, and erythromycin by twofold. According to a time-killing experiment, DPM and erythromycin together showed a more substantial inhibitory effect against drug-resistant *E. coli*. The study of ethidium bromide efflux showed that the addition of DPM significantly reduced dye efflux, indicating that it interferes with the efflux pump [78].

3.1.5. *Campylobacter jejuni* Efflux Pump Inhibitor

Campylobacter jejuni is a Gram-negative and spiral-shaped microaerobic bacterium that is a major cause of foodborne infections in humans in both developed and developing countries [79]. Fluoroquinolones and macrolides are the most commonly used antibiotics to treat *Campylobacter* infections, but the increasing resistance of these bacteria to clinically essential antibiotics is a severe public health problem. The multidrug efflux pump CmeABC stands out among the factors that influence antibiotic resistance in *C. jejuni*. Recent studies have also identified various compounds with significant inhibitory effects on CmeABC.

Phenolic compounds are secondary metabolites of plants; some of them have antibacterial activities against various pathogenic bacteria. Polyphenols are not necessarily the most active in the body, possibly because they are poorly absorbed from the gut, highly metabolised, or rapidly eliminated [80]. Once absorbed through the intestinal barrier, polyphenols are extensively metabolised in tissues [81]. However, polyphenols can affect the bioavailability of many therapeutic drugs by influencing the activity of various enzymes involved in their own metabolism [80], thereby enhancing antibiotic efficacy and antimicrobial action. Oh et al. observed the synergistic antibacterial activity of six phenolic compounds, including five phenolic acids (coumalic acid, sinapic acid, vanillic acid, galluc acid, and caffeic acid) and one flavonoid (taxifolin) (Figure 6). Specifically, 8 μg/mL of each phenolic component decreased the MICs of ciprofloxacin and erythromycin in five distinct *C. jejuni* strains, including three humans and two avian isolates, by roughly 4–32-fold [82]. Caffeic acid has been approved for use in China in the form of tablets as an anti-haemorrhagic agent. Essential oil constituents like the monoterpene (−)-pinene (Figure 6) can stop Gram-negative efflux pumps. (−)-α-Pinene can inhibit the CmeABC efflux pump, thereby increasing the susceptibility of *C. jejuni* to triclosan, ciprofloxacin, and erythromycin [83]. α-Pinene is the main active ingredient in eucalyptol, limonene, and pinene enteric capsules which have been approved for use in China as a respiratory medicine. Martinez et al. investigated the effect of Phe-Arg β-naphthylamide dihydrochloride (MC-207,110) on the function and resistance of CmeABC. The efflux pump inhibitor effectively decreased both inherent and acquired resistance to erythromycin in *C. jejuni* (MIC decreased 64–128-fold), with a fluoroquinolone MIC 2–4–fold lower [84].

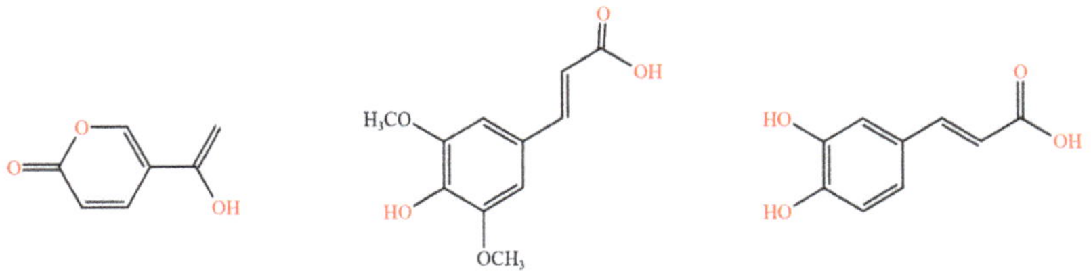

Coumalic acid **Sinapic acid** **Caffeic acid**

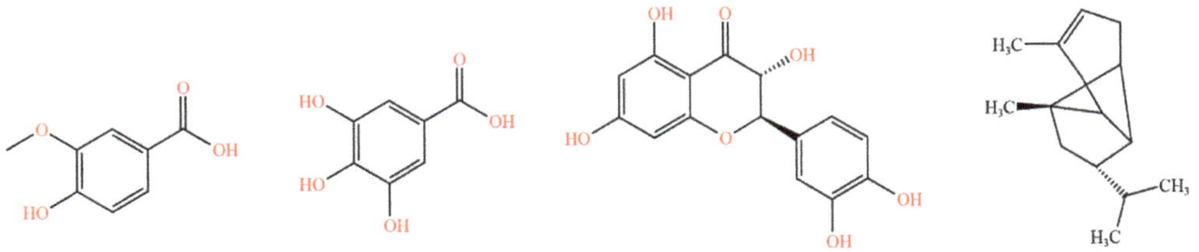

Vanillic acid **Gallic acid** **Taxifolin** **(-)-α-Pinene**

Figure 6. *Campylobacter jejuni* inhibitor structural formulae.

3.2. Screening Methods and Key Technologies of Efflux Pump Inhibitors

A suitable screening method is a prerequisite and an essential guarantee for screening drugs. At present, new drug screening technologies are developing very rapidly. The various screening methods complement and validate each other, providing crucial technical support to screen potentially valuable inhibitors.

Virtual screening is a database search method based on molecular docking technology. It is mainly applied to screen small-molecule inhibitors against enzymes, proteins, and other known three-dimensional-structure receptors [85]. After the docking calculation has been completed for all molecules in the library, the best molecule that binds to the target can be identified by scoring the docked molecules according to their interaction and binding energy [86]. The goals of virtual screening are to discover target compounds from an extensive library of small molecules, to reduce the number of compounds screened in physical experiments, to shorten the study period, and to save research costs.

The checkerboard test, which identifies synergistic combinations of antimicrobial agents, was used to screen for potential efflux pump inhibitors. Successive twofold dilutions of antibiotics and tested compounds were made in bacterial culture plates with different combinations of concentrations [32]. The fractional inhibitory concentration (FIC) index was used to determine the combined bacterial inhibitory effect. An FIC index ≤ 0.5 is a synergistic effect; $0.5 <$ FIC index ≤ 1 is additive effect; $1 <$ FIC index ≤ 4 is irrelevant; and an FIC index > 4 is antagonistic effect.

$$\text{FIC index} = \frac{\text{MIC of A in combination with B}}{\text{MIC of A alone}} + \frac{\text{MIC of B in combination with A}}{\text{MIC of B alone}}$$

A general approach to testing the validity of efflux pump inhibitors is to measure the growth inhibition of antibiotics in different bacterial strains in the absence and presence of efflux pump inhibitors. In experiments, wild-type, knockout, and efflux-pump-overexpression strains are often used. Greater synergy in efflux-pump-overexpressing strains compared with wild-type strains and the lack of growth inhibition in knockout strains are indicators of an efflux pump inhibitor. In this way, researchers can examine the intrinsic antimicrobial activity of potential inhibitors to demonstrate that the reduction in an antibiotic MIC is not the result of other mechanisms of antimicrobial activity [6].

In addition, a fluorescence real-time efflux analysis is an equally important tool for determining the efficacy and specificity of efflux pump inhibitors [32]. The most commonly used compound in efflux analysis is ethidium bromide (EtBr), which fluoresces only when embedded in DNA within the cell. In this test, adding an energy source, such as glucose, to the cells causes a fluorescent dye to extrude. If the efflux is reduced, fluorescence is retained over time. Therefore, an increase in fluorescence with the addition of efflux pump inhibitors is an important indicator of presumed activity. However, severe optical interferences were observed when assessing the efflux pump inhibitory activity of plant extracts and some purified compounds by fluorescence detection due to matrix burst effects in crude extracts and even in some pure compounds. In order to avoid the problem of optical matrix interference, Brown et al. developed a liquid chromatography–mass spectrometry (LC-MS) method for the determination of the concentration of ethidium bromide in fermentation broth filtrates. The LC-MS method produces data with a concentration–response relationship opposite to that produced by fluorescence measurements, i.e., as the concentration of the inhibitor is increased, the inhibitor reduces the exocytosis of EtBr by trapping it in the bacterial cells, and the EtBr concentration in the waste filtrate (measured by LC-MS) decreases [87]. However, EtBr is a DNA insertion agent, and there are concerns about its mutagenicity. One study optimised the method and established a novel $DiOC_3$ 96-well real-time exclusion assay suitable for high-throughput efflux pump inhibitor screening using *S. aureus*. This approach does not use a potentially dangerous DNA insertion agent [88]. The use of virtual screening and physical experiments to complement and validate each other provides essential technical support to screen potentially valuable inhibitors.

4. The Gain Effect of Efflux Pump Inhibitors

4.1. Efflux Pump Inhibitors Inhibit Biofilm Formation

Cells are frequently embedded in self-produced extracellular polymer matrices that are irrevocably connected to the matrix or interface or each other, forming biofilms, which are groups of microorganisms that have been confined. They are usually found in industrial, medicinal, and natural environments on moist surfaces [89]. Numerous environmental cues, such as host-derived signals, nutritional and metabolic cues, mechanical signals, population signals (quorum sensing (QS)), and subinhibitory antimicrobial concentrations, all favour the production of biofilms [90]. Biofilms protect bacterial cells from immune defences and antibiotics, making bacteria highly resistant to antimicrobial drugs.

At least four functions for efflux pumps in the growth of biofilms are possible (Figure 7): (i) biofilm matrix production and QS regulation are both aided by extracellular polymeric substances (EPSs), QS efflux, and population quenching (QQ) molecules; (ii) the indirect control of biofilm-formation-related genes; (iii) the efflux of harmful substances, including antibiotics and metabolic byproducts; and (iv) encouraging or inhibiting adherence to surfaces and other cells to affect aggregation [91]. Antibiotic resistance in biofilms has been linked to the putative multidrug resistance pump YhcQ in *E. coli*. MexAB-OprM and MexCD-OprJ efflux pumps have been implicated in resistance mechanisms specific to biofilms in *P. aeruginosa* [92]. Efflux pumps are actively involved in biofilm formation, so interfering with their function may impair biofilm formation.

Kvist et al. found that efflux pump and drug transporter protein genes are highly upregulated during biofilm growth and demonstrated the value of efflux pump inhibitors as potential anti-biofilm drugs. The effectiveness of three known efflux pump inhibitors, PAβN, azoxystrobin, and 1-(1-naphthyl methyl) piperazine (NMP), was tested against *S. aureus, K. pneumoniae, P. aeruginosa*, and *E. coli*. In almost all strains, the inhibitors significantly reduced biofilm formation [92]. Singh et al. demonstrated that boeravinone B inhibits efflux-pump-mediated fluoroquinolone efflux as well as the formation of an *S. aureus* biofilm. In the presence of boeravinone B (6.25–25 M), there was a concentration-dependent reduction in the minimum biofilm inhibitor concentration (MBIC) of ciprofloxacin against both bacteria. The staining showed a significant decrease in cell viability in the biofilm after the addition of the inhibitor. The combination of ciprofloxacin and the inhibitor showed a better biofilm inhibitory effect [93]. Vitamin C (VC) is a potential antioxidant. Prior research showed that VC may effectively kill Gram-positive organisms by causing oxidative stress and could prevent *E. coli* from forming biofilms. Xuet al. then demonstrated that VC could effectively treat infections caused by carbapenem-resistant hypervirulent *Klebsiella pneumoniae*. the subminimum inhibitory concentration (sub-MIC) of VC functioned as an efflux pump inhibitor, obstructing the passage of EPS to the bacterial cell surface and thus impeding the formation of biofilms. Both confocal laser scanning microscopy and scanning electron microscopy revealed that at lower VC concentrations, the biofilm structure was fragmented and began to disintegrate [94].

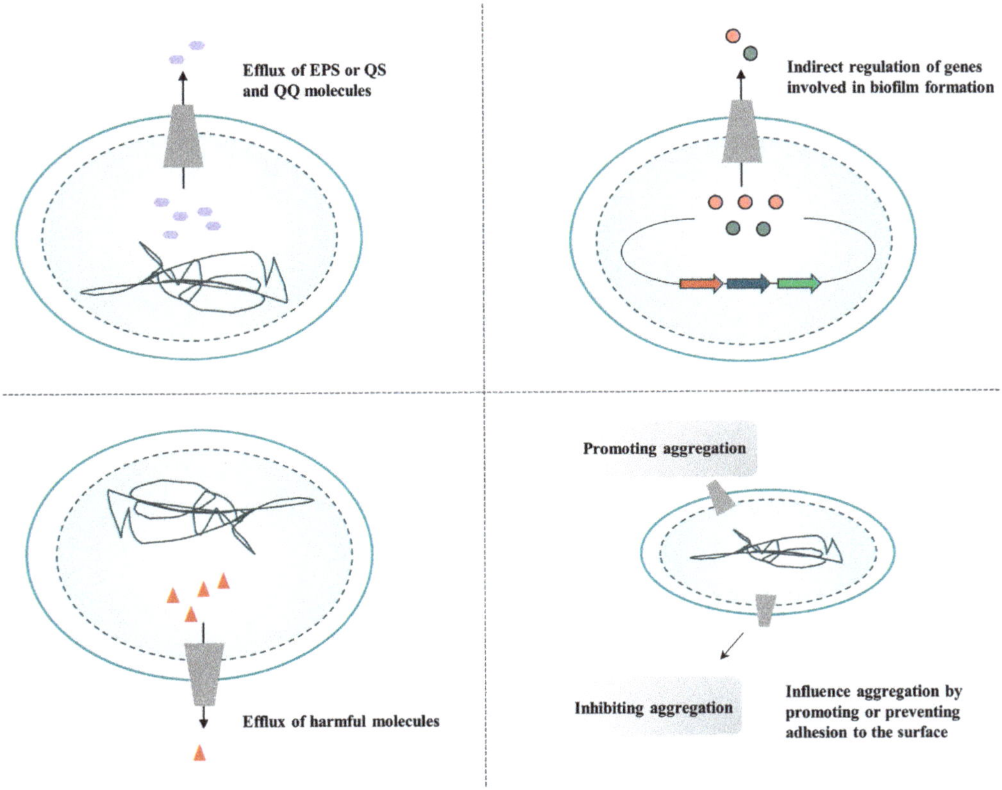

Figure 7. Four different potential roles of efflux pumps in biofilm formation. Adapted from Alav et al. [91].

This suggests to us that due to the intrinsic link between efflux pumps and biofilms, some agents are potent biofilm inhibitors due to their excellent efflux pump inhibition. Efflux pump inhibitors have great potential as anti-biofilm drugs.

4.2. Efflux Pump Inhibitors Reduce Bacterial Virulence

The issue of antibiotic resistance has encouraged the creation of novel therapies, one of which is the discovery of substances that diminish the pathogenic potential of toxic microbes. These compounds, which can be used alone or in combination with antibiotics, can help treat bacterial infections by impairing the ability of bacterial pathogens to proliferate in infected patients [95]. The potential interaction between the antibiotic resistance, virulence, infectivity, and pathogenicity of pathogens mediated by the efflux system emphasises the possible use of efflux pump inhibitors as antivirulence and antiresistance substances.

Under laboratory conditions, researchers have shown that efflux pump inhibitors can impair the virulence of several pathogens. Evidence shows that some RND transport proteins participate in the efflux of virulence-related components of bacteria. PAβN is the most efficient inhibitor of the *P. aeruginosa* RND efflux pump and reduces some of the virulence characteristics of this bacterium. PAβN attenuated the virulence of *P. aeruginosa* PAO1 by inhibiting processes related to motility and micronutrient acquisition (i.e., phosphate and iron). Meanwhile, in the PAO1 strain, PAβN stimulated the generation of N-3-oxododecanoyl-homoserine lactone ($3OC_{12}$-HSL) and pyocyanin, which plays an active role in its virulence [96]. At the transcriptional level, PAβN markedly decreased the relative expression of QS cascades, and the QS-regulated type II secretion gene lasB and toxA (exotoxin A). In addition, PAβN diminished the relative expression of pelA (exopolysaccharides) in clinical virulent isolates. PAβN significantly blocked the QS circuit and inhibited the virulence factors expressed by clinical isolates of *P. aeruginosa*, rendering them less virulent (Figure 8) [97]. In another study, NMP and PAβN reduced the production of cholera toxin and the number of toxin co-regulatory trichomes (TCPs). Various concentrations of PAβN or NMP reduced the expression of the *tcpPH* and *toxT* genes which encode essential regulators of *Vibrio cholerae* virulence genes. They appeared to suppress the expression of other virulence genes independently of RND efflux inhibition. Thus, PAβN and NMP may have additional effects on virulence gene expression [98].

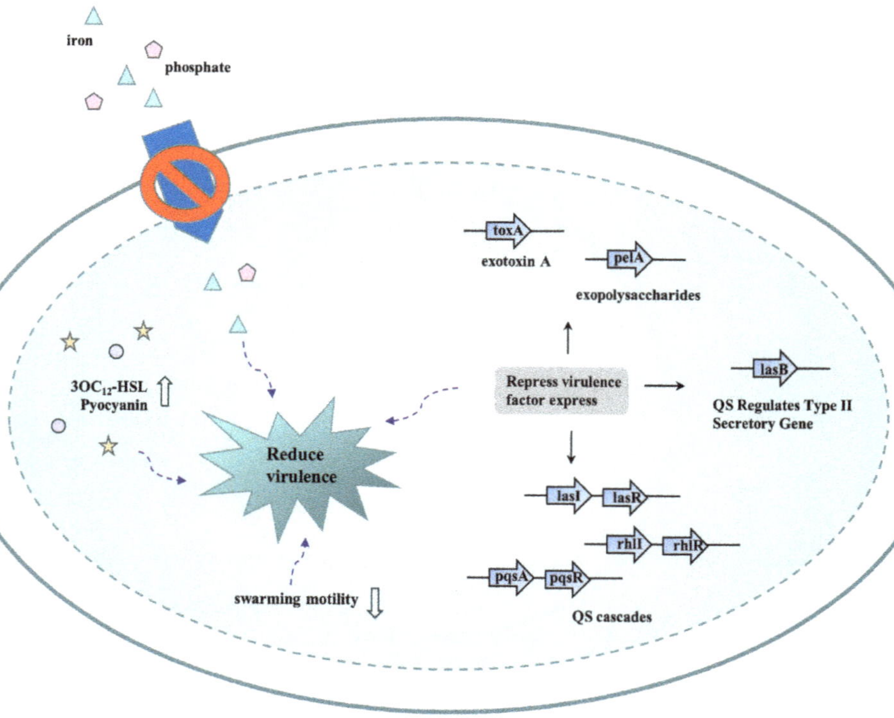

Figure 8. Possible strategies to reduce bacterial virulence.

Overall, pathogen resistance and pathogenicity are decreased by efflux pump inhibitors. Therefore, they can increase the success of antimicrobial therapy when treating infections while potentially reducing the harmful effects of infections on patients [31]. Efflux inhibition may be an essential method to control bacterial virulence and pathogenesis.

4.3. Efflux Pump Inhibitors Reduce the Formation of Bacterial Persister Cells

When it comes to medication therapy, the majority of cells are swiftly eradicated by bactericidal antibiotics. Still, a relatively tiny number of phenotypic variants, known as persister cells, exhibit considerable resistance [99]. The essential difference between persister cells and drug-resistant bacteria in the traditional sense is that the latter result from genetic mutations which prevent the drug from binding to the target in various ways, resulting in the inhibition of the drug's effect. On the other hand, persister cells represent non-genetic phenotypic variation and have a dormant phenotype: almost all life activities have stopped, so even though the drug can bind to the corresponding target, it does not exert a toxic effect on the microorganism. Hence, persister cells tolerate almost all drugs [100].

In persister cells, antibiotic accumulation can be decreased through an increase in efflux, a decrease in cell membrane permeability, or both. Transcriptome sequencing has confirmed that multidrug efflux gene expression is markedly higher in persister cells. Cells with efflux gene upregulation show enhanced persistence, while efflux knockout mutants show an attenuation of persistence, suggesting that efflux pumps play an essential role in bacterial drug resistance. The AcrAB-TolC efflux pump is present in most Gram-negative bacteria. It has a proton-dependent mechanism for releasing hazardous metabolic intermediates and at least nine antibiotics. The lack of AcrB and TolC impairs the persistence of delafloxacin in planktonic stationary phase cultures of *E. coli*. The co-administration of PAβN (an inhibitor of AcrAB-TolC) and delafloxacin resulted in a 40-fold reduction in the survival of stationary phase *E. coli* populations after 7 h of treatment compared with cultures treated with delafloxacin and dimethyl sulphoxide. Targeting the AcrAB-Tol C efflux pump enhanced delafloxacin activity against slow/non-growing *E. coli* during treatment [101]. Pu et al. similarly identified that persisters show higher levels of expression of efflux-related genes. A high level of expression of tolC is essential for promoting persistent cell formation, and persisters combine active efflux and passive dormancy to survive antibiotic attack. They found a significant reduction in persister cells after the addition of efflux pump inhibitors when comparing the use of antibiotics alone with the antibiotics plus PAβN or NMP [99].

Consequently, the combination of efflux pump inhibitors with conventional antibiotics holds promise for the treatment of persistent bacterial disease by reducing intrinsic resistance or acquired multidrug resistance associated with efflux pumps. These compounds assist antibiotics to clear persister cells, thereby ultimately killing the pathogenic microbial population.

4.4. Efflux Pump Inhibitors Limit the Acquisition of Bacterial Resistance

The capacity of bacteria to access genes through horizontal transfer mechanisms is a major cause of the spread of drug resistance. Horizontal gene transfer (HGT) is a driver of antimicrobial resistance gene transmission in bacteria. Drug resistance genes for clinically significant antibiotics are typically carried on plasmids, which are circular DNA segments that can be conjugated between several bacterial species. Studies have shown that even when drugs that restrict gene expression exist, bacteria can still express resistance genes [102]. Researchers revealed the critical role of multidrug efflux complexes in maintaining drug-specific resistance through horizontal gene transfer in the presence of bacterial inhibitors [103]. One reason is that the plasmid delivers both the Tet A resistance protein and the plasmid to the receptor. The transferred Tet A protein can pump out enough tetracycline to allow for translation, which produces more Tet A from the plasmid. AcrAB-TolC is essential for the acquisition of tetracycline resistance via horizontal gene transfer, which may ensure that the newly acquired plasmid is able to synthesise TetA resistance by decreasing the tetracycline content of the cells. When we use antibiotics to treat bacterial infections, it is possible that horizontal gene transfer through the digestive tract may lead to the emergence of new drug-resistant strains, resulting in therapeutic failure. Therefore, in addition to inhibiting bacterial growth through antibiotics, reducing the horizontal transfer of antibiotic resistance may be considered. Antibiotic resistance is mainly attributed to the AcrAB-TolC efflux pump, which raises the possibility that blocking the multidrug efflux system is a significant contributing factor. Efflux pump inhibitors might have an indirect effect by preventing the spread of antibiotic resistance among bacteria during therapy, which would stop the creation of new resistant strains [102].

4.5. *Efflux Pump Inhibitors Improve Mismatch Repair*

Most organisms contain DNA mismatch repair (MMR) systems to prevent mutations and preserve genomic stability. MMR fixes mismatches and insertion–deletion loops caused by DNA synthesis, increasing DNA replication's overall fidelity by up to a thousand times. *MutS* initiates MMR by identifying mismatches in newly replicated DNA [104]. Antibiotic resistance is mainly the result of genetic changes, but bacteria can also resist drug treatment by transiently expressing multidrug efflux pumps. El Meouche et al. focused on transient drug resistance caused by the heterogeneity of AcrAB-TolC efflux pump expression found in various infections. The authors built a two-colour plasmid to record the expression of the *acrAB* and *mutS* promoters in order to study the relationship between *acrAB* and *mutS* expression. Single-cell time-lapse fluorescence microscopy showed that cells with greater levels of *acrAB* expression had much lower levels of *mutS*, making them more susceptible to mutation. Transient antibiotic resistance from increased *acrAB* expression could facilitate spontaneous mutations that lead to permanent antibiotic resistance [105]. This suggests that cells with increased efflux pump expression are more resistant to antibiotics and also have less MMR, making them more susceptible to mutations. We could potentially apply efflux pump inhibitors in combination with antibiotic treatment, which can reduce the expression of efflux pumps and thus increase the probability of MMR to prevent the emergence of drug resistance.

5. Summary and Prospects

Efflux pump inhibitors can restore the antibiotic efficacy of clinically important pathogenic microorganisms and can play a vital role in solving antibiotic resistance. The mechanisms of action of efflux pump inhibitors include competitively blocking the binding sites on the efflux pump, reducing drug efflux by competing with antibiotics for the active site of the bacterial efflux pump [44,106], and the gene-level inhibition of exocytosis pump expression or selective inhibition of the expression of genes encoding exocytosis pumps through antisense nucleotides or small-molecule RNA. The expression of *acrAB* in *E. coli* and the expression of *cmeABC* in multidrug-resistant *C. jejuni* have been inhibited using antisense nucleic acids and oligonucleotides, respectively [107].

The inhibitors we have described in this review include natural and synthetic compounds. They have shown promising efficacy in delaying bacterial resistance when used with antibiotics (Figure 9). These inhibitors have multiple benefits such as facilitating gastrointestinal absorption, improving blood–brain barrier penetration, and increasing drug concentration in mammalian cells to enhance the killing of invasive pathogens [93]. Therefore, developing safe and effective efflux pump inhibitors is essential to improve the activity of antibacterial drugs and the current status of the clinical treatment of bacterial infections.

Currently, most efflux pump inhibitors are still in the preclinical phase or in vitro trials. There are numerous challenges when developing novel antibiotics or clinically effective efflux pump inhibitors that can circumvent multiple drug-resistant bacteria. Due to their poor stability, poor selectivity, and severe cytotoxicity to eukaryotic cells, inhibitors have not yet been employed treating bacterial infections [108]. Efflux pump inhibitors as therapeutic agents present several difficulties, but their value and benefits should not be overlooked. They have long been recognized as drugs that can increase the effectiveness of antibiotics against multidrug-resistant bacteria, and there have been numerous studies backed by data demonstrating their effectiveness. Nevertheless, more investigation is required to fully realize the potential of efflux pump inhibitors.

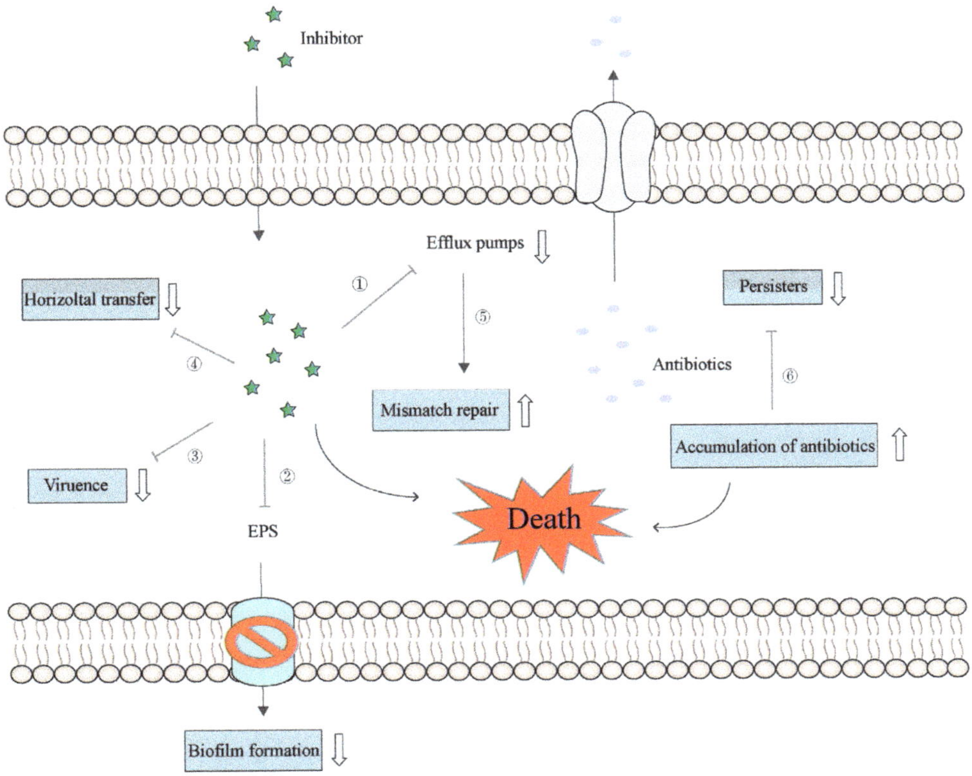

Figure 9. Schematic diagram of the synergistic effects of efflux pump inhibitors and antibiotics. ① Inhibitors promote antibiotic accumulation in cells by inhibiting the multidrug efflux pump function; ② disrupt EPS transport and inhibit biofilm formation; ③ impair bacterial virulence; ④ prevent horizontal transfer of resistance; ⑤ enhance mismatch repair; and ⑥ assist in antibiotic clearance of persister cells.

Author Contributions: Writing—original draft preparation, L.Z., X.T. and L.S.; writing—review and editing, K.M., R.W. and F.G.; project administration, L.H. All authors have read and agreed to the published version of the manuscript.

Funding: This article was supported by the National Key Research and Development Program (grant number 2022YFD1800402).

Institutional Review Board Statement: Not applicable.

Informed Consent Statement: Not applicable.

Data Availability Statement: Not applicable.

Acknowledgments: We thank all the authors for their contributions and Huang for financial support.

Conflicts of Interest: The authors declare that this research was conducted in the absence of any commercial or financial relationships that could be construed as potential conflicts of interest.

References

1. Lei, H.; Wang, C.; Wang, Y.; Wang, C. Single-cell RNA-Seq revealed profound immune alteration in the peripheral blood of patients with bacterial infection. *Int. J. Infect. Dis.* **2021**, *103*, 527–535. [CrossRef]
2. Kovalchuk, O.; Walz, P.; Kovalchuk, I. Does bacterial infection cause genome instability and cancer in the host cell? *Mutat. Res.* **2014**, *761*, 1–14. [CrossRef]
3. García-Castro, M.; Sarabia, F.; Díaz-Morilla, A.; López-Romero, J.M. Approved antibacterial drugs in the last 10 years: From the bench to the clinic. *Explor. Drug Sci.* **2023**, *1*, 180–209. [CrossRef]
4. Allcock, S.; Young, E.H.; Holmes, M.; Gurdasani, D.; Dougan, G.; Sandhu, M.S.; Solomon, L.; Török, M.E. Antimicrobial resistance in human populations: Challenges and opportunities. *Glob. Health Epidemiol. Genom.* **2017**, *2*, e4. [CrossRef]
5. Hancock, R.E. The role of fundamental research and biotechnology in finding solutions to the global problem of antibiotic resistance. *Clin. Infect. Dis.* **1997**, *24* (Suppl. S1), S148–S150. [CrossRef] [PubMed]
6. Lamut, A.; Peterlin Mašič, L.; Kikelj, D.; Tomašič, T. Efflux pump inhibitors of clinically relevant multidrug resistant bacteria. *Med. Res. Rev.* **2019**, *39*, 2460–2504. [CrossRef] [PubMed]
7. Laws, M.; Jin, P.; Rahman, K.M. Efflux pumps in Mycobacterium tuberculosis and their inhibition to tackle antimicrobial resistance. *Trends Microbiol.* **2022**, *30*, 57–68. [CrossRef]
8. Vestergaard, M.; Frees, D.; Ingmer, H. Antibiotic Resistance and the MRSA Problem. *Microbiol. Spectr.* **2019**, *7*, 10–1128. [CrossRef] [PubMed]
9. Guo, Y.; Song, G.; Sun, M.; Wang, J.; Wang, Y. Prevalence and Therapies of Antibiotic-Resistance in Staphylococcus aureus. *Front. Cell. Infect. Microbiol.* **2020**, *10*, 107. [CrossRef] [PubMed]
10. Monteiro, K.L.C.; de Aquino, T.M.; Mendonça Junior, F.J.B. An Update on Staphylococcus aureus NorA Efflux Pump Inhibitors. *Curr. Top. Med. Chem.* **2020**, *20*, 2168–2185. [CrossRef] [PubMed]
11. Liao, W.; Liu, Y.; Zhang, W. Virulence evolution, molecular mechanisms of resistance and prevalence of ST11 carbapenem-resistant Klebsiella pneumoniae in China: A review over the last 10 years. *J. Glob. Antimicrob. Resist.* **2020**, *23*, 174–180. [CrossRef] [PubMed]
12. Pang, Z.; Raudonis, R.; Glick, B.R.; Lin, T.-J.; Cheng, Z. Antibiotic resistance in Pseudomonas aeruginosa: Mechanisms and alternative therapeutic strategies. *Biotechnol. Adv.* **2019**, *37*, 177–192. [CrossRef]
13. von Baum, H.; Marre, R. Antimicrobial resistance of Escherichia coli and therapeutic implications. *Int. J. Med. Microbiol.* **2005**, *295*, 503–511. [CrossRef] [PubMed]
14. Iovine, N.M. Resistance mechanisms in Campylobacter jejuni. *Virulence* **2013**, *4*, 230–240. [CrossRef]
15. Mahmood, H.Y.; Jamshidi, S.; Sutton, J.M.; Rahman, K.M. Current Advances in Developing Inhibitors of Bacterial Multidrug Efflux Pumps. *Curr. Med. Chem.* **2016**, *23*, 1062–1081. [CrossRef] [PubMed]
16. AlMatar, M.; Albarri, O.; Makky, E.A.; Köksal, F. Efflux pump inhibitors: New updates. *Pharmacol. Rep.* **2021**, *73*, 1–16. [CrossRef] [PubMed]
17. Jamshidi, S.; Sutton, J.M.; Rahman, K.M. An overview of bacterial efflux pumps and computational approaches to study efflux pump inhibitors. *Future Med. Chem.* **2016**, *8*, 195–210. [CrossRef] [PubMed]
18. Schindler, B.D.; Kaatz, G.W. Multidrug efflux pumps of Gram-positive bacteria. *Drug Resist. Updat.* **2016**, *27*, 1–13. [CrossRef]
19. Du, D.; Wang-Kan, X.; Neuberger, A.; van Veen, H.W.; Pos, K.M.; Piddock, L.J.V.; Luisi, B.F. Multidrug efflux pumps: Structure, function and regulation. *Nat. Rev. Microbiol.* **2018**, *16*, 523–539. [CrossRef]
20. Marquez, B. Bacterial efflux systems and efflux pumps inhibitors. *Biochimie* **2005**, *87*, 1137–1147. [CrossRef]
21. Sharma, A.; Gupta, V.K.; Pathania, R. Efflux pump inhibitors for bacterial pathogens: From bench to bedside. *Indian J. Med. Res.* **2019**, *149*, 129–145. [CrossRef]
22. Mahey, N.; Tambat, R.; Chandal, N.; Verma, D.K.; Thakur, K.G.; Nandanwar, H. Repurposing Approved Drugs as Fluoroquinolone Potentiators to Overcome Efflux Pump Resistance in Staphylococcus aureus. *Microbiol. Spectr.* **2021**, *9*, e0095121. [CrossRef]
23. Pereira da Cruz, R.; Sampaio de Freitas, T.; Socorro Costa, M.d.; Lucas Dos Santos, A.T.; Ferreira Campina, F.; Pereira, R.L.S.; Bezerra, J.W.A.; Quintans-Júnior, L.J.; De Souza Araújo, A.A.; Júnior, J.P.D.S.; et al. Effect of α-Bisabolol and Its β-Cyclodextrin Complex as TetK and NorA Efflux Pump Inhibitors in Staphylococcus aureus Strains. *Antibiotics* **2020**, *9*, 28. [CrossRef]
24. Zimmermann, S.; Klinger-Strobel, M.; Bohnert, J.A.; Wendler, S.; Rödel, J.; Pletz, M.W.; Löffler, B.; Tuchscherr, L. Clinically Approved Drugs Inhibit the Staphylococcus aureus Multidrug NorA Efflux Pump and Reduce Biofilm Formation. *Front. Microbiol.* **2019**, *10*, 2762. [CrossRef]
25. Xu, Q.; Sheng, Z.; Hao, M.; Jiang, J.; Ye, M.; Chen, Y.; Xu, X.; Guo, Q.; Wang, M. RamA upregulates multidrug resistance efflux pumps AcrAB and OqxAB in Klebsiella pneumoniae. *Int. J. Antimicrob. Agents* **2021**, *57*, 106251. [CrossRef] [PubMed]
26. Blair, J.M.A.; Richmond, G.E.; Piddock, L.J.V. Multidrug efflux pumps in Gram-negative bacteria and their role in antibiotic resistance. *Future Microbiol.* **2014**, *9*, 1165–1177. [CrossRef] [PubMed]
27. Schuster, S.; Vavra, M.; Schweigger, T.M.; Rossen, J.W.A.; Matsumura, Y.; Kern, W.V. Contribution of AcrAB-TolC to multidrug resistance in an Escherichia coli sequence type 131 isolate. *Int. J. Antimicrob. Agents* **2017**, *50*, 477–481. [CrossRef] [PubMed]
28. Guo, B.; Lin, J.; Reynolds, D.L.; Zhang, Q. Contribution of the multidrug efflux transporter CmeABC to antibiotic resistance in different Campylobacter species. *Foodborne Pathog. Dis.* **2010**, *7*, 77–83. [CrossRef]
29. Jeon, B.; Zhang, Q. Sensitization of Campylobacter jejuni to fluoroquinolone and macrolide antibiotics by antisense inhibition of the CmeABC multidrug efflux transporter. *J. Antimicrob. Chemother.* **2009**, *63*, 946–948. [CrossRef] [PubMed]

30. Sharifi, S.; Bakhshi, B.; Najar-Peerayeh, S. Significant contribution of the CmeABC Efflux pump in high-level resistance to ciprofloxacin and tetracycline in Campylobacter jejuni and Campylobacter coli clinical isolates. *Ann. Clin. Microbiol. Antimicrob.* **2021**, *20*, 36. [CrossRef]
31. Blanco, P.; Sanz-García, F.; Hernando-Amado, S.; Martínez, J.L.; Alcalde-Rico, M. The development of efflux pump inhibitors to treat Gram-negative infections. *Expert Opin. Drug Discov.* **2018**, *13*, 919–931. [CrossRef]
32. Stavri, M.; Piddock, L.J.V.; Gibbons, S. Bacterial efflux pump inhibitors from natural sources. *J. Antimicrob. Chemother.* **2007**, *59*, 1247–1260. [CrossRef]
33. Giorgini, G.; Mangiaterra, G.; Cedraro, N.; Laudadio, E.; Sabbatini, G.; Cantarini, M.; Minnelli, C.; Mobbili, G.; Frangipani, E.; Biavasco, F.; et al. Berberine Derivatives as Pseudomonas aeruginosa MexXY-OprM Inhibitors: Activity and In Silico Insights. *Molecules* **2021**, *26*, 6644. [CrossRef]
34. Khan, I.A.; Mirza, Z.M.; Kumar, A.; Verma, V.; Qazi, G.N. Piperine, a phytochemical potentiator of ciprofloxacin against Staphylococcus aureus. *Antimicrob. Agents Chemother.* **2006**, *50*, 810–812. [CrossRef] [PubMed]
35. Mirza, Z.M.; Kumar, A.; Kalia, N.P.; Zargar, A.; Khan, I.A. Piperine as an inhibitor of the MdeA efflux pump of Staphylococcus aureus. *J. Med. Microbiol.* **2011**, *60*, 1472–1478. [CrossRef] [PubMed]
36. Sharma, S.; Kumar, M.; Sharma, S.; Nargotra, A.; Koul, S.; Khan, I.A. Piperine as an inhibitor of Rv1258c, a putative multidrug efflux pump of Mycobacterium tuberculosis. *J. Antimicrob. Chemother.* **2010**, *65*, 1694–1701. [CrossRef]
37. Chan, B.C.L.; Ip, M.; Lau, C.B.S.; Lui, S.L.; Jolivalt, C.; Ganem-Elbaz, C.; Litaudon, M.; Reiner, N.E.; Gong, H.; See, R.H.; et al. Synergistic effects of baicalein with ciprofloxacin against NorA over-expressed methicillin-resistant Staphylococcus aureus (MRSA) and inhibition of MRSA pyruvate kinase. *J. Ethnopharmacol.* **2011**, *137*, 767–773. [CrossRef] [PubMed]
38. Holler, J.G.; Christensen, S.B.; Slotved, H.-C.; Rasmussen, H.B.; Gúzman, A.; Olsen, C.-E.; Petersen, B.; Mølgaard, P. Novel inhibitory activity of the Staphylococcus aureus NorA efflux pump by a kaempferol rhamnoside isolated from Persea lingue Nees. *J. Antimicrob. Chemother.* **2012**, *67*, 1138–1144. [CrossRef] [PubMed]
39. Freitas, T.S.; Xavier, J.C.; Pereira, R.L.S.; Rocha, J.E.; Campina, F.F.; de Araújo Neto, J.B.; Silva, M.M.C.; Barbosa, C.R.S.; Marinho, E.S.; Nogueira, C.E.S.; et al. In vitro and in silico studies of chalcones derived from natural acetophenone inhibitors of NorA and MepA multidrug efflux pumps in Staphylococcus aureus. *Microb. Pathog.* **2021**, *161*, 105286. [CrossRef] [PubMed]
40. Guo, Y.; Huang, C.; Su, H.; Zhang, Z.; Chen, M.; Wang, R.; Zhang, D.; Zhang, L.; Liu, M. Luteolin increases susceptibility to macrolides by inhibiting MsrA efflux pump in Trueperella pyogenes. *Vet. Res.* **2022**, *53*, 3. [CrossRef]
41. Stermitz, F.R.; Lorenz, P.; Tawara, J.N.; Zenewicz, L.A.; Lewis, K. Synergy in a medicinal plant: Antimicrobial action of berberine potentiated by 5′-methoxyhydnocarpin, a multidrug pump inhibitor. *Proc. Natl. Acad. Sci. USA* **2000**, *97*, 1433–1437. [CrossRef] [PubMed]
42. Kanagaratnam, R.; Sheikh, R.; Alharbi, F.; Kwon, D.H. An efflux pump (MexAB-OprM) of Pseudomonas aeruginosa is associated with antibacterial activity of Epigallocatechin-3-gallate (EGCG). *Phytomedicine* **2017**, *36*, 194–200. [CrossRef] [PubMed]
43. Sudano Roccaro, A.; Blanco, A.R.; Giuliano, F.; Rusciano, D.; Enea, V. Epigallocatechin-gallate enhances the activity of tetracycline in staphylococci by inhibiting its efflux from bacterial cells. *Antimicrob. Agents Chemother.* **2004**, *48*, 1968–1973. [CrossRef] [PubMed]
44. Opperman, T.J.; Kwasny, S.M.; Kim, H.-S.; Nguyen, S.T.; Houseweart, C.; D'Souza, S.; Walker, G.C.; Peet, N.P.; Nikaido, H.; Bowlin, T.L. Characterization of a novel pyranopyridine inhibitor of the AcrAB efflux pump of Escherichia coli. *Antimicrob. Agents Chemother.* **2014**, *58*, 722–733. [CrossRef] [PubMed]
45. Tohidpour, A.; Najar Peerayeh, S.; Mehrabadi, J.F.; Rezaei Yazdi, H. Determination of the efflux pump-mediated resistance prevalence in Pseudomonas aeruginosa, using an efflux pump inhibitor. *Curr. Microbiol.* **2009**, *59*, 352–355. [CrossRef]
46. Bailey, A.M.; Paulsen, I.T.; Piddock, L.J.V. RamA confers multidrug resistance in Salmonella enterica via increased expression of acrB, which is inhibited by chlorpromazine. *Antimicrob. Agents Chemother.* **2008**, *52*, 3604–3611. [CrossRef] [PubMed]
47. Pannek, S.; Higgins, P.G.; Steinke, P.; Jonas, D.; Akova, M.; Bohnert, J.A.; Seifert, H.; Kern, W.V. Multidrug efflux inhibition in Acinetobacter baumannii: Comparison between 1-(1-naphthylmethyl)-piperazine and phenyl-arginine-beta-naphthylamide. *J. Antimicrob. Chemother.* **2006**, *57*, 970–974. [CrossRef]
48. Nakashima, R.; Sakurai, K.; Yamasaki, S.; Hayashi, K.; Nagata, C.; Hoshino, K.; Onodera, Y.; Nishino, K.; Yamaguchi, A. Structural basis for the inhibition of bacterial multidrug exporters. *Nature* **2013**, *500*, 102–106. [CrossRef]
49. Kalia, N.P.; Mahajan, P.; Mehra, R.; Nargotra, A.; Sharma, J.P.; Koul, S.; Khan, I.A. Capsaicin, a novel inhibitor of the NorA efflux pump, reduces the intracellular invasion of Staphylococcus aureus. *J. Antimicrob. Chemother.* **2012**, *67*, 2401–2408. [CrossRef]
50. Zhang, Y.; Huang, Z.; Omari-Siaw, E.; Lu, S.; Zhu, Y.; Jiang, D.; Wang, M.; Yu, J.; Xu, X.; Zhang, W. Preparation and In Vitro-In Vivo Evaluation of Sustained-Release Matrix Pellets of Capsaicin to Enhance the Oral Bioavailability. *AAPS PharmSciTech* **2016**, *17*, 339–349. [CrossRef]
51. Naaz, F.; Khan, A.; Kumari, A.; Ali, I.; Ahmad, F.; Ahmad Lone, B.; Ahmad, N.; Ali Khan, I.; Rajput, V.S.; Grover, A.; et al. 1,3,4-oxadiazole conjugates of capsaicin as potent NorA efflux pump inhibitors of Staphylococcus aureus. *Bioorg. Chem.* **2021**, *113*, 105031. [CrossRef] [PubMed]
52. Azam, S.; Park, J.-Y.; Kim, I.-S.; Choi, D.-K. Piperine and Its Metabolite's Pharmacology in Neurodegenerative and Neurological Diseases. *Biomedicines* **2022**, *10*, 154. [CrossRef]

53. Deka, B.; Suri, M.; Sarma, S.; Devi, M.V.; Bora, A.; Sen, T.; Dihingia, A.; Pahari, P.; Singh, A.K. Potentiating the intracellular killing of Staphylococcus aureus by dihydroquinazoline analogues as NorA efflux pump inhibitor. *Bioorg. Med. Chem.* **2022**, *54*, 116580. [CrossRef]
54. Bharatham, N.; Bhowmik, P.; Aoki, M.; Okada, U.; Sharma, S.; Yamashita, E.; Shanbhag, A.P.; Rajagopal, S.; Thomas, T.; Sarma, M.; et al. Structure and function relationship of OqxB efflux pump from Klebsiella pneumoniae. *Nat. Commun.* **2021**, *12*, 5400. [CrossRef]
55. Hasdemir, U.O.; Chevalier, J.; Nordmann, P.; Pagès, J.-M. Detection and prevalence of active drug efflux mechanism in various multidrug-resistant Klebsiella pneumoniae strains from Turkey. *J. Clin. Microbiol.* **2004**, *42*, 2701–2706. [CrossRef] [PubMed]
56. Sun, L.; Sun, L.; Li, X.; Hu, X.; Wang, X.; Nie, T.; Zhang, Y.; You, X. A Novel Tigecycline Adjuvant ML-7 Reverses the Susceptibility of Tigecycline-Resistant Klebsiella pneumoniae. *Front. Cell. Infect. Microbiol.* **2021**, *11*, 809542. [CrossRef] [PubMed]
57. Chevalier, J.; Bredin, J.; Mahamoud, A.; Malléa, M.; Barbe, J.; Pagès, J.-M. Inhibitors of antibiotic efflux in resistant Enterobacter aerogenes and Klebsiella pneumoniae strains. *Antimicrob. Agents Chemother.* **2004**, *48*, 1043–1046. [CrossRef]
58. Mahamoud, A.; Chevalier, J.; Baitiche, M.; Adam, E.; Pagès, J.-M. An alkylaminoquinazoline restores antibiotic activity in Gram-negative resistant isolates. *Microbiology* **2011**, *157*, 566–571. [CrossRef]
59. Chevalier, J.; Mahamoud, A.; Baitiche, M.; Adam, E.; Viveiros, M.; Smarandache, A.; Militaru, A.; Pascu, M.L.; Amaral, L.; Pagès, J.-M. Quinazoline derivatives are efficient chemosensitizers of antibiotic activity in Enterobacter aerogenes, Klebsiella pneumoniae and Pseudomonas aeruginosa resistant strains. *Int. J. Antimicrob. Agents* **2010**, *36*, 164–168. [CrossRef]
60. Yan, S.; Chang, J.; Hao, X.; Liu, J.; Tan, X.; Geng, Z.; Wang, Z. Berberine regulates short-chain fatty acid metabolism and alleviates the colitis-associated colorectal tumorigenesis through remodeling intestinal flora. *Phytomedicine* **2022**, *102*, 154217. [CrossRef]
61. Feng, X.; Wang, K.; Cao, S.; Ding, L.; Qiu, F. Pharmacokinetics and Excretion of Berberine and Its Nine Metabolites in Rats. *Front. Pharmacol.* **2020**, *11*, 594852. [CrossRef]
62. Zhou, X.-Y.; Ye, X.-G.; He, L.-T.; Zhang, S.-R.; Wang, R.-L.; Zhou, J.; He, Z.-S. In vitro characterization and inhibition of the interaction between ciprofloxacin and berberine against multidrug-resistant Klebsiella pneumoniae. *J. Antibiot.* **2016**, *69*, 741–746. [CrossRef] [PubMed]
63. Li, Y.; Yao, J.; Han, C.; Yang, J.; Chaudhry, M.T.; Wang, S.; Liu, H.; Yin, Y. Quercetin, Inflammation and Immunity. *Nutrients* **2016**, *8*, 167. [CrossRef]
64. Pal, A.; Tripathi, A. Quercetin inhibits carbapenemase and efflux pump activities among carbapenem-resistant Gram-negative bacteria. *APMIS* **2020**, *128*, 251–259. [CrossRef]
65. Cheng, Y.; Liu, Y.; Chen, D.; Zhou, Y.; Yu, S.; Lin, H.; Liao, C.K.; Lin, H.; Xu, P.; Huang, M. Dual effects of quercetin on protein digestion and absorption in the digestive tract. *Food Chem.* **2021**, *358*, 129891. [CrossRef]
66. Pokharel, K.; Dawadi, B.R.; Bhatt, C.P.; Gupte, S. Prevalence of Pseudomonas Aeruginosa and its Antibiotic Sensitivity Pattern. *J. Nepal Health Res. Counc.* **2019**, *17*, 109–113. [CrossRef]
67. Sala, A.; Di Ianni, F.; Pelizzone, I.; Bertocchi, M.; Santospirito, D.; Rogato, F.; Flisi, S.; Spadini, C.; Iemmi, T.; Moggia, E.; et al. The prevalence of Pseudomonas aeruginosa and multidrug resistant Pseudomonas aeruginosa in healthy captive ophidian. *PeerJ* **2019**, *7*, e6706. [CrossRef] [PubMed]
68. Gan, R.-Y.; Li, H.-B.; Sui, Z.-Q.; Corke, H. Absorption, metabolism, anti-cancer effect and molecular targets of epigallocatechin gallate (EGCG): An updated review. *Crit. Rev. Food Sci. Nutr.* **2018**, *58*, 924–941. [CrossRef] [PubMed]
69. Zhang, S.; Mao, B.; Cui, S.; Zhang, Q.; Zhao, J.; Tang, X.; Chen, W. Absorption, metabolism, bioactivity, and biotransformation of epigallocatechin gallate. *Crit. Rev. Food Sci. Nutr.* **2023**, 1–21. [CrossRef]
70. He, Y.; Hu, Z.; Li, A.; Zhu, Z.; Yang, N.; Ying, Z.; He, J.; Wang, C.; Yin, S.; Cheng, S. Recent Advances in Biotransformation of Saponins. *Molecules* **2019**, *24*, 2365. [CrossRef]
71. Hazlett, L.D.; Ekanayaka, S.A.; McClellan, S.A.; Francis, R. Glycyrrhizin Use for Multi-Drug Resistant Pseudomonas aeruginosa: In Vitro and In Vivo Studies. *Investig. Ophthalmol. Vis. Sci.* **2019**, *60*, 2978–2989. [CrossRef]
72. Tambat, R.; Mahey, N.; Chandal, N.; Verma, D.K.; Jangra, M.; Thakur, K.G.; Nandanwar, H. A Microbe-Derived Efflux Pump Inhibitor of the Resistance-Nodulation-Cell Division Protein Restores Antibiotic Susceptibility in Escherichia coli and Pseudomonas aeruginosa. *ACS Infect. Dis.* **2022**, *8*, 255–270. [CrossRef]
73. Yuan, Y.; Rosado-Lugo, J.D.; Zhang, Y.; Datta, P.; Sun, Y.; Cao, Y.; Banerjee, A.; Parhi, A.K. Evaluation of Heterocyclic Carboxamides as Potential Efflux Pump Inhibitors in Pseudomonas aeruginosa. *Antibiotics* **2021**, *11*, 30. [CrossRef] [PubMed]
74. Zhang, Y.; Rosado-Lugo, J.D.; Datta, P.; Sun, Y.; Cao, Y.; Banerjee, A.; Yuan, Y.; Parhi, A.K. Evaluation of a Conformationally Constrained Indole Carboxamide as a Potential Efflux Pump Inhibitor in Pseudomonas aeruginosa. *Antibiotics* **2022**, *11*, 716. [CrossRef] [PubMed]
75. Abd El-Tawab, A.A.; Ammar, A.M.; Ahmed, H.A.; Hefny, A.A. Efflux Pump Inhibitors, Alpha-Tocopherol and Aspirin: Role in Campylobacter jejuni and Campylobacter coli Fluoroquinolone Resistance. *Microb. Drug Resist.* **2019**, *25*, 203–211. [CrossRef] [PubMed]
76. Zeng, B.; Wang, H.; Zou, L.; Zhang, A.; Yang, X.; Guan, Z. Evaluation and target validation of indole derivatives as inhibitors of the AcrAB-TolC efflux pump. *Biosci. Biotechnol. Biochem.* **2010**, *74*, 2237–2241. [CrossRef] [PubMed]
77. Cui, Z.-H.; He, H.-L.; Zheng, Z.-J.; Yuan, Z.-Q.; Chen, Y.; Huang, X.-Y.; Ren, H.; Zhou, Y.-F.; Zhao, D.-H.; Fang, L.-X.; et al. Phentolamine Significantly Enhances Macrolide Antibiotic Antibacterial Activity against MDR Gram-Negative Bacteria. *Antibiotics* **2023**, *12*, 760. [CrossRef]

78. Lu, W.-J.; Hsu, P.-H.; Chang, C.-J.; Su, C.-K.; Huang, Y.-J.; Lin, H.-J.; Lai, M.; Ooi, G.-X.; Dai, J.-Y.; Lin, H.-T.V. Identified Seaweed Compound Diphenylmethane Serves as an Efflux Pump Inhibitor in Drug-Resistant Escherichia coli. *Antibiotics* **2021**, *10*, 1378. [CrossRef]
79. García-Sánchez, L.; Melero, B.; Jaime, I.; Rossi, M.; Ortega, I.; Rovira, J. Biofilm formation, virulence and antimicrobial resistance of different Campylobacter jejuni isolates from a poultry slaughterhouse. *Food Microbiol.* **2019**, *83*, 193–199. [CrossRef] [PubMed]
80. Manach, C.; Scalbert, A.; Morand, C.; Rémésy, C.; Jiménez, L. Polyphenols: Food sources and bioavailability. *Am. J. Clin. Nutr.* **2004**, *79*, 727–747. [CrossRef]
81. Scalbert, A.; Morand, C.; Manach, C.; Rémésy, C. Absorption and metabolism of polyphenols in the gut and impact on health. *Biomed. Pharmacother.* **2002**, *56*, 276–282. [CrossRef]
82. Oh, E.; Jeon, B. Synergistic anti-Campylobacter jejuni activity of fluoroquinolone and macrolide antibiotics with phenolic compounds. *Front. Microbiol.* **2015**, *6*, 1129. [CrossRef]
83. Kovač, J.; Šimunović, K.; Wu, Z.; Klančnik, A.; Bucar, F.; Zhang, Q.; Možina, S.S. Antibiotic resistance modulation and modes of action of (-)-α-pinene in Campylobacter jejuni. *PLoS ONE* **2015**, *10*, e0122871. [CrossRef]
84. Martinez, A.L.; Lin, J. Effect of an efflux pump inhibitor on the function of the multidrug efflux pump CmeABC and antimicrobial resistance in Campylobacter. *Foodborne Pathog. Dis.* **2006**, *3*, 393–402. [CrossRef]
85. Kumar, A.; Ito, A.; Hirohama, M.; Yoshida, M.; Zhang, K.Y.J. Identification of new SUMO activating enzyme 1 inhibitors using virtual screening and scaffold hopping. *Bioorg. Med. Chem. Lett.* **2016**, *26*, 1218–1223. [CrossRef]
86. Ke, Y.-Y.; Coumar, M.S.; Shiao, H.-Y.; Wang, W.-C.; Chen, C.-W.; Song, J.-S.; Chen, C.-H.; Lin, W.-H.; Wu, S.-H.; Hsu, J.T.A.; et al. Ligand efficiency based approach for efficient virtual screening of compound libraries. *Eur. J. Med. Chem.* **2014**, *83*, 226–235. [CrossRef]
87. Brown, A.R.; Ettefagh, K.A.; Todd, D.; Cole, P.S.; Egan, J.M.; Foil, D.H.; Graf, T.N.; Schindler, B.D.; Kaatz, G.W.; Cech, N.B. A mass spectrometry-based assay for improved quantitative measurements of efflux pump inhibition. *PLoS ONE* **2015**, *10*, e0124814. [CrossRef] [PubMed]
88. Elhidar, N.; Nafis, A.; Goehler, A.; Abbad, A.; Hassani, L.; Mezrioui, N.-E.; Bohnert, J.A. Novel DiOC3 96-well real-time efflux assay for discovery of NorA efflux pump inhibitors in Staphylococcus aureus. *J. Microbiol. Methods* **2021**, *181*, 106128. [CrossRef] [PubMed]
89. Høiby, N.; Bjarnsholt, T.; Givskov, M.; Molin, S.; Ciofu, O. Antibiotic resistance of bacterial biofilms. *Int. J. Antimicrob. Agents* **2010**, *35*, 322–332. [CrossRef] [PubMed]
90. Reza, A.; Sutton, J.M.; Rahman, K.M. Effectiveness of Efflux Pump Inhibitors as Biofilm Disruptors and Resistance Breakers in Gram-Negative (ESKAPEE) Bacteria. *Antibiotics* **2019**, *8*, 229. [CrossRef]
91. Alav, I.; Sutton, J.M.; Rahman, K.M. Role of bacterial efflux pumps in biofilm formation. *J. Antimicrob. Chemother.* **2018**, *73*, 2003–2020. [CrossRef]
92. Kvist, M.; Hancock, V.; Klemm, P. Inactivation of efflux pumps abolishes bacterial biofilm formation. *Appl. Environ. Microbiol.* **2008**, *74*, 7376–7382. [CrossRef]
93. Singh, S.; Kalia, N.P.; Joshi, P.; Kumar, A.; Sharma, P.R.; Kumar, A.; Bharate, S.B.; Khan, I.A. Boeravinone B, A Novel Dual Inhibitor of NorA Bacterial Efflux Pump of Staphylococcus aureus and Human P-Glycoprotein, Reduces the Biofilm Formation and Intracellular Invasion of Bacteria. *Front. Microbiol.* **2017**, *8*, 1868. [CrossRef]
94. Xu, C.; Dong, N.; Chen, K.; Yang, X.; Zeng, P.; Hou, C.; Chi Chan, E.W.; Yao, X.; Chen, S. Bactericidal, anti-biofilm, and anti-virulence activity of vitamin C against carbapenem-resistant hypervirulent Klebsiella pneumoniae. *iScience* **2022**, *25*, 103894. [CrossRef]
95. Brannon, J.R.; Hadjifrangiskou, M. The arsenal of pathogens and antivirulence therapeutic strategies for disarming them. *Drug Des. Devel Ther.* **2016**, *10*, 1795–1806. [CrossRef]
96. Rampioni, G.; Pillai, C.R.; Longo, F.; Bondì, R.; Baldelli, V.; Messina, M.; Imperi, F.; Visca, P.; Leoni, L. Effect of efflux pump inhibition on Pseudomonas aeruginosa transcriptome and virulence. *Sci. Rep.* **2017**, *7*, 11392. [CrossRef] [PubMed]
97. El-Shaer, S.; Shaaban, M.; Barwa, R.; Hassan, R. Control of quorum sensing and virulence factors of Pseudomonas aeruginosa using phenylalanine arginyl β-naphthylamide. *J. Med. Microbiol.* **2016**, *65*, 1194–1204. [CrossRef] [PubMed]
98. Bina, X.R.; Philippart, J.A.; Bina, J.E. Effect of the efflux inhibitors 1-(1-naphthylmethyl)-piperazine and phenyl-arginine-beta-naphthylamide on antimicrobial susceptibility and virulence factor production in Vibrio cholerae. *J. Antimicrob. Chemother.* **2009**, *63*, 103–108. [CrossRef] [PubMed]
99. Pu, Y.; Zhao, Z.; Li, Y.; Zou, J.; Ma, Q.; Zhao, Y.; Ke, Y.; Zhu, Y.; Chen, H.; Baker, M.A.B.; et al. Enhanced Efflux Activity Facilitates Drug Tolerance in Dormant Bacterial Cells. *Mol. Cell.* **2016**, *62*, 284–294. [CrossRef]
100. Chang, J.; Lee, R.-E.; Lee, W. A pursuit of Staphylococcus aureus continues: A role of persister cells. *Arch. Pharm. Res.* **2020**, *43*, 630–638. [CrossRef] [PubMed]
101. Byrd, B.A.; Zenick, B.; Rocha-Granados, M.C.; Englander, H.E.; Hare, P.J.; LaGree, T.J.; DeMarco, A.M.; Mok, W.W.K. The AcrAB-TolC Efflux Pump Impacts Persistence and Resistance Development in Stationary-Phase Escherichia coli following Delafloxacin Treatment. *Antimicrob. Agents Chemother.* **2021**, *65*, e0028121. [CrossRef] [PubMed]
102. Povolo, V.R.; Ackermann, M. Disseminating antibiotic resistance during treatment. *Science* **2019**, *364*, 737–738. [CrossRef]
103. Nolivos, S.; Cayron, J.; Dedieu, A.; Page, A.; Delolme, F.; Lesterlin, C. Role of AcrAB-TolC multidrug efflux pump in drug-resistance acquisition by plasmid transfer. *Science* **2019**, *364*, 778–782. [CrossRef]

104. Busseti, M.I.I.; Margara, L.M.; Castell, S.D.; Fernández, M.M.; Malchiodi, E.L.; Montich, G.G.; Miguel, V.; Argaraña, C.E.; Monti, M.R. MutS recognition of mismatches within primed DNA replication intermediates. *DNA Repair* **2022**, *119*, 103392. [CrossRef]
105. El Meouche, I.; Dunlop, M.J. Heterogeneity in efflux pump expression predisposes antibiotic-resistant cells to mutation. *Science* **2018**, *362*, 686–690. [CrossRef] [PubMed]
106. Sjuts, H.; Vargiu, A.V.; Kwasny, S.M.; Nguyen, S.T.; Kim, H.-S.; Ding, X.; Ornik, A.R.; Ruggerone, P.; Bowlin, T.L.; Nikaido, H.; et al. Molecular basis for inhibition of AcrB multidrug efflux pump by novel and powerful pyranopyridine derivatives. *Proc. Natl. Acad. Sci. USA* **2016**, *113*, 3509–3514. [CrossRef]
107. Ayhan, D.H.; Tamer, Y.T.; Akbar, M.; Bailey, S.M.; Wong, M.; Daly, S.M.; Greenberg, D.E.; Toprak, E. Sequence-Specific Targeting of Bacterial Resistance Genes Increases Antibiotic Efficacy. *PLoS Biol.* **2016**, *14*, e1002552. [CrossRef]
108. Spengler, G.; Kincses, A.; Gajdács, M.; Amaral, L. New Roads Leading to Old Destinations: Efflux Pumps as Targets to Reverse Multidrug Resistance in Bacteria. *Molecules* **2017**, *22*, 468. [CrossRef] [PubMed]

pharmaceutics

MDPI

Review

Comprehensive Therapeutic Approaches to Tuberculous Meningitis: Pharmacokinetics, Combined Dosing, and Advanced Intrathecal Therapies

Ahmad Khalid Madadi [1] and Moon-Jun Sohn [1,2,*]

[1] Department of Biomedical Science, Graduate School of Medicine, Inje University, 75, Bokji-ro, Busanjin-gu, Busan 47392, Republic of Korea; khalidmadadi@yahoo.com

[2] Department of Neurosurgery, Neuroscience & Radiosurgery Hybrid Research Center, College of Medicine, Inje University Ilsan Paik Hospital, 170, Juhwa-ro, Ilsanseo-gu, Goyang City 10380, Republic of Korea

* Correspondence: mjsohn@paik.ac.kr; Tel.: +82-31-910-7730

Abstract: Tuberculous meningitis (TBM) presents a critical neurologic emergency characterized by high mortality and morbidity rates, necessitating immediate therapeutic intervention, often ahead of definitive microbiological and molecular diagnoses. The primary hurdle in effective TBM treatment is the blood–brain barrier (BBB), which significantly restricts the delivery of anti-tuberculous medications to the central nervous system (CNS), leading to subtherapeutic drug levels and poor treatment outcomes. The standard regimen for initial TBM treatment frequently falls short, followed by adverse side effects, vasculitis, and hydrocephalus, driving the condition toward a refractory state. To overcome this obstacle, intrathecal (IT) sustained release of anti-TB medication emerges as a promising approach. This method enables a steady, uninterrupted, and prolonged release of medication directly into the cerebrospinal fluid (CSF), thus preventing systemic side effects by limiting drug exposure to the rest of the body. Our review diligently investigates the existing literature and treatment methodologies, aiming to highlight their shortcomings. As part of our enhanced strategy for sustained IT anti-TB delivery, we particularly seek to explore the utilization of nanoparticle-infused hydrogels containing isoniazid (INH) and rifampicin (RIF), alongside osmotic pump usage, as innovative treatments for TBM. This comprehensive review delineates an optimized framework for the management of TBM, including an integrated approach that combines pharmacokinetic insights, concomitant drug administration strategies, and the latest advancements in IT and intraventricular (IVT) therapy for CNS infections. By proposing a multifaceted treatment strategy, this analysis aims to enhance the clinical outcomes for TBM patients, highlighting the critical role of targeted drug delivery in overcoming the formidable challenges presented by the blood–brain barrier and the complex pathophysiology of TBM.

Keywords: intrathecal; intraventricular; anti-TB drugs; tuberculous meningitis; prolonged drug delivery

1. Introduction

Tuberculous meningitis (TBM), the most severe manifestation of extrapulmonary tuberculosis (TB), is rare but carries a significant global impact [1]. It affects 1–5% of TB cases worldwide, constituting approximately 13.91% of meningitis cases (WHO Global Tuberculosis Report 2020) [2–4]. In 2019 alone, TBM resulted in an estimated 78,200 (95% UI; 52,300–104,000) adult deaths, accounting for 48–50% of incident cases globally, representing a significant proportion of incident TBM cases [5,6]. Vulnerable populations, including children under 5 and immunocompromised individuals (HIV-1 co-infected patients), face increased risks [7–9]. Delayed diagnosis and treatment can lead to fatality rates exceeding 50% and severe neurological complications [10]. Despite its lower prevalence, TBM carries significant mortality and morbidity rates, particularly among HIV-1 co-infected individuals

(50%) and children (19.3%) [11–14]. TBM, accounting for about 20% of childhood TB mortality, often results in neurological complications in over half of survivors [11,15].

TBM poses significant challenges in treatment due to the limited penetration of drugs into the central nervous system (CNS) and the emergence of drug-resistant strains. Our emphasis lies on prolonged intrathecal (IT) drug administration to bypass the barriers and mitigate systemic side effects and risks linked to repetitive IT injections while improving therapeutic outcomes by ensuring sufficient anti-TB drug concentrations at the site of infection. Current TBM treatment strategies involve long-term and complex antimicrobial therapy, often resulting in adverse effects and patient non-compliance [4,12,16]. Determining the optimal combinations, durations, doses, and frequencies of drug regimens for TBM remains uncertain, with insufficient evidence to guide empiric treatment regimens for TBM [11]. The blood–brain barrier (BBB) complicates matters by impeding drug delivery to the CNS, necessitating alternative administration methods to ensure optimal drug concentration in the cerebrospinal fluid (CSF) and brain [17].

The IT drug delivery, involving the direct injection of substances into the thecal sac housing CSF, represents a promising approach for drug administration. This strategy enables the attainment of elevated concentrations of therapeutic agents within the CNS, concurrently mitigating off-target exposure and the associated toxicity [18,19]. Despite its potential, IT therapy presents risks and limitations, encompassing chemical ventriculitis or meningitis, seizures, local adverse events, and infections [20,21]. Historically, IT/intraventricular (IVT) administration of isoniazid (INH) and rifampicin (RIF) for TBM, documented since 1955 and 1976, respectively, shows promise in refractory cases resistant to oral treatments, allowing for direct CNS targeting, enabling reduced dosages, minimizing side effects, and improving therapeutic outcomes without reported serious side effects. The initial documented use of IVT-RIF involved a TBM patient experiencing relapse with tuberculoma complications [22]. Cases involving IVT-RIF and systemic anti-TB drugs demonstrate high efficacy and safety without documented toxicity [23,24]. The treatment with IT amikacin (AMK) and levofloxacin (LVX) resulted in successful clinical and microbiological outcomes even in the case of multidrug-resistant TBM (MDR-TBM) [25]. The current WHO guidelines for treating TBM are derived from those established for pulmonary tuberculosis. These guidelines recommend a treatment regimen comprising two months of RIF, INH, pyrazinamide (PZA), and ethambutol (ETB), followed by a continuation phase of up to ten months consisting of RIF and INH for all TBM patients [26]. However, determining the optimal duration of treatment lacks a consensus based on evidence, resulting in variations in treatment duration observed between different countries [27,28]. The global prevalence of rifampicin/multidrug-resistant TBM (RR/MDR-TBM) is exhibiting an upward trend, leading to catastrophic prognostic implications, including a mortality rate exceeding 80% [29–31]. Therefore, early diagnosis of TBM and evaluation of its drug resistance and customization of treatment regimens to achieve a high concentration of anti-tuberculosis (RIF or INH) in CSF are essential for effectively managing RR/MDR-TBM [32].

This review meticulously synthesizes clinical experiences, case reports, and the pharmacokinetics–pharmacodynamics (PK-PD) of RIF and INH within the CSF, spotlighting the critical gaps in achieving sustained drug concentrations, exploring alternative drug formulations, and refining treatment regimens. The review emphasizes the potential of the potential of innovative IT drug delivery methods, such as osmotic pumps and nanoparticles (NP)-laden hydrogels, which promise enhanced and prolonged drug release directly into the CSF. These advancements aim to circumvent the limitations posed by traditional administration routes, offering further research to optimize dosage regimens and improve TBM treatment strategies. TBM persists as a formidable challenge within the spectrum of infectious diseases, primarily due to its severe impact on the central CNS and the intricate barriers to effective treatment. Among the myriad of hurdles, the PK dynamics of anti-TB drugs within the CSF stand out as a critical determinant of therapeutic success. This comprehensive review investigates the physiological and pharmacological factors influencing PK of anti-TB drugs in the CSF, emphasizing the theoretical approach and strategy required to optimize treatment for this devastating condition.

1.1. Physiology of BBB and CSF

A thorough grasp of the CSF system is vital for understanding how substances within it behave and the clinical implications for drug delivery. CSF serves essential roles in brain protection, maintaining balance, and regulating neuronal function. Numerous studies have demonstrated the critical importance of normal CSF production, circulation, and absorption for typical brain development and function.

In adults, CSF volume is approximately 150 mL, with about 25 mL in brain ventricles, 50 mL in cerebral subarachnoid space, and 75 mL in spinal subarachnoid space. Daily CSF production ranges from 400–600 mL, predominantly by the choroid plexuses of lateral ventricles. CSF composition includes water (99%), proteins at low concentrations, ions, neurotransmitters, and glucose. Its production is finely controlled by the autonomic nervous system and neuropeptides like dopamine and atrial natriuretic peptides. Sympathetic nervous system activity reduces CSF secretion, while cholinergic system activation enhances it. Following production in lateral ventricles, CSF flows through the interventricular foramen of Monro and aqueduct into the third and fourth ventricles, respectively. Subsequently, it passes through the median (foramen of Magendie) and lateral (foramina of Luschka) apertures in the fourth ventricle into the subarachnoid space at the base of the brain, then over the brain's convexity and along the spinal cord [33].

Recent studies have unveiled additional pathways involved in CSF and solute movement throughout the CNS. In an experimental study by Iliff et al., conducted in mice, it was observed that CSF traverses intracortical periarterial spaces akin to blood flow. Subsequently, CSF moves through perivascular astrocytic endfeet mediated by aquaporin-4 (AQP4) water channels, entering the parenchyma where it merges with interstitial fluid. CSF then exits the parenchyma through AQP4 channels, returning to the subarachnoid space via perivenous spaces. From there, it can follow two routes: drainage through arachnoid granulations into venous sinuses or exit via the lymphatic system. This insight was gleaned from injecting fluorescently labeled tracers through the cisterna magna of mice and tracking flow in real-time with in vivo two-photon microscopy. Both small (3 kilodaltons, kDa) and large (2000 kDa) tracers could traverse the perivascular space of penetrating arteries and arterioles. However, solute entry into the interstitial space was size-dependent, with small molecules dispersing quickly throughout the brain interstitium and larger molecules concentrating in the perivascular space. This discrepancy is attributed to the "sieving effect" of perivascular astrocytic endfeet, acting as a physical barrier with gap widths of approximately 20 nanometers. Moreover, clearance of solutes occurs from the interstitial space along perivenous spaces to the subarachnoid space and beyond. Polarized AQP4 water channels on astrocyte endfeet facilitate CSF flow into and through the interstitium while clearing fluid and solutes from the interstitial space [34]. AQP4 constitutes a crucial component of the glymphatic system, characterized by 0.3- to 0.6-nanometer pores selectively permeable to water, widely expressed on capillary-facing astrocyte endfeet along perivascular spaces. However, the precise mechanism by which AQP4 facilitates solute elimination and bidirectional delivery to the parenchyma remains poorly understood [35]. These experiments introduced the concept of the "glymphatic system", a waste clearance system of the CNS utilizing perivascular channels formed by astroglial endfeet, responsible for removing amyloid β and other waste products from the CNS [34]. Traditionally, CSF absorption was thought to occur primarily in arachnoid villi along superior sagittal and intracranial venous sinuses and around spinal nerve roots. However, recent studies propose additional arachnoid pathways as major sites of CSF absorption, including passage through brain parenchyma, lymphatics near the cribriform plate, or perineural sheaths of cranial nerves [33].

1.2. Physiological Factors Influencing IT-PK

The blood–brain and blood–CSF barriers: The main challenge in treating TBM lies in the BBB and the blood–CSF barrier (BCSFB), intricate networks of lipid layers designed to shield the CNS from potential threats. These barriers, while vital for neuroprotection, significantly impede the entry of anti-TB drugs to the sites of infection within the CNS, necessitating higher systemic doses that often come with increased toxicity risks. The delicate balance between protecting the brain and allowing therapeutic agents access is a critical consideration in developing effective TBM treatments.

The efficacy of anti-TB drugs in treating TBM is contingent upon achieving a specific exposure level at the infection site, particularly within the CSF. Determining the appropriate dosing regimen for IT administration necessitates a consideration of various factors. These factors include the BBB and BCSFB, the size of the CSF system, CSF volume, the site of antibiotic administration (IT or IVT), and drainage volume. The BBB and BCSFB encompass lipid layers enveloping the CNS to protect the brain from harmful agents circulating in the bloodstream. These barriers play a crucial role in restricting the entry of antibiotics to the infection site during the treatment of meningitis or ventriculitis.

The role of inflammation in drug permeability: Inflammation, a characteristic response to TBM, disrupts the BBB's integrity, variably enhancing the CSF's permeability to antibiotics, especially those with hydrophilic properties [36–38]. This disruption, though potentially beneficial in increasing drug penetration, introduces a layer of unpredictability in treatment outcomes. This effect is more pronounced in meningitis than in ventriculitis. Neurosurgical procedures may further disrupt the BBB, potentially increasing systemic antibiotic penetration [39]. However, the extent of penetration during inflammation is unpredictable. The extent of BBB disruption and its impact on drug delivery efficacy underscore the need for adaptable and precise dosing regimens that can account for the dynamic nature of the disease.

The CSF compartment: A challenge of volume and flow. The CSF compartment's complexity adds another layer of difficulty in achieving optimal drug delivery. Variability in the compartment's size, influenced by factors such as age and disease, affects the distribution volume of administered drugs. Furthermore, the non-uniform distribution and directional flow of CSF necessitate careful consideration of the site of antibiotic administration. The choice between IT and IVT dosing can significantly impact the distribution and efficacy of anti-TB drugs within the CNS.

The CSF compartment's size, crucial for determining the distribution volume post-BBB penetration, shows significant interindividual variability, influenced by age, ventricular and subarachnoid space dimensions, and underlying diseases. The dimension of the CSF compartment, a pivotal determinant for the volume of distribution (V_{CSF}) in IT dosing following BBB penetration, demonstrates notable variability. The V_{CSF} encompasses the four ventricles, an aqueduct, basal cisterns, and the subarachnoid space. Interindividual variations in V_{CSF} size are influenced by factors such as age, widths of ventricles and subarachnoid space, dimensions of the spinal canal, and underlying diseases, all contributing to the diversity in this critical parameter [40]. Studies have documented varying mean CSF volumes across different conditions, with the presence of clots potentially reducing this volume. Numerous studies have reported mean CSF sizes in healthy adults ranging from 250 to 326 mL [41–43], in communicating hydrocephalus patients at 488 mL, and in non-communicating hydrocephalus patients at 593 mL [43]. The presence of clots in ventricles or basal cisterns may result in a reduction in V_{CSF} size. Despite the intricate nature of the CSF compartment, its distribution exhibits non-uniformity [40,44]. Approximately two-thirds of the CSF is generated by the choroid plexus, with the remaining one-third originating from the extracellular space of the brain and spinal cord. CSF equilibration relies on oscillations induced by heartbeat and respiration. The overall flow of CSF occurs from the ventricles to the cisterna magna, cerebral convexities, and the spinal canal. The concentrations of antibiotics obtained through lumbar puncture and extra ventricular drain sampling may vary. Following entry into the CSF, antibiotics can diffuse into the extracellular fluid of the brain and spinal cord due to the absence of a tight barrier, contrasting with the bulk flow of CSF [40].

PK considerations for effective drug delivery: Effective management of TBM requires not only overcoming physiological barriers but also considering the PK landscape of drug administration. The drainage volume of CSF, influenced by factors such as meningitis-induced production changes and external drainage systems, plays a crucial role in the clearance of antibiotics from the CSF. Understanding these dynamics is essential for timing IT doses to maximize therapeutic efficacy while minimizing the risk of drug resistance.

Given the complexity of physiological factors affecting IT-PK, it becomes increasingly clear that a one-size-fits-all approach to TBM treatment is insufficient. The advent of advanced drug delivery mechanisms, such as nanoparticle-infused hydrogels and osmotic pumps, offers promising avenues for achieving sustained drug release within the CNS. These innovations, coupled with a deeper understanding of the physiological and pharmacological factors at play, pave the way for tailored, effective treatment strategies that can navigate the complex terrain of TBM therapy.

Selecting the administration site for IT doses is of paramount importance due to the non-homogeneous characteristics of the CSF compartment and the directional flow of CSF from the ventricles. Discrepancies in drug concentrations are noted across the ventricular, cisternal, and lumbar regions within the CSF compartment [45]. As IT doses are frequently employed for ventriculitis, ensuring optimal antibiotic exposure in the ventricles becomes essential, making IVT administration preferable. In contrast to intravenous (IV) administration, where lumbar CSF concentrations generally surpass ventricular CSF concentrations for most drugs [36], IVT dosing guarantees distribution throughout the CSF compartment unless impeded by factors such as bleeding. The intraluminal or intraspinal administration of antibiotics leads to decreased concentrations in ventricular CSF attributed to the net flow of CSF originating from the ventricles [46]. This phenomenon results in an uneven distribution within the CSF space and the possibility of insufficient drug levels in the ventricles. PK data endorse IVT dosing, especially in the management of ventriculitis, despite the absence of current clinical evidence supporting this preference [38]. The drainage volume, indicating the daily CSF elimination in mL/day, significantly influences the clearance of antibiotics from the CSF. Keeping the external drain at a consistent level facilitates CSF excretion during elevated intracranial pressure surpassing drain pressure, and any changes in drain levels may impact drainage volume. Research highlights the crucial role of drainage volume in determining IT antibiotic doses [47–49]. CSF production, exhibiting circadian variation, reaches a minimum of 12 ± 7 mL/h around 6 p.m. and a nightly peak of 42 ± 2 mL/h around 2 a.m. in both healthy volunteers and patients with external ventriculostomies [50,51]. Meningeal inflammation reduces CSF production, affecting drainage volume and antibiotic clearance, emphasizing the importance of timing IT doses [36].

In summary, the strategy for optimizing TBM treatment is fraught with challenges, from considering physiological barriers to understanding the intricacies of ITPK. By embracing a multifaceted approach that combines PK insights with innovative drug delivery systems (DDS), we can move closer to turning the tide against this devastating disease, offering hope for improved outcomes in the fight against TBM.

1.3. Advancing IT Therapy for TBM: Beyond Physiological Barriers to PK Optimization

PK factors affecting IT therapy: The journey of anti-TB drugs from the systemic circulation to the CSF is influenced by a myriad of PK variables. The clearance of drugs from CSF to blood is intricately affected by the drugs' inherent characteristics. This process is governed by bulk flow, retrograde diffusion across the blood–CSF and BBB, and active transport mechanisms. The dynamics of these processes underscore the nuanced approach needed to ensure that therapeutic agents reach their target sites within the CNS effectively.

The IT-PKs of anti-TB drugs are greatly impacted by the drugs' intrinsic characteristics and their systemic administration. The process of clearing drugs from the CSF to the blood, indicated as CLCSF to Blood, is profoundly influenced by the drugs' properties. The elimination process is dependent on several drug-specific factors. Key elements that influence CLCSF to Blood include (1) bulk flow, which is equivalent to the rate of CSF production, (2) retrograde diffusion across the blood–CSF and BBB, and (3) active transport mechanisms [52]. Bulk flow primarily facilitates the removal of large hydrophilic molecules, leading to decreased clearance rates in patients with hydrocephalus [40,53]. In contrast, small to moderately lipophilic molecules predominantly are cleared through passive elimination via retrograde diffusion across the barriers [54].

The Role of Drug Properties in CSF Penetration: The PK behavior of anti-TB drugs in the CSF is significantly dictated by their hydrophilic or lipophilic nature, molecular weight (MW), and protein binding affinity. Hydrophilic antibiotics face challenges in penetrating the CSF, highlighting the necessity for strategic dosing to circumvent these limitations. Conversely, lipophilic drugs, with their capacity for easier equilibration and membrane binding, present a different set of considerations for achieving optimal CSF distribution. The impact of drug characteristics on treatment efficacy illuminates the path toward tailored therapeutic strategies that account for these PK characteristics.

Variations in BBB permeability and CSF volume distribution are linked to dissimilarities in the characteristics of hydrophilic and lipophilic antibiotics. Hydrophilic antibiotics demonstrate restricted entry into the CSF, with the CSF volume of distribution determined by the combination of CSF volume distribution and the portion of brain extracellular space equilibrating with CSF [55,56]. Conversely, lipophilic drugs, facilitated by easier equilibration and membrane binding, generally exhibit a greater CSF volume of distribution when compared to hydrophilic drugs. In instances of meningitis, inflammation-induced BBB penetration is more prominent for hydrophilic antibiotics, whereas lipophilic antibiotics exert a minimal impact [38]. The entry into the CSF is markedly affected by both molecular weight and protein binding [57]. In instances where the barrier is intact, only the unbound fraction can traverse the CSF, as binding proteins generally exhibit minimal crossing of the barriers [57]. Compounds characterized by low protein binding display an increased unbound fraction, thereby promoting improved penetration into the CNS, with CSF protein binding generally being lower in comparison to serum/plasma [58]. Furthermore, molecular weight serves as a decisive factor in BBB penetration, with substances featuring larger molecular weights typically experiencing restricted penetration.

Optimizing treatment through combined IT and IV administration: The strategic combination of IT and IV therapies has been identified as a crucial approach for overcoming the challenges posed by the BBB and enhancing drug levels within the CSF. This integrated method not only facilitates higher antibiotic concentrations in the CSF but also mitigates the risks of bacterial resistance and disease relapse. The synergy between systemic and local drug administration, leveraging the strengths of both approaches, represents a forward-thinking strategy in the management of MDR-CNS infections.

The combined use of IT and IV treatments results in slightly higher antibiotic levels in CSF compared to IVT therapy alone, which is crucial for managing MDR-CNS infections [59]. Simultaneous IVT and IV administration of antibiotics prevent low concentrations in brain compartments, reducing the likelihood of resistant bacteria selection and relapse [60]. Concurrent systemic antibiotic therapy, utilizing the same or alternative effective agents, is strongly endorsed and backed by numerous studies showing successful IT antibiotic administration alongside IV therapy [40]. Simultaneous IV and local administrations are deemed optimal, considering CSF circulation and obstruction [48]. However, the effect of simultaneous IV dosing on CSF-PK for all antibiotics remains uncertain [40,52]. While most patients receive a combination of IV and IT doses, evidence supporting a shift to IT-only regimens is lacking. The lack of data solely focusing on IT drug administration is attributed to the necessity of combining IT and IV therapy for optimal therapeutic outcomes [40]. In CNS infections caused by MDR pathogens, the combination of IVT and systemic antimicrobial therapy is acknowledged as potentially life-saving [40].

The imperative for PK research and innovation: Despite the advancements in understanding the PKs of IT therapy, significant gaps remain in our knowledge, particularly regarding the optimal integration of systemic and local drug administration. The lack of comprehensive data focused solely on IT drug delivery underscores the urgent need for research aimed at refining dosage regimens and exploring novel therapeutic interventions. The potential of emerging technologies, such as nanoparticle-infused hydrogels and osmotic pumps, in achieving sustained drug release within the CNS offers a promising avenue for future exploration.

Pioneering a new era in TBM treatment: The effective management of TBM at the intersection of PKs and advanced drug delivery technologies presents a multifaceted challenge. However, it also offers an opportunity to significantly improve patient outcomes. By harnessing insights into drug properties, systemic administration, and innovative delivery methods, we can develop a comprehensive framework for TBM treatment that transcends traditional barriers. This integrated approach not only promises to enhance the precision and efficacy of therapy but also marks a step towards a new era in the battle against TBM, one that is characterized by informed, patient-centric, and outcome-driven strategies.

2. Materials and Method

This comprehensive review synthesizes findings from electronic scientific databases, including PubMed, Google Scholar, and Science Direct, focusing on English-language articles related to the use of "Intrathecal or Intraventricular Rifampicin" and "Intrathecal or Intraventricular Isoniazid" in the treatment of TBM. Our initial search yielded 3277 articles, from which 169 were selected through database searches. This pool was further enriched by manually adding references found in recent research articles, systematic reviews, meta-analyses, clinical trials, randomized controlled trials, cohort studies, clinical experiences, and case reports dating from 1958 to the current day. After a thorough screening for title relevance and the removal of duplicate entries, 3108 articles were excluded. The references of the remaining literature were manually checked for completeness, ensuring the inclusion of all pertinent information regarding the IT and IVT treatment modalities for TBM. The inclusion criteria were English-language publications and original research, excluding citations, patents, non-English articles, and abstracts lacking data.

The methodology of this review follows a thematic structure outlined as follows: First, the introduction is presented, followed by an examination of the physiological factors influencing IT-PKs, and subsequently, the advancing IT therapy for TBM. Next, the results obtained from the literature search are detailed. Following this, an analysis of the PKs of systemic anti-TB drugs in CSF is conducted. Additionally, a comparative analysis is undertaken between combined systemic dosing and IT/IVT administration alone. The general aspects of IT/IVT drug administration are then discussed, followed by an exploration of therapeutic drug monitoring. Furthermore, clinical experiences concerning IT/IVT anti-TB treatment are reviewed, along with an evaluation of adverse events and the associated risks of IT/IVT drug administration. Strategies for optimizing therapeutic regimens for IT anti-TB therapy are explored, including discussions on strategies for IT-prolonged drug delivery, such as the clinical application of an osmotic pump and nanoparticle-laden hydrogel.

3. Results

3.1. *PK of Systemic Anti-TB Drug in CSF: General Overview, Importance, and Challenges*

Understanding the PKs of first-line anti-TB drugs is essential for the effective management of TB, highlighting the significance of absorption, distribution, metabolism, and excretion processes. Accurate PK knowledge is critical to prevent suboptimal drug levels, thus reducing the risk of treatment failure, mortality [61], and adverse effects, such as drug-induced liver injury (DILI) [62]. The application of PK principles can significantly enhance TB treatment outcomes, especially in cases involving multidrug resistance [63]. Despite advancements, significant gaps remain in PK data for anti-TB drugs, notably among pediatric patients with TBM [61].

TBM represents one of the most severe extrapulmonary manifestations of TB, necessitating a nuanced approach to treatment that hinges on the effective penetration of anti-TB drugs into the CNS. Among the arsenal of first-line anti-TB drugs, INH and RIF stand out for their critical roles in TBM treatment, albeit with contrasting PK profiles that influence their efficacy and therapeutic strategies.

An essential element of TB management is the ability of drugs to cross the BBB, typically facilitated by lipophilic drugs with a molecular weight under 400 g/mol and fewer than eight hydrogen bonds, allowing lipid-mediated free diffusion across the BBB [64]. However, the penetrative efficacy of anti-TB medications in the CNS varies significantly. RIF, for instance, whether administered orally or intravenously, often results in suboptimal CSF concentrations due to limited CNS penetration [61]. In contrast, INH effectively penetrates the CNS, achieving peak CSF concentrations comparable to plasma levels in both children and adults [65–67]. This variation in drug penetration underlines the critical need for dose optimization strategies, particularly for the treatment of extrapulmonary TB (EPTB) manifestations, such as meningitis and bone/joint disease, to ensure therapeutic efficacy and improve patient outcomes. A review analyzing CSF concentrations of INH after administration revealed significant variability across different dosages and methodologies, with INH concentrations in CSF ranging from 0.55 to 14.1 μg/mL after 3 h of administration at a dosage of 1.5 to 20 mg/kg in both children and adults [67]. This variability underscores the complexity of achieving optimal therapeutic levels in the treatment of TBM.

The PK-PD metrics crucial for evaluating the efficacy of anti-TB are the ratio of C_{max} to MIC (minimal inhibitory concentration) and the ratio of AUC (area under curve) at the end of the dosing interval relative to MIC (AUC_{0-24}/MIC). These measures highlight the critical role of MIC values in determining drug efficacy against *Mycobacterium tuberculosis* [63]. In a comprehensive study, MICs for 14 anti-TB drugs were assessed against *Mycobacterium tuberculosis* strains. Remarkably, MICs for these drugs consistently remained within a narrow range, indicating susceptibility across all tested strains. The MIC values, reported in μg/mL, demonstrated variability across the drugs: INH (0.02 to 0.04), RIF (0.2 to 0.4), ETB and streptomycin (0.5–2.0), ethionamide (0.25–0.5), D-cycloserine (25–75), capreomycin (CM) (1–2), kanamycin (KAN) (2–4), amikacin (AMK) (0.5–1.0), clofazimine (0.1–0.4), ofloxacin (0.5–1.0), ciprofloxacin (0.25–1.0), and sparfloxacin (0.1–0.4) [68]. Intriguingly, the study revealed that RIF, even at a minimal concentration of 0.2 μg/mL, was effective in inhibiting the metabolism of susceptible bacterial strains. Conversely, at concentrations up to 32 μg/mL, RIF showed minimal impact on drug-resistant strains of *M. tuberculosis*, highlighting the challenge of combating resistance and the necessity for precise dosing strategies [68].

A prospective observational study conducted in Indonesia involving 20 TBM patients provided insights into the PK of INH, RIF, and PZA. This study uncovered that RIF displayed suboptimal RIF AUC_{0-24} levels for TBM treatment and notably low concentrations in CSF. Additionally, a significant correlation was identified between elevated AUC_{0-24} and C_{max} values for INH, RIF, and PZA and the incidence of DILI, emphasizing the need for meticulous drug monitoring. Moreover, within the initial four weeks, four patients encountered grade 2–3 DILI, leading to a temporary discontinuation in drug therapy. However, upon the drugs' reintroduction, no subsequent cases of DILI were noted. Notably, patients who developed DILI exhibited higher AUC_{0-24} for INH, RIF, and PZA, as well as elevated C_{max} values for INH and PZA on day 10 relative to those who did not experience DILI ($p < 0.05$), indicating a potential link between drug exposure levels and the risk of DILI [61].

RIF encounters significant barriers in achieving therapeutic CSF levels for TBM treatment due to its high molecular weight, impeding its BBB passage. The effectiveness of RIF in treating TBM critically depends on exceeding specific MIC levels in the CSF, a criterion not met by current treatment protocols. This often results in sub-therapeutic drug concentrations in the majority of TBM cases, underscoring the imperative for precise dosage adjustments to amplify RIF's therapeutic effects. An in-depth examination of RIF's PKs, including its absorption, metabolism, and excretion, alongside a comparative analysis of oral versus IV administration, reveals the complexity involved in optimizing CNS drug delivery. These insights strongly advocate for customized dosing strategies to enhance TBM treatment efficacy, highlighting the challenges and proposing potential solutions for attaining therapeutic drug concentrations within the CSF.

In a PKs assessment involving 20 participants, INH demonstrated an $AUC_{0–24}$ of 18.5 (5.1–47.4) h·mg/L, with a C_{max} of 4.6 (1.0–10.0) mg/L. Concentration in CSF varied across time intervals, ranging from 1.4 (0.5–6.1) to 1.6 (1.2–2.5) mg/L in the first two hours, and from 1.3 (1.2–4.3) to 2.3 (1.9–2.8) mg/L between six to eight hours. RIF exhibited an $AUC_{0–24}$ of 66.9 (21.7–118.6) h·mg/L and a plasma C_{max} of 9.4 (2.9–23.7) mg/L. However, its concentration in CSF was notably lower, ranging from 0.2 (0.1–0.4) to 0.4 (0.1–1.4) mg/L across different intervals. PZA showed the highest $AUC_{0–24}$ of 315.5 (100.6–599.0) h·mg/L and a C_{max} of 37.7 (15.9–61.7) mg/L. Concentration in CSF ranged from 24.4 (11.1–54.9) to 19.6 (7.2–37.7) mg/L across various intervals. A subsequent evaluation with 12 participants confirmed these patterns, with no significant variances in PK parameters observed between the assessments [61].

3.1.1. PK of RIF in CSF

RIF, a first-line anti-TB medication [69], is notably less effective at reaching therapeutic CSF levels due to its substantial molecular weight (Mw 822.9 g/mol), impeding BBB permeability [70]. Such limitation is exacerbated by plasma protein binding, with only ~10–20% of RIF reaching the CSF [71]. Crucially, for RIF to be effective in TBM treatment, CSF MIC values must exceed 15 μg/mL [22], a condition unmet by existing treatment guidelines, leading to sub-therapeutic RIF levels in up to 89% of TBM patients [72]. This dilemma highlights the necessity of carefully tailored dosage adjustments. RIF shows a dose-dependent elevation in serum concentration [73–77], underscoring the importance of meticulous dosage adjustments to enhance its efficacy. In the treatment of bacterial meningitis, it is crucial to rapidly eliminate bacteria from the CSF to ensure survival without neurological impairment [78]. Achieving therapeutic levels of antimicrobial drugs within the CSF is essential for eradicating bacteria in TBM, influenced by both drug PKs and the strain's susceptibility [79].

RIF metabolism occurs in the liver, with hepatic esterase transforming RIF into deacetyl-RIF [80]. Both RIF and its metabolite undergo biliary excretion [80,81], with approximately 17% of a 600 mg RIF dose recovered unchanged in urine [81]. Liver excretion saturates between 300–450 mg doses, leading to higher serum concentrations. With repeated doses, the body's self-metabolism increases during the initial treatment phase, causing serum concentration and half-life to decrease [82]. Roughly 80% of RIF binds to albumin, dispersing across tissues [82]. The elimination half-life of RIF in CSF is extended compared to serum, ranging from 9.1 to 21 h with "uninflamed" meninges, with a median of 14.5 h, $n = 5$). This is notably longer than the serum half-life of RIF, which ranges from 2.2 to 5.8 h, with a median of 3.6 h; ($n = 7$) [83]. The distribution of RIF within the CNS is restricted, impacting its efficacy. In a study using a standard dosing regimen, a large proportion of TBM patients exhibit undetectable levels of RIF in their CSF [84], rarely exceeding 1 μg/mL [67,77,85] as approximately only about ~5% of plasma RIF reaches the CSF due to high protein binding, indicating poor penetration into the CNS, falling below the MIC required for the eradication of *Mycobacterium tuberculosis* and potentially leading to suboptimal dosing and treatment failure [72].

A study examining CSF concentrations of RIF reported variability at different time intervals after administration, with doses ranging from 3 mg/kg to 20 mg/kg in both children and adults. Notably, CSF RIF levels varied from 0.14 to 1.7 μg/mL three hours after administration, indicating the challenges of achieving therapeutic CSF levels with oral RIF in TBM treatment [67]. Such PK insights necessitate the development of optimized dosing regimens, potentially necessitating higher doses for children and adults alike to effectively combat TBM, reduce mortality rates, and address the critical challenges in CNS drug delivery.

This comprehensive analysis underscores the complexity of achieving adequate therapeutic levels of RIF within the CSF for TBM treatment and the urgent need for innovative dosing strategies to improve patient outcomes.

Upon oral administration, RIF is rapidly and completely absorbed in the intestines, especially when taken on an empty stomach. A single 600 mg dose achieves a peak serum concentration of 10 μg/mL within 2 h, corresponding to a half-life of 2.5 h [82]. RIF is metabolized in the liver through deacetylation by hepatic esterase, forming diacetyl-RIF, a more polar and microbiologically active metabolite. It is excreted in similar quantities in both bile and urine, though the presence of diacetyl-RIF is more likely to be found in bile, at a ratio four times higher than its presence in urine [82]. Given its potential for DILI with long-term or high-dose usage, cautious application of RIF is advised for patients with renal and hepatic impairment [86].

A comparative study of oral and IV RIF administration indicated that higher oral doses (750 mg or 16.7 mg/kg and 900 mg or 18 mg/kg) achieve similar total plasma RIF levels as a 600 mg (13.3 mg/kg) IV dose within 1.5 h. IV administration, which has been associated with reduced mortality in TBM, results in higher peak plasma concentrations [85]. In African patients with HIV-associated TBM, the plasma RIF $AUC_{0–24}$ was greater for an oral dose of 35 mg/kg than for an IV dose of 20 mg/kg, though C_{max} values remained comparable [87].

A review of 18 studies revealed that only seven reported RIF CSF concentrations exceeding 1.0 μg/mL [67]. Notably, higher CSF concentrations of RIF were observed in four cases where children received dosages ranging from 15 to 20 mg/kg. Moreover, one study documented mean RIF CSF concentrations of 2.01 μg/mL (range 0.8–3.5 μg/mL) six hours post-dosing in adults with TBM. Another report on seven adults requiring ventriculostomy indicated median ventricular CSF RIF concentrations of 0.73 μg/mL (range 0.57–1.24 μg/mL). Generally, recorded RIF CSF concentrations did not consistently surpass the MIC for *M. tuberculosis*, suggesting suboptimal therapeutic efficacy.

Although some efficacy might persist at these low concentrations, the situation cannot be considered optimal. The highest CSF concentration of RIF is typically observed in children receiving a 20 mg/kg dosage of RIF. Given that children generally exhibit lower serum concentrations of RIF compared to adults at standard dosages (8–12 mg/kg), advocating higher dosages for TBM management in children seems reasonable. Dosages of at least 20 mg/kg in young children and infants weighing less than 10 kg, and at least 15 mg/kg in those weighing between 10 and 20 kg, are suggested. The optimal CSF concentration of RIF was typically seen in children at a 20 mg/kg dosage, suggesting a need for higher dosage recommendations for TBM management in pediatric populations [67].

Ellard and colleagues emphasize the vital importance of RIF in the treatment of TBM, despite its limited penetration into the CSF. The significance of RIF, especially in addressing drug-resistant TB strains and managing severe disease symptoms, including widespread lesions and the need for an effective therapeutic response, is highlighted. This focus emphasizes the necessity for comprehensive treatment strategies, particularly in the face of INH resistance, to ensure an effective response against TB. This emphasis sheds light on the necessity for robust treatment approaches, particularly in scenarios where INH resistance is present, to secure an effective therapeutic response against TB [88].

In a study involving 27 Chinese TBM patients, the PK of RIF following oral dosage (11 mg/kg) revealed slow penetration of RIF concentrations in serum and CSF from 2 to 6 h. Serum concentrations were initiated at 11.5 mg/L (2 h) and declined to 4.7 mg/L (6 h), maintaining levels slightly above the MIC for *M. tuberculosis* (0.3 mg/L). Poor CSF penetration was observed, with CSF/serum ratios increasing from 0.04 (2 h) to 0.11 (6 h) [88]. This highlights challenges in achieving therapeutic CSF levels with oral RIF in TBM treatment. PK targets, such as RIF AUC_{24} (at least 116 μg·h/mL (equivalent to AUC6 of 70 μg·h/mL) and C_{max} (22 μg/mL) have been recommended for optimizing TBM treatment, emphasizing the need for adequate drug exposure in the CSF [89]. Peak CSF concentrations, observed 0–8 h (median = 1 h) post-infusion completion, varied between 0.57 to 1.24 mg/L (median = 0.73 mg/L) [83]. PK modeling studies indicate that for childhood TBM, RIF doses of at least 30 mg/kg orally or 15 mg/kg intravenously may be necessary to attain adequate CSF drug concentrations and effectively eradicate bacteria, thus lowering mortality rates. However, it remains unclear whether similar dosing regimens are applicable to adults [90].

IV administration of RIF has shown promise, achieving higher plasma and CSF concentrations compared to oral dosing, potentially offering a pathway to optimizing CNS drug delivery. IV RIF demonstrates higher plasma and CSF concentrations compared to oral dosing, suggesting potential benefits for optimizing CNS drug delivery [84]. Addressing drug penetration challenges and optimizing dosing strategies are crucial for improving TBM treatment outcomes.

Despite high-dose IV RIF administration, mortality rates in TBM remain high, with a 35% mortality rate (at 13 mg/kg), with a comparatively higher rate of 65% observed in standard doses reported by Ruslami et al. [84]. In a study of 26 patients receiving IV (600 mg) and oral (450 mg) doses, IV administration significantly increased plasma AUC_{0-6} (78.7 mg.h/L) compared to oral (26.0 mg.h/L), with a ratio of 3.0. IV C_{max} in plasma was markedly higher (22.1 mg/L) than oral (6.3 mg/L), with a ratio of 3.5. Median T_{max} for both routes was 2 h, but IV administration showed a significantly narrower range (1–2 h) compared to oral (1–6 h, $p = 0.048$). In CSF, IV C_{max} (0.60 mg/L) was significantly higher than oral (0.21 mg/L), with a ratio of 2.92 ($p < 0.0001$) [84]. Table 1 details the systemic therapeutic regimen, CSF penetration and side effects of anti-TB in the treatment of TBM.

Table 1. Systemic therapeutic regimen of anti-TB drugs used in TBM [11,91], WHO Guidelines.

Category	Drug	Adult Dose (WHO) in (mg/kg)	Child Dose (WHO) in (mg/kg)	Duration (WHO)	CSF Penetration (CSF/Plasma%)	Adverse Effects
First-line drugs for DS-TBM	RIF	15 (10–20); max. 600 mg	10 (8–12); max. 600 mg	12 months	10–20%	Hepatotoxicity, orange urine, many drug interactions
	INH	10 (7–15); max. 300 mg	5 (4–6); max. 300 mg	12 months	80–90%	Hepatotoxicity, peripheral neuropathy, lupus-like syndrome, confusion, seizures
	PZA	35 (30–40) mg/k	25 (20–30) mg/kg	First 2 months	90–100%	Hepatotoxicity, arthralgia, gout
	ETB	20 (15–25) mg/kg	15 (15–20) mg/kg	First 2 months	20–30%	Dose-related retrobulbar neuritis, more common in renal impaired
	STM	15–30; max. 1 g IV/ IM	15 (12–18); max. 1 g	First 2 months	10–20%	Nephrotoxicity and ototoxicity
Core second-line drugs for MDR-TBM	LVX	<5 Ys: 15–20, ≥5 Ys: 10–15	10–15 (mg/kg)	During treatment	70–80%	Nausea, headache, tremor, confusion, tendon rupture (rare)
	MXF	10–20: max. 400 mg NWE	400 mg/d	During treatment	70–80%	Nausea, headache, tremor, confusion, tendon rupture (rare)
	AMK	15–30 mg/kg; max. 1 g IV or IM	15; max. 1 g IV or IM	Intensive phase only	10–20%	Nephrotoxicity and ototoxicity
	KAN	15–30 mg/kg; max. 1 g IV or IM	15; max. 1 g IV or IM	Intensive phase only	10–20%	Nephrotoxicity and ototoxicity
	CM	15–30 mg/kg; max. 1 g IV or IM	15; max. 1 g IV or IM	Intensive phase only	No Data	Nephrotoxicity and ototoxicity
	ETO	15–20 mg/kg; max. 1 g	15–20; max. 1 g	During treatment	80–90%	Anorexia, nausea, vomiting, gynaecomastia, hypothyroidism, confusion, seizures
	CYC	10–20 mg/kg; max. 1 g	10–15; max. 1 g	During treatment	80–90%	CNS toxicity
	LNZ	10 mg/kg; max. 600 mg	600 mg/d	During treatment	30–70%	Myelosuppression, optic neuropathy; use with pyridoxine.

Table 1. *Cont.*

Category	Drug	Adult Dose (WHO) in (mg/kg)	Child Dose (WHO) in (mg/kg)	Duration (WHO)	CSF Penetration (CSF/Plasma%)	Adverse Effects
Other drugs for MDR-TB, in TBM	CFZ	1 mg/kg	100–200 mg/d	NA	No Data	skin discoloration (orange/red) and sun sensitivity.
		100 mg/d	-	NA	0 [92]	NA
	PAS	200–300 mg/kg	8–12 g	NA	No Data	Vomiting, diarrhea, reversible hypothyroidism (increased risk with ethionamide)
	BDQ	Not determined	400 mg 1x/d for 2 wks, then 200 mg 3x/wks for 22 wks	NA	No Data	Nausea, vomiting, arthralgia, QT prolongation
	Dlm	Not determined	200 mg	NA	No Data	Nausea, vomiting, and dizziness rarely; QT prolongation

Abbreviations: PAS, Para-aminosalicylic acid; RIF, Rifampin; INH, Isoniazid; RFB, Rifabutin; LVX, Levofloxacin; AMK, Amikacin; CYC, Cycloserine; LNZ, Linezolid; CFZ, Clofazimine; BDQ, Bedaquiline; Dlm, Delamanid; CM, Capreomycin; KAN, Kanamycin; ETO, Ethionamide; NEW, not well established; DS, drug sensitive; Ys, years, wks, weeks.

In summary, clinical observations and PK modeling studies suggest that higher doses of RIF may be necessary to attain adequate CSF drug concentrations, a strategy that might help reduce mortality rates in TBM. However, the risk of DILI and the PKs of RIF, including its metabolism and excretion, demand careful consideration.

3.1.2. PK of INH in CSF

INH, a hydrophilic antimicrobial, exemplifies the potential for effective BBB penetration despite the challenges associated with delivering drugs to the CNS. Unlike lipophilic agents that cross the BBB via lipid-mediated diffusion, INH's water solubility enables it to traverse the BBB paracellularly, achieving significant concentrations within the CSF. This property is critical for combating *Mycobacterium tuberculosis* within a compartment notoriously difficult to reach. Clinical PK analysis demonstrates that INH attains steady-state concentrations in both plasma and CSF, with peak levels occurring shortly after administration and maintaining bactericidal activity within the CNS. The WHO's recommendation for higher INH dosages in the treatment of MDR/XDR-TBM/PTB acknowledges this drug's capacity for high CNS penetration and underlines the importance of achieving therapeutic CSF concentrations to combat resistant TB strains effectively.

The PKs of INH, another important first-line TB drug, in CSF are important for the effective treatment of TBM [93]. Unlike lipophilic agents, hydrophilic antimicrobials like INH and PZA cross the BBB paracellularly due to their water solubility [93]. Optimal BBB penetration occurs when drug log *p*-values range from 1.5 to 2.7, with a mean value of 2.1 [94]. INH, a hydrophilic drug (Mw 137.14 g/mol), freely penetrates the BBB (80–90% into CSF), exhibiting potent bactericidal activity [95–97]. CSF penetration is predicted by the AUC_{csf}/AUC_{serum} ratio, with INH showing an AUC_{CSF}/AUC_{Serum} close to 1.0, making it valuable for treating CNS infections [36,98]. The MIC of INH in liquid media ranges from 0.02 to 0.04 µg/mL [68]. In confirmed TBM patients, a PK analysis of INH transfer into CSF demonstrated steady-state concentrations in both plasma and CSF. Peak plasma levels (4.17–21.5 µg/mL) occurred 0.25 to 3 h post a multiple INH dose (600 mg/day). Parameters included a terminal half-life of 1.42 ± 0.41 h, total clearance (CI/F) of 0.47 ± 0.22 L/kg/h, and volume of distribution (Vd/F) of 0.93 ± 0.48 L/kg. In CSF, INH concentrations were highest at 3 h (Mean, 4.18 µg/mL) and 0.54 ± 0.21 µg/mL at 12 h post the last dose of INH 10 mg/kg/day. Using a modified PK/PD model, the disposition rate constant from CSF-to-plasma and CSF/plasma partitioning ratio of INH was estimated to be 0.39–1 h and 1.17, respectively [99]. Similarly, in a study involving 27 Chinese TBM patients, INH exhibited rapid diffusion into the CSF after oral administration of INH (9 mg/kg) [88] and attained peak concentrations surpassing 3 mg/L, exceeding its MIC against *Mycobacterium tuberculosis* by over 30 times, achieved within 4 h [28]. Mean serum concentrations peaked at 4.4 mg/L within 2 h, decreasing to about 1 mg/L by 6 h. CSF concentrations reached 1.9 mg/L within 2 h, and increased to 3.2 mg/L by 4 h, surpassing 30 times its MIC against *Mycobacterium tuberculosis*. CSF/serum INH ratios rose from approximately 0.5 at 2 h to 2.1 at 6 h, unaffected by steroid administration [88].

Considering the good absorption and penetration of INH in CSF, achieving therapeutic concentrations in CSF is feasible. However, to prevent systemic side effects, IT administration of INH is preferred. Typically, a 24 h dosing interval is sufficient for moderately lipophilic antibiotics (MW > 1000 g/mole) and hydrophilic drugs (MW > 400 g/mole). Therapeutic drug monitoring (TDM) is necessary only in cases of treatment failure [40].

For the treatment of MDR/XDR-TBM/PTB, the WHO recommends 15–20 mg/kg INH, recognized for its high CNS penetration [65,66]. Utilizing a high dose (800 mg/d), peak CSF concentration (11.57 μg/mL) was observed after 6 h, indicating nearly 100% brain penetration and effective bactericidal activity [61]. Long-term use of INH may lead to peripheral neuropathy at the sixth month, with reported neuritis [92]. The MIC of RIF for pre-XDR-TBM found in a case study > 16 μg/mL and for INH is 0.06 μg/mL while the reference value is <1 and <0.25, respectively [92].

In summary, the ability of INH to penetrate the BBB and reach concentrations above the MIC underscores its potential effectiveness in treating TBM. High doses of INH can enhance BBB penetration, leading to higher concentrations in the CSF. However, prolonged use of INH is linked to side effects such as peripheral neuropathy and neuritis.

3.1.3. Comparative PK with Other Anti-TB Drugs in CSF

The treatment of TBM, particularly MDR forms, necessitates a nuanced understanding of the PK of various anti-TB drugs to ensure effective CNS penetration and therapeutic outcomes. This editorial examines the PK profiles of linezolid, bedaquiline, and clofazimine, comparing their efficacy and challenges in treating TBM.

Linezolid: a promising option for MDR-TB. Linezolid, an antibiotic effective against MDR-TB, demonstrates significant promise due to its favorable PK profile in both serum and CSF. Administered 600 mg twice daily via IV infusion, linezolid achieves a mean C_{max} of 18.6 μg/mL in serum and 10.8 μg/mL in CSF, with a CSF-to-serum penetration ratio of 0.66. Notably, linezolid's elimination half-life in the CSF (19.1 h) surpasses that in serum (6.5 h), suggesting sustained activity within the CNS [100]. This extended half-life in the CSF is particularly advantageous for treating TBM, allowing for more consistent antimicrobial pressure against *Mycobacterium tuberculosis*.

Bedaquiline: challenges and potential. Bedaquiline, another critical component in the treatment of MDR-TB, presents a more complex PK profile. While plasma concentrations fall within expected ranges, indicating adequate systemic absorption, CSF penetration appears limited. Bedaquiline's C_{max} in the CSF is significantly lower than in plasma, reflecting a CSF-to-plasma concentration ratio of merely 0.12%. PKs were studied with thirty-eight plasma and seven CSF samples. Plasma concentrations were within the expected range, with a bedaquiline C_{max} of 1368.1 ng/mL and AUC last of 19,825.9 ng·h/mL; M2 reached a C_{max} of 217.3 ng/mL and AUC last of 4134.6 ng·h/mL. In CSF, bedaquiline had a C_{max} of 3.790 ng/mL, while M_2 had a C_{max} of 1.400 ng/mL. CSF-to-plasma concentration ratios were 0.12% for bedaquiline and 0.3% for M_2, aligning with estimated plasma unbound fractions [101]. Despite this, the inclusion of bedaquiline in intensified treatment regimens has contributed to improved clinical outcomes in MDR-TBM, suggesting that even limited CNS penetration can be clinically beneficial when combined with other effective anti-TB agents.

Clofazimine: limited CNS penetration. Clofazimine, despite its efficacy in treating MDR-TB, exhibits limited benefits for TBM due to its poor CNS penetration. In a case report, plasma concentrations reached peak levels post-administration, while CSF levels remained undetectable. The plasma concentration peaked at 0.35 μg/mL 6 h post-administration, while CSF levels were undetectable. Despite its lipophilicity, the drug's high plasma protein binding rate (>85%) may limit CNS penetration. Human efflux transporters could further impede brain and CSF access. Findings suggest limited benefits of clofazimine in treating MDR- or pre-XDR-TBM [92]. The drug's high plasma protein binding rate and potential interaction with human efflux transporters may impede its CNS access, highlighting the challenges of achieving therapeutic CSF concentrations with certain anti-TB drugs.

Intensified treatment regimens: a path forward. An intensified treatment regimen, including fluoroquinolone, injectables (AMK/KAN/CM), linezolid, carbapenem, RIF, INH, PZA, clofazimine, delamanid, and bedaquiline, administered for 361 days in drug-susceptible TBM and 486 days in MDR-TBM, resulted in a 9% mortality rate in the MDR group [102]. Neurological outcomes improved over time, with 18% reporting a Modified Rankin Scale (MRS) score of 0 at 1 month, increasing to 85% at 6 months, and further to 94% at 12 months. Disability rates also showed improvement, with 36% having low Glasgow Outcome Scale (GOS) disability at 1 month, 80% at 6 months, and 86% at 12 months. These findings underscore the positive impact of intensified regimens on mortality and neurological outcomes in TBM. Increased use of drugs that penetrate well into the CSF, including LZD, may be one reason for our favorable outcomes, given the concern for limited CSF penetration of first-line anti-TB agents (reported by a prospective cohort study conducted in the National Center for TB and Lung Diseases, Georgia) [102].

The success of intensified treatment regimens, incorporating drugs like linezolid and bedaquiline alongside standard anti-TB medications, underscores the importance of optimizing drug selection based on PK profiles and CNS penetration capabilities. Such regimens have shown a significant positive impact on mortality and neurological outcomes in TBM, particularly in MDR cases. The careful combination of drugs with varying CSF penetration efficiencies allows for a comprehensive approach to combating TB within the CNS, addressing the limitations posed by individual drugs.

The comparative PK of linezolid, bedaquiline, and clofazimine highlights the complexities of treating TBM, especially in the context of drug resistance. While challenges such as limited CNS penetration persist, the strategic use of these drugs within intensified treatment regimens offers hope for improved outcomes. Future research should continue to focus on understanding the PK properties of anti-TB drugs, exploring new delivery methods, and developing regimens that maximize CNS penetration and therapeutic efficacy against TBM. In summary, the PK of anti-TB drugs is critical for treatment success, with challenges like limited CNS penetration. Key drugs, such as RIF, face difficulties in reaching therapeutic levels in CSF. However, INH demonstrates better CNS penetration. An intensified treatment regimen, administered long-term (361 days for drug-susceptible TBM and 486 days for MDR-TBM), yields improved clinical outcomes. Linezolid, with enhanced CSF penetration, holds promise for better results, addressing concerns about limited penetration of first-line anti-TB agents. IT/IVT anti-TB treatments show superior efficacy with short durations and minimal side effects. Table 2 presents the PK indices of the anti-TB treatments.

Table 2. Anti-TB PK in serum/CSF, systemic administration for TBM.

Drug	Adm. Route	Sample #	Dose	Pathogen Type and Drug Susc.	MIC in μg/mL	Ser Conc. in μg/mL (Mean ± SD)	CSF Conc. in μg/mL (Mean ± SD)	T_{max} in CSF	CSF/Serum Ratio	References
RIF.	PO	1	NA	INH-R, BDP RIF-R	1	NA	NA	NA	NA	[103]
	PO	19	10.7 ± 0.5 mg/kg	NA	NA	11.5 ± 1.0	0.39 ± 0.06	NA	0.04 ± 0.01	[88]
	PO	10	11.1 ± 0.5 mg/kg	NA	NA	10.6 ± 1.4	0.38 ± 0.06	NA	0.04 ± 0.01	[88]
	PO	7	10.1 ± 0.6 mg/kg	NA	NA	10.1 ± 1.1	0.78 ± 0.13	NA	0.08 ± 0.02	[88]
	PO	7	10.5 ± 0.8 mg/kg	NA	NA	4.7 ± 0.6	0.47 ± 0.06	NA	0.11 ± 0.03	[88]
	PO	26	450 mg/d	NA	NA	C_{max} (6.3 mg/L)	C_{max} (0.21 mg/L)	2	NA	[84]
	IV	26	600 mg/d	NA	NA	C_{max} (22.1 mg/L)	C_{max} (0.60 mg/L)	2	NA	[84]

Table 2. *Cont.*

Drug	Adm. Route	Sample #	Dose	Pathogen Type and Drug Susc.	MIC in μg/mL	Ser Conc. in μg/mL (Mean ± SD)	CSF Conc. in μg/mL (Mean ± SD)	T_{max} in CSF	CSF/Serum Ratio	References
RIF.	PO	1	NA	Pre-XDR-TBM, R	>16	NA	NA	NA	NA	[92]
	In Vitro	-	NA	S. Strain	0.2–0.4	NA	NA	NA	NA	[68]
	PO	20	NA	NA	NA	9.4 (2.9–23.7)	0.2 (0.1–0.4) $CCSF_{0-2}$	NA	NA	[61]
	PO	20	NA	NA	NA	9.4 (2.9–23.7)	0.4 (0.1–1.4) $CCSF_{0-8}$	NA	NA	[61]
INH	PO	1	NA	Mod. RS to INH	2–4	NA	NA	NA	NA	[103]
	PO	19	8.5 ± 0.4 mg/kg	NA	NA	4.4 ± 0.5	1.9 ± 0.3	NA	0.47 ± 0.04	[88]
	PO	8	9.1 ± 0.6 mg/kg	NA	NA	2.6 ± 0.8	3.2 ± 0.8	NA	1.31 ± 0.13	[88]
	PO	9	9.0 ± 0.8 mg/kg	NA	NA	2.1 ± 0.6	1.8 ± 0.5	NA	1.03 ± 0.14	[88]
	PO	8	7.5 ± 0.9 mg/kg	NA	NA	1.0 ± 0.3	1.8 ± 0.5	NA	2.12 ± 0.25	[88]
	PO	1	600 mg/d	Pre-XDR-TBM, R	0.06	9.86 C_{max}	11.57 C_{max}	NA	1.51	[92]
	In Vitro		NA	S. Strain	0.02–0.04	NA	NA	NA	NA	[68]
	PO	6 Child	2.5–3.3 mg/kg tid–qid	S. Strain	-	4.84 ± 2.31	3.18 ± 1.27	-	-	[65,104]
	PO	20	NA	NA	NA	4.6 (1.0–10.0) 4.7 (2.5–13.6)	1.4 (0.5–6.1) 1.3 (1.2–4.3)	NA	NA	[61]
RFB	PO	1	NA	RFB sus.	≤0.250–0.5	NA	NA	NA	NA	[103]
	PO	1	NA	Pre-XDR-TBM, R	8	NA	NA	NA	NA	[92]
LVX	PO	1	10–15 mg/kg	NA	NA	NA	NA	NA	NA	[11]
OFX	In Vitro	-	-	-	(0.5–1.0)	-	-	-	-	[68]
AMK	IV or IM	1	15 mg/kg	NA	NA	NA	10–20% of Ser conc.	NA	NA	[11]
	IV or IM	1	-	-	0.5	-	-	-	-	[92]
Mfx	PO	1	400 mg	NA	NA	NA	70–80% of Ser conc.	NA	NA	[11]
	PO	1	400 mg/d	Pre-XDR-TBM, R	4	2.11	0.62	-	0.48	[92]
CYC	PO	1	10–15 mg/kg	NA	NA	NA	80–90% of Ser Conc.	NA	NA	[11]
	PO	1	500 mg/d	Pre-XDR-TBM, R	16	36.28 C_{max}	20.62 C_{max}	-	-	[92]
LNZ	PO	1	600 mg	NA	NA	NA	30–70% of Ser conc.	NA	NA	[11]
	PO	14	600 mg 2x/d	NA	NA	18.6 ± 9.6	10.8 ± 5.7	101.6 ± 59.6 μg·h/mL	0.66	[100]
	PO	1	600 mg 2x/d	Pre-XDR-TBM, R	-	31.81	15.72	-	-	[92]
CFZ	PO	1	100–200 mg	NA	NA	NA	Probably low	NA	NA	[11]
	PO	1	100 mg/d	Pre-XDR-TBM, R	-	0.35	0	-	-	[92]
BDQ	PO	7	400 mg 1x/d for 2 wks, then 200 mg 3x/wk for 22 wks.	NA	>30 ng/mL	1.1442 (C_{max} = 1.832)	0.00149 (C_{max} = 0.00379)	5 h	0.12	[101]
	PO	1	400 mg 1x/d for 2 wks then 200 mg 3x/wk for 22 wks.	NA	NA	NA	Probably very low	NA	NA	[11]

Abbreviations: RIF, Rifampin; INH, Isoniazid; RFB, Rifabutin; LVX, Levofloxacin; AMK, Amikacin; CYC, Cycloserine; LNZ, Linezolid; CFZ, Clofazimine; BDQ, Bedaquiline; R, Resistant; Mod., Moderate; Sus., Susceptible; BL, Borderline; d, daily; wks, weeks.

3.2. Clinical Experience and Evidence Supports Concomitant IT/IVT Anti-TB Effectiveness and Safety

Administering drugs directly into the spinal canal offers benefits, including precise targeting of the CNS, reduced overall doses, and improved treatment outcomes. However, the absence of standardized treatment regimens underscores the necessity of leveraging existing data to develop consistent drug protocols. Analysis of published literature, encompassing case reports, cohort studies, meta-analyses, and other documents, has yielded compelling evidence supporting the effective clinical use of RIF and INH via IT/IVT administration. Our literature review examined the use of RIF, INH, rifabutin, and fluoroquinolones administered via IT or IVT for treating TBM and meningitis caused by *Flavobacterium meningosepticum*. Extensive documentation on INH and RIF, dating back to 1955 and 1976, respectively [105,106], reveals promising outcomes even in cases of refractory TBM, without serious side effects. Despite limited PK data, evidence supports the efficacy of IT/IVT administration in managing TBM.

Analyzing the published literature, including case reports, cohort studies, meta-analyses, and additional literature pieces, we extracted compelling evidence supporting the effective clinical use of RIF and INH through IT/IVT administration.

IT/IVT INH: Historically, in 1955, the treatment of twelve cases of TBM with IT-INH doses ranging from 10 to 20 mg per day led to significant clinical improvement in ten patients. Follow-up assessments, conducted over a period of up to 14 months, reported no toxicity or side effects associated with this treatment regimen [105]. In another instance, 18 patients treated with IT-INH demonstrated the infection eradicated in 17 cases, but two deaths occurred. Patients received streptomycin and oral INH, with doses ranging from 25 to 75 mg by weight for streptomycin and 5–7 mg/lb daily for oral INH, with IT doses of 25–50 mg. Most patients fully recovered, some with neurological improvement. IT-INH injection at approximately 0.5 mg/lb was safe and well-tolerated, but doses below 0.3 mg/lb showed a slower treatment response. Side effects were minimal, including non-specific meningeal irritation in one infant and major epileptic attacks in an adult with pre-existing conditions. This approach, combined with systemic therapy, was effective in reducing the need for frequent streptomycin injections, indicating most patients recovered fully or showed significant neurological improvement [107].

Similarly, a patient developing hepatitis from high oral doses of INH (1000 mg/day) showed clinical improvement upon gradual reintroduction of IT and oral INH. Consequently, INH was discontinued and later re-administered in gradually increasing IT and subsequently oral doses, up to the final dose of 400 mg/day. The patient was treated with IT-INH at a dosage of 35–37 mg every 3 days, resulting in an improvement of clinical symptoms [108]. A study involving 23 TBM cases demonstrated successful treatment with 100 mg of INH and 2 mg dexamethasone administered directly into the lateral ventricle, showing positive outcomes with IT administration [24]. A case of refractory TBM treated with IT-INH at 100 mg three times per week demonstrated significant symptom improvement without any observed toxicity. Following the commencement of IT therapy, there was an immediate improvement in the patient's consciousness disturbance and CSF findings. One month later, a ventriculoperitoneal shunt operation was performed to address hydrocephalus, leading to further clinical improvement [109]. In another study, 12 pediatric cases with severe TBM, hydrocephalus, and altered consciousness received IT INH and dexamethasone injections. From day 2 onwards, doses of 5 to 20 mg of INH and 1 mg of dexamethasone in 5 mL of saline were administered into the ventricle through the Ommaya reservoir using a syringe, with injections repeated once every 2 to 3 days. Successful recovery was observed in nine out of twelve cases [110]. A 30-year-old Vietnamese woman, 19 weeks pregnant, presented with acute cerebral infarction, left middle cerebral artery stenosis, TBM, and miliary tuberculosis. Treatment with heparin, quadruple anti-TB therapy, and dexamethasone led to rapid symptomatic improvement, but she experienced a stillbirth, followed by recurrent acute cerebral infarction with LMCA occlusion, sinus thrombosis, and cranial base inflammation. Thrice-weekly 100 mg IT-INH improved meningeal inflammation signs. Discharged ambulatory after 7 months, suggesting multimodal therapy with IT-INH and steroids for refractory TBM, as reported by Nakatani et al. [111]. In treating 10 ventricular TB patients with IT-INH (0.1 g) and dexamethasone (2.5–5 mg) alongside multidrug anti-TB therapy, patients showed positive clinical outcomes. MRI findings demonstrated complete resolution or significant reduction in the size of tubercular lesions in seven out of eight patients. These results suggest the potential effectiveness of combining IT INH and corticosteroids for intracranial TB treatment [112].

The effectiveness of INH is significantly influenced by the activity of the N-acetyltransferase 2 (NAT2) enzyme in the liver, which is categorized into slow, intermediate, or rapid acetylation rates. These rates affect the metabolism of the drug, thereby impacting its efficacy in treatment. The presence of variations in NAT2 genotypes necessitates the adjustment of INH dosages to optimize treatment outcomes, particularly in populations with a high prevalence of rapid acetylators. Rapid acetylators metabolize INH more quickly, leading to lower concentrations of the drug in the plasma and an increased risk of treatment failure [113,114]. Given the notable prevalence of rapid acetylators in the Chinese population, higher doses of INH might be essential to improve treatment outcomes [115]. Additionally, recent studies suggest the potential benefits of administering high doses of INH for managing cases of MDR-TB [116,117]. Administering high doses of INH, especially through IT routes, holds promise for treating MDR-TB. This approach can potentially increase the concentration of the drug in the CSF while minimizing the adverse effects associated with higher systemic doses [108,118].

IT/IVT-RIF: In meningitis caused by *Flavobacterium meningosepticum*, RIF was administered intravenously and directly into the cerebral ventricles for neonatal cases. IVT dosage varied from 2 to 5 mg/day for 10 days in two cases and for 10 weeks in one case, in conjunction with systemic therapy. This approach led to rapid bacterial clearance from the CSF. Notably, jaundice was observed as a side effect, and the majority of infants developed hydrocephalus, necessitating shunt placement. Nevertheless, two infants achieved normal neurological development following treatment [106]. In a review of treatment strategies from 1977, a series of seven meningitis cases attributed to the same bacterium were managed with IVT-RIF, dosed between 2 to 5 mg/day until achieving CSF sterilization. Treatment was complemented with intramuscular RIF at 20 mg/kg every 12 h. Transient jaundice was a noted side effect, with ventricular antibiotic concentrations evaluated 20 to 24 h post-administration [119]. Further reports highlighted a treatment regimen consisting of concurrent intravenous RIF at 40 mg/kg per day and IVT-RIF at 5 mg daily for 22 days, resulting in CSF sterilization without reported adverse effects or toxicity [120]. An evaluation of nine meningitis cases treated with either IT or IVT-RIF at 3 mg/kg/day for a minimum of 10 days showed that five patients achieved CSF sterilization and survival. While two deaths were due to other causes, two other patients did not survive the infection. The combined use of intravenous RIF, moxalactam, and piperacillin emerged as a promising antibiotic regimen for treating patients with normal or mildly dilated ventricles. In cases where significant ventriculomegaly was present, the simultaneous intravenous and IVT administration of antibiotics, tailored to the organism's sensitivity, proved necessary for eradicating the infection. Remarkably, no side effects or toxicity were reported throughout these treatment courses [121]. Additionally, two infections treated with IVT-RIF at a daily dose of 5 mg for 7–10 days were documented, successfully leading to CSF sterilization without any reported side effects or toxicity [122]. These meningitis patients were also treated with a systemic regimen of clindamycin, RIF, and cefotaxime, complemented by IVT administration of RIF.

A case of TBM involved a patient experiencing complications and a relapse after oral anti-TB treatment, who was then treated with IVT-RIF. The treatment regimen comprised an initial daily dose of 5 mg RIF for one week, followed by 3 mg every other day for three additional weeks. Despite a brief interruption due to technical issues and patient non-compliance, treatment resumed after drain placement and continued for another month, aiming to achieve a CSF concentration exceeding 15 μg/mL. This integrated approach successfully effectively managed the meningeal infection, leading to significant clinical improvement without adverse effects [22]. IVT-RIF was highly effective in severe cases of

TB meningoencephalitis. In one instance, a patient with concurrent hepatic and renal failure underwent successful treatment with daily 5 mg IVT-RIF administered via an Ommaya reservoir. The 50-day course of IVT-RIF treatment was well-tolerated and demonstrated high efficacy in conjunction with systemic therapy [23]. Furthermore, an accidental high-dose administration of IT-RIF at 600 mg in a postoperative spine infection case showed no immediate or 6-month adverse effects, indicating the potential safety margin of the drug in acute settings [123].

IT/IVT fluoroquinolones in the treatment of MDR-TBM: The application of IT fluoroquinolones has been observed in the treatment of TBM. Notable, a case of MDR-TBM was effectively managed using a flexible regimen. The regimen comprised alternating daily IT administration of LVX at a dose of 1.5 mg and AMK at 5 mg with IV doses of LVX at 500 mg and AMK at 1200 mg. This was in conjunction with oral anti-TB medications and prednisolone at a dosage of 25 mg. Adjustments to the regimen were made to accommodate the patient's tolerance and the CSF concentration, leading to positive clinical and microbiological results. The targeted CSF concentrations (LVX: 8–10 µg/mL; AMK: 40 µg/mL) were successfully achieved [25].

In summary, the evaluation of IT/IVT anti-TB (RIF, INH, and fluoroquinolones) in managing TBM yields promising outcomes, supported by existing data demonstrating their efficacy and safety, particularly in clinical cases and even in refractory TBM. The summarized literature under the subtitle "clinical experiences and evidence supports IT/IVT anti-TB effectiveness and safety" is provided in Table 3, affirming the high effectiveness and safety of IT/IVT anti-TB drugs.

3.3. Optimization of IT Therapeutic Regimens of RIF and INH in the Treatment of TBM

Despite the absence of an optimized IT/IVT therapeutic regimen for anti-TB treatment in TBM, our analysis of published data aims to propose a regimen to enhance treatment strategies, ultimately improving patient outcomes and reducing associated morbidity and mortality. We have synthesized data from case reports, cohort studies, and literature articles to optimize therapeutic regimens for IT/IVT administration of anti-TB drugs in TBM management.

RIF: In seven published literature, IT/IVT-RIF was given at doses of 2–5 mg for 7–50 days in seven studies. This regimen proved highly effective, especially in complex cases and when oral drugs failed, with no significant side effects [22,23,106,120–123]. The targeted concentration of RIF in the CSF was 15 mg/L. See Table 3 for dosage and treatment duration details [22].

INH: Similarly, IT/IVT administration of INH, typically given thrice weekly at dosages between 5 to 100, proves highly effective with a mild side effect. Details of dosage and treatment duration are provided in Table 3 [24,105,107–112].

LVX and AMK: The recommended dosage for IT fluoroquinolones (LVX and AMK) in treating TBM is LVX 1.5 mg and Am 5 mg, supplemented with IV-LVX 500 mg and Am 1200 mg, oral anti-TB drugs, and 25 mg prednisolone. This flexible regimen, tailored to patient tolerance and CSF concentration, has yielded positive clinical and microbiological outcomes, achieving target CSF levels (LVX, 8–10 g/mL; AMK, 40 µg/mL) [25].

Figure 1 depicts the mean and median doses of IT/IVT-RIF, INH, and fluoroquinolones for TBM treatment, sourced from the published literature.

The synthesis of data from various sources informs optimized therapeutic regimens for IT/IVT administration of anti-TB drugs in TBM management, emphasizing efficacy and minimal side effects.

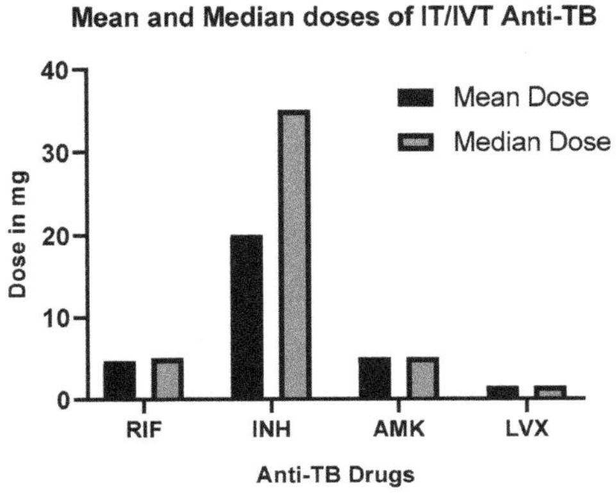

Figure 1. As demonstrated, the mean dose of IT/IVT-RIF is: 4.71 mg/day, INH: 20 mg, LVX: 5 mg, AMK: 1.5 mg, and the median dose of IT/IVT-RIF: is 5 mg, INH: 3 mg, LVX: 5 mg and AMK: 1.5 mg.

3.4. Comparative Analysis of Systemic Treatment versus Concomitant IT-Anti-TB Therapy

A comprehensive assessment comparing systemic anti-TB therapy to concomitant IT-Anti-TB therapy is essential. However, there is a notable lack of studies that thoroughly compare these approaches across all anti-TB medications in the treatment of TBM. In evaluating combined systemic and IT/IVT dosing, understanding the central volume of distribution is critical, as it correlates with CSF concentrations in both compartments, potentially influencing their effectiveness [52].

Comparing IT-INH alongside systemic therapy vs. systemic INH alone: A propensity-matched cohort study compared IT-INH alongside systemic therapy to systemic anti-TB therapy alone, highlighting the advantages of the concomitant treatment approach. This study showed that integrating IT therapy with systemic anti-TB therapy was more effective than systemic therapy alone in patients with TBM over long-term follow-up [124]. In the overall cohort, there were 36 patients in the IT group and 162 patients in the systemic therapy group. Among the propensity-matched cohort comprising 30 patients in each group, the analysis revealed noteworthy findings. Specifically, the odds ratio (OR) for death in the propensity-matched cohort was 0.266 (95% CI: 0.073–0.964; $p = 0.037$), indicating a significantly lower risk of mortality in the IT therapy group compared to the systemic therapy alone group. Furthermore, the OR for poor outcomes was 0.386 (95% CI: 0.136–1.094; $p = 0.073$), suggesting a trend towards improved outcomes in the IT therapy group, although statistical significance was not reached. These findings underscore the potential benefit of incorporating IT therapy with systemic anti-TB in the management of TBM, particularly in reducing mortality rates [124].

Table 3 outlines the doses, regimens, and efficacy of RIF, INH, LVX, and AMK when administered via IT and IVT administration routes for the treatment of TBM. The data spans from 1955 to the present and encompasses all relevant literature on IT/IVT anti-TB therapies.

In conclusion, while there is a scarcity of data to directly compare the clinical outcomes of all anti-TB drugs between concomitant IT/IVT therapy and systemic anti-TB drugs alone for treating TBM, clinical observations and case studies, including published case reports and cohort studies on TBM patients, suggest and support the practice of administering IT/IVT alongside systemic anti-TB therapies concurrently.

Table 3. IT/IVT Administration of RIF, INH, LVX, and AMK in Anti-TB Therapy.

Drugs	Disease	Pts	Dosage and Regimens	Efficacy	Toxicity/Side Effect	References
INH	TBM	12	10–20 mg/d IVT	Cured	Convulsion 1, partial optic atrophy 1	Sifontes, J. et al., 1955 [105]
	TBM	18	25–50 mg IT for 1~119 days	Cured	Hemiplegia 2, etc.	Swift, P. et al., 1956 [107]
	TBM	1	35–37 mg IT every 3 days	Cured	Hepatotoxicity	Daielides, I., 1983 [108]
	TBM	23	100 mg/d or qod IT based on the severity	Cured	8 pts mild brain hernia rapidly improved	Chen, Y. et al., 1996 [24,125]
	TBM	1	100 mg, 3x/wk IT.	Cured	No	Takahashi, T., 2003 [109]
	TBMH	12	5–20 mg one time every 2–3 days	9 cured, 1 Pt died, 2 Pts not cured	2 pts disability	Lin, J. et al., 2012 [110],
	TBM	1	100 mg, 3x/wk, tapered to 1x/wk upon symptom improvement.	Cured	Mild aphasia, 1 pts hemi-paresis	Nakatani, Y et al., 2017 [111]
	TBM	10	100 mg/d IT	Cured	One patient, hemiparesis	Li, D. et al., 2017 [112]
RIF	F.M	3	2–5 mg/d, IVT for 10 days (2 patients) or 10 wks (others).	Cured	Transient Jaundice	Lee, E. et al., 1976 [106]
	F.M	7	2–5 mg/d, IVT	Cured	Transient Jaundice	Lee, E. et al., 1977 [119]
	F.M	1	5 mg 1x/d, IVT/ for 22 days	Cured	No	Rios, I., 1978 [120]
	TBM	1	5 mg/d, IVT for 7 days, then 3 mg qod for a total of 63 days.	Cured	No	Dajez, P et al., 1981 [22]
	F.M	9	3 mg/kg/d, IVT for at least 10 days	5 cured, 2 died of other DS, and 2 succumbed.	No	Boo, N. et al., 1989 [121]
	F.M	2	5 mg/kg/d, IVT for 7–10 days	Cured	No	Bruun, B. et al., 1989 [122]
	TBM	1	5 mg/d IVT for 50 days	Cured	No	Vincken, W et al., 1992 [23]
	TBM	1	600 mg infusion 1x for 4 h, IT	Inadvertent Inj.	No	Senbaga, N et al., 2005 [123]
LVX/ AMK	MDR-TBM	1	1.5 mg LVX and 5 mg AMK qod/IT	Cured	Mild; Insomnia and Myalgia	Berning, S.E. et al., 2001 [25]

Note: Ds, disease; Pts, number of patients; MIC, minimum inhibitory concentration; RIF, rifampicin; INH, isoniazid; LVX, levofloxacin; AMK, amikacin; F.M, *flavobacterium meningosepticum*; TBM, tuberculosis meningitis; /d, per day; qod, every other day, wk, week.

3.5. Adverse Events and Risks of IT/IVT Administration of Anti-TB Drugs

Similar to other clinical procedures, the IT/IVT administration of anti-TB drugs, while minimally invasive, is not without risks and adverse effects. Data from various case reports and studies indicate the use of IT/IVT-RIF for *Flavobacterium meningosepticum* and *Mycobacterium tuberculosis* in TBM treatment, typically administered at 2 to 5 mg/day over 7 to 50 days. Reversible jaundice is reported in some cases, demonstrated in Table 3 [22,23,106,120–123]. Notably, even an accidental IT infusion of 600 mg RIF over 4 h showed no adverse events [123]. IT/IVT administration of INH, given thrice weekly at dosages between 5 to 100 mg, is associated with side effects like hemiplegia, quadriplegia, convulsion, partial optic atrophy, and hydrocephalus. Caution is advised based on reported cases, as demonstrated in Table 3 [24,105,107–112]. The total side effects and risks of IT/IVT INH and RIF in our review include convulsions, partial optic atrophy, hemiplegia, hepatotoxicity, mild brain hernia that rapidly improved in one patient, mild aphasia, transient jaundice, insomnia, and myalgia, and no significant side effects in most patients noted, as demonstrated in Table 3.

4. Discussion

This narrative review indicates that most anti-TB drugs, specifically RIF and clofazimine, exhibit poor CSF penetration when administered systemically, leading to suboptimal concentrations and ineffective therapeutic outcomes. To overcome these issues, healthcare professionals are considering the IT and IVT administration of INH and RIF. This approach is inspired by the successful use of IT analgesics [126], and the treatment of Gram-negative bacterial ventriculitis with IVT antibiotics, including RIF [119].

4.1. PK Properties of Anti-TB Drugs

The PK properties of anti-TB drugs within the CSF are pivotal in determining the efficacy of treating TBM and optimizing dosage. This review article addresses the challenges linked with conventional treatments for CNS infections, with a specific focus on TBM. It explores the potential of IT administration of RIF and INH as a targeted drug delivery strategy to mitigate systemic exposure, bypass the BBB, achieve elevated or adequate concentrations of anti-TB drugs in the CSF surpassing the MIC for a sufficient duration, and minimize side effects. A main obstacle identified is the lack of adequate PK data for the direct IT/IVT administration of these drugs. Nonetheless, the findings underscore the efficacy of IT/IVT administration of RIF and INH, demonstrating favorable clinical outcomes without serious side effects, utilizing short durations and small doses. Physiological and PK factors influencing IT anti-TB therapy are dissected, emphasizing the need for therapeutic regimen optimization. Barriers such as the BBB and BCSFB hinder drug entry, although infection-induced meningeal inflammation may enhance penetration for certain antibiotics [36,37].

In this study, we analyzed and summarized the differences in the concentration ratios of drugs in the CSF compared to serum concentrations following oral and IV administration, with a focus on RIF and INH. The comparison between oral and IV administration of RIF revealed distinct differences in drug concentration levels in the CSF. Our analysis, presented in Table 2, showed that oral administration resulted in a mean CSF concentration of 0.49 μg/mL, with a C_{max} of 0.21 mg/L following a 450 mg dose. Conversely, IV administration achieved a higher mean CSF concentration of 0.60 mg/L after a 600 mg dose, indicating more effective penetration into the CSF. These findings indicate that only 5 to 20% of the plasma concentration can reach the CSF through the two different administration routes. However, the evidence suggests that the difference is dose-dependent rather than being solely determined by the route of administration. In contrast, for INH, oral administration resulted in an average CSF concentration of 3.55 μg/mL, ranging from 1.3 to 11.57 μg/mL. The data presented significant differences between RIF and INH regarding their PKs and efficacy against MTB in the treatment of CNS infections. INH showed a much higher average CSF concentration (3.55 μg/mL) compared to RIF (0.49 μg/mL for oral administration), indicating better penetration of INH into the CNS.

The CSF/serum ratio significantly differs between the two drugs, with INH showing a much higher ratio (0.47–2.12) than RIF (0.04–0.11). This suggests that INH is more readily transported from the serum into the CSF compared to RIF. Both drugs exhibited a T_{max} of 2 h post-oral administration for RIF, but INH has a broader range of T_{max} (2 to 4 h), suggesting a slower absorption into the bloodstream in some cases. RIF had a longer half-life (9.1 to 21 h) compared to INH (approximately 1.42 h), indicating that RIF remained in the CSF longer than INH, potentially extending its antimicrobial activity within the CNS.

Regarding MIC values, for RIF, MIC against susceptible MTB ranged from 0.02 to 0.04 μg/mL but was >16 μg/mL for pre-XDR MTB, and varied from >8 to >32 μg/mL for multidrug-resistant (MDR) and rifampicin-resistant (RR) MTB. For INH, MIC against susceptible MTB ranged from 0.02 to 0.04 μg/mL, 0.06 μg/mL for pre-XDR MTB, and from 0.4 μg/mL to >4 μg/mL for MDR- and INH-resistant MTB. RIF and INH both had varying MIC values against different strains of MTB, with INH generally requiring lower concentrations to inhibit susceptible MTB. However, RIF shows efficacy over a broader range of MTB strains, including pre-XDR and MDR/RR MTB, albeit at higher concentrations.

Overall, these differences highlighted the unique PK properties and effectiveness of RIF and INH in treating CNS infections, underscoring the need for tailored therapeutic strategies based on the drug's ability to penetrate the CNS and its activity against various MTB strains. The findings suggested that IT or IVT administration may offer better treatment outcomes for CNS infections. Nevertheless, the choice of administration route should also take into consideration patient-specific factors, such as condition, drug tolerance, and potential side effects.

4.2. The Clinical Significance of Concurrent IT/IVT Therapy

4.2.1. Concurrent IT/IVT Therapy of Anti-TB Drugs

Concurrent therapy and its clinical outcomes of anti-TB drugs: Concurrent IT and IV therapy alongside systemic treatment is crucial for preventing bacterial resistance and reducing relapse risk, especially in challenging cases like MDR-TBM. Simultaneous administration of RIF and INH has shown effective outcomes with minimal side effects, utilizing short durations and small doses. The clinical outcomes of concomitant IT/IVT anti-TB with systemic anti-TB in the treatment of TBM show promise. Clinical evidence supports the efficacy and safety of IT/IVT administration, particularly for RIF and INH, even in refractory TBM cases [22–24,105–112,120–123]. Some adverse events, such as infections, nerve damage, transient jaundice, seizures, etc., have been observed following the IT or IVT administration of RIF and INH [22–24,105–112,120–123]. However, from our perspective, it is challenging to ascertain whether these side effects are directly attributable to RIF and INH or if they stem from disease progression or the specific procedure and route of administration. These events necessitate careful consideration and monitoring. Comparing concurrent administration of IT/IVT anti-TB alongside systemic therapy of anti-TB in the treatment of TBM, there is a lack of data, although one study compared the concomitant IT INH versus systemic alone and showed superior concomitant therapy [124].

4.2.2. Concurrent IT/IVT Therapy of Non-Anti-TB Antibiotics

The IT/IVT administration of non-anti-TB antibiotics such as vancomycin, colistin, tigecycline, and others, plus systemic therapy, has been studied in several papers, and we have documented them here. Concurrent administration of IT/IVT antibiotics alongside systemic treatment is essential in preventing bacterial resistance and reducing the risk of relapse, particularly in cases involving MDR pathogens where systemic options are limited, as acknowledged in CNS infections caused by MDR pathogens and postoperative meningitis or ventriculitis, which highlight the potential life-saving benefits of combining IT/IVT and systemic antimicrobial therapy [40,127]. This strategy allows for achieving high concentrations of antimicrobial agents within the CNS while minimizing systemic side effects, influenced by drug properties and concurrent systemic dosing [56,59,128–130].

The benefit of elevated CSF drug levels becomes apparent when IT and IV treatments are combined, potentially leading to slightly higher antibiotic concentrations in the CSF compared to IVT therapy alone [59]. Although this complexity presents challenges in pharmacokinetic analysis, it proves advantageous in managing MDR-CNS infections. Extensive research supports the efficacy of combining IT administration of antibiotics, such as colistin, vancomycin, tigecycline, and others, in the treatment of CNS infections, including meningitis and ventriculitis, as well as in addressing MDR *Acinetobacter baumannii* ventriculostomy-related infections, when administered concurrently with IV therapy [40,56,59,128–130]. Thus, the integration of systemic antibiotic therapy, whether using the same antimicrobial agent or a different active ingredient, is strongly advocated.

Upon examining CSF concentrations following IT administration, the consensus is that the influence of plasma concentrations on CSF kinetics can be overlooked. This simplification is accepted due to the scarcity of studies involving human subjects treated exclusively with IVT anti-infectives [40]. However, the impact of simultaneous IV dosing on CSF-PK for all antibiotics remains uncertain [40,52]. While most patients receive a combination of IV and IT antibiotic doses, the absence of data exclusively focused on IT drug delivery underscores the imperative need for combining IT and systemic (IV) therapy to achieve optimal therapeutic effects [40]. A review paper highlighted that the IT/IVT administration of antibiotics plus IV in the treatment of postsurgical meningitis or ventriculitis decreased mortality 7.09 times compared to IV therapy alone (OR 0.27 [95% CI 0.15–0.49] $p \leq 0.00001$) with low heterogeneity ($Chi^2 = 7.2$ df $= 7$, $I^2 = 3\%$), with no significant differences in reinfection rate or poor functional outcome [127]. The suitable volume ranges for IT injection ranged from 0.5–5 mL [131]. IT and IVT administration of antibiotics have been documented for over a century [132], emphasizing the key advantage of direct antibiotic delivery, bypassing the BBB, leading to higher effective concentrations at the infection site (CNS) and potentially minimizing systemic side effects [40,133–136]. IT administration injects substances into the thecal sac with CSF, ensuring high therapeutic CNS concentration while minimizing off-target exposure and toxicity [18].

Regarding the uniformity of drug distribution within the CSF following IT administration, we have cited studies that highlight challenges in achieving adequate antibiotic concentrations in the ventricles. The distribution of drugs in the CSF following IT (Intralumbar) administration compared to IVT administration is subject to debate. One study noted that drug distribution after intralumbar dosing is not homogeneous and often fails to attain adequate antibiotic concentrations in the ventricles [44].

4.2.3. TDM and Adverse Effect of IT/IVT Drug Administration

The importance of TDM in CSF for patients who exhibit slow response, treatment failure, or risk of drug interactions is emphasized. The decision-making process for performing TDM was influenced by factors such as limited understanding of PK-PD relationships and desired target concentrations in CSF [52]. Despite the potential benefits of TDM, the text acknowledges several challenges, including limited data on CSF drug levels, the need for individualized monitoring, and the risks associated with frequent CSF sampling, such as infections related to ventriculostomy [40,137–139]. TDM might be considered in specific cases, like when there is persistent positivity in CSF cultures and evidence of therapeutic failure [40,140]. The proposal to use positron emission tomography for tracking drug distribution in TBM patients indicates a potential shift for future research [92,141].

The risks and adverse effects associated with the procedure of IT/IVT administration of a drug, while minimally invasive, are not negligible. The side effects following IT/IVT dosing of other antibiotics for treatment of Gram-negative meningitis (ventriculitis/meningitis) by gentamycin, polymyxin-B, colistin, tobramycin, AMI, etc. have been observed. The common side effects include chemical ventriculitis, meningitis, seizures, local events, or infection [20,21]. A comprehensive meta-analysis involving 23 studies (229 patients) reported a 13% overall complication rate, with chemical meningitis and seizures as predominant complications [21]. The reported complication rates may not show the true number of issues because patients getting IT treatment are usually in critical health. However, similar adverse effects are also seen with IV dosing alone, without IT administration [142]. Side effects associated with IT therapy may encompass aseptic meningitis, seizures, nerve root irritation, and even brain herniation [125], emphasizing the importance of careful consideration of seizures, chemical ventriculitis, and drug-related toxicities [143]. Although drawbacks associated with these administration routes encompass limited data for determining optimal dosage, the risk of infection due to access to the subarachnoid space and ventricles, and adverse events and risks specific to IT administration [144,145], the IT/IVT delivery of antibiotics has demonstrated enhanced survival and cure rates in individuals with postsurgical meningitis or ventriculitis [127].

4.3. Strategies for Prolonged IT DDS in TBM

While the efficacy of IT or IVT administration of RIF and INH has been substantiated by the literature we have reviewed, the frequent injection of drugs via IT routes is invasive and may result in patient non-compliance and procedure-associated risk. In light of this, we have explored prolonged drug delivery methods for IT administration and suggest that utilizing an implantable osmotic and infusion pump or nanoparticle-laden hydrogel could be advantageous for achieving prolonged drug delivery to the CNS. IT-prolonged drug delivery facilitates the direct administration of drugs, enhancing drug concentration in the CNS. Implantable pumps (like Duro's technology) or NP-laden hydrogel are recommended for sustaining antibiotic levels in the CSF, minimizing the need for frequent injections.

They have demonstrated efficacy in delivering medications subcutaneously, intrathecally, and via other routes, for managing prostate cancer, pain, and various medical conditions. Given our understanding that osmotic pumps and NP-laden hydrogels serve as means of prolonged drug delivery and have already been applied for drug delivery in humans, we believe that they could offer potential benefits for the IT delivery of anti-TB drugs.

4.3.1. Implantable Device for Prolonged Drug Delivery

Implantable infusion pumps, utilized for prolonged drug delivery directly into the body, varied in their operation based on the underlying pumping mechanism. These devices were divided into two main categories: passive pumps, which include osmotic pumps and those driven by fluorocarbon propellants, and active pumps, which are typically powered electrically. An ideal implantable drug delivery pump was designed to maintain precise delivery rates over its operational lifespan, fulfilling both standard and increased drug delivery requirements as determined by clinical conditions. It should support various delivery modes, including bolus, continuous, and adjustable-rate delivery, to accommodate different clinical needs. While both implantable osmotic pump devices and battery-powered programmable implantable drug infusion devices were employed for IT drug delivery, they were based on distinct operational principles and possess unique features, making them suitable for various therapeutic applications [146].

Battery-powered programmable implantable drug infusion devices for IT drug delivery: A sophisticated implantable programmable drug delivery device for IT usage represents a significant advancement in targeted drug delivery, particularly for delivering medications directly into the space surrounding the spinal cord. This technology enables precise medication delivery to CNS, significantly reducing the potential for systemic side effects. The device consists of a drug reservoir and a programmable pump, both of which are surgically implanted under the skin. It is connected by a catheter to the IT space, where it can release specific doses of medication at predetermined intervals. This method is especially beneficial for managing chronic conditions such as severe spasticity, chronic pain, and certain types of cancer pain, as it ensures consistent medication levels, minimizes side effects, and enhances the quality of life for patients. Such technology marks a leap forward in personalized medicine and targeted drug delivery, offering new hope for patients with conditions that are challenging to manage with traditional treatment methods. Commercially available drug infusion pumps, including the Medtronic SynchroMed II, Medtronic IsoMed, Codman 3000, and Codman Midstream, provide a range of specifications designed to meet specific medical requirements. For instance, the SynchroMed™ II infusion pump is a battery-powered, programmable, implantable device that delivers drugs to the IT space via an implanted catheter. These pumps can be programmed and managed using the Control Workflow approach, a strategy aimed at reducing the need for systemic opioids and achieving effective pain relief, highlighting the personalized and efficient nature of this DDS [147].

Implantable osmotic pump devices for IT drug delivery: Osmotic pumps are a form of membrane-controlled release DDS, driven by osmotic pressure to administer medication at a steady rate. The core principle of their operation involves water permeating through a semipermeable membrane into the pump. This action allows water to dissolve the drug formulation inside the pump while keeping the active agent itself from passing through the membrane. As the internal solution becomes diluted, osmotic pressure builds up, pushing the dissolved content out of the pump and into the body, thus delivering the active agent at a controlled rate [148]. The delivery rate of osmotic pumps is constant and determined by the device's design before implantation; however, unlike some other DDS, it cannot be adjusted or programmed after the pump has been implanted. This characteristic makes osmotic pumps suitable for applications where a steady, continuous drug release is required over a specific period. The development of the osmotic pump for drug delivery dates back to 1955, involving a design with three chambers primarily intended for pharmacological research, though it was never patented. The Rose–Nelson pump, consisting of three chambers, was designed for pharmacological research but was never patented [148]. An osmotic pump is recognized as a reliable means of controlled drug delivery, dispensing active agents through osmotic pressure [146]. The pump offered predictable release rates independent of the active agent, suitable for various therapeutic agents. Typically, these pumps exhibit a zero-order release profile following an initial lag, capable of achieving higher release rates compared to conventional diffusion-based systems [149]. The implantable osmotic pump systems come in various designs, such as the Elementary Osmotic Pump (EOP), Rose–Nelson Pump, Push–pull, Push–stick, Controlled Porosity, and Single Composition Osmotic Tablet. Among these, the most commonly utilized is the EOP [146].

Several implantable osmotic pumps for DDS have been developed, with two products from a company gaining popularity among others. The preclinical implantable osmotic pump, developed by Alza and intended for preclinical implantation in laboratory animals, has been used across various anatomical sites and species. Documented in over 10,000 scientific publications, its application has enabled the exploration of innovative therapies with potential implications for human treatment [150]. The ALZET® system has delivered a broad spectrum of active agents, ranging from amino acids to gastrointestinal modulators (e.g., amino acids, anesthetics, antibiotics, antibodies, anticancer agents, anticoagulants, anti-epileptics, antigens, antihypertensives, anti-parasitic agents, anti-Parkinson agents, ascitic fluid, catecholamines, chelators, cholinergic, CNS acting agents, enzyme inhibitors, and gastrointestinal modulators) [148]. Currently, the zero-order DDS osmotic implantable pump, developed by Alza and known as the DUROS® system, represents a significant breakthrough as the first implantable osmotic pump used in humans. The implantable pump, with its compact design incorporating a titanium alloy cylinder and a semi-permeable membrane activated by body fluids, facilitates continuous, zero-order drug release. In vitro studies have highlighted its capability to provide consistent zero-order release for up to one year, demonstrating its potential for long-term therapeutic applications [148]. This pump technology has been widely investigated for its use in various human therapies, offering delivery durations ranging from several months up to a year, making it particularly effective for potent drugs that require precise, extended release. A notable example is the Viadur® system, which employs DUROS technology and has received FDA approval for the continuous delivery of leuprolide in the treatment of advanced prostate cancer, achieving a 12-month release cycle at a daily rate of 125 µg [149,151–153].

Development is ongoing for other systems such as the Chronogesic® system for sufentanil delivery in chronic pain management, the Omega system for omega interferon in hepatitis C treatment, and catheterized osmotic pump systems for IT opioid delivery and local chemotherapy for brain tumors [150]. The technology is versatile, capable of administering various compounds irrespective of the drug's properties, from the leuprolide implant for prostate cancer to salmon calcitonin for osteoporosis and Paget's disease, typically implanted subcutaneously in the upper arm. This ensures a constant osmotic gradient for effective drug delivery over periods ranging from 3 to 12 months, significantly overcoming the limitations associated with bolus injections [154,155]. The pump's mechanism relies on osmosis to draw water through a semi-permeable membrane, which expands the osmotic agent, displaces a piston, and dispenses precise drug quantities [155]. User satisfaction is high, with a 96% reimplantation rate and more than 90% of users reporting satisfaction with the comfort and convenience of the device at 24- and 52-week intervals post-implantation [156].

4.3.2. Application of NP-Laden Hydrogel for IT-Prolonged Drug Release

The application of NP-laden hydrogels for IT-prolonged drug release has gained significant attention for its potential to enhance the delivery of nanomedicines in both preclinical and clinical settings. Notably, liposomal cytarabine (DepoCyt®) was approved by the FDA in 1999 for the IT treatment of lymphomatous meningitis and has shown effectiveness in various cancer treatments, demonstrating improved PK over free drug administration with cytotoxic levels maintained in the CSF for up to 14 days post-IVT injection and nearly 7 days after IT lumbar injection [19,157–160]. Furthermore, the transition of several IT medications, including morphine, ziconotide, baclofen, and nusinersen, to clinical trials or FDA approval underscores the significance of nanoparticle-based drug delivery in the subarachnoid space [161]. Additionally, NP-laden hydrogels have gained attention for their potential to improve drug release kinetics and enhance bioavailability [162]. These NP-laden hydrogels, which incorporate nanoparticles within 3D polymer matrices, offer targeted and controlled drug release to specific sites, such as the injured CNS [163–166]. The broad spread of NPs along the neuraxis underscores their pivotal role in efficient nanomedicine delivery [167]. This technology, incorporating NPs into hydrogels, addressed issues like burst release and low encapsulation efficiency, presenting a promising avenue to improve drug delivery efficiency [162]. Despite their potential and promising preclinical studies, the clinical adoption of NP-laden hydrogels is still emerging, with ongoing research focused on translating experimental findings into clinical practice [168]. Studies have demonstrated the efficacy of thermosensitive gels in delivering drugs like amphotericin-B, reducing dosing frequency and neurotoxicity while enhancing antifungal efficacy in animal models [169].

The availability of IT drug delivery via osmotic pump therapy or NP-hydrogel composite may vary slightly between first-world and third-world countries. While the cost of osmotic pumps compared to battery-powered pumps is lower, accessibility differs based on healthcare infrastructure and regulatory frameworks. In first-world nations, advanced medical facilities and skilled personnel make this therapy more accessible, whereas third-world countries encounter challenges such as limited healthcare resources, medical professional shortages, and regulatory barriers which may affect access to these treatments.

The evolving landscape of IT anti-TB therapy for TBM is paving the way for the development of optimized therapeutic regimens. This progress is driven by an in-depth understanding of various factors including physiological barriers, PKs, drug properties, administration routes, clinical outcomes, and strategies for prolonged drug delivery. Such comprehensive insights are crucial for enhancing the efficacy and safety of treatments for TBM, ultimately improving patient outcomes.

5. Conclusions

Understanding the PK-PD parameters of anti-TB drugs is key to optimizing dosages for TBM treatment. Despite facing challenges, IT administration of RIF and INH promising for increasing CSF drug concentration and enhancing clinical outcomes while minimizing systemic side effects. The efficacy and safety of IT/IVT-RIF and INH are supported by case reports, cohort studies, and the literature, even in refractory TBM cases. Osmotic pumps offer a practical solution for sustained IT drug delivery, and nanoparticle-laden hydrogels hold potential for the future, awaiting clinical validation.

Based on CSF-PK data following oral and IV administration of anti-TB drugs, a daily release of 5 mg of RIF and 20 mg of INH via an IT osmotic pump, supplemented by an oral regimen, may effectively eradicate MTB and yield positive clinical outcomes. This approach aims to prevent neurological damage caused by the infection. Despite the hurdles in dose optimization, the current literature supports the concept of optimized regimens for TBM treatment, illustrating a hopeful path forward in the fight against this severe condition.

Author Contributions: Conceptualization, M.-J.S.; methodology, A.K.M. and M.-J.S.; formal analysis, A.K.M.; investigation, A.K.M.; writing—original draft preparation, A.K.M.; writing—review and editing, A.K.M. and M.-J.S.; visualization, A.K.M. and M.-J.S.; supervision, M.-J.S.; project administration, M.-J.S.; funding acquisition, M.-J.S. All authors have read and agreed to the published version of the manuscript.

Funding: This research was supported by the Basic Science Research Program through the National Research Foundation of Korea (NRF) funded by the Ministry of Education (NRF-2020R1I1A3073930).

Institutional Review Board Statement: Not applicable.

Informed Consent Statement: Not applicable.

Data Availability Statement: Not applicable (due to this review summarizing the published literature).

Conflicts of Interest: The authors declare no conflicts of interest.

List of Abbreviations

TBM	Tuberculous Meningitis
BBB	Blood–Brain Barrier
CNS	Central Nervous System
IT	Intrathecal
IVT	Intraventricular
CSF	Cerebrospinal Fluid
TB	Tuberculosis
PK	Pharmacokinetics
PD	Pharmacodynamics
NP	Nanoparticles
BCSFB	Blood–CSF Barrier
UI	Uncertainty Interval
HIV-1	Human Immunodeficiency Virus type 1
RR/MDR-TBM	Rifampicin/Multidrug-Resistant Tuberculous Meningitis
IV	Intravenous
MDR	Multidrug-Resistant
DDS	Drug Delivery Systems
MW	Molecular Weight
MIC	Minimal Inhibitory Concentration
AUC	Area Under Curve
DILI	Drug-Induced Liver Injury
EPTB	Extrapulmonary Tuberculosis
HIV	Human Immunodeficiency Virus
INH	Isoniazid
RIF	Rifampicin
LVX	Levofloxacin
AMK	Amikacin
PZA	Pyrazinamide
ETB	Ethambutol
PAS	Para-aminosalicylic acid
RFB	Rifabutin
CYC	Cycloserine
LNZ	Linezolid
CFZ	Clofazimine
BDQ	Bedaquiline
Dlm	Delamanid
CM	Capreomycin
KAN	Kanamycin
ETO	Ethionamide
NEW	Not well established
DS	Drug sensitive
Ys	Years
XDR	Extensively drug-resistant
CI/F	Clearance over bioavailability
Vd/F	Volume of distribution over bioavailability
GOS	Glasgow Outcome Scale
MRS	Modified Rankin Scale
R	Resistant
Mod.	Moderate
Sus.	Susceptible
BL	Borderline
NAT2	N-acetyltransferase 2

MDR-TB	Multidrug-Resistant Tuberculosis
LMCA	Left Middle Cerebral Artery
BCSFB	Blood-Cerebrospinal Fluid Barrier
XDR	Extensively Drug-Resistant
TDM	Therapeutic Drug Monitoring
PK-PD	Pharmacokinetic-Pharmacodynamic
FDA	Food and Drug Administration
MTB	*Mycobacterium tuberculosis*

References

1. Marais, S.; Pepper, D.J.; Schutz, C.; Wilkinson, R.J.; Meintjes, G. Presentation and outcome of tuberculous meningitis in a high HIV prevalence setting. *PLoS ONE* **2011**, *6*, e20077. [CrossRef]
2. Mezochow, A.; Thakur, K.; Vinnard, C. Tuberculous Meningitis in Children and Adults: New Insights for an Ancient Foe. *Curr. Neurol. Neurosci. Rep.* **2017**, *17*, 85. [CrossRef] [PubMed]
3. Bourgi, K.; Fiske, C.; Sterling, T.R. Tuberculosis Meningitis. *Curr. Infect. Dis. Rep.* **2017**, *19*, 39. [CrossRef]
4. Navarro-Flores, A.; Fernandez-Chinguel, J.E.; Pacheco-Barrios, N.; Soriano-Moreno, D.R.; Pacheco-Barrios, K. Global morbidity and mortality of central nervous system tuberculosis: A systematic review and meta-analysis. *J. Neurol.* **2022**, *269*, 3482–3494. [CrossRef] [PubMed]
5. Dodd, P.J.; Osman, M.; Cresswell, F.V.; Stadelman, A.M.; Lan, N.H.; Thuong, N.T.T.; Muzyamba, M.; Glaser, L.; Dlamini, S.S.; Seddon, J.A. The global burden of tuberculous meningitis in adults: A modelling study. *PLoS Glob. Public. Health* **2021**, *1*, e0000069. [CrossRef] [PubMed]
6. Manyelo, C.M.; Solomons, R.S.; Walzl, G.; Chegou, N.N. Tuberculous Meningitis: Pathogenesis, Immune Responses, Diagnostic Challenges, and the Potential of Biomarker-Based Approaches. *J. Clin. Microbiol.* **2021**, *59*, 10–1128. [CrossRef]
7. du Preez, K.; Seddon, J.A.; Schaaf, H.S.; Hesseling, A.C.; Starke, J.R.; Osman, M.; Lombard, C.J.; Solomons, R. Global shortages of BCG vaccine and tuberculous meningitis in children. *Lancet Glob. Health* **2019**, *7*, e28–e29. [CrossRef]
8. Cecchini, D.; Ambrosioni, J.; Brezzo, C.; Corti, M.; Rybko, A.; Perez, M.; Poggi, S.; Ambroggi, M. Tuberculous meningitis in HIV-infected and non-infected patients: Comparison of cerebrospinal fluid findings. *Int. J. Tuberc. Lung Dis.* **2009**, *13*, 269–271. [PubMed]
9. Thwaites, G.E.; Duc Bang, N.; Huy Dung, N.; Thi Quy, H.; Thi Tuong Oanh, D.; Thi Cam Thoa, N.; Quang Hien, N.; Tri Thuc, N.; Ngoc Hai, N.; Thi Ngoc Lan, N.; et al. The influence of HIV infection on clinical presentation, response to treatment, and outcome in adults with Tuberculous meningitis. *J. Infect. Dis.* **2005**, *192*, 2134–2141. [CrossRef]
10. Soria, J.; Metcalf, T.; Mori, N.; Newby, R.E.; Montano, S.M.; Huaroto, L.; Ticona, E.; Zunt, J.R. Mortality in hospitalized patients with tuberculous meningitis. *BMC Infect. Dis.* **2019**, *19*, 9. [CrossRef]
11. Wilkinson, R.J.; Rohlwink, U.; Misra, U.K.; van Crevel, R.; Mai, N.T.H.; Dooley, K.E.; Caws, M.; Figaji, A.; Savic, R.; Solomons, R.; et al. Tuberculous meningitis. *Nat. Rev. Neurol.* **2017**, *13*, 581–598. [CrossRef] [PubMed]
12. Cherian, A.; Ajitha, K.C.; Iype, T.; Divya, K.P. Neurotuberculosis: An update. *Acta Neurol. Belg.* **2021**, *121*, 11–21. [CrossRef] [PubMed]
13. Davis, A.G.; Rohlwink, U.K.; Proust, A.; Figaji, A.A.; Wilkinson, R.J. The pathogenesis of tuberculous meningitis. *J. Leukoc. Biol.* **2019**, *105*, 267–280. [CrossRef] [PubMed]
14. Nataprawira, H.M.; Gafar, F.; Risan, N.A.; Wulandari, D.A.; Sudarwati, S.; Marais, B.J.; Stevens, J.; Alffenaar, J.C.; Ruslami, R. Treatment Outcomes of Childhood Tuberculous Meningitis in a Real-World Retrospective Cohort, Bandung, Indonesia. *Emerg. Infect. Dis.* **2022**, *28*, 660–671. [CrossRef] [PubMed]
15. Chiang, S.S.; Khan, F.A.; Milstein, M.B.; Tolman, A.W.; Benedetti, A.; Starke, J.R.; Becerra, M.C. Treatment outcomes of childhood tuberculous meningitis: A systematic review and meta-analysis. *Lancet Infect. Dis.* **2014**, *14*, 947–957. [CrossRef]
16. Dian, S.; Ganiem, A.R.; van Laarhoven, A. Central nervous system tuberculosis. *Curr. Opin. Neurol.* **2021**, *34*, 396–402. [CrossRef]
17. da Silva Leite, J.M.; Patriota, Y.B.G.; de La Roca, M.F.; Soares-Sobrinho, J.L. New Perspectives in Drug Delivery Systems for the Treatment of Tuberculosis. *Curr. Med. Chem.* **2022**, *29*, 1936–1958. [CrossRef]
18. Calias, P.; Banks, W.A.; Begley, D.; Scarpa, M.; Dickson, P. Intrathecal delivery of protein therapeutics to the brain: A critical reassessment. *Pharmacol. Ther.* **2014**, *144*, 114–122. [CrossRef]
19. Fowler, M.J.; Cotter, J.D.; Knight, B.E.; Sevick-Muraca, E.M.; Sandberg, D.I.; Sirianni, R.W. Intrathecal drug delivery in the era of nanomedicine. *Adv. Drug Deliv. Rev.* **2020**, *165–166*, 77–95. [CrossRef] [PubMed]
20. Khan, S.A.; Waqas, M.; Siddiqui, U.T.; Shamim, M.S.; Nathani, K.R.; Jooma, R.; Mehmood, F. Intrathecal and intraventricular antibiotics for postoperative Gram-negative meningitis and ventriculitis. *Surg. Neurol. Int.* **2017**, *8*, 226. [CrossRef]
21. Brotis, A.G.; Churis, I.; Karvouniaris, M. Local complications of adjunct intrathecal antibiotics for nosocomial meningitis associated with gram-negative pathogens: A meta-analysis. *Neurosurg. Rev.* **2021**, *44*, 139–152. [CrossRef] [PubMed]
22. Dajez, P.; Vincken, W.; Lambelin, D.; Noterman, J.; Yourassowski, E.; Telerman-Toppet, N. Intraventricular administration of rifampin for tuberculous meningitis. *J. Neurol.* **1981**, *225*, 153–156. [CrossRef] [PubMed]
23. Vincken, W.; Meysman, M.; Verbeelen, D.; Lauwers, S.; D'Haens, J. Intraventricular rifampicin in severe tuberculous meningo-encephalitis. *Eur. Respir. J.* **1992**, *5*, 891–893. [CrossRef] [PubMed]

24. Chen, Y.; Chang, F.; Wang, Z. Treatment of the secondary hydrocephalus of tuberculous meningitis by lateral ventricular drainage and drug injection. *Zhonghua Jie He He Hu Xi Za Zhi* **1996**, *19*, 297–299. [PubMed]
25. Berning, S.E.; Cherry, T.A.; Iseman, M.D. Novel treatment of meningitis caused by multidrug-resistant *Mycobacterium tuberculosis* with intrathecal levofloxacin and amikacin: Case report. *Clin. Infect. Dis.* **2001**, *32*, 643–646. [CrossRef] [PubMed]
26. World Health Organization. WHO Guidelines Approved by the Guidelines Review Committee. In *Treatment of Tuberculosis: Guidelines*; World Health Organization: Geneva, Switzerland, 2010.
27. Nahid, P.; Dorman, S.E.; Alipanah, N.; Barry, P.M.; Brozek, J.L.; Cattamanchi, A.; Chaisson, L.H.; Chaisson, R.E.; Daley, C.L.; Grzemska, M.; et al. Official American Thoracic Society/Centers for Disease Control and Prevention/Infectious Diseases Society of America Clinical Practice Guidelines: Treatment of Drug-Susceptible Tuberculosis. *Clin. Infect. Dis.* **2016**, *63*, e147–e195. [CrossRef] [PubMed]
28. Thwaites, G.E.; van Toorn, R.; Schoeman, J. Tuberculous meningitis: More questions, still too few answers. *Lancet Neurol.* **2013**, *12*, 999–1010. [CrossRef] [PubMed]
29. Thwaites, G.E.; Lan, N.T.; Dung, N.H.; Quy, H.T.; Oanh, D.T.; Thoa, N.T.; Hien, N.Q.; Thuc, N.T.; Hai, N.N.; Bang, N.D.; et al. Effect of antituberculosis drug resistance on response to treatment and outcome in adults with tuberculous meningitis. *J. Infect. Dis.* **2005**, *192*, 79–88. [CrossRef]
30. Thwaites, G.E.; Nguyen, D.B.; Nguyen, H.D.; Hoang, T.Q.; Do, T.T.; Nguyen, T.C.; Nguyen, Q.H.; Nguyen, T.T.; Nguyen, N.H.; Nguyen, T.N.; et al. Dexamethasone for the treatment of tuberculous meningitis in adolescents and adults. *N. Engl. J. Med.* **2004**, *351*, 1741–1751. [CrossRef]
31. Seddon, J.A.; Visser, D.H.; Bartens, M.; Jordaan, A.M.; Victor, T.C.; van Furth, A.M.; Schoeman, J.F.; Schaaf, H.S. Impact of drug resistance on clinical outcome in children with tuberculous meningitis. *Pediatr. Infect. Dis. J.* **2012**, *31*, 711–716. [CrossRef]
32. Fang, M.-T.; Su, Y.-F.; An, H.-R.; Zhang, P.-Z.; Deng, G.-F.; Liu, H.-M.; Mao, Z.; Zeng, J.-F.; Li, G.; Yang, Q.-T.; et al. Decreased mortality seen in rifampicin/multidrug-resistant tuberculous meningitis treated with linezolid in Shenzhen, China. *BMC Infect. Dis.* **2021**, *21*, 1015. [CrossRef] [PubMed]
33. Shapey, J.; Toma, A.; Saeed, S.R. Physiology of cerebrospinal fluid circulation. *Curr. Opin. Otolaryngol. Head. Neck Surg.* **2019**, *27*, 326–333. [CrossRef] [PubMed]
34. Iliff, J.J.; Wang, M.; Liao, Y.; Plogg, B.A.; Peng, W.; Gundersen, G.A.; Benveniste, H.; Vates, G.E.; Deane, R.; Goldman, S.A.; et al. A paravascular pathway facilitates CSF flow through the brain parenchyma and the clearance of interstitial solutes, including amyloid β. *Sci. Transl. Med.* **2012**, *4*, 147ra111. [CrossRef] [PubMed]
35. Naseri Kouzehgarani, G.; Feldsien, T.; Engelhard, H.H.; Mirakhur, K.K.; Phipps, C.; Nimmrich, V.; Clausznitzer, D.; Lefebvre, D.R. Harnessing cerebrospinal fluid circulation for drug delivery to brain tissues. *Adv. Drug Deliv. Rev.* **2021**, *173*, 20–59. [CrossRef] [PubMed]
36. Nau, R.; Sörgel, F.; Eiffert, H. Penetration of drugs through the blood-cerebrospinal fluid/blood-brain barrier for treatment of central nervous system infections. *Clin. Microbiol. Rev.* **2010**, *23*, 858–883. [CrossRef] [PubMed]
37. Albanèse, J.; Léone, M.; Bruguerolle, B.; Ayem, M.L.; Lacarelle, B.; Martin, C. Cerebrospinal fluid penetration and pharmacokinetics of vancomycin administered by continuous infusion to mechanically ventilated patients in an intensive care unit. *Antimicrob. Agents Chemother.* **2000**, *44*, 1356–1358. [CrossRef] [PubMed]
38. Lutsar, I.; McCracken, G.H., Jr.; Friedland, I.R. Antibiotic Pharmacodynamics in Cerebrospinal Fluid. *Clin. Infect. Dis.* **1998**, *27*, 1117–1129. [CrossRef] [PubMed]
39. Lin Wu, F.L.; Liu, S.S.; Yang, T.Y.; Win, M.F.; Lin, S.W.; Huang, C.F.; Wang, K.C.; Shen, L.J. A Larger Dose of Vancomycin Is Required in Adult Neurosurgical Intensive Care Unit Patients Due to Augmented Clearance. *Ther. Drug Monit.* **2015**, *37*, 609–618. [CrossRef]
40. Nau, R.; Blei, C.; Eiffert, H. Intrathecal Antibacterial and Antifungal Therapies. *Clin. Microbiol. Rev.* **2020**, *33*, 10–1128. [CrossRef]
41. Courchesne, E.; Chisum, H.J.; Townsend, J.; Cowles, A.; Covington, J.; Egaas, B.; Harwood, M.; Hinds, S.; Press, G.A. Normal brain development and aging: Quantitative analysis at in vivo MR imaging in healthy volunteers. *Radiology* **2000**, *216*, 672–682. [CrossRef]
42. Fleischhack, G.; Jaehde, U.; Bode, U. Pharmacokinetics following intraventricular administration of chemotherapy in patients with neoplastic meningitis. *Clin. Pharmacokinet.* **2005**, *44*, 1–31. [CrossRef] [PubMed]
43. Lebret, A.; Hodel, J.; Rahmouni, A.; Decq, P.; Petit, E. Cerebrospinal fluid volume analysis for hydrocephalus diagnosis and clinical research. *Comput. Med. Imaging Graph.* **2013**, *37*, 224–233. [CrossRef] [PubMed]
44. Gerber, J.; Tumani, H.; Kolenda, H.; Nau, R. Lumbar and ventricular CSF protein, leukocytes, and lactate in suspected bacterial CNS infections. *Neurology* **1998**, *51*, 1710–1714. [CrossRef] [PubMed]
45. Shapiro, W.R.; Young, D.F.; Mehta, B.M. Methotrexate: Distribution in cerebrospinal fluid after intravenous, ventricular and lumbar injections. *N. Engl. J. Med.* **1975**, *293*, 161–166. [CrossRef] [PubMed]
46. Kaiser, A.B.; McGee, Z.A. Aminoglycoside therapy of gram-negative bacillary meningitis. *N. Engl. J. Med.* **1975**, *293*, 1215–1220. [CrossRef] [PubMed]
47. Ichie, T.; Urano, K.; Suzuki, D.; Okada, T.; Kobayashi, N.; Hayashi, H.; Sugiura, Y.; Yamamura, K.; Sugiyama, T. Influence of cerebral fluid drainage on the pharmacokinetics of vancomycin in neurosurgical patients. *Pharmazie* **2015**, *70*, 404–409. [PubMed]

48. Li, X.; Wang, X.; Wu, Y.; Sun, S.; Chen, K.; Lu, Y.; Wang, Q.; Zhao, Z. Plasma and cerebrospinal fluid population pharmacokinetic modeling and simulation of meropenem after intravenous and intrathecal administration in postoperative neurosurgical patients. *Diagn. Microbiol. Infect. Dis.* **2019**, *93*, 386–392. [CrossRef]
49. Li, X.; Sun, S.; Wang, Q.; Zhao, Z. Population Pharmacokinetics of Combined Intravenous and Local Intrathecal Administration of Meropenem in Aneurysm Patients with Suspected Intracranial Infections After Craniotomy. *Eur. J. Drug Metab. Pharmacokinet.* **2018**, *43*, 45–53. [CrossRef] [PubMed]
50. Nilsson, C.; Ståhlberg, F.; Thomsen, C.; Henriksen, O.; Herning, M.; Owman, C. Circadian variation in human cerebrospinal fluid production measured by magnetic resonance imaging. *Am. J. Physiol.* **1992**, *262*, R20–R24. [CrossRef]
51. Klein, O.; Demoulin, B.; Jean Auque, R.T.; Audibert, G.; Sainte-Rose, C.; Marchal, J.C.; Marchal, F. Cerebrospinal fluid outflow and intracranial pressure in hydrocephalic patients with external ventricular drainage. *Acta Neurol. Scand.* **2010**, *122*, 140–147. [CrossRef]
52. Muller, A.E.; van Vliet, P.; Koch, B.C.P. Clinical Experience with Off-Label Intrathecal Administration of Selected Antibiotics in Adults: An Overview with Pharmacometric Considerations. *Antibiotics* **2023**, *12*, 1291. [CrossRef] [PubMed]
53. Wright, P.F.; Kaiser, A.B.; Bowman, C.M.; McKee, K.T., Jr.; Trujillo, H.; McGee, Z.A. The pharmacokinetics and efficacy of an aminoglycoside administered into the cerebral ventricles in neonates: Implications for further evaluation of this route of therapy in meningitis. *J. Infect. Dis.* **1981**, *143*, 141–147. [CrossRef] [PubMed]
54. Ooie, T.; Suzuki, H.; Terasaki, T.; Sugiyama, Y. Kinetics of quinolone antibiotics in rats: Efflux from cerebrospinal fluid to the circulation. *Pharm. Res.* **1996**, *13*, 1065–1068. [CrossRef] [PubMed]
55. Reesor, C.; Chow, A.W.; Kureishi, A.; Jewesson, P.J. Kinetics of intraventricular vancomycin in infections of cerebrospinal fluid shunts. *J. Infect. Dis.* **1988**, *158*, 1142–1143. [CrossRef] [PubMed]
56. Imberti, R.; Cusato, M.; Accetta, G.; Marinò, V.; Procaccio, F.; Del Gaudio, A.; Iotti, G.A.; Regazzi, M. Pharmacokinetics of colistin in cerebrospinal fluid after intraventricular administration of colistin methanesulfonate. *Antimicrob. Agents Chemother.* **2012**, *56*, 4416–4421. [CrossRef] [PubMed]
57. Norrby, S.R. Role of cephalosporins in the treatment of bacterial meningitis in adults. Overview with special emphasis on ceftazidime. *Am. J. Med.* **1985**, *79*, 56–61. [CrossRef] [PubMed]
58. Jongmans, C.; Muller, A.E.; Van Den Broek, P.; Cruz De Almeida, B.M.; Van Den Berg, C.; Van Oldenrijk, J.; Bos, P.K.; Koch, B.C.P. An Overview of the Protein Binding of Cephalosporins in Human Body Fluids: A Systematic Review. *Front. Pharmacol.* **2022**, *13*, 900551. [CrossRef] [PubMed]
59. Ziaka, M.; Markantonis, S.L.; Fousteri, M.; Zygoulis, P.; Panidis, D.; Karvouniaris, M.; Makris, D.; Zakynthinos, E. Combined intravenous and intraventricular administration of colistin methanesulfonate in critically ill patients with central nervous system infection. *Antimicrob. Agents Chemother.* **2013**, *57*, 1938–1940. [CrossRef]
60. Nau, R.; Seele, J.; Djukic, M.; Eiffert, H. Pharmacokinetics and pharmacodynamics of antibiotics in central nervous system infections. *Curr. Opin. Infect. Dis.* **2018**, *31*, 57–68. [CrossRef]
61. Ruslami, R.; Gafar, F.; Yunivita, V.; Parwati, I.; Ganiem, A.R.; Aarnoutse, R.E.; Wilffert, B.; Alffenaar, J.C.; Nataprawira, H.M. Pharmacokinetics and safety/tolerability of isoniazid, rifampicin and pyrazinamide in children and adolescents treated for tuberculous meningitis. *Arch. Dis. Child.* **2022**, *107*, 70–77. [CrossRef]
62. Donald, P.R. Antituberculosis drug-induced hepatotoxicity in children. *Pediatr. Rep.* **2011**, *3*, e16. [CrossRef] [PubMed]
63. Reynolds, J.; Heysell, S.K. Understanding pharmacokinetics to improve tuberculosis treatment outcome. *Expert. Opin. Drug Metab. Toxicol.* **2014**, *10*, 813–823. [CrossRef]
64. Pardridge, W.M. Drug transport across the blood-brain barrier. *J. Cereb. Blood Flow. Metab.* **2012**, *32*, 1959–1972. [CrossRef]
65. Sullins, A.K.; Abdel-Rahman, S.M. Pharmacokinetics of Antibacterial Agents in the CSF of Children and Adolescents. *Pediatric Drugs* **2013**, *15*, 93–117. [CrossRef] [PubMed]
66. Peloquin, C.A.; Jaresko, G.S.; Yong, C.L.; Keung, A.C.; Bulpitt, A.E.; Jelliffe, R.W. Population pharmacokinetic modeling of isoniazid, rifampin, and pyrazinamide. *Antimicrob. Agents Chemother.* **1997**, *41*, 2670–2679. [CrossRef] [PubMed]
67. Donald, P.R. Cerebrospinal fluid concentrations of antituberculosis agents in adults and children. *Tuberculosis* **2010**, *90*, 279–292. [CrossRef]
68. Rastogi, N.; Labrousse, V.; Goh, K.S. In vitro activities of fourteen antimicrobial agents against drug susceptible and resistant clinical isolates of *Mycobacterium tuberculosis* and comparative intracellular activities against the virulent H37Rv strain in human macrophages. *Curr. Microbiol.* **1996**, *33*, 167–175. [CrossRef]
69. Kohli, M.; Schiller, I.; Dendukuri, N.; Dheda, K.; Denkinger, C.M.; Schumacher, S.G.; Steingart, K.R. Xpert(®) MTB/RIF assay for extrapulmonary tuberculosis and rifampicin resistance. *Cochrane Database Syst. Rev.* **2018**, *8*, Cd012768. [CrossRef]
70. Rifampin. *Tuberculosis* **2008**, *88*, 151–154. [CrossRef]
71. Sousa, M.; Pozniak, A.; Boffito, M. Pharmacokinetics and pharmacodynamics of drug interactions involving rifampicin, rifabutin and antimalarial drugs. *J. Antimicrob. Chemother.* **2008**, *62*, 872–878. [CrossRef]
72. Cresswell, F.V.; Meya, D.B.; Kagimu, E.; Grint, D.; Te Brake, L.; Kasibante, J.; Martyn, E.; Rutakingirwa, M.; Quinn, C.M.; Okirwoth, M.; et al. High-Dose Oral and Intravenous Rifampicin for the Treatment of Tuberculous Meningitis in Predominantly Human Immunodeficiency Virus (HIV)-Positive Ugandan Adults: A Phase II Open-Label Randomized Controlled Trial. *Clin. Infect. Dis.* **2021**, *73*, 876–884. [CrossRef] [PubMed]

73. Boeree, M.J.; Diacon, A.H.; Dawson, R.; Narunsky, K.; du Bois, J.; Venter, A.; Phillips, P.P.; Gillespie, S.H.; McHugh, T.D.; Hoelscher, M.; et al. A dose-ranging trial to optimize the dose of rifampin in the treatment of tuberculosis. *Am. J. Respir. Crit. Care Med.* **2015**, *191*, 1058–1065. [CrossRef] [PubMed]
74. Curci, G.; Bergamini, N.; Delli Veneri, F.; Ninni, A.; Nitti, V. Half-life of rifampicin after repeated administration of different doses in humans. *Chemotherapy* **1972**, *17*, 373–381. [CrossRef] [PubMed]
75. Nitti, V. Antituberculosis activity of rifampin. Report of studies performed and in progress (1966–1971). *Chest* **1972**, *61*, 589–598. [CrossRef] [PubMed]
76. Svensson, R.J.; Aarnoutse, R.E.; Diacon, A.H.; Dawson, R.; Gillespie, S.H.; Boeree, M.J.; Simonsson, U.S.H. A Population Pharmacokinetic Model Incorporating Saturable Pharmacokinetics and Autoinduction for High Rifampicin Doses. *Clin. Pharmacol. Ther.* **2018**, *103*, 674–683. [CrossRef] [PubMed]
77. Furesz, S.; Scotti, R.; Pallanza, R.; Mapelli, E. Rifampicin: A new rifamycin. 3. Absorption, distribution, and elimination in man. *Arzneimittelforschung* **1967**, *17*, 534–537. [PubMed]
78. Nau, R.; Sörgel, F.; Prange, H.W. Pharmacokinetic optimisation of the treatment of bacterial central nervous system infections. *Clin. Pharmacokinet.* **1998**, *35*, 223–246. [CrossRef]
79. Mezochow, A.; Thakur, K.T.; Zentner, I.; Subbian, S.; Kagan, L.; Vinnard, C. Attainment of target rifampicin concentrations in cerebrospinal fluid during treatment of tuberculous meningitis. *Int. J. Infect. Dis.* **2019**, *84*, 15–21. [CrossRef] [PubMed]
80. Sloan, D.J.; McCallum, A.D.; Schipani, A.; Egan, D.; Mwandumba, H.C.; Ward, S.A.; Waterhouse, D.; Banda, G.; Allain, T.J.; Owen, A.; et al. Genetic Determinants of the Pharmacokinetic Variability of Rifampin in Malawian Adults with Pulmonary Tuberculosis. *Antimicrob. Agents Chemother.* **2017**, *61*, 10–1128. [CrossRef]
81. Jeanes, C.W.; Jessamine, A.G.; Eidus, L. Treatment of chronic drug-resistant pulmonary tuberculosis with rifampin and ethambutol. *Can. Med. Assoc. J.* **1972**, *106*, 884–888.
82. Acocella, G. Clinical pharmacokinetics of rifampicin. *Clin. Pharmacokinet.* **1978**, *3*, 108–127. [CrossRef] [PubMed]
83. Nau, R.; Prange, H.W.; Menck, S.; Kolenda, H.; Visser, K.; Seydel, J.K. Penetration of rifampicin into the cerebrospinal fluid of adults with uninflamed meninges. *J. Antimicrob. Chemother.* **1992**, *29*, 719–724. [CrossRef] [PubMed]
84. Ruslami, R.; Ganiem, A.R.; Dian, S.; Apriani, L.; Achmad, T.H.; van der Ven, A.J.; Borm, G.; Aarnoutse, R.E.; van Crevel, R. Intensified regimen containing rifampicin and moxifloxacin for tuberculous meningitis: An open-label, randomised controlled phase 2 trial. *Lancet Infect. Dis.* **2013**, *13*, 27–35. [CrossRef] [PubMed]
85. Yunivita, V.; Dian, S.; Ganiem, A.R.; Hayati, E.; Hanggono Achmad, T.; Purnama Dewi, A.; Teulen, M.; Meijerhof-Jager, P.; van Crevel, R.; Aarnoutse, R.; et al. Pharmacokinetics and safety/tolerability of higher oral and intravenous doses of rifampicin in adult tuberculous meningitis patients. *Int. J. Antimicrob. Agents* **2016**, *48*, 415–421. [CrossRef] [PubMed]
86. Shang, P.; Xia, Y.; Liu, F.; Wang, X.; Yuan, Y.; Hu, D.; Tu, D.; Chen, Y.; Deng, P.; Cheng, S.; et al. Incidence, clinical features and impact on anti-tuberculosis treatment of anti-tuberculosis drug induced liver injury (ATLI) in China. *PLoS ONE* **2011**, *6*, e21836. [CrossRef] [PubMed]
87. Wasserman, S.; Davis, A.; Stek, C.; Chirehwa, M.; Botha, S.; Daroowala, R.; Bremer, M.; Maxebengula, M.; Koekemoer, S.; Goliath, R.; et al. Plasma Pharmacokinetics of High-Dose Oral versus Intravenous Rifampicin in Patients with Tuberculous Meningitis: A Randomized Controlled Trial. *Antimicrob. Agents Chemother.* **2021**, *65*, e0014021. [CrossRef]
88. Ellard, G.A.; Humphries, M.J.; Allen, B.W. Cerebrospinal fluid drug concentrations and the treatment of tuberculous meningitis. *Am. Rev. Respir. Dis.* **1993**, *148*, 650–655. [CrossRef] [PubMed]
89. Te Brake, L.; Dian, S.; Ganiem, A.R.; Ruesen, C.; Burger, D.; Donders, R.; Ruslami, R.; van Crevel, R.; Aarnoutse, R. Pharmacokinetic/pharmacodynamic analysis of an intensified regimen containing rifampicin and moxifloxacin for tuberculous meningitis. *Int. J. Antimicrob. Agents* **2015**, *45*, 496–503. [CrossRef]
90. Savic, R.M.; Ruslami, R.; Hibma, J.E.; Hesseling, A.; Ramachandran, G.; Ganiem, A.R.; Swaminathan, S.; McIlleron, H.; Gupta, A.; Thakur, K.; et al. Pediatric tuberculous meningitis: Model-based approach to determining optimal doses of the anti-tuberculosis drugs rifampin and levofloxacin for children. *Clin. Pharmacol. Ther.* **2015**, *98*, 622–629. [CrossRef]
91. Davis, A.; Meintjes, G.; Wilkinson, R.J. Treatment of Tuberculous Meningitis and Its Complications in Adults. *Curr. Treat. Options Neurol.* **2018**, *20*, 5. [CrossRef]
92. Liang, Z.; Liao, W.; Chen, Q.; Li, H.; Ye, M.; Zou, J.; Deng, G.; Zhang, P. Pharmacokinetics of Antituberculosis Drugs in Plasma and Cerebrospinal Fluid in a Patient with Pre-Extensive Drug Resistant Tuberculosis Meningitis. *Infect. Drug Resist.* **2023**, *16*, 1669–1676. [CrossRef] [PubMed]
93. Azarmi, M.; Maleki, H.; Nikkam, N.; Malekinejad, H. Transcellular brain drug delivery: A review on recent advancements. *Int. J. Pharm.* **2020**, *586*, 119582. [CrossRef] [PubMed]
94. Pajouhesh, H.; Lenz, G.R. Medicinal chemical properties of successful central nervous system drugs. *NeuroRx* **2005**, *2*, 541–553. [CrossRef] [PubMed]
95. Isoniazid. *Tuberculosis* **2008**, *88*, 112–116. [CrossRef] [PubMed]
96. Lakshminarayana, S.B.; Huat, T.B.; Ho, P.C.; Manjunatha, U.H.; Dartois, V.; Dick, T.; Rao, S.P. Comprehensive physicochemical, pharmacokinetic and activity profiling of anti-TB agents. *J. Antimicrob. Chemother.* **2015**, *70*, 857–867. [CrossRef] [PubMed]
97. Mitchison, D.A. Role of individual drugs in the chemotherapy of tuberculosis. *Int. J. Tuberc. Lung Dis.* **2000**, *4*, 796–806.
98. Djukic, M.; Munz, M.; Sörgel, F.; Holzgrabe, U.; Eiffert, H.; Nau, R. Overton's rule helps to estimate the penetration of anti-infectives into patients' cerebrospinal fluid. *Antimicrob. Agents Chemother.* **2012**, *56*, 979–988. [CrossRef]

99. Shin, S.G.; Roh, J.K.; Lee, N.S.; Shin, J.G.; Jang, I.J.; Park, C.W.; Myung, H.J. Kinetics of isoniazid transfer into cerebrospinal fluid in patients with tuberculous meningitis. *J. Korean Med. Sci.* **1990**, *5*, 39–45. [CrossRef] [PubMed]
100. Myrianthefs, P.; Markantonis, S.L.; Vlachos, K.; Anagnostaki, M.; Boutzouka, E.; Panidis, D.; Baltopoulos, G. Serum and cerebrospinal fluid concentrations of linezolid in neurosurgical patients. *Antimicrob. Agents Chemother.* **2006**, *50*, 3971–3976. [CrossRef]
101. Upton, C.M.; Steele, C.I.; Maartens, G.; Diacon, A.H.; Wiesner, L.; Dooley, K.E. Pharmacokinetics of bedaquiline in cerebrospinal fluid (CSF) in patients with pulmonary tuberculosis (TB). *J. Antimicrob. Chemother.* **2022**, *77*, 1720–1724. [CrossRef]
102. Smith, A.G.C.; Gujabidze, M.; Avaliani, T.; Blumberg, H.M.; Collins, J.M.; Sabanadze, S.; Bakuradze, T.; Avaliani, Z.; Kempker, R.R.; Kipiani, M. Clinical outcomes among patients with tuberculous meningitis receiving intensified treatment regimens. *Int. J. Tuberc. Lung Dis.* **2021**, *25*, 632–639. [CrossRef] [PubMed]
103. Getahun, M.; Blumberg, H.M.; Ameni, G.; Beyene, D.; Kempker, R.R. Minimum inhibitory concentrations of rifampin and isoniazid among multidrug and isoniazid resistant *Mycobacterium tuberculosis* in Ethiopia. *PLoS ONE* **2022**, *17*, e0274426. [CrossRef] [PubMed]
104. Pellegrino, E.D.; Petrik, F.G.; Horton, R. The treatment of tuberculous meningitis in infants with streptomycin and isonicotinic acid hydrazide (isoniazid); a preliminary report of six patients under the age of two years treated without intrathecal medication. *Dis. Chest* **1954**, *26*, 146–165. [CrossRef] [PubMed]
105. Sifontes, J.E.; Berio, C.L.; Rivera, K.R.; Grana, J. Isoniazid by the intrathecal route in the therapy of tuberculous meningitis. *Bol. De La Asoc. Medica De Puerto Rico* **1955**, *47*, 1–19.
106. Lee, E.L.; Robinson, M.J.; Thong, M.L.; Puthucheary, S.D. Rifamycin in neonatal flavobacteria meningitis. *Arch. Dis. Child.* **1976**, *51*, 209–213. [CrossRef] [PubMed]
107. Swift, P.N. Intrathecal isoniazid in tuberculous meningitis. *Arch. Dis. Child.* **1956**, *31*, 328–332. [CrossRef]
108. Danielides, I.C.; Constantoulakis, M.; Daikos, G.K. Hepatitis on high dose isoniazid: Reintroduction of the drug in severe tuberculous meningitis. *Am. J. Gastroenterol.* **1983**, *78*, 378.
109. Takahashi, T.; Ogawa, K.; Sawada, S.; Nakayama, T.; Mizutani, T. A case of refractory tuberculous meningitis markedly improved by intrathecal administration of isoniazid (INH). *Rinsho Shinkeigaku* **2003**, *43*, 20–25. [PubMed]
110. Lin, J.; Zhou, H.; Zhang, N.; Yin, B.; Sheng, H.S. Effects of the implantation of Ommaya reservoir in children with tuberculous meningitis hydrocephalus: A preliminary study. *Childs Nerv. Syst.* **2012**, *28*, 1003–1008. [CrossRef]
111. Nakatani, Y.; Suto, Y.; Fukuma, K.; Yamawaki, M.; Sakata, R.; Takahashi, S.; Nakayasu, H.; Nakashima, K. Intrathecal Isoniazid for Refractory Tuberculous Meningitis with Cerebral Infarction. *Intern. Med.* **2017**, *56*, 953–957. [CrossRef]
112. Li, D.; Lv, P.; Lv, Y.; Ma, D.; Yang, J. Magnetic resonance imaging characteristics and treatment aspects of ventricular tuberculosis in adult patients. *Acta Radiol.* **2017**, *58*, 91–97. [CrossRef] [PubMed]
113. Azuma, J.; Ohno, M.; Kubota, R.; Yokota, S.; Nagai, T.; Tsuyuguchi, K.; Okuda, Y.; Takashima, T.; Kamimura, S.; Fujio, Y.; et al. NAT2 genotype guided regimen reduces isoniazid-induced liver injury and early treatment failure in the 6-month four-drug standard treatment of tuberculosis: A randomized controlled trial for pharmacogenetics-based therapy. *Eur. J. Clin. Pharmacol.* **2013**, *69*, 1091–1101. [CrossRef] [PubMed]
114. Motta, I.; Calcagno, A.; Bonora, S. Pharmacokinetics and pharmacogenetics of anti-tubercular drugs: A tool for treatment optimization? *Expert. Opin. Drug Metab. Toxicol.* **2018**, *14*, 59–82. [CrossRef] [PubMed]
115. Jing, W.; Zong, Z.; Tang, B.; Wang, J.; Zhang, T.; Wen, S.; Xue, Y.; Chu, N.; Zhao, W.; Huang, H. Population Pharmacokinetic Analysis of Isoniazid among Pulmonary Tuberculosis Patients from China. *Antimicrob. Agents Chemother.* **2020**, *64*, e01736-19. [CrossRef] [PubMed]
116. Walsh, K.F.; Vilbrun, S.C.; Souroutzidis, A.; Delva, S.; Joissaint, G.; Mathurin, L.; Ocheretina, O.; Cremieux, P.; Pape, J.W.; Koenig, S.P. Improved Outcomes With High-dose Isoniazid in Multidrug-resistant Tuberculosis Treatment in Haiti. *Clin. Infect. Dis.* **2019**, *69*, 717–719. [CrossRef] [PubMed]
117. Jacobson, K.R.; Theron, D.; Victor, T.C.; Streicher, E.M.; Warren, R.M.; Murray, M.B. Treatment outcomes of isoniazid-resistant tuberculosis patients, Western Cape Province, South Africa. *Clin. Infect. Dis.* **2011**, *53*, 369–372. [CrossRef]
118. Begley, D.J. Delivery of therapeutic agents to the central nervous system: The problems and the possibilities. *Pharmacol. Ther.* **2004**, *104*, 29–45. [CrossRef]
119. Lee, E.L.; Robinson, M.J.; Thong, M.L.; Puthucheary, S.D.; Ong, T.H.; Ng, K.K. Intraventricular chemotherapy in neonatal meningitis. *J. Pediatr.* **1977**, *91*, 991–995. [CrossRef] [PubMed]
120. Rios, I.; Klimek, J.J.; Maderazo, E.; Quintiliani, R. *Flavobacterium meningosepticum* meningitis: Report of selected aspects. *Antimicrob. Agents Chemother.* **1978**, *14*, 444–447. [CrossRef]
121. Boo, N.Y.; Lim, V.K.; Yakin, F.M.; Sakijan, A.S. Management of Flavobacterium meningitis in the neonates: Experience with 18 consecutive cases. *Singapore Med. J.* **1989**, *30*, 177–183.
122. Bruun, B.; Jensen, E.T.; Lundstrøm, K.; Andersen, G.E. *Flavobacterium meningosepticum* infection in a neonatal ward. *Eur. J. Clin. Microbiol. Infect. Dis.* **1989**, *8*, 509–514. [CrossRef] [PubMed]
123. Senbaga, N.; Davies, E.M. Inadvertent intrathecal administration of rifampicin. *Br. J. Clin. Pharmacol.* **2005**, *60*, 116. [CrossRef] [PubMed]
124. Li, K.; Wang, L.; Wen, L.; Wang, J.; Li, M. Intrathecal therapy for tuberculous meningitis: Propensity-matched cohort study. *Neurol. Sci.* **2022**, *43*, 2693–2698. [CrossRef] [PubMed]

125. Mrowczynski, O.D.; Langan, S.T.; Rizk, E.B. Intra-cerebrospinal fluid antibiotics to treat central nervous system infections: A review and update. *Clin. Neurol. Neurosurg.* **2018**, *170*, 140–158. [CrossRef] [PubMed]
126. Engström, A.J.E.; Ivarsson, C.A.; Aldergård, Å.; Henningsson, R.N. Analgesic effects of intrathecal morphine and bupivacaine during robotic-assisted surgery: A prospective randomized controlled study. *Pain. Pract.* **2023**, *23*, 631–638. [CrossRef] [PubMed]
127. Florez-Perdomo, W.A.; Escobar-Cardona, D.; Janjua, T.; Agrawal, A.; Vasquez, H.; Lozada-Martinez, I.D.; Moscote-Salazar, L.R. Antibiotic therapy by intrathecal or intraventricular approach for postsurgical meningitis or ventriculitis: A systematic review and meta-analysis. *Egypt. J. Neurosurg.* **2023**, *38*, 20. [CrossRef]
128. Lewin, J.J., 3rd; Cook, A.M.; Gonzales, C.; Merola, D.; Neyens, R.; Peppard, W.J.; Brophy, G.M.; Kurczewski, L.; Giarratano, M.; Makii, J.; et al. Current Practices of Intraventricular Antibiotic Therapy in the Treatment of Meningitis and Ventriculitis: Results from a Multicenter Retrospective Cohort Study. *Neurocrit Care* **2019**, *30*, 609–616. [CrossRef]
129. Li, X.; Sun, S.; Ling, X.; Chen, K.; Wang, Q.; Zhao, Z. Plasma and cerebrospinal fluid population pharmacokinetics of vancomycin in postoperative neurosurgical patients after combined intravenous and intraventricular administration. *Eur. J. Clin. Pharmacol.* **2017**, *73*, 1599–1607. [CrossRef]
130. Perier, F.; Couffin, S.; Martin, M.; Bardon, J.; Cook, F.; Mounier, R. Multidrug-Resistant Acinetobacter baumannii Ventriculostomy-Related Infection, Treated by a Colistin, Tigecycline, and Intraventricular Fibrinolysis. *World Neurosurg.* **2019**, *121*, 111–116. [CrossRef]
131. Dodou, K. Intrathecal route of drug delivery can save lives or improve quality of life. *Pharm. J.* **2012**, *289*, 501.
132. Clifford, H.E.; Stewart, G.T. Intraventricular administration of a new derivative of polymyxin B in meningitis due to Ps. pyocyanea. *Lancet* **1961**, *2*, 177–180. [CrossRef] [PubMed]
133. Rodríguez Guardado, A.; Blanco, A.; Asensi, V.; Pérez, F.; Rial, J.C.; Pintado, V.; Bustillo, E.; Lantero, M.; Tenza, E.; Alvarez, M.; et al. Multidrug-resistant Acinetobacter meningitis in neurosurgical patients with intraventricular catheters: Assessment of different treatments. *J. Antimicrob. Chemother.* **2008**, *61*, 908–913. [CrossRef] [PubMed]
134. Allinson, R.R.; Stach, P.E. Intrathecal Drug Therapy. *Drug Intell. Clin. Pharm.* **1978**, *12*, 347–359. [CrossRef]
135. Cook, A.M.; Mieure, K.D.; Owen, R.D.; Pesaturo, A.B.; Hatton, J. Intracerebroventricular administration of drugs. *Pharmacotherapy* **2009**, *29*, 832–845. [CrossRef] [PubMed]
136. Atkinson, A.J., Jr. Intracerebroventricular drug administration. *Transl. Clin. Pharmacol.* **2017**, *25*, 117–124. [CrossRef] [PubMed]
137. Huynh, L.; Agossah, C.; Lelong-Boulouard, V.; Marie, J.; Brossier, D.; Goyer, I. Therapeutic drug monitoring of intravenous anti-tuberculous therapy: Management of an 8-month-old child with tuberculous meningitis. *Paediatr. Int. Child Health* **2021**, *41*, 285–290. [CrossRef] [PubMed]
138. Peloquin, C.A. Therapeutic drug monitoring in the treatment of tuberculosis. *Drugs* **2002**, *62*, 2169–2183. [CrossRef]
139. Peloquin, C. The Role of Therapeutic Drug Monitoring in Mycobacterial Infections. *Microbiol. Spectr.* **2017**, *5*, 10–1128. [CrossRef] [PubMed]
140. LeBras, M.; Chow, I.; Mabasa, V.H.; Ensom, M.H. Systematic Review of Efficacy, Pharmacokinetics, and Administration of Intraventricular Aminoglycosides in Adults. *Neurocrit Care* **2016**, *25*, 492–507. [CrossRef]
141. Tucker, E.W.; Guglieri-Lopez, B.; Ordonez, A.A.; Ritchie, B.; Klunk, M.H.; Sharma, R.; Chang, Y.S.; Sanchez-Bautista, J.; Frey, S.; Lodge, M.A.; et al. Noninvasive (11)C-rifampin positron emission tomography reveals drug biodistribution in tuberculous meningitis. *Sci. Transl. Med.* **2018**, *10*, eaau0965. [CrossRef]
142. Chusri, S.; Sakarunchai, I.; Kositpantawong, N.; Panthuwong, S.; Santimaleeworagun, W.; Pattharachayakul, S.; Singkhamanan, K.; Doi, Y. Outcomes of adjunctive therapy with intrathecal or intraventricular administration of colistin for post-neurosurgical meningitis and ventriculitis due to carbapenem-resistant acinetobacter baumannii. *Int. J. Antimicrob. Agents* **2018**, *51*, 646–650. [CrossRef] [PubMed]
143. Khawcharoenporn, T.; Apisarnthanarak, A.; Mundy, L.M. Intrathecal colistin for drug-resistant Acinetobacter baumannii central nervous system infection: A case series and systematic review. *Clin. Microbiol. Infect.* **2010**, *16*, 888–894. [CrossRef] [PubMed]
144. Roitberg, B.Z.; Khan, N.; Alp, M.S.; Hersonskey, T.; Charbel, F.T.; Ausman, J.I. Bedside external ventricular drain placement for the treatment of acute hydrocephalus. *Br. J. Neurosurg.* **2001**, *15*, 324–327. [CrossRef] [PubMed]
145. Wiesel, J.; Rose, D.N.; Silver, A.L.; Sacks, H.S.; Bernstein, R.H. Lumbar puncture in asymptomatic late syphilis. An analysis of the benefits and risks. *Arch. Intern. Med.* **1985**, *145*, 465–468. [CrossRef] [PubMed]
146. Almoshari, Y. Osmotic Pump Drug Delivery Systems-A Comprehensive Review. *Pharmaceuticals* **2022**, *15*, 1430. [CrossRef] [PubMed]
147. Meng, E.; Hoang, T. MEMS-enabled implantable drug infusion pumps for laboratory animal research, preclinical, and clinical applications. *Adv. Drug Deliv. Rev.* **2012**, *64*, 1628–1638. [CrossRef] [PubMed]
148. 6-Drug delivery systems. In *Strategies to Modify the Drug Release from Pharmaceutical Systems*; Bruschi, M.L. Ed.; Woodhead Publishing: Sawston, UK, 2015; pp. 87–194.
149. Wright, J.C.; Tao Leonard, S.; Stevenson, C.L.; Beck, J.C.; Chen, G.; Jao, R.M.; Johnson, P.A.; Leonard, J.; Skowronski, R.J. An in vivo/in vitro comparison with a leuprolide osmotic implant for the treatment of prostate cancer. *J. Control Release* **2001**, *75*, 1–10. [CrossRef]
150. Wright, J.C. Critical variables associated with nonbiodegradable osmotically controlled implants. *AAPS J.* **2010**, *12*, 437–442. [CrossRef]

151. Stevenson, C.L.; Theeuwes, F.; Wright, J.C. Osmotic implantable delivery systems. In *Handbook of Pharmaceutical Controlled Release Technology*; CRC Press: Boca Raton, FL, USA, 2000; pp. 225–254.
152. Stevenson, C.; Fereira, P.; Tan, M.; Ayer, R.; Dehnad, H.; Berry, S. Glucagon like peptide suspension stability and delivery from a DUROS® implant. *Proc. Int. Symp. Control. Rel. Bioact. Mater.* **2000**, *27*, 986–987.
153. Dehnad, H.; Berry, S.; Fereira, P.; Ayer, R.; O'Brien, A.; Chen, G.; Priebe, D.; Matsuura, J.; Muchnik, A.; Kang, L. Functionality of osmotically driven, implantable therapeutic systems (DUROS® implants): Effect of osmotic pressure and delivery of suspension formulations of peptides and proteins. *Proc. Int. Symp. Control. Rel. Bioact. Mater.* **2000**, *27*, 1016–1017.
154. Rohloff, C.M.; Alessi, T.R.; Yang, B.; Dahms, J.; Carr, J.P.; Lautenbach, S.D. DUROS technology delivers peptides and proteins at consistent rate continuously for 3 to 12 months. *J. Diabetes Sci. Technol.* **2008**, *2*, 461–467. [CrossRef] [PubMed]
155. Thorat, M.S.; Sapkale, A.; Prasad, V.; Singh, C. Overview of past and current osmotic drug delivery systems. *Int. J. Pharm. Chem. Sci.* **2012**, *1*, 743–753.
156. Fowler, J.E., Jr. Patient-reported experience with the Viadur 12-month leuprolide implant for prostate cancer. *Urology* **2001**, *58*, 430–434. [CrossRef] [PubMed]
157. Bomgaars, L.; Geyer, J.R.; Franklin, J.; Dahl, G.; Park, J.; Winick, N.J.; Klenke, R.; Berg, S.L.; Blaney, S.M. Phase I trial of intrathecal liposomal cytarabine in children with neoplastic meningitis. *J. Clin. Oncol.* **2004**, *22*, 3916–3921. [CrossRef]
158. Phuphanich, S.; Maria, B.; Braeckman, R.; Chamberlain, M. A pharmacokinetic study of intra-CSF administered encapsulated cytarabine (DepoCyt) for the treatment of neoplastic meningitis in patients with leukemia, lymphoma, or solid tumors as part of a phase III study. *J. Neurooncol.* **2007**, *81*, 201–208. [CrossRef] [PubMed]
159. Peyrl, A.; Sauermann, R.; Traunmueller, F.; Azizi, A.A.; Gruber-Olipitz, M.; Gupper, A.; Slavc, I. Pharmacokinetics and safety of intrathecal liposomal cytarabine in children aged <3 years. *Clin. Pharmacokinet.* **2009**, *48*, 265–271. [CrossRef] [PubMed]
160. Peyrl, A.; Sauermann, R.; Chocholous, M.; Azizi, A.A.; Jäger, W.; Höferl, M.; Slavc, I. Pharmacokinetics and toxicity of intrathecal liposomal cytarabine in children and adolescents following age-adapted dosing. *Clin. Pharmacokinet.* **2014**, *53*, 165–173. [CrossRef]
161. Khang, M.; Bindra, R.S.; Mark Saltzman, W. Intrathecal delivery and its applications in leptomeningeal disease. *Adv. Drug Deliv. Rev.* **2022**, *186*, 114338. [CrossRef]
162. Gao, W.; Zhang, Y.; Zhang, Q.; Zhang, L. Nanoparticle-Hydrogel: A Hybrid Biomaterial System for Localized Drug Delivery. *Ann. Biomed. Eng.* **2016**, *44*, 2049–2061. [CrossRef]
163. Baumann, M.D.; Kang, C.E.; Tator, C.H.; Shoichet, M.S. Intrathecal delivery of a polymeric nanocomposite hydrogel after spinal cord injury. *Biomaterials* **2010**, *31*, 7631–7639. [CrossRef]
164. Chvatal, S.A.; Kim, Y.T.; Bratt-Leal, A.M.; Lee, H.; Bellamkonda, R.V. Spatial distribution and acute anti-inflammatory effects of Methylprednisolone after sustained local delivery to the contused spinal cord. *Biomaterials* **2008**, *29*, 1967–1975. [CrossRef] [PubMed]
165. Ramer, L.M.; Ramer, M.S.; Steeves, J.D. Setting the stage for functional repair of spinal cord injuries: A cast of thousands. *Spinal Cord.* **2005**, *43*, 134–161. [CrossRef] [PubMed]
166. Romero, M.I.; Rangappa, N.; Garry, M.G.; Smith, G.M. Functional regeneration of chronically injured sensory afferents into adult spinal cord after neurotrophin gene therapy. *J. Neurosci.* **2001**, *21*, 8408–8416. [CrossRef] [PubMed]
167. Householder, K.T.; Dharmaraj, S.; Sandberg, D.I.; Wechsler-Reya, R.J.; Sirianni, R.W. Fate of nanoparticles in the central nervous system after intrathecal injection in healthy mice. *Sci. Rep.* **2019**, *9*, 12587. [CrossRef] [PubMed]
168. Khaing, Z.Z.; Ehsanipour, A.; Hofstetter, C.P.; Seidlits, S.K. Injectable Hydrogels for Spinal Cord Repair: A Focus on Swelling and Intraspinal Pressure. *Cells Tissues Organs* **2016**, *202*, 67–84. [CrossRef]
169. Lin, W.; Xu, T.; Wang, Z.; Chen, J. Sustained intrathecal delivery of amphotericin B using an injectable and biodegradable thermogel. *Drug Deliv.* **2021**, *28*, 499–509. [CrossRef]

pharmaceuticals

Article

Discovery of Biofilm-Inhibiting Compounds to Enhance Antibiotic Effectiveness Against *M. abscessus* Infections

Elizaveta Dzalamidze [1], Mylene Gorzynski [1], Rebecca Vande Voorde [1], Dylan Nelson [2] and Lia Danelishvili [1,3,*]

1 Department of Biomedical Sciences, Carlson College of Veterinary Medicine, Oregon State University, Corvallis, OR 97331, USA
2 Department of Pharmaceutical Sciences, College of Pharmacy, Oregon State University, Corvallis, OR 97331, USA
3 Department of Microbiology, College of Science, Oregon State University, Corvallis, OR 97331, USA
* Correspondence: lia.danelishvili@oregonstate.edu; Tel.: +1-(541)-737-6544; Fax: +1-(541)-737-2730

Abstract: Background/Objectives: *Mycobacterium abscessus* (MAB) is a highly resilient pathogen that causes difficult-to-treat pulmonary infections, particularly in individuals with cystic fibrosis (CF) and other underlying conditions. Its ability to form robust biofilms within the CF lung environment is a major factor contributing to its resistance to antibiotics and evasion of the host immune response, making conventional treatments largely ineffective. These biofilms, encased in an extracellular matrix, enhance drug tolerance and facilitate metabolic adaptations in hypoxic conditions, driving the bacteria into a persistent, non-replicative state that further exacerbates antimicrobial resistance. Treatment options remain limited, with multidrug regimens showing low success rates, highlighting the urgent need for more effective therapeutic strategies. **Methods**: In this study, we employed artificial sputum media to simulate the CF lung environment and conducted high-throughput screening of 24,000 compounds from diverse chemical libraries to identify inhibitors of MAB biofilm formation, using the Crystal Violet (CV) assay. **Results**: The screen established 17 hits with ≥30% biofilm inhibitory activity in mycobacteria. Six of these compounds inhibited MAB biofilm formation by over 60%, disrupted established biofilms by ≥40%, and significantly impaired bacterial viability within the biofilms, as confirmed by reduced CFU counts. In conformational assays, select compounds showed potent inhibitory activity in biofilms formed by clinical isolates of both MAB and *Mycobacterium avium* subsp. *hominissuis* (MAH). Key compounds, including ethacridine, phenothiazine, and fluorene derivatives, demonstrated potent activity against pre- and post-biofilm conditions, enhanced antibiotic efficacy, and reduced intracellular bacterial loads in macrophages. **Conclusions**: This study results underscore the potential of these compounds to target biofilm-associated resistance mechanisms, making them valuable candidates for use as adjuncts to existing therapies. These findings also emphasize the need for further investigations, including the initiation of a medicinal chemistry campaign to leverage structure–activity relationship studies and optimize the biological activity of these underexplored class of compounds against nontuberculous mycobacterial (NTM) strains.

Keywords: *Mycobacterium abscessus*; nontuberculous mycobacteria; NTM; high-throughput screening; small molecules; biofilm; intracellular killing; macrophage

1. Introduction

Mycobacterium abscessus belongs to the group of clinically important nontuberculous mycobacteria that is associated with infections in both immunocompromised and healthy individuals [1,2]. MAB is a rapidly growing environmental organism freely found in soil, dust, and moving and still bodies of water [3] and causes severe lung infections in CF patients or individuals with other underlying pulmonary diseases such as chronic obstructive pulmonary disease (COPD) or bronchiectasis [4]. While MAB is a commonly isolated organism from respiratory specimens, it can also lead to a wide range of infections in the skin, post-surgery or post-trauma soft tissues, the central nervous system, and bacteremia [5–7]. The incidence of MAB lung infections has been steadily increasing over the past 25 years [1,8–10], often leading to chronic and, in many cases, fatal disease due to the natural resistance of this pathogen to most antibiotics [1].

The ability of MAB to form biofilms in the bronchial airways of patients from which invasive variants emerge is an important aspect of MAB pathogenesis and contributes to the clinical manifestations of infection [9]. MAB forms significantly more and robust extracellular biofilm matrices in synthetic cystic fibrosis sputum medium (SCFM), suggesting a favorable host environmental condition for bacteria to establish biofilms [11] while minimizing the efficacy of antibiotics [12–14]. The nature and structure of biofilms, including nutrient and oxygen availability to bacteria, contributes to intrinsic and acquired resistance in mycobacteria. For example, in nutrient-depleted areas in biofilms, bacteria enter a stationary phase that make pathogens insensible to antibiotics [12]. Also, the heterogeneous metabolic states of bacteria within biofilm matrices stimulate variation in the antibiotic susceptibility of facultative anaerobes such as mycobacteria and can lead to the failure of antibiotic treatment [15–17]. Mycobacterial biofilms are composed of bacterial cells, exopolysaccharides, glycopeptidolipids, eDNA, proteins, carbohydrates, free mycolic acids, and other surrounding materials which can vary depending on location [18]. Biofilms protect bacteria against the host immune system by impairing the activation of phagocytes and the complement system as well as promoting premature TNF-α-dependent apoptosis in phagocytes [19], while stimulating better invasion through airway epithelial cells [20]. Mature biofilms can prevent phagocytosis [19], but microaggregates with pre-biofilm phenotypes have been demonstrated to increase the bacterial uptake process [21]. Furthermore, biofilms create a physical barrier, increasing resistance against conventional antibiotics by up to 1000 folds [22]. The disruption of biofilm eDNA by treating mycobacterial matrices with DNase I has been reported to reduce established biofilms *in vitro*, while also diminishing mycobacterial tolerance to antimicrobials and improving treatment outcomes with various antibiotics [23].

Furthermore, intercellular signaling through the quorum-sensing phenomenon has a critical role in biofilm matrices' structural integrity and regulates the bacterial division of labor, metabolic shift, and cooperative activities between cells [24,25]. This mutual interaction adds another level of complexity to biofilms and, in a mammalian host, benefits to pathogen colonization, increased virulence, and adaptations with varied host environments. The infections that emerge from biofilms are commonly resistant to the highest deliverable levels of antibiotics and cannot be treated with standard antibacterial strategies [15,16,26], often resulting in tissue damage and chronic suffering in patients.

Current strategies to combat bacterial biofilms are very limited and antibiotic treatment remains the only effective measure to control biofilm infections in NTM patients. Although many antimicrobial molecules and substances have been discovered and shown to effectively inhibit biofilms and eliminate viable bacteria *in vitro* [27,28], eradicating biofilm infections in vivo remains challenging due to the higher doses required, which often lead to increased toxicity and side effects. In recent years, new concepts have emerged to control bacterial biofilms by disturbing bacterial quorum-sensing factors [29], targeting c-di-GMP nucleotide signaling which is implicated in the pathogenicity-like quorum-sensing process, and inhibiting bacterial amyloids, helping pathogens in their adhesion to surfaces [30,31].

These studies highlight the importance of targeting relevant bacterial factors that are expressed in biofilm conditions and enable pathogens to become phenotypically resistant to antimicrobial treatment [15,16,32,33].

While numerous high-throughput compound screening assays have been conducted, leading to the identification of various classes of antibacterial compounds through the use of mycobacteria grown *in vitro* in diverse culture media and conditions [34,35], this approach often overlooks compounds that might target bacterial factors exclusively expressed during the physiological stages of biofilm formation or biofilm maintenance. Targeting these factors may have a significant impact on mycobacterial biofilms, which are difficult to eliminate due to their enhanced resistance to antibiotics and inherent persistence to host immune responses. Consequently, environments that more closely mimic *in vivo* conditions, such as synthetic cystic fibrosis sputum media, represent more biologically relevant conditions for drug discovery compared to standard growth media or buffers. These more complex environments may better reveal compounds with novel mechanisms of action that target bacterial pathogenicity factors essential for the integrity of non-tuberculous mycobacterial biofilms. Therefore, in this study, we utilized artificial sputum media and performed high-throughput screening to identify effective antibiofilm compounds in mycobacteria. We aimed to establish compounds with activity to inhibit biofilm formation as well as compromise established biofilms with an intention to use active compounds in combination with current anti-NTM antibiotics and improve treatment outcomes for biofilm infections.

2. Results

2.1. Optimization of High-Throughput Screening (HTS) Assay

The HTS assay was optimized using a 384-well plate (384WP) format in 24 technical replicates and a 96-well plate (96WP) format in 12 technical replicates by varying inoculum concentrations (Figure 1). The 384WP format was utilized for primary screening to identify active compounds, while the 96WP format was employed for secondary validation. This optimization aimed to determine the bacterial concentration, time points, and control antibiotic concentrations that promoted optimal mycobacterial biofilm formation and effectively assessed biofilm inhibition. MAB biofilms were generated under biologically relevant conditions using SCFM and evaluated over the course of a week using CV staining (Figure 1A). Furthermore, the antibiotic treatment groups alongside the DMSO control group were tested to identify parameters that produced the most significant differences in biofilm inhibition by day five, as measured by the percentage of inhibition between DMSO- and antibiotic-treated control wells (Figure 1B). Optimal results were achieved with biofilms containing 10^7 CFU/mL followed by a five-day clarithromycin (CLA) treatment at a 70 μg/mL concentration. The MAB biofilms formed from 10^6 CFU/mL inoculums were weak, resulting in high variability in CV staining and antibiotic treatment outcomes. In contrast, biofilms formed from 10^8 CFU/mL inoculums showed minimal susceptibility to antibiotics (Figure 1B). Under selected conditions, the MAB biofilms and viability were assessed in the 96WP format. As shown in Figure 1C,D, a consistent and significant difference was observed between the bacterial growth control group (DMSO-treated wells) and the negative control group (CLA treatment). Optimal results were accomplished in a 96WP with biofilms generated using a 10^7 CFU/mL inoculum and subjected to CLA treatment at a concentration of 70 μg/mL for five days. These parameters were selected later for the secondary confirmational assay.

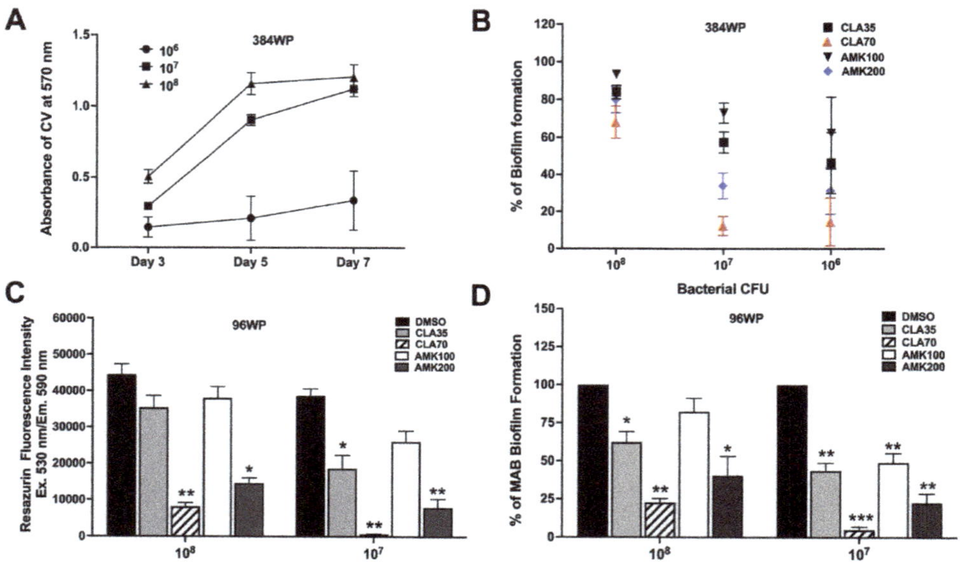

Figure 1. The development of the HTS assay for the identification of compounds with potency against mycobacterial biofilms. (**A**) The CV absorbance readings were recorded for MAB grown in SCFM over time and across a range of concentrations in the 384WP format. (**B**) The percentage of biofilm formation on day five was calculated as described in Section 4 using DMSO- and antibiotic-treated wells as controls for bacterial growth and inhibition in the 384WP format. (**C**) MAB viability was recorded as fluorescent readings for the control and antibiotic-treated groups, based on resazurin reduction rates in the 96WP format. (**D**) The percentage of MAB biofilm formation on day 5 was determined based on the CV assay. Data are expressed as the means ± standard deviations (SD) of three independent experiments. The statistical significance between the DMSO and antibiotic-treated groups is indicated as follows: * $p < 0.05$; ** $p < 0.01$; and *** $p < 0.001$.

2.2. *Identification of Active Compounds Against Mycobacteria Biofilms*

In the primary screen, a total of 24,000 compounds were tested, including 14,400 from the Maybridge compound library, 9000 from the ChemBridge DIVERSet compound library, and 600 from the Spectrum compound library, all sourced from the OSU/College of Pharmacy High-Throughput Screening Services Laboratory. MAB inoculum (10^7 CFU/mL) was added to a 384WP preloaded with compounds at 10 µM in SCFM and, after staining the biofilm mass on day five, 354 compounds were identified as hits showing ≥ 30% biofilm inhibition. This corresponded to a primary screening hit rate of 1.48%. A secondary screening was conducted in a 96WP format to validate the hits. This stage identified 17 active compounds with ≥30% biofilm inhibition, yielding a hit rate of 4.8%. The selected hit compounds, listed in Table 1, were subsequently purchased from MolPort for the third confirmatory screening, where the biofilm-inhibitory activity was verified in three biological replicates, each performed with four technical replicates (Figure 2).

The antibacterial activity of 17 selected compounds was assessed against MAB under both pre- and post-biofilm formation conditions in SCFM (Figure 2A,B). Alongside this, bacterial viably was also measured using resazurin assay to determine whether biofilm inhibition resulted from bacterial killing or interference with the active mechanisms of biofilm formation. The results indicate that six compounds (**1**, **5**, **7**, **11**, **14**, and **16**) demonstrated high activity against MAB biofilms, achieving inhibition rates of over 60% and 40% in pre- and post-formation conditions, respectively. Of these six compounds, five also reduced bacterial viability by 40% or more under both conditions.

Table 1. The list of active compounds identified in this study.

Compound	Structure	SMILES Molecular Formula	Name
1	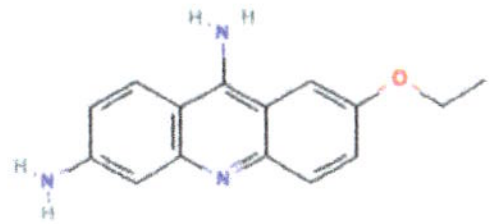	CCOc1ccc2nc3cc(ccc3c(c2c1)N)N $C_{15}H_{15}N_3O$	Ethacridine
2	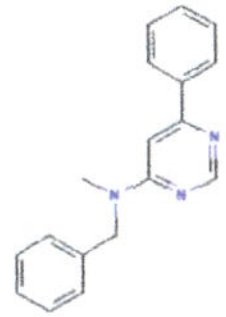	CN(Cc1ccccc1)c1cc(ncn1)-c1ccccc1 $C_{18}H_{17}N_3$	N-benzyl-N-methyl-6-phenylpyrimidin-4-amine
3	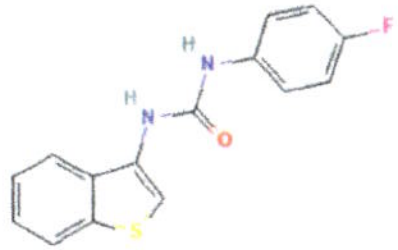	Fc1ccc(cc1)NC(=O)Nc1csc2ccccc12 $C_{15}H_{11}FN_2OS$	N-(1-benzothiophen-3-yl)-N′-(4-fluorophenyl)urea
4	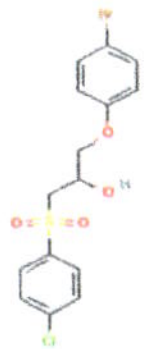	OC(COc1ccc(cc1)Br)CS(=O)(=O)c1ccc(cc1)Cl $C_{15}H_{14}BrClO_4S$	1-(4-Bromophenoxy)-3-(4-chlorophenyl)sulfonylpropan-2-ol
5	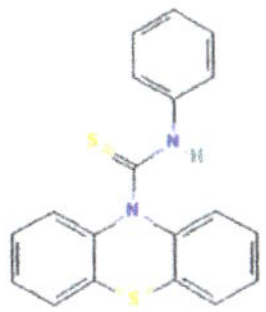	S=C(Nc1ccccc1)N1c2ccccc2Sc2ccccc21 $C_{19}H_{14}N_2S_2$	N-phenylphenothiazine-10-carbothioamide

Table 1. *Cont.*

Compound	Structure	SMILES Molecular Formula	Name
6	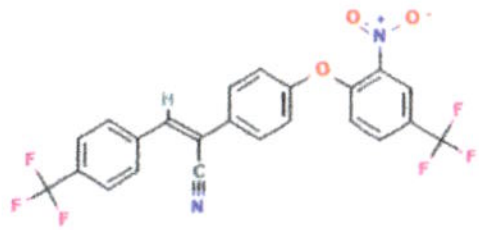	[O-][N+](=O)c1cc(ccc1Oc1ccc(cc1)/C(=C/c1ccc(cc1)C(F)(F)F)C#N)C(F)(F)F $C_{23}H_{12}F_6N_2O_3$	(Z)-2-[4-[2-nitro-4-(trifluoromethyl)phenoxy]phenyl]-3-[4-(trifluoromethyl)phenyl]prop-2-enenitrile
7	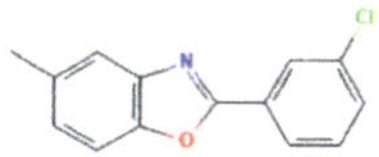	Cc1ccc2oc(nc2c1)-c1cccc(c1)Cl $C_{14}H_{10}ClNO$	2-(3-Chlorophenyl)-5-methylbenzo[d]oxazole
8	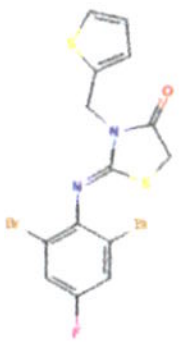	Fc1cc(c(c(c1)Br)/N=C/1\SCC(=O)N1Cc1cccs1)Br $C_{14}H_9Br_2FN_2OS_2$	2-(2,6-Dibromo-4-fluorophenyl)imino-3-(thiophen-2-ylmethyl)-1,3-thiazolidin-4-one
9	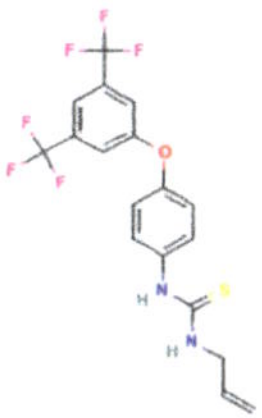	FC(F)(F)c1cc(cc(c1)C(F)(F)F)Oc1ccc(cc1)NC(=S)NCC=C $C_{18}H_{14}F_6N_2OS$	1-[4-[3,5-Bis(trifluoromethyl)phenoxy]phenyl]-3-prop-2-enylthiourea
10	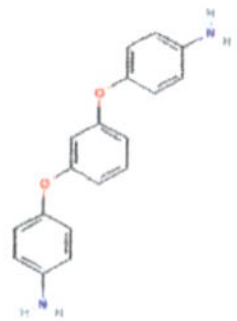	Nc1ccc(cc1)Oc1cccc(c1)Oc1ccc(cc1)N $C_{18}H_{16}N_2O_2$	1,3-Bis(4-aminophenoxy)benzene

Table 1. *Cont.*

Compound	Structure	SMILES Molecular Formula	Name
11	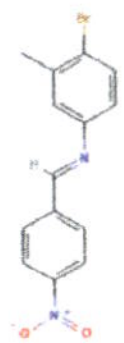	Cc1cc(ccc1Br)/N=C/c1ccc(cc1)[N+](=O)[O-] $C_{14}H_{11}BrN_2O_2$	N-(4-bromo-3-methylphenyl)-1-(4-nitrophenyl)methanimine
12	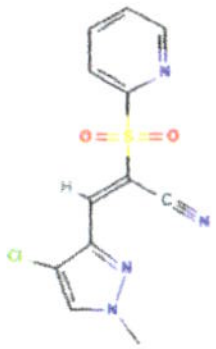	Cn1cc(c(n1)/C=C(\C#N)S(=O)(=O)c1ccccn1)Cl $C_{12}H_9ClN_4O_2S$	(E)-3-(4-chloro-1-methylpyrazol-3-yl)-2-pyridin-2-ylsulfonylprop-2-enenitrile
13	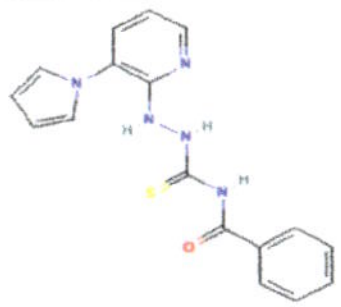	O=C(NC(=S)NNc1ncccc1-n1cccc1)c1ccccc1 $C_{17}H_{15}N_5OS$	Cyto10B8
14	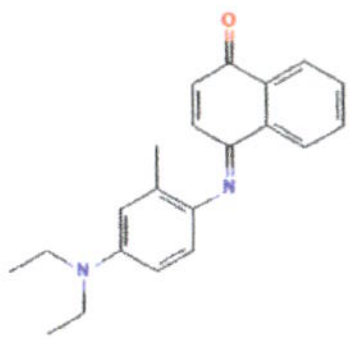	CCN(CC)c1ccc(c(c1)C)/N=C\1/C=CC(=O)c2ccccc21 $C_{21}H_{22}N_2O$	4-((4-(Diethylamino)-2-methylphenyl)imino)naphthalen-1(4H)-one

Table 1. *Cont.*

Compound	Structure	SMILES Molecular Formula	Name
15	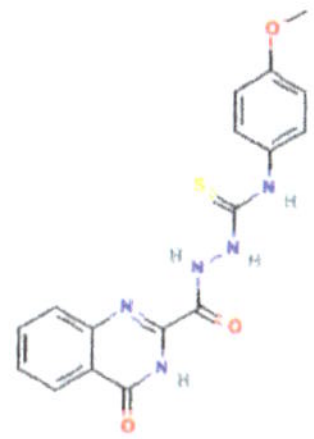	COc1ccc(cc1)NC(=S)NNC(=O)C1=Nc2ccccc2C(=O)N1 $C_{17}H_{15}N_5O_3S$	1-(4-methoxyphenyl)-3-[(4-oxo-3H-quinazoline-2-carbonyl)amino]thiourea
16	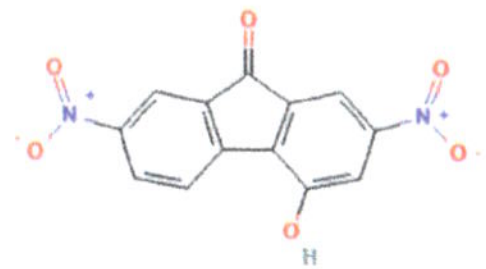	Oc1cc(cc2c1-c1ccc(cc1C2=O)[N+](=O)[O-])[N+](=O)[O-] $C_{13}H_6N_2O_6$	4-Hydroxy-2,7-dinitrofluoren-9-one
17	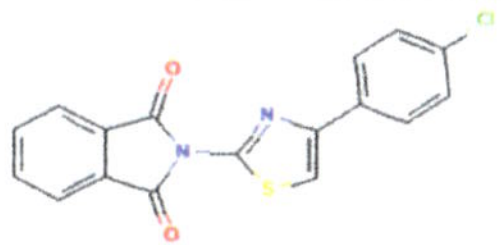	Clc1ccc(cc1)-c1csc(n1)N1C(=O)c2ccccc2C1=O $C_{17}H_9ClN_2O_2S$	2-(4-(4-Chlorophenyl)thiazol-2-yl)isoindoline-1,3-dione

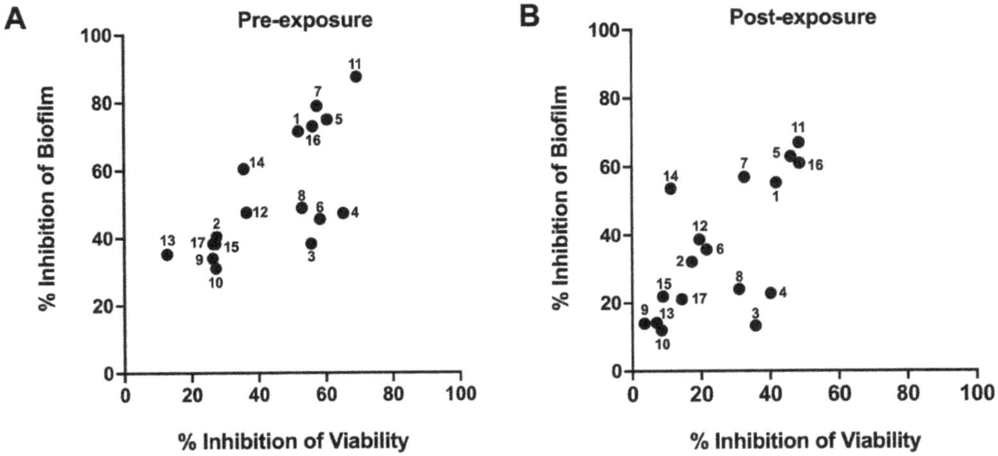

Figure 2. Activity of selected hit compounds against biofilms of MAB19977 under pre- and post-formation conditions. (**A**) Prevention of biofilm formation (pre-exposure condition). (**B**) Inhibition of 24 h pre-formed biofilms (post-exposure condition). Data are presented as percentage of inhibition in bacterial viability in biofilms or biomass relative to untreated control biofilms, based on three independent biological replicates.

The quantification of bacterial biomass confirmed the effectiveness of select hit compounds in inhibiting biofilm formation in NTM clinical strains. We conducted additional investigations to evaluate the effectiveness of six selected compounds (**1**, **5**, **7**, **11**, **14**, and **16**) in preventing biofilm formation by MAB and MAH clinical isolates. As shown in Figure 3, all tested compounds significantly reduced biofilm formation relative to untreated controls. Notably, compounds **5** and **16** demonstrated the most substantial reduction in biofilm biomass across all tested isolates, highlighting their potential as effective agents against biofilm development in these NTM strains.

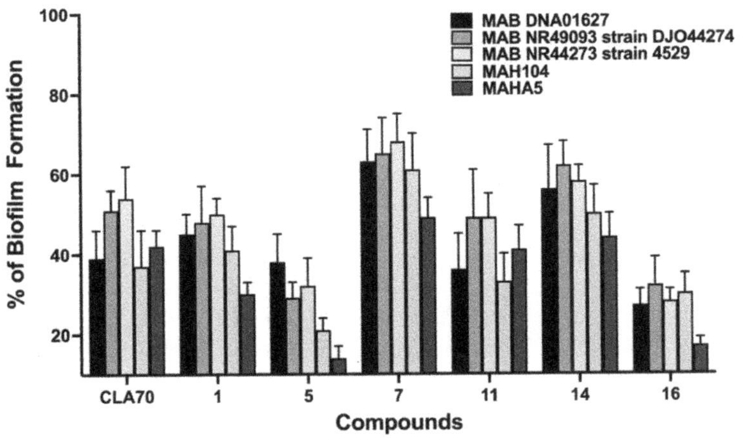

Figure 3. The screening of antibiofilm activity across NTM strains. Biofilm biomass was quantified using crystal violet staining under pre-treatment conditions with select hit compounds at a 100 µM concentration, as described in the Materials and Methods. DMSO and CLA70 served as growth and inhibition controls, respectively. Data are presented as the mean ± standard deviation from three independent replicates.

2.3. The Hit Compounds Exhibited Antimicrobial Activity in Replicating Mycobacteria In Vitro

To determine whether the compounds with antibiofilm activity also exhibited efficacy against replicating bacteria *in vitro*, we conducted antimicrobial susceptibility assays using mid-log-phase cultures of MAB19977 and MAH104. Two-fold serial dilutions of each compound were prepared in 1% DMSO and added to cultures grown in 7H9 broth. Bacterial viability was assessed by measuring the optical density at 600 nm (OD_{600}) and by evaluating resazurin substrate oxidation rates on day 3 for MAB and day 4 for MAH. To further assess the efficacy of the hit compounds, we expanded this study to include clinical isolates of MAB and MAH obtained from CF patients. The MAB isolates included DNA01627, NR49093 (strain DJO44274), and NR44273 (strain 4529). For MAH, we tested the MAHA5 clinical isolate, a strain known for its robust biofilm formation and resilience to antimicrobial agents [20]. The 50% inhibitory concentrations of the compounds against these strains are summarized in Table 2. While the IC_{50} values varied between the two NTM species, the majority of the tested compounds demonstrated consistent *in vitro* activity across all tested strains.

Table 2. *In vitro* activity of hit compounds against NTM strains.

Compound	IC_{50} [μM]					
	MAB19977	DNA 01627	NR49093 Strain DJO44274	NR44273 Strain 4529	MAH104	MAHA5
1	3	10	10	3	10	10
2	10	10	10	10	10	10
3	32	32	32	32	100	100
4	100	100	32	32	-	-
5	32	32	32	32	10	10
6	3	3	10	10	3	3
7	100	100	100	100	32	32
8	3	3	3	3	10	10
9	10	-	-	-	32	100
10	10	10	32	32	-	-
11	10	10	32	32	100	100
12	10	10	-	-	100	100
13	32	32	10	32	10	10
14	10	10	10	10	10	10
15	100	-	-	-	32	100
16	3	3	3	3	10	10
17	10	-	-	-	10	-

In addition, we conducted bacterial survival studies using selected hit compounds that demonstrated *in vitro* activity against both MAB and MAH and exhibited antibiofilm properties. These compounds were tested in combination with the frontline antibiotic amikacin (AMK) to assess their potential to enhance antibiotic efficacy. The mid-log-phase cultures of MAB19977 and MAH104 were treated with either the compound alone at the corresponding IC_{50}, AMK alone at two times the MIC, or a combination of the compound and AMK. The goal was to determine whether the hit compounds, potentially acting through novel mechanisms, could improve the effectiveness of AMK and enhance bacterial clearance compared to AMK alone. Our findings demonstrate the potential of selected hit compounds (**1**, **5**, and **16**) to exhibit synergistic effects with AMK, improving treatment efficacy against both mycobacterial strains (Figure 4).

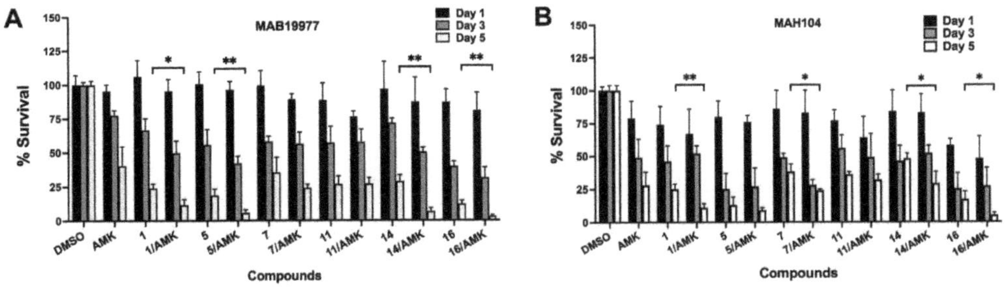

Figure 4. The in vitro time-kill dynamics of mycobacteria during treatment with select hit compounds and in combination with an antibiotic. (**A**) MAB19977 survival rates in AMK, compound, and compound–AMK treatment groups over 5 days in 7H9 broth. (**B**) MAH104 survival rates in AMK, compound, and compound–AMK combination treatment groups over 5 days in 7H9 broth. Bacterial CFUs were recorded after treatment with the antibiotic at 2× the MIC and/or compounds at IC_{50} concentrations. Survival percentages were calculated relative to the DMSO growth control (untreated). Antimicrobials were added at time zero and then supplemented every other day throughout the duration of the experiment. Data are presented as means ± standard deviations (SD) from two independent experiments performed in triplicate. The statistical significance between the compound group alone and antibiotic–compound combination treatment groups on day 5 is indicated as * $p < 0.05$ and ** $p < 0.01$.

2.4. Select Hit Compounds Demonstrated Potency Against Intracellular Mycobacteria Within Phagocytic Cells

To evaluate the potential of the six select hit compounds (**1**, **5**, **7**, **11**, **14**, and **16**) against intracellular mycobacteria, their cytotoxicity was assessed initially in differentiated THP-1 human macrophages. Using a resazurin oxidation assay and visual inspection of the cell monolayers, compounds were tested across a concentration range of 0.1 μM to 100 μM. The highest nontoxic concentrations for THP-1 cells were identified for each compound, as shown in Table 3. Subsequently, bacterial survival assays were performed at these nontoxic concentrations to determine the efficacy of the compounds against intracellular mycobacteria.

Table 3. Activity of select hit compounds in THP-1 macrophages.

Compound	THP-1 Cytotoxicity [μM]	Intracellular Killing in THP-1 Cells	
		MAB19977	MAH104
1	32	Yes	Yes
5	100	Yes	Yes
7	100	*	*
11	32	*	Yes
14	10	*	*
16	32	Yes	Yes

Note: for active compounds marked with an asterisk, bacterial infection increased the toxicity to THP-1 cells at concentrations that were initially nontoxic.

Among the six tested compounds, three demonstrated a significant reduction in the intracellular growth of MAB19977, measured as colony-forming units, while four compounds effectively decreased the viable MAH104 loads over a 5-day infection period (Figure 5). Compounds **1**, **5**, and **16** exhibited activities against both NTM strains, whereas compound **11** demonstrated potency in reducing MAH intracellular growth. It is important to note that certain compounds, when combined with bacterial infection, caused increased toxicity to THP-1 cells at previously identified nontoxic concentrations. These compounds are marked with asterisks in Table 3, and due to this toxicity, no data could be recorded for their efficacy in reducing intracellular bacterial loads.

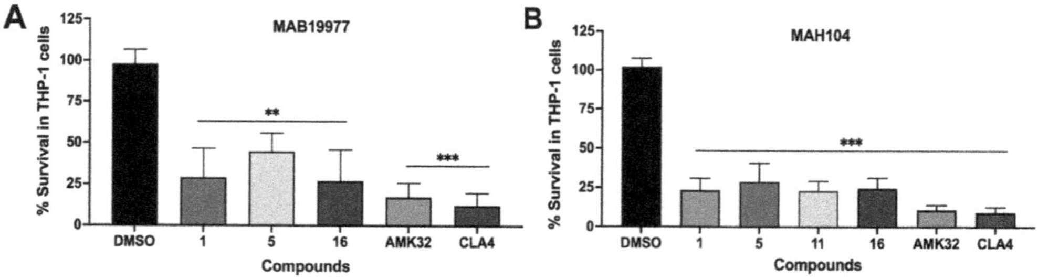

Figure 5. The quantification of bacterial CFUs for establishing the intracellular potency of hit compounds in infected THP-1 macrophages. Compounds at their highest nontoxic concentrations (as listed in Table 3) were added to (**A**) MAB19977-infected or (**B**) MAH104-infected THP-1 cell monolayers at 2 h post-infection and, subsequently, every other day throughout the duration of the experiment. Bacterial CFUs were determined by lysing the cells with 0.1% Triton X-100 on day 5 for both MAB19977 and MAH104, followed by plating serial dilutions on 7H10 agar plates. The percentage of surviving bacteria was calculated by dividing the CFUs per well for each treatment group by the CFUs in the DMSO growth control, which represented a 100% bacterial survival. AMK and CLA at a bactericidal concentration of 32 μg/mL and 4 μg/mL, respectively, were used as controls to assess the growth inhibition of MAB19977 and MAH104 within macrophages. The tested compounds demonstrated a statistically significant reduction in intracellular bacterial survival as determined by one-way ANOVA with multiple comparisons between the DMSO control and compound-treated groups. The significance between the DMSO control group and compound treatment groups on day 5 is indicated as ** $p < 0.01$ and *** $p < 0.001$.

3. Discussion

Mycobacterium abscessus causes one of the most challenging lung infections, particularly in patients with CF, due to its capacity to form resilient biofilms in the bronchial space [13]. These biofilms, also observed in the lung cavity of patients with COPD [36], are embedded in an extracellular matrix that enhances drug tolerance and results in poor treatment outcomes. Additionally, the dense cell aggregation within the biofilms, combined with the mucus-rich lung environment, restricts oxygen diffusion. This creates a hypoxic environment which, along with other lung-specific factors, promotes the metabolic remodeling of the pathogen, driving its transition into a persistent, non-replicative state and further exacerbating its resistance to antimicrobial therapy and complicating therapeutic interventions [17,37]. Current treatment options remain limited, with no standardized regimens or consistently effective therapies [5,38,39]. Multidrug regimens, often involving aminoglycosides, macrolides, and other antibiotics, offer partial relief but demonstrate poor success rates, particularly in the presence of resistance genes [40,41].

In this study, we employed artificial sputum media to simulate the cystic fibrosis lung environment and conducted high-throughput screening to identify compounds with antibiofilm activity against MAB. Our goal was to identify novel compounds under clinically relevant conditions with potent antibiofilm activity to enhance the efficacy of antibiotics against biofilm-associated bacteria. A total of 24,000 compounds were screened from the Maybridge, ChemBridge DIVERSet, and Spectrum compound libraries, resulting in the identification of 354 potential hits with ≥30% biofilm inhibition in the primary screen. After further validation in secondary and confirmatory screens, 17 active compounds were selected. Six compounds exhibited robust biofilm inhibition, with over 60% inhibition in pre-formation assays and over 40% in post-formation assays. Notably, five of these compounds also reduced bacterial viability by at least 40%, highlighting their dual action in preventing biofilm formation and reducing bacterial survival. The *in vitro* antimicrobial susceptibility tests showed that several of these compounds had consistent activity against both MAB and MAH clinical isolates, particularly compounds **1**, **5**, **11**, and **16**. Additionally, combining the hit compounds with AMK, one of the most important antibiotics in the aminoglycoside class for treating pulmonary NTM infections, enhanced bacterial clearance *in vitro*, highlighting potential synergistic effects. The further evaluation of select compounds for their ability to reduce intracellular mycobacterial loads in human macrophages revealed that compounds **1**, **5**, and **16** significantly decreased the intracellular survival of both MAB and MAH. We also observed that some compounds induced toxicity at nontoxic concentrations during infection, highlighting the need for the careful evaluation of their therapeutic potential. Overall, our findings suggest that certain antibiofilm compounds could be valuable adjuncts for improving current antibiotic effectiveness in treating MAB and other NTM infections.

In this study, we identified compounds belonging to various clusters, including pyrimidinone, pyrimidine, benzothiazole, sulfonylphenyl, phenothiazine, benzoxazole, pyrazolopyridine, quinazoline, thiourea, thiazole, and others. Some of these compound classes have been previously identified through the high-throughput screening of large chemical libraries and are known for their antimicrobial activities, including against *M. tuberculosis (Mtb)*. For instance, ethacridine (compound **1**), known for its broad-spectrum antimicrobial properties and effectiveness against various bacteria [42], including *Mtb* [43,44], and within biofilms of Staphylococcus aureus [45], was found in our study to effectively disrupt biofilm formation in MAB. This FDA-approved drug primarily targets the MtopI enzyme, which is essential for DNA replication, transcription, recombination, and chromosome condensation in Mtb, making it a promising candidate for disrupting bacterial viability. Also, its ability to enhance the activity of moxifloxacin against Mtb suggests a potential synergistic effect, offering promise for treating multidrug-resistant tuberculosis (MDR-TB) [46]. With its mechanism of action, involving interference with DNA topology, ethacridine could be a valuable compound for both preventing and treating biofilm-related non-tuberculous mycobacteria infections.

Compounds **5** and **16** exhibited significant antibacterial activity by not only inhibiting bacterial viability but also demonstrating potent efficacy against pre-formed biofilms. Additionally, these compounds maintained their activity under diverse conditions, including *in vitro* replicative environments and intracellular states. This suggests that the compounds target a critical mechanism in NTM that is universally expressed and indispensable across all stages of bacterial growth and survival, making them promising candidates for broad-spectrum therapeutic applications. Compound **5** belongs to the phenothiazine group which has been recognized for its ability to inhibit multidrug efflux pump activity in *S. aureus* [47] and type IV pilus production in multiple Gram-negative bacterial species [48], as well as for being active against *Enterococcus faecalis* and *Enterococcus faecium* human and animal clinical isolates [49] while effectively reducing biofilm mass in *Enterococcus faecalis* [50]. Phenothiazines have also demonstrated both *in vitro* and *in vivo* activity against Mtb, including MDR-TB [51]. These compounds disrupt biofilms by targeting the quorum-sensing systems of sessile bacterial populations [52], which are often more resistant to antibiotics, thereby enhancing overall antimicrobial efficacy. The synergistic interaction between phenothiazines and other antimicrobial agents further suggests their potential as therapeutic agents for chronic infections [53,54].

Furthermore, Compound **16**, a derivative of fluoren, a tricyclic aromatic hydrocarbon commonly found in various pharmaceutical compounds, was found in our study to exhibit high *in vitro*, antibiofilm, and intracellular activities against MAB. Numerous fluorene derivatives have demonstrated broad-spectrum antimicrobial activity against a wide range of microorganisms, including both Gram-positive and Gram-negative bacterial strains as well as yeasts, with some outperforming standard antibiotics [55]. Additionally, research highlights that certain fluorene-based compounds exhibit significant activity against mycobacteria, particularly in Mtb and *M. bovis*, including drug-resistant strains, showcasing their potential as novel anti-tuberculosis drugs [56]. Studies have explored structural modifications of fluorene with diverse functional groups to enhance antimycobacterial efficacy and elucidate their mechanisms of action. Notably, some derivatives show promising activity against both replicating and non-replicating bacterial populations, including intracellular Mtb within infected macrophages [57–59]. These broad-spectrum antimicrobial properties and the ability to inhibit biofilm formation underscore their potential for further investigation in the treatment of infections caused by NTM.

Compound **2**, N-benzyl-N-methyl-6-phenylpyrimidin-4-amine, is a pyrimidine derivative, a class of compounds frequently investigated for their ability to inhibit enzymes involved in DNA synthesis [60,61]. Given its chemical structure and mechanism of action, this compound aligns with other pyrimidine analogs that have been studied for their potential efficacy against antibiotic-resistant strains and biofilm formation, which are major challenges in treating persistent infections [62,63]. Compound **3**, N-(1-benzothiophen-3-yl)-N′-(4-fluorophenyl)urea, which belongs to benzothiazole cluster, demonstrated notable potential for MAB biofilm disruption. Benzothiazole-based compounds have shown promising antibacterial activity against a variety of bacteria, including Gram-positive and Gram-negative bacteria, and Mtb. They can bind to different targets in bacterial cells, such as enzymes involved in cell-wall synthesis, cell division, and DNA replication [64,65]. Interestingly, benzothiazole–urea hybrids have demonstrated antibacterial and anti-biofilm activity against *S. aureus*, including MRSA being almost as effective as vancomycin in reducing bacterial load in a mouse model of abdominal infection [66]. Other compounds in this study, including sulfonylphenyl and benzoxazole derivatives, are not widely reported for their anti-mycobacterial or antibiofilm activities. However, based on their known ability to interact with bacterial membranes and metabolic pathways, they may possess properties that suggest potential for disrupting biofilm integrity. Compounds **13** and **14**, which are fluorescent dyes, contrast with compound **17**, a thiazole-based compound identified as 2-(4-(4-Chlorophenyl)-1,3-thiazol-2-yl)isoindoline-1,3-dione. It is well documented that thiazole, a dominant pharmacophore widely present in antibacterial drugs, can exhibit significant antibacterial activity. For instance, novel thiazole nortopsentin analogs were recently synthesized and evaluated for their ability to inhibit biofilm formation in clinically relevant Gram-positive and Gram-negative pathogens. These compounds were found to interfere with the initial step of biofilm formation in a dose-dependent manner, showing selective activity against staphylococcal strains [67].

4. Materials and Methods

4.1. Bacterial Strains and Growth Culture

Mycobacterium abscessus subsp. *abscessus* (MAB) strain 19977 was obtained from the American Type Culture Collection (ATCC). *Mycobacterium avium* subsp. *hominissuis* 104 (MAH104) and *Mycobacterium avium* subsp. *hominissuis* A5 (MAHA5) were isolated from the blood of a patient with AIDS [68]. Clinical isolates of MAB, including DNA01627, NR49093 strain DJO44274, and NR44273 strain 4529, were provided in collaboration with the Cystic Fibrosis Research and Development Program at National Jewish Health, Denver, CO, USA. All studies were conducted in compliance with Oregon State University Institutional Biosafety Committee approval, protocol #3623. Bacteria were cultured on Middlebrook 7H10 agar or in Middlebrook 7H9 broth (Difco Laboratories, Detroit, MI, USA) supplemented with 10% oleic acid, albumin, dextrose, and catalase (OADC; Hardy

Diagnostics, Santa Maria, CA, USA), along with glycerol. To prepare bacterial suspensions, mid-log-phase cultures grown on 7H10 agar plates were resuspended in Hanks' Balanced Salt Solution (HBSS; VWR, Visalia, CA, USA) and adjusted to a McFarland standard of 3.0 (~9×10^8 cells/mL). The suspensions were sonicated for 1 min in a Branson ultrasonic bath and passed through a 22-gauge syringe 10 times to minimize clumping. Following this, the inoculums were allowed to settle for 15 min, and the top portion of the suspension was transferred to a new tube. The optical density at 600 nm (OD_{600}) was measured (1.0 OD_{600}~3×10^8 cells/mL), and the suspensions were further diluted in SCFM or RPMI-1640 mammalian cell culture media to the desired concentrations. Additionally, the sonicated inoculums were serially diluted and plated onto 7H10 agar plates to ensure precise inoculum concentrations.

4.2. Synthetic Cystic Fibrosis Sputum Media

SCFM was formulated based on the average concentrations of ions, free amino acids, glucose, and lactate observed in the CF sputum samples [69] and is outlined in Table 4. Amino acids were prepared as 100 mM stock solutions in deionized water and stored in the dark at 4 °C. Specific amino acids, such as tyrosine, aspartate, and tryptophan, were dissolved in 1.0 M, 0.5 M, and 0.2 M NaOH, respectively, to enhance solubility. Lactate stock solutions were adjusted to a pH of 7.0 using NaOH. All components were combined sequentially, and the final solution was adjusted to a pH of 6.8. The medium was then sterilized by filtration using 0.2 µm pore-size filtration flasks (Corning, Glendale, CA, USA) and stored at 4 °C until use.

Table 4. Synthetic cystic fibrosis media ingredient list.

Order	Reagent Name	Stock Solution	Volume for 1 L
1	NaH_2PO_4	0.2 M	6.5 mL
2	Na_2HPO_4	0.2 M	6.25 mL
3	KNO_3	1 M	0.348 mL
4	NH_4Cl	-	0.122 g
5	KCl	-	1.114 g
6	NaCl	-	3.03 g
7	MOPS	10 mM	10 mL
8	DI H2O	-	779.6 mL
9	l-aspartate,	0.5 M in NaOH	8.27 mL
10	l-threonine	100mM	10.72 mL
11	l-serine	100 mM	14.46 mL
12	l-glutamate·HCl	100 mM	15.49 mL
13	l-proline	100 mM	16.61 mL
14	l-glycine	100 mM	12.03 mL
15	l-alanine	100 mM	17.8 mL
16	l-cysteine·HCl	100 mM	1.6 mL
17	l-valine	100 mM	11.17 mL
18	l-methionine	100 mM	6.33 mL
19	l-isoleucine	100 mM	11.2 mL
20	l-leucine	100 mM	16.09 mL
21	l-tyrosine	1 M in NaOH	8.02 mL
22	l-phenylalanine	100 mM	5.3 mL
23	l-ornithine·HCl,	100 mM	6.76 mL
24	l-lysine·HCl,	100 mM	21.28 mL
25	l-histidine·HCl,	100 mM	5.19 mL
26	l-tryptophan	0.2 M in NaOH	0.13 mL
27	l-arginine·HCl	100 mM	3.06 mL
28	$CaCl_2$	1 M	1.754 mL

Table 4. *Cont.*

Order	Reagent Name	Stock Solution	Volume for 1 L
29	$MgCl_2$	1 M	0.606 mL
30	$FeSO_4 \cdot 7H_2O$	3.6 mM	1 mL
31	d-glucose	1 M	3 mL
32	l-lactate	1 M	9.3 mL

4.3. Chemicals and Compound Libraries

The screening process involved a total of 24,000 compounds, sourced from three distinct libraries. Of these, 14,400 compounds were from the Maybridge compound library, 9000 from the Cambridge compound library, and 600 from the Spectrum compound library. All compounds were obtained from the Oregon State University/College of Pharmacy High-Throughput Screening Services Laboratory. The selection of compounds was designed to maximize chemical diversity, thereby increasing the likelihood of identifying biologically active candidates. The compounds were supplied in two formats: (1) as a single-dose preparation at a concentration of 10 μM in SCFM dispensed in a 384WP with a total volume of 40 μL per well and (2) as a stock solution at a concentration of 100 μM in 1% DMSO used as the solvent carrier, provided in a 96WP.

Antibiotics included amikacin (AMK), purchased from Research Products International (Mt. Prospect, IL, USA), and clarithromycin (CLA), obtained from Tokyo Chemical Industry (Tokyo, Japan). Crystal violet, used for biofilm staining, was purchased from Sigma-Aldrich. Additional reagents included Phorbol 12-myristate 13-acetate (PMA) and resazurin, both sourced from Sigma-Aldrich (St. Louis, MO, USA). All compounds and reagents were prepared and stored according to the manufacturers' specifications to ensure optimal stability and consistency throughout the experiment.

4.4. Compound Screening

Ten microliters of MAB inoculum, prepared at a concentration of 9×10^8 CFU/mL in SCFM, were added to each well of a 384WP pre-loaded with compounds in 40 μL of SCFM at a final concentration of 10 μM. The plates were incubated at 37 °C for up to five days. As a control for 100% inhibition, the last two columns of wells received 70 μg/mL CLA. Additionally, DMSO was included as a control to assess any potential effects on bacterial growth and/or biofilm formation.

Compounds that inhibited bacterial biofilm formation by more than 30%, assessed by crystal violet staining, described below, were identified as potential hits. These compounds were subsequently cherry-picked from the stock library of 100 μM in 1% DMSO and subjected to a secondary assay in 96-well round-bottom plates, performed in triplicate under the same conditions as the primary screen. A total of 100 μL of bacterial inoculum at a concentration of 3×10^8 CFU/mL in SCFM was dispensed into each well of a 96WP. Compounds were added to each well at a final concentration of 10 μM, and the plates were incubated at 37 °C for up to five days. Compounds that demonstrated consistent activity in both the primary and secondary assays were purchased from MolPort and further validated as hit compounds through three independent biological replicates, each performed in triplicate using a 96WP format.

4.5. Bacterial Viability Using Resazurin Staining

Resazurin staining was conducted prior to biofilm visualization only in the secondary and confirmation assay with purchased compounds. These assays focused on selected hit compounds, incorporating both bacterial growth and killing controls to ensure accurate evaluation. Briefly, the planktonic solution was gently removed to preserve biofilm integrity.

The biofilms were washed once with phosphate-buffered saline (PBS) and incubated with resazurin at 5 μg/well, protected from light, at room temperature for up to 2–3 h or until the appropriate color developed in the bacterial growth control (without compound exposure). Fluorescence was measured at an excitation/emission wavelength of 530 nm/590 nm using a Tecan Infinite F200 Fluorescent Microplate Reader. In addition, compound activity was assessed in both planktonic and biofilm-associated bacteria by serially diluting the supernatants and biofilm bacteria, followed by the quantification of colony-forming units on 7H10 agar plates. This assay was conducted exclusively with the hit compounds.

4.6. Crystal Violet (CV) Staining

To measure the biomass, briefly, the resazurin solution was gently removed from the plates. The wells were fixed with 95% ethanol (40 μL for 384WP and 200 μL for 96WP) for 10 min at room temperature and then air-dried. Next, the wells were stained with 0.1% crystal violet (Sigma-Aldrich, St. Louis, MO, USA) for 10 min. After staining, the wells were washed twice with deionized water and allowed to air-dry for 10 min. To quantify biofilm formation, the stained biofilms were solubilized in 30% acetic acid for 30 min and transferred to 384- or 96-well optically clear flat-bottom plates. The absorbance was measured using an Epoch Microplate Spectrophotometer (BioTek, Winooski, VT, USA) at a wavelength of 570 nm.

The percentage of biofilm inhibition was calculated using the following formula:

$$\%\,\text{inhibition} = 100 \times \left[1 - \frac{(\text{Drug treated well average}) - (\text{Negative control well average})}{(\text{Positive control well average}) - (\text{Negative control well average})}\right]$$

where the drug-treated well average is the absorbance value from wells treated with the test compound. The positive control well average is the absorbance value from wells containing only the bacterial culture and no compound. The negative control well average is the absorbance value from wells treated with the antibiotic CLA70.

4.7. Compound Activity in Pre- and Post-Biofilm Formation

The activity of the compounds was additionally assessed against clinical isolates of MAB and MAH in two contexts: the inhibition of biofilm formation (pre-biofilm condition) and disruption of formed biofilms (post-biofilm condition). For the inhibition of biofilms, the assay was carried out with the selected hit compounds, as detailed above. Biofilm formation was quantified using CV staining, while bacterial viability was assessed through the redox indicator resazurin dye, with assessments conducted for both assays on day 5 following compound addition.

The activity of the selected hit compound was evaluated with post-formed biofilms under the following conditions: A total of 100 μL of bacterial inoculum at a concentration of 3×10^8 CFU/mL in SCFM was dispensed into each well of a 96WP and incubated at 37 °C for three days to allow biofilm formation. Subsequently, the compound was added to each well at a final concentration of 10 μM. The bacterial viability and biofilm mass were then evaluated on day 5 using resazurin and CV staining, respectively, as detailed above.

4.8. Assay for 50% Inhibitory Concentration (IC50)

To establish the minimum inhibitory concentration (MIC) required to inhibit 50% of the pathogen, bacterial inoculums of MAB19977, MAH104, or clinical isolates were prepared in 7H9 media supplemented with OADC, 10% glycerol, and 0.1% Tween 80. The bacterial suspension was adjusted to an OD_{600} of 0.5 and diluted to 0.03–0.05 in 7H9 media prior to compound addition. Compounds were prepared in DMSO at concentrations ranging from 10 μM to 0.01 μM by half-log serial dilutions. CLA at 70 μg/mL and 1%

DMSO served as positive and negative controls, respectively. The plates were incubated statically for 72 h for MAB19977 and 96 h for MAH104. Resazurin was then added to the plates at a final concentration of 5 μg/well and incubated at 37 °C until a color change was observed. Fluorescence was measured at 530 nm/590 nm using a Tecan plate reader, and the percentage of inhibition was calculated using the formula described above. The experiment was conducted in two biological replicates.

4.9. Assay for 50% Cytotoxic Concentration (CC50)

Human-derived THP-1 monocytes (ATCC TIB-202) were cultured in RPMI-1640 medium (Cellgro, Manassas, VA, USA) supplemented with 10% heat-inactivated fetal bovine serum, 2 mM L-glutamine, and 25 mM HEPES and maintained at 37 °C with 5% CO_2. To differentiate the THP-1 cells into macrophages, they were plated at 1×10^5 cells per well in a clear, flat-bottom 96WP with culture media and treated overnight with 30 ng/mL of phorbol-12-myristate-13-acetate (PMA). The following day, the wells were replenished with fresh media, and the cells were allowed to rest for an additional two days at 37 °C with 5% CO_2. The hit compounds were serially diluted in DMSO, ranging from 10 μM to 0.01 μM, using half-log dilutions. To determine the cytotoxic concentration inducing 50% cell death (IC_{50}), 4 μL of each compound dilution was added to 196 μL of RPMI-1640 medium and dispensed onto the cell monolayers in each well of a 96WP. Cyclosporine A (50 μg/mL) was used as a positive control for cytotoxicity, while 1% DMSO served as the negative control. After incubating the plates at 37 °C with 5% CO_2 for 72 h, cytotoxicity was assessed using the resazurin assay by adding 10 μL of 50 μg/mL resazurin in PBS to each well. The plates were incubated for 2–3 h, and fluorescence was measured at 530 nm/590 nm using a Tecan plate reader. Additionally, cell monolayers were visually examined under an inverted microscope to detect any signs of cytotoxicity. The cytotoxic concentration was identified as the lowest compound concentration that neither caused a color change in resazurin nor induced visible damage to the cell monolayer. This concentration was used to calculate the IC_{50}, representing the concentration at which 50% of the host cells were killed.

4.10. Compound Activity in Macrophages

The THP-1 monolayers were established in a 96WP, as previously described, and bacterial infections were conducted with a multiplicity of infection (MOI) of 1 for MAB19977 and a MOI of 5 for MAH104. After a 2 h infection period, cells were washed with PBS and treated with 200 μg/mL AMK for 2 h to kill any extracellular bacteria. The wells were then replenished with fresh media containing the tested compounds at their highest nontoxic concentrations and, subsequently, every other day throughout the duration of the experiment. Additionally, antibiotics were added at bactericidal concentrations of 32 μg/mL AMK and 4 μg/mL CLA for both pathogens. After 5 days of incubation at 37 °C with 5% CO_2, monolayers were lysed 0.1% Triton-X-100 and viable bacteria were enumerated by plating serial dilutions on 7H10 agar plates. Data were calculated as the percentage of surviving bacteria relative to the DMSO-only control.

4.11. Statistical Analysis

Statistical analyses were performed using Student's *t*-test for comparisons between two groups and one-way analysis of variance (ANOVA) for comparisons across multiple groups, followed by post hoc multiple-comparison tests to assess pairwise differences between the DMSO control and experimental treatment groups. Results were derived from three biological replicates, each performed in three technical replicates unless otherwise stated. Statistical analyses and graphical visualizations were carried out using GraphPad Prism (version 10.0). A *p*-value of less than 0.05 was considered statistically significant (* $p < 0.05$; ** $p < 0.01$; *** $p < 0.001$), with specific values reported in figure legends and indicated by asterisks in graphical outputs. All data are presented as the mean ± standard deviation (SD), with error bars included in graphical representations to reflect variability within replicates.

5. Conclusions

This study aimed to identify compounds with anti-NTM biofilm activity through the high-throughput screening of a diverse compound library under conditions simulating a CF patient's sputum. Our findings revealed several promising candidates with varying degrees of biofilm inhibition and antimicrobial activity against clinical NTM strains. Notably, compounds such as ethacridine, phenothiazine, and fluorene derivatives demonstrated significant potential by not only preventing biofilm formation but also enhancing antimicrobial efficacy and markedly reducing intracellular MAB loads in macrophages. These results suggest that the identified compounds likely target a critical, universally expressed mechanism in NTM, essential for bacterial growth and survival across different stages. This positions them as promising candidates for broad-spectrum therapeutic applications. However, the precise molecular mechanisms underlying their activity remain unclear and require further elucidation through molecular and biochemical studies. While these findings highlight their potential as valuable candidates for further development as adjuncts to existing antibiotics, it is challenging to assess their draggability at this stage without *in vivo* studies. Also, some chemical modifications may be necessary to optimize their activity and minimize toxicity for future therapeutic use. Notably, certain compounds, as highlighted in this paper, exhibited increased toxicity when tested in the phagocytic cell infection model. These findings underscore the importance of *in vivo* studies to better understand the clinical relevance and safety of these compounds.

Author Contributions: Conceptualization, L.D.; Methodology, E.D., M.G., R.V.V., D.N. and L.D.; Validation, E.D., M.G. and L.D.; Formal analysis, E.D. and L.D.; Investigation, E.D., M.G., R.V.V., D.N. and L.D.; Resources, L.D.; Data curation, E.D., M.G. and L.D.; Writing—original draft, E.D.; Writing—review and editing, L.D.; Visualization, E.D., M.G. and L.D.; Project administration, L.D.; Funding acquisition, L.D. All authors have read and agreed to the published version of the manuscript.

Funding: This work was supported by the NIH/NIAID grant 1R15AI142609-01A1 (L. Danelishvili).

Institutional Review Board Statement: Not applicable.

Informed Consent Statement: Not applicable.

Data Availability Statement: The original contributions presented in this study are included in the article. Further inquiries can be directed to the corresponding author.

Conflicts of Interest: The authors declare no conflicts of interest.

References

1. Lee, M.R.; Sheng, W.H.; Hung, C.C.; Yu, C.J.; Lee, L.N.; Hsueh, P.R. *Mycobacterium abscessus* Complex Infections in Humans. *Emerg. Infect. Dis.* **2015**, *21*, 1638–1646. [CrossRef]
2. Lopeman, R.C.; Harrison, J.; Desai, M.; Cox, J.A.G. *Mycobacterium abscessus*: Environmental Bacterium Turned Clinical Nightmare. *Microorganisms* **2019**, *7*, 90. [CrossRef]
3. Honda, J.R.; Virdi, R.; Chan, E.D. Global Environmental Nontuberculous Mycobacteria and Their Contemporaneous Man-Made and Natural Niches. *Front. Microbiol.* **2018**, *9*, 2029. [CrossRef]
4. Schuurbiers, M.M.F.; Bruno, M.; Zweijpfenning, S.M.H.; Magis-Escurra, C.; Boeree, M.; Netea, M.G.; van Ingen, J.; van de Veerdonk, F.; Hoefsloot, W. Immune defects in patients with pulmonary *Mycobacterium abscessus* disease without cystic fibrosis. *ERJ Open Res.* **2020**, *6*, 00590-2020. [CrossRef]
5. Griffith, D.E.; Aksamit, T.; Brown-Elliott, B.A.; Catanzaro, A.; Daley, C.; Gordin, F.; Holland, S.M.; Horsburgh, R.; Huitt, G.; Iademarco, M.F.; et al. An official ATS/IDSA statement: Diagnosis, treatment, and prevention of nontuberculous mycobacterial diseases. *Am. J. Respir. Crit. Care Med.* **2007**, *175*, 367–416. [CrossRef]

6. Nessar, R.; Cambau, E.; Reyrat, J.M.; Murray, A.; Gicquel, B. *Mycobacterium abscessus*: A new antibiotic nightmare. *J. Antimicrob Chemother.* **2012**, *67*, 810–818. [CrossRef] [PubMed]
7. Zhang, X.; Liu, W.; Liu, W.; Jiang, H.; Zong, W.; Zhang, G.; Jin, P.; Wang, H. Cutaneous infections caused by rapidly growing mycobacteria: Case reports and review of clinical and laboratory aspects. *Acta Derm. Venereol.* **2015**, *95*, 985–989. [CrossRef]
8. Kendall, B.A.; Winthrop, K.L. Update on the epidemiology of pulmonary nontuberculous mycobacterial infections. *Semin. Respir. Crit. Care Med.* **2013**, *34*, 87–94. [CrossRef]
9. Ryan, K.; Byrd, T.F. *Mycobacterium abscessus*: Shapeshifter of the Mycobacterial World. *Front. Microbiol.* **2018**, *9*, 2642. [CrossRef] [PubMed]
10. Petrini, B. *Mycobacterium abscessus*: An emerging rapid-growing potential pathogen. *APMIS* **2006**, *114*, 319–328. [CrossRef]
11. Keefe, B.F.; Bermudez, L.E. Environment in the lung of cystic fibrosis patients stimulates the expression of biofilm phenotype in *Mycobacterium abscessus*. *J. Med. Microbiol.* **2022**, *71*, 001467. [CrossRef] [PubMed]
12. Greendyke, R.; Byrd, T.F. Differential antibiotic susceptibility of *Mycobacterium abscessus* variants in biofilms and macrophages compared to that of planktonic bacteria. *Antimicrob. Agents Chemother.* **2008**, *52*, 2019–2026. [CrossRef]
13. Qvist, T.; Eickhardt, S.; Kragh, K.N.; Andersen, C.B.; Iversen, M.; Hoiby, N.; Bjarnsholt, T. Chronic pulmonary disease with *Mycobacterium abscessus* complex is a biofilm infection. *Eur. Respir. J.* **2015**, *46*, 1823–1826. [CrossRef] [PubMed]
14. Hunt-Serracin, A.C.; Parks, B.J.; Boll, J.; Boutte, C.C. *Mycobacterium abscessus* Cells Have Altered Antibiotic Tolerance and Surface Glycolipids in Artificial Cystic Fibrosis Sputum Medium. *Antimicrob. Agents Chemother.* **2019**, *63*. [CrossRef] [PubMed]
15. Rojony, R.; Danelishvili, L.; Campeau, A.; Wozniak, J.M.; Gonzalez, D.J.; Bermudez, L.E. Exposure of *Mycobacterium abscessus* to Environmental Stress and Clinically Used Antibiotics Reveals Common Proteome Response among Pathogenic Mycobacteria. *Microorganisms* **2020**, *8*, 698. [CrossRef]
16. Rojony, R.; Martin, M.; Campeau, A.; Wozniak, J.M.; Gonzalez, D.J.; Jaiswal, P.; Danelishvili, L.; Bermudez, L.E. Quantitative analysis of *Mycobacterium avium* subsp. hominissuis proteome in response to antibiotics and during exposure to different environmental conditions. *Clin. Proteomics* **2019**, *16*, 39. [CrossRef]
17. Kurbatfinski, N.; Hill, P.J.; Tobin, N.; Kramer, C.N.; Wickham, J.; Goodman, S.D.; Hall-Stoodley, L.; Bakaletz, L.O. Disruption of nontuberculous mycobacteria biofilms induces a highly vulnerable to antibiotic killing phenotype. *Biofilm* **2023**, *6*, 100166. [CrossRef]
18. Chakraborty, P.; Kumar, A. The extracellular matrix of mycobacterial biofilms: Could we shorten the treatment of mycobacterial infections? *Microb. Cell* **2019**, *6*, 105–122. [CrossRef] [PubMed]
19. Rose, S.J.; Bermudez, L.E. *Mycobacterium avium* biofilm attenuates mononuclear phagocyte function by triggering hyperstimulation and apoptosis during early infection. *Infect. Immun.* **2014**, *82*, 405–412. [CrossRef] [PubMed]
20. Yamazaki, Y.; Danelishvili, L.; Wu, M.; Hidaka, E.; Katsuyama, T.; Stang, B.; Petrofsky, M.; Bildfell, R.; Bermudez, L.E. The ability to form biofilm influences *Mycobacterium avium* invasion and translocation of bronchial epithelial cells. *Cell Microbiol.* **2006**, *8*, 806–814. [CrossRef] [PubMed]
21. Babrak, L.; Danelishvili, L.; Rose, S.J.; Kornberg, T.; Bermudez, L.E. The environment of "*Mycobacterium avium* subsp. hominissuis" microaggregates induces synthesis of small proteins associated with efficient infection of respiratory epithelial cells. *Infect. Immun.* **2015**, *83*, 625–636. [CrossRef] [PubMed]
22. Uruen, C.; Chopo-Escuin, G.; Tommassen, J.; Mainar-Jaime, R.C.; Arenas, J. Biofilms as Promoters of Bacterial Antibiotic Resistance and Tolerance. *Antibiotics* **2020**, *10*, 3. [CrossRef] [PubMed]
23. Aung, T.T.; Yam, J.K.; Lin, S.; Salleh, S.M.; Givskov, M.; Liu, S.; Lwin, N.C.; Yang, L.; Beuerman, R.W. Biofilms of Pathogenic Nontuberculous Mycobacteria Targeted by New Therapeutic Approaches. *Antimicrob. Agents Chemother.* **2016**, *60*, 24–35. [CrossRef] [PubMed]
24. Antunes, L.C.M.; Ferreira, R.B.R.; Buckner, M.M.C.; Finlay, B.B. Quorum sensing in bacterial virulence. *Microbiology* **2010**, *156*, 2271–2282. [CrossRef] [PubMed]
25. Boopathi, S.; Ramasamy, S.; Haridevamuthu, B.; Murugan, R.; Veerabadhran, M.; Jia, A.Q.; Arockiaraj, J. Intercellular communication and social behaviors in mycobacteria. *Front. Microbiol.* **2022**, *13*, 943278. [CrossRef]
26. Sharma, S.; Mohler, J.; Mahajan, S.D.; Schwartz, S.A.; Bruggemann, L.; Aalinkeel, R. Microbial Biofilm: A Review on Formation, Infection, Antibiotic Resistance, Control Measures, and Innovative Treatment. *Microorganisms* **2023**, *11*, 1614. [CrossRef] [PubMed]
27. Roy, R.; Tiwari, M.; Donelli, G.; Tiwari, V. Strategies for combating bacterial biofilms: A focus on anti-biofilm agents and their mechanisms of action. *Virulence* **2018**, *9*, 522–554. [CrossRef] [PubMed]
28. Hawas, S.; Verderosa, A.D.; Totsika, M. Combination Therapies for Biofilm Inhibition and Eradication: A Comparative Review of Laboratory and Preclinical Studies. *Front. Cell Infect. Microbiol.* **2022**, *12*, 850030. [CrossRef]
29. Mitra, A. Combatting biofilm-mediated infections in clinical settings by targeting quorum sensing. *Cell Surf.* **2024**, *12*, 100133. [CrossRef]

30. Tursi, S.A.; Puligedda, R.D.; Szabo, P.; Nicastro, L.K.; Miller, A.L.; Qiu, C.; Gallucci, S.; Relkin, N.R.; Buttaro, B.A.; Dessain, S.K.; et al. Salmonella Typhimurium biofilm disruption by a human antibody that binds a pan-amyloid epitope on curli. *Nat. Commun.* **2020**, *11*, 1007. [CrossRef] [PubMed]
31. Cegelski, L.; Pinkner, J.S.; Hammer, N.D.; Cusumano, C.K.; Hung, C.S.; Chorell, E.; Aberg, V.; Walker, J.N.; Seed, P.C.; Almqvist, F.; et al. Small-molecule inhibitors target Escherichia coli amyloid biogenesis and biofilm formation. *Nat. Chem. Biol.* **2009**, *5*, 913–919. [CrossRef] [PubMed]
32. Silva, C.; Rojony, R.; Bermudez, L.E.; Danelishvili, L. Short-Chain Fatty Acids Promote *Mycobacterium avium* subsp. hominissuis Growth in Nutrient-Limited Environments and Influence Susceptibility to Antibiotics. *Pathogens* **2020**, *9*, 700. [CrossRef]
33. Danelishvili, L.; Shulzhenko, N.; Chinison, J.J.J.; Babrak, L.; Hu, J.; Morgun, A.; Burrows, G.; Bermudez, L.E. Mycobacterium tuberculosis Proteome Response to Antituberculosis Compounds Reveals Metabolic "Escape" Pathways That Prolong Bacterial Survival. *Antimicrob. Agents Chemother.* **2017**, *61*. [CrossRef] [PubMed]
34. Gupta, R.; Netherton, M.; Byrd, T.F.; Rohde, K.H. Reporter-Based Assays for High-Throughput Drug Screening against *Mycobacterium abscessus*. *Front. Microbiol.* **2017**, *8*, 2204. [CrossRef]
35. Richter, A.; Strauch, A.; Chao, J.; Ko, M.; Av-Gay, Y. Screening of Preselected Libraries Targeting *Mycobacterium abscessus* for Drug Discovery. *Antimicrob. Agents Chemother.* **2018**, *62*. [CrossRef] [PubMed]
36. Fennelly, K.P.; Ojano-Dirain, C.; Yang, Q.; Liu, L.; Lu, L.; Progulske-Fox, A.; Wang, G.P.; Antonelli, P.; Schultz, G. Biofilm Formation by *Mycobacterium abscessus* in a Lung Cavity. *Am. J. Respir. Crit. Care Med.* **2016**, *193*, 692–693. [CrossRef] [PubMed]
37. Kolpen, M.; Jensen, P.O.; Qvist, T.; Kragh, K.N.; Ravnholt, C.; Fritz, B.G.; Johansen, U.R.; Bjarnsholt, T.; Hoiby, N. Biofilms of *Mycobacterium abscessus* Complex Can Be Sensitized to Antibiotics by Disaggregation and Oxygenation. *Antimicrob. Agents Chemother.* **2020**, *64*, e01212-19. [CrossRef]
38. Nathavitharana, R.R.; Strnad, L.; Lederer, P.A.; Shah, M.; Hurtado, R.M. Top Questions in the Diagnosis and Treatment of Pulmonary *M. abscessus* Disease. *Open Forum Infect. Dis.* **2019**, *6*, ofz221. [CrossRef] [PubMed]
39. Novosad, S.A.; Beekmann, S.E.; Polgreen, P.M.; Mackey, K.; Winthrop, K.L.; Team, M.a.S. Treatment of *Mycobacterium abscessus* Infection. *Emerg. Infect. Dis.* **2016**, *22*, 511–514. [CrossRef]
40. Nash, K.A.; Brown-Elliott, B.A.; Wallace, R.J., Jr. A novel gene, erm(41), confers inducible macrolide resistance to clinical isolates of *Mycobacterium abscessus* but is absent from Mycobacterium chelonae. *Antimicrob. Agents Chemother.* **2009**, *53*, 1367–1376. [CrossRef] [PubMed]
41. Maurer, F.P.; Castelberg, C.; Quiblier, C.; Bottger, E.C.; Somoskovi, A. Erm(41)-dependent inducible resistance to azithromycin and clarithromycin in clinical isolates of *Mycobacterium abscessus*. *J. Antimicrob. Chemother.* **2014**, *69*, 1559–1563. [CrossRef] [PubMed]
42. Jabri, T.; Khan, N.A.; Makhlouf, Z.; Akbar, N.; Gul, J.; Shah, M.R.; Siddiqui, R. Antibacterial Properties of Ethacridine Lactate and Sulfmethoxazole Loaded Functionalized Graphene Oxide Nanocomposites. *Antibiotics* **2023**, *12*, 755. [CrossRef]
43. Bunalema, L.; Ocan, M.; Bbosa, G.S.; Lubega, A.; Lanyero, H.; Nahereza, I.T. Anti-Mycobacterial and Synergistic Activity of Synthetic Acridine-9-Carboraldehyde and 9-Hydroxy-4-Methoxy Acridine Alkaloids against Clinical Isolates of Multidrug-Resistant Mycobacterium Tuberculosis. *J. Pharm. Pharmacol. Res.* **2022**, *6*, 25–34. [CrossRef]
44. Petrikaite, V.; Tarasevicius, E.; Pavilonis, A. New ethacridine derivatives as the potential antifungal and antibacterial preparations. *Medicina* **2007**, *43*, 657. [CrossRef]
45. Junka, A.; Bartoszewicz, M.; Smutnicka, D.; Secewicz, A.; Szymczyk, P. Efficacy of antiseptics containing povidone-iodine, octenidine dihydrochloride and ethacridine lactate against biofilm formed by Pseudomonas aeruginosa and Staphylococcus aureus measured with the novel biofilm-oriented antiseptics test. *Int. Wound J.* **2014**, *11*, 730–734. [CrossRef] [PubMed]
46. Garcia-Moreno, P.K. *Mycobacterium tuberculosis* inhibitors: Action and resistance. Ph.D. Thesis, Florida International University, Miami, FL, USA, 2018. [CrossRef]
47. Kaatz, G.W.; Moudgal, V.V.; Seo, S.M.; Kristiansen, J.E. Phenothiazines and thioxanthenes inhibit multidrug efflux pump activity in Staphylococcus aureus. *Antimicrob. Agents Chemother.* **2003**, *47*, 719–726. [CrossRef] [PubMed]
48. Vo, N.; Sidner, B.S.; Yu, Y.; Piepenbrink, K.H. Type IV Pilus-Mediated Inhibition of Acinetobacter baumannii Biofilm Formation by Phenothiazine Compounds. *Microbiol. Spectr.* **2023**, *11*, e0102323. [CrossRef] [PubMed]
49. Hendricks, O.; Molnar, A.; Butterworth, T.S.; Butaye, P.; Kolmos, H.J.; Christensen, J.B.; Kristiansen, J.E. In vitro activity of phenothiazine derivatives in Enterococcus faecalis and Enterococcus faecium. *Basic. Clin. Pharmacol. Toxicol.* **2005**, *96*, 33–36. [CrossRef]
50. Fiallos, N.M.; Ribeiro Aguiar, A.L.; da Silva, B.N.; Pergentino, M.L.M.; Rocha, M.F.G.; Sidrim, J.J.C.; Maia, D.; Cordeiro, R.A. The Potential of Phenothiazines against Endodontic Pathogens: A Focus on Enterococcus-Candida Dual-Species Biofilm. *Antibiotics* **2022**, *11*, 1562. [CrossRef]
51. Warman, A.J.; Rito, T.S.; Fisher, N.E.; Moss, D.M.; Berry, N.G.; O'Neill, P.M.; Ward, S.A.; Biagini, G.A. Antitubercular pharmacodynamics of phenothiazines. *J. Antimicrob. Chemother.* **2013**, *68*, 869–880. [CrossRef] [PubMed]
52. Gad, A.I.; El-Ganiny, A.M.; Eissa, A.G.; Noureldin, N.A.; Nazeih, S.I. Miconazole and phenothiazine hinder the quorum sensing regulated virulence in Pseudomonas aeruginosa. *J. Antibiot.* **2024**, *77*, 454–465. [CrossRef] [PubMed]

53. Chan, Y.Y.; Ong, Y.M.; Chua, K.L. Synergistic interaction between phenothiazines and antimicrobial agents against Burkholderia pseudomallei. *Antimicrob. Agents Chemother.* **2007**, *51*, 623–630. [CrossRef] [PubMed]
54. Aguilar-Vega, L.; Lopez-Jacome, L.E.; Franco, B.; Munoz-Carranza, S.; Vargas-Maya, N.; Franco-Cendejas, R.; Hernandez-Duran M.; Otero-Zuniga, M.; Campo-Beleno, C.; Jimenez-Cortes, J.G.; et al. Antibacterial properties of phenothiazine derivatives against multidrug-resistant Acinetobacter baumannii strains. *J. Appl. Microbiol.* **2021**, *131*, 2235–2243. [CrossRef] [PubMed]
55. Vlad, I.M.; Nuta, D.C.; Ancuceanu, R.V.; Caproiou, M.T.; Dumitrascu, F.; Marinas, I.C.; Chifiriuc, M.C.; Marutescu, L.G.; Zarafu I.; Papacocea, I.R.; et al. New O-Aryl-Carbamoyl-Oxymino-Fluorene Derivatives with MI-Crobicidal and Antibiofilm Activity Enhanced by Combination with Iron Oxide Nanoparticles. *Molecules* **2021**, *26*, 3002. [CrossRef] [PubMed]
56. Rezayan, A.H.; Hariri, S.; Azerang, P.; Ghavami, G.; Portugal, I.; Sardari, S. Synthesis of Novel Fluorene Bisamide Derivatives via Ugi Reaction and Evaluation their Biological Activity against Mycobacterium species. *Iran. J. Pharm. Res.* **2017**, *16*, 745–755.
57. Eke, I.E.; Williams, J.T.; Haiderer, E.R.; Albrecht, V.J.; Murdoch, H.M.; Abdalla, B.J.; Abramovitch, R.B. Discovery and characterization of antimycobacterial nitro-containing compounds with distinct mechanisms of action and in vivo efficacy. *Antimicrob Agents Chemother.* **2023**, *67*, e0047423. [CrossRef]
58. Chollet, A.; Mori, G.; Menendez, C.; Rodriguez, F.; Fabing, I.; Pasca, M.R.; Madacki, J.; Kordulakova, J.; Constant, P.; Quemard, A. et al. Design, synthesis and evaluation of new GEQ derivatives as inhibitors of InhA enzyme and Mycobacterium tuberculosis growth. *Eur. J. Med. Chem.* **2015**, *101*, 218–235. [CrossRef]
59. Ahmed, S.; Manning, A.; Flint, L.; Awasthi, D.; Ovechkina, Y.; Parish, T. Identification of Novel Chemical Scaffolds that Inhibit the Growth of Mycobacterium tuberculosis in Macrophages. *Front. Pharmacol.* **2021**, *12*, 790583. [CrossRef] [PubMed]
60. El-Gharably, A.A.; Nassar, A.A.; El-Ganzory, N.M.; Saad-Allah, K.M.; El-Barbary, A.A. Sulfoxidation of pyrimidine thioate derivatives and study their biological activities. *Sci. Rep.* **2025**, *15*, 1024. [CrossRef]
61. Guo, M.J.; Hildbrand, S.; Leumann, C.J.; McLaughlin, L.W.; Waring, M.J. Inhibition of DNA polymerase reactions by pyrimidine nucleotide analogues lacking the 2-keto group. *Nucleic Acids Res.* **1998**, *26*, 1863–1869. [CrossRef] [PubMed]
62. Garg, S.; Kumar, R.; Kunimoto, D.; Rayat, G.R. Anti-Mycobacterial Activity of Flavonoid and Pyrimidine Compounds. *Molecules* **2022**, *27*, 6714. [CrossRef]
63. Cui, Q.; Shin, W.S.; Luo, Y.; Tian, J.; Cui, H.; Yin, D. Thymidylate kinase: An old topic brings new perspectives. *Curr. Med. Chem.* **2013**, *20*, 1286–1305. [CrossRef]
64. Gjorgjieva, M.; Tomasic, T.; Kikelj, D.; Masic, L.P. Benzothiazole-based Compounds in Antibacterial Drug Discovery. *Curr. Med. Chem.* **2018**, *25*, 5218–5236. [CrossRef] [PubMed]
65. Kashyap, P.; Verma, S.; Gupta, P.; Narang, R.; Lal, S.; Devgun, M. Recent insights into antibacterial potential of benzothiazole derivatives. *Med. Chem. Res.* **2023**, *32*, 1543–1573. [CrossRef]
66. Zha, L.; Xie, Y.; Wu, C.; Lei, M.; Lu, X.; Tang, W.; Zhang, J. Novel benzothiazole–urea hybrids: Design, synthesis and biological activity as potent anti-bacterial agents against MRSA. *Eur. J. Med. Chem.* **2022**, *236*, 114333. [CrossRef]
67. Carbone, A.; Parrino, B.; Cusimano, M.G.; Spano, V.; Montalbano, A.; Barraja, P.; Schillaci, D.; Cirrincione, G.; Diana, P.; Cascioferro, S. New Thiazole Nortopsentin Analogues Inhibit Bacterial Biofilm Formation. *Mar. Drugs* **2018**, *16*, 274. [CrossRef] [PubMed]
68. Bermudez, L.E.; Petrofsky, M.; Kolonoski, P.; Young, L.S. An animal model of *Mycobacterium avium* complex disseminated infection after colonization of the intestinal tract. *J. Infect. Dis.* **1992**, *165*, 75–79. [CrossRef]
69. Palmer, K.L.; Aye, L.M.; Whiteley, M. Nutritional cues control Pseudomonas aeruginosa multicellular behavior in cystic fibrosis sputum. *J. Bacteriol.* **2007**, *189*, 8079–8087. [CrossRef] [PubMed]

MDPI AG
Grosspeteranlage 5
4052 Basel
Switzerland
Tel.: +41 61 683 77 34

MDPI Books Editorial Office
E-mail: books@mdpi.com
www.mdpi.com/books

www.ingramcontent.com/pod-product-compliance
Lightning Source LLC
Chambersburg PA
CBHW061810210926
51850CB00006B/662
* 9 7 8 3 7 2 5 8 7 3 3 4 0 *